Lecture Notes in Computer Science

Lecture Notes in Artificial Intelligence 15442

Founding Editor

Jörg Siekmann

The series Lecture Notes in Artificial Intelligence (LNAI) was established in 1988 as a topical subseries of LNCS devoted to artificial intelligence.

The series publishes state-of-the-art research results at a high level. As with the LNCS mother series, the mission of the series is to serve the international R & D community by providing an invaluable service, mainly focused on the publication of conference and workshop proceedings and postproceedings.

Mingming Gong · Yiliao Song · Yun Sing Koh ·
Wei Xiang · Derui Wang
Editors

AI 2024: Advances in Artificial Intelligence

37th Australasian Joint Conference on Artificial Intelligence, AI 2024
Melbourne, VIC, Australia, November 25–29, 2024
Proceedings, Part I

Springer

Editors
Mingming Gong
The University of Melbourne
Parkville, VIC, Australia

Yiliao Song
The University of Adelaide
Adelaide, SA, Australia

Yun Sing Koh
The University of Auckland
Auckland, New Zealand

Wei Xiang
La Trobe University
Bundoora, VIC, Australia

Derui Wang
CSIRO's Data61
Clayton, VIC, Australia

ISSN 0302-9743 ISSN 1611-3349 (electronic)
Lecture Notes in Artificial Intelligence
ISBN 978-981-96-0347-3 ISBN 978-981-96-0348-0 (eBook)
https://doi.org/10.1007/978-981-96-0348-0

LNCS Sublibrary: SL7 – Artificial Intelligence

This Springer imprint is published by the registered company Springer Nature Singapore Pte Ltd.
The registered company address is: 152 Beach Road, #21-01/04 Gateway East, Singapore 189721, Singapore

Preface

This volume contains the papers presented at the 37th Australasian Joint Conference on Artificial Intelligence (AJCAI 2024). The conference was held during November 25 – November 29, 2024, and was hosted by the University of Melbourne and the Royal Melbourne Institute of Technology in Melbourne, Australia. This annual conference is one of the longest running conferences in artificial intelligence, with the first conference held in Sydney in 1987. The conference remains the premier event for artificial intelligence in Australasia, offering a forum for researchers and practitioners across all subfields of artificial intelligence to meet and discuss recent advances.

AJCAI 2024 received 108 submissions, and each submission was reviewed by at least two Program Committee (PC) members or external reviewers in a double-blind process. After a thorough discussion and rigorous scrutiny by the reviewers, 62 papers were accepted for long oral presentations. These 62 submissions were accepted for publication as full papers in these proceedings with an acceptance rate of 58%. AJCAI 2024 had two keynote talks by the following distinguished scientists: Liming Zhu, Flora Salim.

The following are notable aspects of the AJCAI 2024 conference:

- AJCAI 2024 was held jointly with the Defence Artificial Intelligence 2024 Symposium (November 26, 2024). The Defence Artificial Intelligence Symposium is an exciting opportunity for Defence and AI researchers to come together and explore priorities, opportunities and commonalities.
- The conference included an Industry Day on November 28. The focus of the Industry Day was on discussion of the recent Australian Government voluntary AI safety standard, and consultations on mandatory AI guardrails for high-risk settings.
- The conference included a Special Session on November 29. The focus was on AI Bias and Ethics. Invited speakers at the session were Rita Arrigo, Geoff Webb, Anthony McCosker, Jeannie Paterson and Rahil Garnavi.
- The AJCAI 2024 program included one workshop, held on November 25: Trustworthy Federated Intelligence by Guodong Long.
- The AJCAI 2024 program included six tutorials, held on November 25: Towards Safe and Controlled Large Language Models by Hung Le, Instance Space Analysis for Rigorous and Insightful Algorithm Testing (ISA) by Kate Smith-Miles and Mario Andres Munoz, Continual Learning for Large Language Models by Tongtong Wu and Trang Vu, Towards Autonomous Machine Learning: Evolution of AutoML, Roles Of Humans, and Related Topics by Bogdan Gabrys and Thanh Tung Khuat, Theory and Practice of AI Safety by Andrew Cullen and Hanxun Huang, and Functional Bayesian Deep Learning: Beyond Function Approximation to Function Distribution Approximation by Junyu Xuan.
- The AJCAI 2024 program included a PhD Forum, held on November 26, to mentor and assist postgraduate students in developing their research, with mentorship provided by research leaders.

- The AJCAI 2024 program included an Encore track, allowing speakers to present papers that were previously accepted at top-tier journals and conferences.

We especially appreciate the work of the members of the Program Committee and the external reviewers for their expertise and tireless effort in assessing the papers within a strict timeline. We are also very grateful to the members of the Organising Committee for their efforts in the preparation, promotion and organisation of the conference, especially the general chairs, James Bailey and Sarah Monazam Erfani, for coordinating the whole event.

Lastly, we thank the National Committee for Artificial Intelligence of the Australian Computer Society, Springer, for the professional service provided by the Lecture Notes in Artificial Intelligence editorial and publishing teams, and our conference sponsors: Melbourne Connect, the School of Computing Information and Systems at the University of Melbourne, the School of Computing Technologies at RMIT, the School of Science, Computing and Engineering Technologies at Swinburne University, Pioneer, Yep AI, Defence Artificial Intelligence Research Network (DAIRNET), Australian Institute of Machine Learning at the University of Adelaide, and Australian Computer Society.

October 2024

Mingming Gong
Yiliao Song
Yun Sing Koh
Wei Xiang
Derui Wang

Organization

General Chairs

James Bailey	University of Melbourne, Australia
Sarah Erfani	University of Melbourne, Australia

Program Committee Chairs

Mingming Gong	University of Melbourne, Australia
Yiliao Song	University of Adelaide, Australia
Yun Sing Koh	University of Auckland, New Zealand

Advisory Board

Sally Cripps	University of Technology Sydney, Australia
Tongliang Liu	University of Sydney, Australia
Jie Lu	University of Technology Sydney, Australia
Guodong Long	University of Technology Sydney, Australia
Abdul Sattar	Griffith University, Australia
Andy Song	RMIT University, Australia
Dacheng Tao	Nanyang Technological University, Singapore
Geoff Webb	Monash University, Australia
Miao Xu	University of Queensland, Australia
Chengqi Zhang	University of Technology Sydney, Australia
Mengjie Zhang	Victoria University of Wellington, New Zealand

Senior Program Committee

Feng Liu	University of Melbourne, Australia
Runnan Chen	University of Sydney, Australia
Sasha Rubin	University of Sydney, Australia
Jingfeng Zhang	University of Auckland, New Zealand
Yanbin Liu	Auckland University of Technology, New Zealand
Xin Yu	University of Queensland, Australia
Xiaogang Zhu	University of Adelaide, Australia

Susan Wei	University of Melbourne, Australia
Lingqiao Liu	University of Adelaide, Australia
Yu Yao	University of Sydney, Australia
Yuxia Wu	Singapore Management University, Singapore
Shan Xue	Macquarie University, Australia
Miaomiao Liu	Australian National University, Australia
Yuxuan Du	JD Explore Academy, China
Haytham M. Fayek	RMIT University, Australia
Liam Hodgkinson	University of Melbourne, Australia
Xuequan Lu	La Trobe University, Australia
Junyu Xuan	University of Technology Sydney, Australia
Miao Xu	University of Queensland, Australia
Yadan Luo	University of Queensland, Australia
Lizhen Qu	Monash University, Australia
Weidong Cai	University of Sydney, Australia

Program Committee

En Yu	University of Technology Sydney, Australia
Huaxi Huang	Shanghai Artificial Intelligence Laboratory, China
Wenqin Liu	University of Melbourne, Australia
Jinhao Li	University of Melbourne, Australia
Tong Wu	Auckland University of Technology, New Zealand
Xiangyu Sun	University of Sydney, Australia
Changlu Chen	University of Technology Sydney, Australia
HaoChuan Xu	University of Auckland, New Zealand
Yimin Deng	Xi'an Jiaotong University, China
Jiyang Zheng	University of Sydney, Australia
Brendon J. Woodford	University of Otago, New Zealand
Fangfang Zhang	Victoria University of Wellington, New Zealand
Lynn Miller	Monash University, Australia
Ying Bi	Zhengzhou University, China
Bach Hoai Nguyen	Victoria University of Wellington, New Zealand
Ran Wang	University of Technology Sydney, Australia
Ziye Chen	University of Melbourne, Australia
Jiaxin Huang	Mohamed bin Zayed University of Artificial Intelligence, United Arab Emirates
Marcus Gallagher	University of Queensland, Australia
Guohang Zeng	University of Technology Sydney, Australia
Feng Zhu	University of Technology Sydney, Australia
Jinghe Yang	University of Melbourne, Australia

Huiqiang Chen	University of Technology Sydney, Australia
Angus Dempster	Monash University, Australia
Chengyi Cai	University of Melbourne, Australia
Ickjai Lee	James Cook University of North Queensland, Australia
Yanjun Shu	Harbin Institute of Technology, China
Kun Wang	University of Technology Sydney, Australia
Liangwei Nathan Zheng	University of Adelaide, Australia
Jiahao Ma	Australian National University, Australia
Shaofei Shen	University of Queensland, Australia
Yeliz Yesilada	University of Manchester, UK
Yi Mei	Victoria University of Wellington, New Zealand
Kun Han	University of Queensland, Australia
Xin Guo	University of Queensland, Australia
M. A. Hakim Newton	University of Newcastle, Australia
Huan Huo	University of Technology Sydney, Australia
Yixuan Qiu	University of Queensland, Australia
Jing Teng	North China Electric Power University, China
Jianglin Qiao	University of South Australia, Australia
Markus Wagner	University of Adelaide, Australia
Shanshan Ye	University of Technology Sydney, Australia
Kairui Guo	University of Technology Sydney, Australia
Stephen Chen	York University, Canada
Ningyuan Zhang	University of Technology Sydney, Australia
Peter Baumgartner	Commonwealth Scientific and Industrial Research Organisation, Australia
Wenjie Wang	University of Melbourne, Australia
Jun Wang	University of Sydney, Australia
Wenhua Zhang	University of Hong Kong, China
Dongting Hu	University of Melbourne, Australia
Jean Lee	University of Sydney, Australia
Pengqian Lu	University of Technology Sydney, Australia
Aoqi Zuo	University of Melbourne, Australia
Zehong Cao	University of South Australia, Australia
Yuhao Wu	University of Sydney, Australia
Qiang Qu	University of Sydney, Australia
Rina Carines Cabral	University of Sydney, Australia
Changqin Huang	Zhejiang Normal University, China
Dung Ngoc Nguyen	Commonwealth Scientific and Industrial Research Organisation, Australia
Weijia Zhang	University of Newcastle, Australia
Chang George Dong	University of Adelaide, Australia

Zesheng Ye	University of Melbourne, Australia
Yiqiao Li	University of Technology Sydney, Australia
Matthew Damigos	Ionian University, Greece
Haodong Chen	University of Sydney, Australia
Ruijiang Dong	University of Melbourne, Australia
Muxing Li	University of Melbourne, Australia
Jianhua Yang	University of Western Sydney, Australia
Cong Lei	University of Sydney, Australia
Wei Duan	University of Technology Sydney, Australia
Tim Miller	University of Queensland, Australia
Yexiong Lin	University of Sydney, Australia
Yawen Zhao	University of Queensland, Australia
Jiale Liu	Pennsylvania State University, USA
Maurice Pagnucco	University of New South Wales, Australia
Jiacheng Zhang	University of Melbourne, Australia
Muyang Li	University of Sydney, Australia
Yuan Liu	Hong Kong University of Science and Technology, China
Daokun Zhang	University of Nottingham Ningbo China, China
Sung-Bae Cho	Yonsei University, South Korea
Youquan Liu	Hochschule Bremerhaven, Germany
Dharmender Salian	Avco Consulting, USA
Zihe Liu	University of Technology Sydney, Australia
Xueping Peng	University of Technology Sydney, Australia
Yuanyuan Wang	University of Melbourne, Australia
Jiyang Zheng	University of Sydney, Australia
Chenhao Zhang	University of Queensland, Australia

Sponsors

Contents – Part I

Trustworthy and Explainable AI

Machine Learning and Data Mining

Contents – Part II

Computer Vision

Knowledge Representation and NLP

DELA: Dual Embedding Using LSTM and Attention for Asset Tag Inference in Industrial Automation Systems

Zhen Zhao[1,3](✉), Brian Kenneth Erickson[2], Shantanu Chakraborty[1], and Wei Liu[3]

[1] AVEVA Group plc (Australia), Sydney, NSW, Australia
{zhen.zhao,shantanu.chakraborty}@aveva.com,
zhen.zhao-1@student.uts.edu.au
[2] AVEVA Group plc (USA), Lake Forest, CA, USA
brian.erickson@aveva.com
[3] University of Technology Sydney, Sydney, NSW, Australia
wei.liu@uts.edu.au

Abstract. Artificial Intelligence (AI) is a key driver of the Industry 4.0 revolution. In industrial automation systems, data points of assets are represented by globally unique identifiers known as "Tags," which often contain abbreviated asset and attribute information. These abbreviations need translation into concrete names to map data points to their corresponding assets. In this paper, we introduce DELA (Dual Embedding using LSTM and Attention), an innovative deep learning approach that uses two neural networks to classify "Attribute" and "Asset" for tag-to-asset mapping. The models are trained on real-world industrial standard datasets from the automation industry. To evaluate the generalization of our models, our experiments included a testing dataset with numerous abbreviations not present in the training set. This setup ensures that DELA can handle data with uncommon naming conventions. Our extensive experiments show that DELA efficiently achieves surpassing performance over current state-of-the-art approaches.

Keywords: Artificial Intelligence · Industry 4.0 · Information Extraction · Tag Inference · LSTM · Attention Mechanisms

1 Introduction

The Fourth Industrial Revolution (Industry 4.0) is transforming global production and supply chain operations through advanced technologies like Artificial Intelligence (AI) and the Internet of Things (IoT). It promises smarter, more flexible, and efficient industrial activities. A key feature of this new era is self-monitoring technology, enabling machines to diagnose issues automatically, reducing the need for human intervention. This advancement is expected to enhance production capacity, improve product quality, and lower costs. Motivated by Industry 4.0, industries and academics are increasingly integrating neural networks (NN) based machine learning (ML) methods into automation

M. Gong et al. (Eds.): AI 2024, LNAI 15442, pp. 3–15, 2025.
https://doi.org/10.1007/978-981-96-0348-0_1

control systems. These systems produce vast daily data, including industrial asset configurations and state changes. The data are valuable for ML models to address issues like anomaly detection [1] and recognizing underperforming assets [2].

However, the massiveness of data introduced new challenges. Applying ML in manufacturing and process industries requires data scientists with industrial knowledge to manage and interpret extensive data, identifying relevant data points for each asset during feature and target variable selection. A comprehensive knowledge base (KB) that groups data points, assets, processes, and subsystems is crucial. This KB aids engineers in performance analysis, issue diagnosis, system maintenance, and simplifies feature engineering for ML solutions.

Most real-world automation systems use tag-based databases, where a tag represents the value of a data point associated with an asset. These databases typically employ a flat namespace, so engineers often incorporate the system's hierarchy - such as asset type, location, function, and other relevant properties - into tag names for readability. As a result, a tag's name can provide information about the associated asset and its attributes. Naming a tag involves two primary elements: the structure of the tag and the abbreviations used for assets and attributes. In legacy automation systems, tags often require manual intervention by engineers and data scientists to decipher abbreviations and map them to physical assets, creating huge human effects and labour costs.

To address this problem, in this paper, we design a novel interactive neural network framework, DELA (Dual Embedding using LSTM (Long Short-Term Memory) and Attention), with a dual embedding strategy for automatic mapping of unstructured tags and resolution of abbreviations. Through experiments conducted in real-world industrial settings, DELA has demonstrated superior performance compared to other advanced approaches.

2 Background

In this section, we underlay the current state of arts in the legacy automation system and the potential problems in using them. We also discuss the distinctive requirements of industrial automation and explain why Large Language Models (LLMs) are not optimal for solving these problems, despite their proven effectiveness in many other domains.

2.1 Data Representation in the Legacy Automation Systems

The structure of a tag is how a system's hierarchy is represented in the tag name. Figure 1 is an example that follows a structure $\{Asset\ Code\}\{Area\ ID\}\{Sub\ Area\ ID\}\{Asset\ ID\}\_\{Attribute\ Code\}$.

While structure recognition (which relies on case sensitivity, delimiters, and other character patterns typically handled by rule-based methods like regular expressions) falls beyond this research's scope, resolving abbreviations remains crucial. Abbreviations are commonly used in tag names to shorten them and

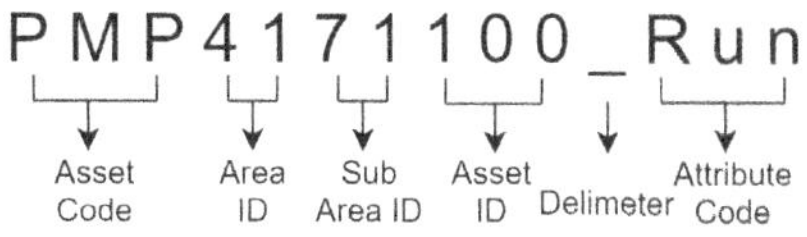

Fig. 1. A Tag Represents the Running Status of a Pump, Using 'PMP' as the Code for the Asset and 'Run' for the Attribute.

save time in coding. For example, `PMP4171100_Run` indicates whether the pump with ID 4171100 is running. However, due to the lack of standardization in early industry 3.0, interpreting tags and linking them to physical devices can be challenging without sufficient knowledge or reference documents. This challenge motivated the development of a generic model capable of extracting asset and attribute knowledge from tags.

Tag inference faces challenges due to varying naming conventions between training and real-world datasets, leading to a high number of out-of-vocabulary (OOV) tokens and reduced inference accuracy. Despite unpredictable naming standards, abbreviations for the same asset or attribute tend to be similar, typically consisting of a subset of letters from the original words in a specific order. From an NLP (Natural Language Processing) perspective, these abbreviations share the same semantics.

Traditional methods for semantic similarity involve word embedding, defining similarity metrics, and searching for nearest neighbors. Existing work [3] improved word similarity measurement significantly by combining source words with their nearest neighbors. In fields like biomedicine, where abbreviations are prevalent, ambiguities introduced by abbreviations complicate information extraction. Methods to address these ambiguities include clustering techniques [4], bidirectional LSTM NNs [5], and attention NNs [6].

2.2 Distinctive Requirements and Features of Industrial Automation

Legacy industrial automation systems typically operate on isolated networks, making it challenging to utilize cloud-based LLM services [18]. Due to the sensitive nature of the information, on-premises data processing is preferable. Processing vast amounts of tags and assets in large control systems requires significant computational resources, especially if models are complex. LLMs, designed for natural language tasks, may not be optimized for industrial data, necessitating customization for unique naming conventions. When systems are upgraded, models need fine-tuning or retraining to adapt. Active learning sometimes is also needed, as it allows for intermediate corrections by system engineers and domain experts. Thus, customized models and on-premises solutions can be more cost-effective.

3 Our Proposed Method for Tag Inference

In this section, we introduce our proposed method, DELA (Dual Embedding using LSTM and Attention), for inferring tags from abbreviations in tag names and properties such as data type, category, and engineering units.

3.1 Basic Model Architectures

We first introduce two basic model structures before formally introducing our DELA model. The first basic architecture is structured with six distinct layers, beginning with an embedding layer. The embedding layer utilizes a trainable Keras embedding (TKE) [20]. This layer takes indices of words in the input sequence, where each index corresponds to a word in a vocabulary dictionary that has been constructed from the training data.

Followed by the embedding layer is a Bidirectional Recurrent NN (Bi-LSTM) Layer. The collaboration of the input, forget, and output gates enables LSTM layers to process sequential data by remembering long-term dependencies [13].

The Bi-LSTM layer concatenates the outputs of the LSTMs from both directions to include more information from the original sequence.

Followed by the Bi-LSTM layer is an attention layer. The attention layer helps with identifying the parts of the sequences that are most influential for the classification task. An attention mechanism can be written in mathematical format as:

$$Attention(Q, K, V) = softmax(\frac{QK^T}{\sqrt{d_k}})V \tag{1}$$

Where Q, K, and V represent the query matrix, the key matrix, and the value matrix respectively, and d_k is the dimensionality of the keys [12].

Our work employs self-attention in which,

$$Q = O_{Bi-LSTM} \times W_Q \tag{2}$$

$$K = O_{Bi-LSTM} \times W_K \tag{3}$$

$$V = O_{Bi-LSTM} \times W_V \tag{4}$$

Where $O_{Bi-LSTM}$ is the output from Bi-LSTM layer and W_Q, W_K and W_V are trainable weight matrix.

The Bi-LSTM and attention layers convert the initial vectors generated by the embedding layer to contextual vectors. The new vectors are then processed by three fully connected dense layers, in which the first two use Rectified Linear Unit (ReLU) as the activation function, and the third uses softmax, to produce the probability distribution for the concrete name of the asset or attribute.

Considering the effectiveness of categorical cross entropy (CEE) on error measurement when training deep NN for classification problems, we employed it as our loss function.

Anticipating low accuracy due to the aforementioned OOV issue, we propose a second basic model using subword tokenization, which splits words into

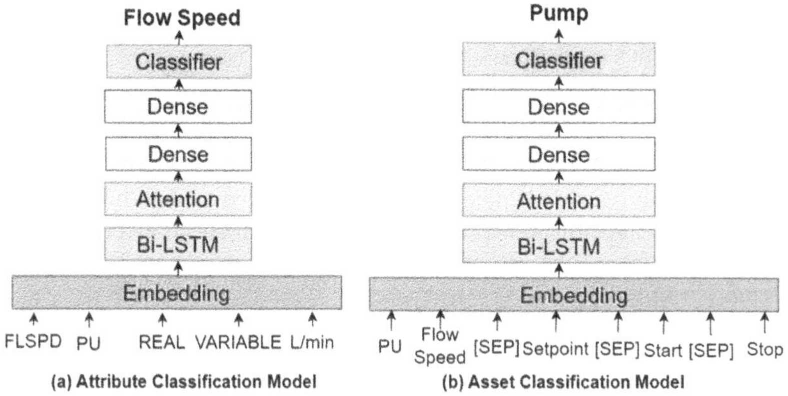

Fig. 2. The NN Architectures of Attribute Classification Model (a) and Asset Classification Model (b) Using Attentitive Bi-LSTM.

smaller chunks. We will utilize BERT (Bidirectional Encoder Representations from Transformers), BPE (Byte Pair Encoding) and Unigram Language Models [15]. This experiment aims to determine if subword tokenization improves accuracy over word tokenization for Tag inference. The architectures of the basic models are shown in Fig. 2.

In the following two sections, we are going to discuss the two novel embedding models designed for converting asset and attribute abbreviations to numerical representations. We propose combining character embeddings for asset and attribute abbreviations with word embeddings for other tokens to effectively interpret diverse elements within tag definitions. The updated model replaces word/subword embeddings with attribute sequence embeddings and asset sequence embeddings in the Attribute and Asset Classification Models, respectively. This requires character-level tokenization for asset and attribute abbreviations and word-level tokenization for other sequences, termed Character level Representation for Abbreviations (CRA). The first embedding model draws inspiration from the Transformers architecture [12], in which positional encoding and multi-head attention mechanisms are employed. In the second model, LSTM layers coupled with attention layers are used.

3.2 Dual Embedding Using Positional Encoding and Multi-head Attention (DEPA)

In this design, in the Attribute Classification Model, the input comprises abbreviations which are initially processed through a character embedding layer that transforms each character in the input into a vector. Following this, positional encoding, that imbues the sequence with information about the order of the characters, is applied to the vectors produced by the embedding layer. In the project, the same sine and cosine functions used in [12] are applied.

By applying positional encoding, the vector generated from embedding layer is converted to a new vector, by:

$$O_{pos} = O_{embedding} + V_{pe} \tag{5}$$

Where $O_{embedding}$ is the output from the embedding layer and V_{pe} is the same output positionally encoded. The positionally-encoded embeddings, O_{pos}, are then directed to a multi-head attention layer. The multi-head attention concatenates the output from each attention, called head, to a new vector, with linear transformation applied:

$$O_{multihead} = Concat(Head_1, ..., Head_k)W^o \tag{6}$$

Where k is the number of heads and W^o is the weight matrix for applying linear transformation to the concatenation. Multi-head attention allows the model to focus on different parts of the input sequence, enabling it to capture various aspects of the input simultaneously. The output from this attention layer is then combined with the original positionally encoded vectors using an addition operation:

$$O_{add} = O_{pos} + O_{multihead} \tag{7}$$

The resulting vectors undergo normalization in a layer normalization step, in order to stabilize the learning process and help in faster convergence.

The normalized vectors are subsequently fed into a dense layer. For the asset code and attribute code parts of the input, their respective outputs from the dense layer are concatenated. This concatenated output then passes through another dense layer, further refining the representation.

Simultaneously, the remaining parts of the sequence - the data type, category, and engineering unit values - go through a similar process of embedding, positional encoding, attention, and dense layers. The outputs from these layers are then appended to the previous concatenated asset and attribute code vectors.

The dense layers employ a linear activation function. The final output of this entire process is a comprehensive embedding that represents all components of the input. This embedding is then utilized in attribute inference, by the classification model to make predictions (Fig. 3a).

Similarly, in Asset Classification Model, the abbreviation used as the asset code and attributes of the asset go through the same process to generate a numerical representation of the features describing an asset for the inferencing task (Fig. 3b).

3.3 Dual Embedding Using LSTM and Attention (DELA)

In our second design, as illustrated in Fig. 4a, the attribute resolving process begins with abbreviations used as asset and attribute being fed into a character embedding layer. This layer transforms each abbreviation into numerical vectors. The vectors then pass through an LSTM layer. The LSTM layer is adept at

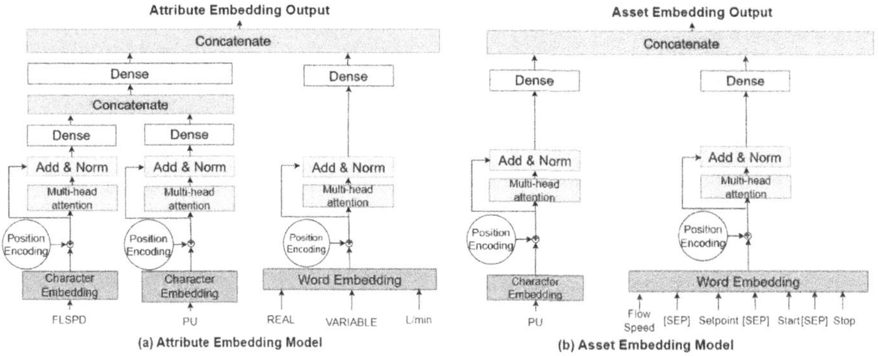

Fig. 3. The Architecture of DEPA for Generating Attribute and Asset Sequence Embeddings by Utilizing Positional Encoding and Multi-head Attention.

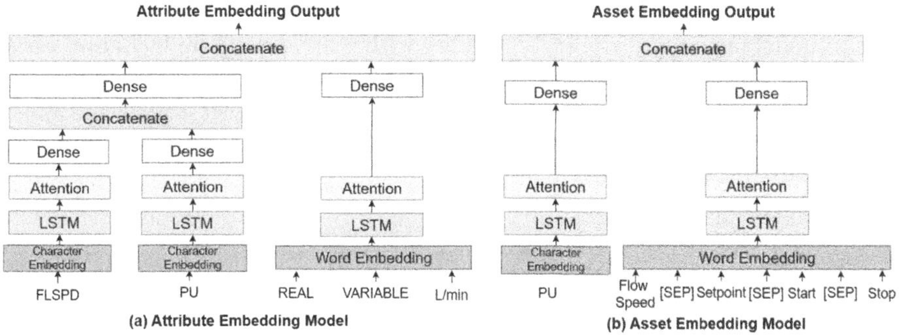

Fig. 4. The Architecture of DELA for Generating Attribute and Asset Sequence Embeddings by Utilizing LSTM and Self-attention.

handling sequences by capturing both short-term and long-term dependencies within the input. Following the LSTM layer, an attention mechanism is applied. This attention layer enables the model to focus on the most relevant parts of the sequence, enhancing the model's ability to capture important features from the LSTM output. The output of the attention layer is then processed through a dense layer, which serves to convert the attention-modulated features into a more suitable representation for the subsequent stages of the model.

The dense layers are equipped with a linear activation function. The outputs of the dense layers for both asset and attribute abbreviations are concatenated to form a new vector. This concatenated vector is a comprehensive representation that combines information from both asset and attribute abbreviations. In parallel, the model processes additional properties of the attribute such as data type, category, and engineering unit, which undergo a similar sequence of embedding, that involves LSTM, attention, and dense layers. The resulting vector from this process is then appended to the previously obtained concatenated vector.

The combined vector, now representing all aspects of the attribute including its abbreviations and properties, is used as the final embedding for the attribute inferencing task. Asset abbreviation resolving utilizes a similar process for embedding that involves character and word embeddings, LSTM layers, attention layers and dense layers (Fig. 4b). The output encapsulates the essential characteristics of the asset and its attributes, making it suitable for the asset inferencing task.

4 Experiment Results

Our model aims to infer the full names of assets and their attributes from tags. The training set consists of 23,050 tags from 2,474 asset instances and 904 labels combining asset and attribute full names. These were prepared based on industry standards such as ISO14224:2016 [7], ISO14617-1:2005 [8], and IEC61850 [9], and real-world samples.

The model faces four primary challenges with real-world data, including varying naming conventions within the same dataset (e.g., Pump A as PU and Pump B as PMP), ambiguity in abbreviations (e.g., PU could refer to either Pump, or Power Unit, or Process Unit), uncommon naming practices, such as inconsistent use of letters from the asset's full name, and context-specific prefixes or suffixes in attribute names that may confuse the model. The testing dataset was designed to reflect these challenges, containing 52,907 tags for evaluation. Of these, 35,810 tags have unseen asset or attribute abbreviations, meaning over 67% of the test data includes at least one novel abbreviation (i.e., not contained in the training data at all). Additionally, more than 14% of the test data features instances where both abbreviations are new to the model.

4.1 Experiment Design

The problem involves inferring the asset type and attribute from a tag definition, including its name, data type, category, and unit.

It is similar to Morphosyntactic Tagging in NLP, typically addressed using word embedding and sequence classification models [11]. We propose using these models with supplementary techniques.

An asset instance often has multiple tags, each representing an attribute. Identifying assets and their attributes requires understanding abbreviations and their context. Our experiment decouples asset and attribute inference into two steps, emphasizing the importance of token order and meaning representation. During attribute inference, raw tags are converted to sequences in the format: "{*Attribute Code*} {*Asset Code*} {*Data Type*} {*Category*} {*Unit*}". For instance, given a tag "PU130_FLSPD" with attribute code *FLSPD*, asset code *PU*, data type *REAL*, category *VARIABLE*, and engineering unit *L/min*, the input sequence can be formulated as "*FLSPD PU REAL VARIABLE L/min*". These sequences are tokenized, embedded, and input into a classification model to predict the attribute name.

In the asset inferencing step, tags linked to the same asset instance are grouped. For each group, a sequence is composed by appending a sorted list of inferred attributes to the asset code. To ensure clarity, especially in cases where an attribute name contains multiple words, separators are inserted between attributes. The format used is: "{*Asset Code*} {*Attribute 1*} {*Separator*} {*Attribute 2*} {*Separator*} {*Attribute 3*} ...". To predict the asset name, the sequence constructed from an asset type that uses *PU* as the code and contains *Flow Speed*, *Setpoint*, *Start* and *Stop* attributes would be: "*PU Flow Speed [SEP] Setpoint [SEP] Start [SEP] Stop*". These sequences are also tokenized, embedded, and input into a classification model to determine the asset name.

Each tag is then labeled with the predicted asset and attribute names. Tag inference is treated as a multi-class classification problem, with performance evaluated using precision, recall, F1 score, and accuracy.

4.2 Benchmarks

In our experiments, we establish benchmarks using BERT models, incorporating a pre-trained tokenizer and an embedding technique that combines token, segment, and positional embeddings [19]. BERT is a well-known NN architecture widely used for solving various NLP tasks, including resolving abbreviations, due to its state-of-the-art performance. We fine-tune the bert-base-uncased model with our training data and evaluate its effectiveness on our testing dataset. Additionally, we develop custom BERT models trained exclusively on our dataset, utilizing the same tokenizer and embedding methods. The model achieving the highest performance on the testing dataset will serve as the benchmark for comparison with subsequent algorithms and models.

4.3 Result Analysis

The classification models were evaluated on a dataset where over 67% of samples included at least one unfamiliar abbreviation. Initial results showed low accuracy, highlighting the inadequacy of word and subword embeddings for inconsistent tag naming standards. However, the proposed embedding algorithms improved abbreviation resolution. Specifically, DEPA enhanced accuracy to 0.65, while DELA achieved an accuracy of 0.81, outperforming both fine-tuned BERT and custom BERT models. Detailed performance differences are presented in Table 1.

Because of the highest accuracy that DELA delivers, it is selected as our preferred embedding model for the Tag Inference task. To train the models utilizing DELA, Adam optimizer [16] was adopted with $\beta_1 = 0.9$, $\beta_2 = 0.999$, $\epsilon = 10^{-7}$, and $\alpha = 10^{-3}$.

Table 1. Performance of Classification NN (Fig. 2) Using Different Embedding Techniques Compared to Fine-tuned BERT and Custom BERT models.

Category	Model	Accuracy	Precision	Recall	F1 Score
Benchmarks	Fine-tuned BERT (bert-base-uncased)	0.61	0.75	0.61	0.63
	Custom BERT using BERT Embedding (bert-base-uncased)	0.55	0.76	0.55	0.61
	Custom BERT using BERT Embedding (bert-large-uncased)	0.54	0.75	0.54	0.60
Basic Methods	Classification NN using Word Embedding	0.34	0.81	0.34	0.43
	Classification NN using Subword Embedding (BPE)	0.49	0.77	0.49	0.56
	Classification NN using Subword Embedding (Unigram)	0.49	0.65	0.49	0.52
	Classification NN using BERT Embedding (bert-base-uncased)	0.57	0.74	0.57	0.61
	Classification NN using BERT Embedding (bert-large-uncased)	0.49	0.71	0.49	0.54
	Classification NN using BERT Embedding (CRA)	0.69	0.77	0.69	0.70
Advanced Methods	Classification NN using DEPA	0.65	0.75	0.65	0.65
	Classification NN using **DELA**	**0.81**	**0.87**	**0.81**	**0.82**

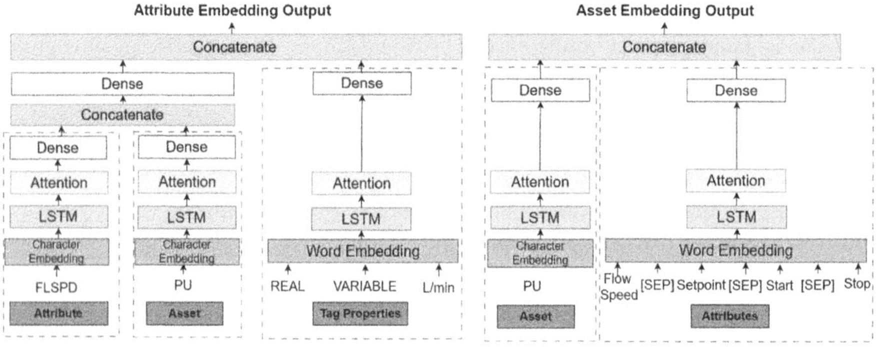

(a) Attribute Embedding Model (b) Asset Embedding Model

Fig. 5. Components of DELA in Attribute and Asset Classification Models for Ablation Study.

4.4 Ablation Study Against DELA

To explore potential optimizations, an ablation study was conducted. We monitored changes in classification performance after omitting components from either the Attribute Embedding Model or the Asset Embedding Model. Figure 5a illustrates the three components within the Attribute Embedding Model: one for processing attribute abbreviations, another for asset abbreviations, and the third for handling tag properties. Figure 5b shows the two components in the Asset Embedding Model, dedicated to processing asset abbreviations and attributes, respectively.

The results suggest that all components positively impact performance. Omitting the Attribute Abbreviation from the Attribute Embedding Model

Table 2. Ablation Study Results Against DELA. The Delta (δ) Columns Report the Changes in the Current Performance Metrics When a Component Is Removed From the Model.

Model	Omitted Component	Accuracy	δ	Precision	δ	Recall	δ	F1 Score	δ
Attribute Embedding	Attribute	0.14	−0.67	0.09	−0.78	0.14	−0.67	0.09	−0.73
	Asset	0.64	−0.17	0.78	−0.09	0.64	−0.17	0.68	−0.14
	Tag Properties	0.75	−0.06	0.83	−0.04	0.75	−0.06	0.77	−0.05
Asset Embedding	Asset	0.27	−0.54	0.31	−0.56	0.27	−0.54	0.26	−0.56
	Attributes	0.76	−0.05	0.84	−0.03	0.76	−0.05	0.77	−0.05

drops accuracy from 0.81 to 0.14, and omitting the Asset Abbreviation from Asset Embedding Model reduces accuracy to 0.27. Although removing other components does not significantly reduce classification performance, they still help resolve data ambiguities, as shown in Table 2.

5 Conclusion

In this paper, we propose an integrative approach to solve the tag-to-asset mapping problem using a fusion of two specially designed NNs. The initial architecture included a Keras embedding layer, a bidirectional Recurrent NN (Bi-LSTM) layer, an attention layer, and three dense layers for both attribute and asset inferences. While LSTM and attention mechanisms effectively resolved ambiguities such as identical abbreviations for different assets or attributes, they were inefficient with tags containing out-of-vocabulary (OOV) abbreviations not present in the training dataset.

To address the OOV problem, we first tested the subword approach, commonly used in NLP solutions. Although it provided slight performance improvement, it was not entirely satisfactory. Therefore, we designed and tested two embedding models employing an innovative dual embedding strategy to replace standard embedding layers. This strategy tokenizes abbreviations into characters while maintaining word tokenization for other features. The first model, DEPA (Dual Embedding using Positional Encoding and Multi-head Attention), uses positional encoding and multi-head attention, along with dense layers, to transform sequences into embeddings. The second model, DELA (Dual Embedding using LSTM and Attention), leverages LSTM, attention, and dense layers for the same purpose.

The training data were prepared based on industrial standards and real-world automation system configurations. The test set included tags with novel abbreviations, simulating scenarios with non-standard naming conventions. Our models outperformed both fine-tuned and custom BERT models in this domain. To further enhance this solution, we plan to integrate active learning into the framework in the future.

References

1. Nazir, S., Patel, S., Patel, D.: Autoencoder based anomaly detection for SCADA networks. Int. J. Artif. Intell. Mach. Learn. **11**, 17 (2021)
2. Teoh, Y., Gill, S., Parlikad, A.: IOT and Fog computing based predictive maintenance model for effective asset management in Industry 4.0 using machine learning. IEEE Internet Things J. **10**, 2087–2094 (2021)
3. Li, C., Ma, T., Zhou, Y., Cheng, J., Xu, B.: Measuring word semantic similarity based on transferred vectors. In: International Conference on Neural Information Processing, pp. 326–335 (2017)
4. Xu, H., Wu, Y., Elhadad, N., Stetson, P., Friedman, C.: A new clustering method for detecting rare senses of abbreviations in clinical notes. J. Biomed. Inform. **45**(6), 1075–1083 (2012)
5. Zhang, C., Biś, D., Liu, X., He, Z.: Biomedical word sense disambiguation with bidirectional long short-term memory and attention-based neural networks. BMC Bioinf. **20**(1), 502 (2019)
6. Zhang, C., Pang, S., Gao, X., Liu, J., Yu, B.: Attention neural network for biomedical word sense disambiguation. Discret. Dyn. Nat. Soc. **2022**, 1–14 (2022)
7. ISO 14224:2016, ISO - International Organization for Standardization. https://www.iso.org/standard/64076.html. Accessed 25 Oct 2023
8. ISO 14617-1:2005, ISO - International Organization for Standardization. https://www.iso.org/standard/41838.html. Accessed 25 Oct 2023
9. IEC 61850:2022 SER Series, IEC - International Electrotechnical Commission. https://webstore.iec.ch/publication/6028. Accessed 25 Oct 2023
10. Sokolova, M., Lapalme, G.: A systematic analysis of performance measures for classification tasks. Info. Process. Manag. **45**(4), 427–437 (2009)
11. Bohnet, B., McDonald, R., Simões, G., Andor, D., Pitler, E., Maynez, J.: Morphosyntactic tagging with a meta-BiLSTM model over context sensitive token encodings. In: Proceedings of the 56th Annual Meeting of the Association for Computational Linguistics (Volume 1: Long Papers), pp. 2642–2652 (2018)
12. Vaswani, A., et al.: Attention is all you need. Adv. Neural Info. Process. Syst. **30** (2017)
13. Xu, G., Meng, Y., Qiu, X., Yu, Z., Wu, X.: Sentiment analysis of comment texts based on BiLSTM. IEEE Access **7**, 51522–51532 (2019)
14. Rusiecki, A.: Trimmed categorical cross-entropy for deep learning with label noise. Electron. Lett. **55**(6), 319–320 (2019)
15. Patel, R., Domeniconi, C.: Estimator vectors: OOV word embeddings based on subword and context clue estimates. In: 2020 International Joint Conference on Neural Networks (IJCNN), pp. 1–8 (2020)
16. Kingma, D.P., Ba, J.: Adam: a method for stochastic optimization. In: 3rd International Conference for Learning Representations (2015)
17. Schuster, M., Paliwal, K.K.: Bidirectional recurrent neural networks. IEEE Trans. Sig. Process. **45**(11), 2673–2681 (1997)
18. Yadav, G., Paul, K.: Architecture and security of SCADA systems: a review. In: Proceedings of the Khosla School of Information Technology, IIT Delhi, India, pp. 1–10 (2024)

19. Devlin, J., Chang, M.-W., Lee, K., Toutanova, K.: BERT: Pre-training of deep bidirectional transformers for language understanding. In: Proceedings of the 2019 Conference of the North American Chapter of the Association for Computational Linguistics: Human Language Technologies, Volume 1 (Long and Short Papers), pp. 4171–4186, Minneapolis, Minnesota. Association for Computational Linguistics (2019)
20. Chollet, F.: Keras. In: Deep Learning with Python, pp. 301–304. Manning Publications Co. (2017)

Combined Change Operators for Trust and Belief

Aaron Hunter(✉)

British Columbia Institute of Technology, Burnaby, Canada
aaron_hunter@bcit.ca

Abstract. We present a formal framework for representing and reasoning about simultaneous changes of trust and belief. We introduce trust states, and we demonstrate how trust states can be explicitly updated by a new class of trust change operators. Trust change postulates are introduced, and a representation result is presented for these operators. We then demonstrate how we can use trust change operators to make implicit updates to trust in other agents, based on the accuracy of the reports provided by other agents. Broadly, agents are more strongly trusted when they provide reports that agree with observation and they are less strongly trusted when they provide reports that conflict with observation. We define combined trust-belief change operators that allow an agent to simultaneously update their trust in other agents while also revising their beliefs. Applications and implementations are considered.

Keywords: Trust · Belief Revision · Knowledge Representation

1 Introduction

Belief revision refers to the process in which an agent incorporates new information with their pre-existing beliefs. In formal models of belief change, the new information is a logical formula that must be believed following revision. This is not always appropriate when information comes from both reports and observations. In these cases, there are actually two different concerns related to trust. First, trust impacts the likelihood that we will believe reports from other agents. This problem has been addressed to some extent in the literature. The second concern is related to *trust change*. Our level of trust in other agents should change, depending on how often their reports agree with our observations. This problem has not been addressed extensively in the context of belief revision operators. In this paper, we introduce a formal approach to modeling the simultaneous dynamics of trust and belief.

Our approach is the following. We first introduce *trust states*, along with a set of postulates for trust change; these postulates specify basic conditions that we expect to hold when we determine that some agent should be more (or less) trusted. We then prove a representation result for operators satisfying these postulates. Finally, we introduce new *combined* change operators, which specify

M. Gong et al. (Eds.): AI 2024, LNAI 15442, pp. 16–28, 2025.
https://doi.org/10.1007/978-981-96-0348-0_2

both how beliefs and trust should change when an agent receives a sequence of reports followed by an observation.

This paper makes several contributions to the literature on belief change. First, *trust change operators* have not been explored in detail in the theory of belief change. As such, both the set of trust-change postulates and the representation result are new contributions in the area. Another contribution is the introduction of combined change operators for trust and belief. Our approach makes the relationship between belief change and trust change explicit, framed in the context of a classical model of belief revision. Finally, the work here opens up opportunities for practical applications in reasoning about reputation systems.

2 Preliminaries

2.1 Belief Revision

The dominant formal approach to belief revision is the AGM approach [1]. Assume a set of propositional variables V. A *state* is a propositional interpretation over V, and a *belief state* is a set of states. A belief revision operator is a function $*$ that maps a belief state K and a formula ϕ to a new belief state $K * \phi$. An AGM revision operator must also satisfy the so-called AGM postulates. Every AGM revision operator can characterized by mapping each belief state to a total pre-order over states [9]. The states believed after revision are the minimal states in this ordering, which are consistent with the formula for revision.

One problem with AGM revision is the fact that it does not support iteration. The most influential approach to iterated belief revision is the DP approach [4]. In this setting, an *epistemic state* $\mathbb{E}$ is a structure that includes a total pre-order $\preceq_{\mathbb{E}}$ over states. A DP revision operator maps an epistemic state $\mathbb{E}$ and a formula ϕ to a new epistemic state $\mathbb{E} * \phi$. Note that the result of DP revision includes the ordering on states which is required for further revision. We let $B(\mathbb{E})$ denote the minimal elements of $\preceq_{\mathbb{E}}$, which are understood to represent the belief state.

2.2 Trust

Trust has been studied extensively in distributed systems and network communication [13,14]. However, it is not a considered in many formal models of belief revision, where new information must be believed following revision. This is, of course, not reasonable if the new information is a report from another agent.

We distinguish between knowledge-based trust and reliability-based trust. Knowledge-based trust is concerned with the domain expertise of a reporting agent. For example, a doctor is trusted on medicine; they may not be trusted on other topics. Knowledge-based trust has been used as a means for ranking search results on the Internet [5]. There has also been work on knowledge-based trust in formal models of belief change [3,11]. However, knowledge-based trust is not the focus of this paper.

We are concerned with reliability-based trust. An agent is reliable if their reports agree with known facts or direct observation. If an agent provides inaccurate reports, they will not be trusted. This has been addressed in [7], where a notion of *conflict* is introduced to determine which reports should be ignored. On the other hand, an agent that is not initially trusted may earn trust by continually providing accurate reports. This problem has not been directly addressed in connection with belief revision. We remark that honesty is one factor related to reliability, but we do not assume agents are lying when a report is incorrect.

3 Trust Change

3.1 Motivating Example

Suppose we are investigating a crime scene and we can receive reports from two agents: Juan(J) and Alma(A). Juan is considered to be trustworthy, whereas Alma is not.

We are initially unsure if the door was forced open (F), and we believe that there is no crowbar in the house($\neg C$). So if our initial epistemic state is $\mathbb{E}$, then $B(\mathbb{E})$ should be the set of models of $(F \wedge \neg C) \vee (\neg F \wedge \neg C)$. Now suppose that we receive a report from Alma that the door was forced open and there is a crowbar in the house. Since Alma is not trusted, this report does not initially trigger a belief change. Juan then reports that the door was not forced open and that there is no crowbar in the house. The most plausible states in our new epistemic state $\mathbb{E}'$ are the models of $\neg F \wedge \neg C$. Suppose that we now observe a crowbar is in the living room. What should be believe now, and who should we trust?

It seems like we should decrease our trust Juan, since he has provided incorrect information. We also need to revisit our trust in Alma. She has provided a report that is consistent with our observation, so our trust in her should increase. We might even want to retroactively believe her initial report.

Reasoning about this kind of problem requires a model that keeps track of beliefs as well as trust in reporting agents. As information is acquired, we not only need to revise our beliefs - we also need to increase (resp. decrease) trust in agents that have provided accurate (resp. inaccurate) reports. In this paper, we introduce a family of combined trust-belief change operators for this purpose. We remark that this kind of reasoning does not only occur in commonsense problems; it is also the basis for trust in reputation systems [6,8].

3.2 Graded Trust Change Operators

We introduce a model of trust change that is defined with respect to a set of agents $\mathbf{A}$. Belief change will be added in Sect. 4. We first define *trust states*, which are ranking functions that capture the trust held in other agents.

Definition 1. *A trust state T is a function $T : \mathbf{A} \to \mathbb{Z}$. We write $\alpha(T) = \{A \mid T(A) \leq 0\}$, and we refer to this as the set of* trusted agents.

Informally, if $T(A) < T(B)$, then A is trusted more than B. The set $\alpha(T)$ is similar to the set of "believed" states in an epistemic state, but there is an important difference. Although the agents in $\alpha(T)$ are all trusted, we still can rank them and to determine which agents are most *strongly* trusted.

We now introduce a simple class of *trust change operators*. In the following definition, an *agent literal* is either A or -A, where $A \in \mathbf{A}$.

Definition 2. *A basic trust change operator is a function $\star$ that maps a trust state T and an agent literal L to a new trust state $T \star L$.*

Intuitively, $T \star A$ is the operation that occurs when A has done something that causes them to be more trusted. For example, if an agent provides a report that is consistent with direct observation, then we will increase trust in that agent. On the other hand, $T \star \text{-}A$ captures the situation where an agent becomes less trusted. This would occur, for example, when the agent has provided a report that is inconsistent with direct observation.

We give some desirable properties for basic trust change operators. The following postulates are all implicitly universally quantified over trust states T, T' and agents A, B. For clarity, we use square brackets to write $[T \star A](A)$, which is the value assigned to A by the trust state $T \star A$.

R1. $[T \star A](A) < T(A)$.
R2. $[T \star A](\text{-}A) > T(A)$.
R3. If $B \neq A$, then $T(B) = [T \star A](B)$ and $T(B) = [T \star \text{-}A](B)$.

Postulate $R1$ says that, when an agent A becomes more trusted, the T-ranking for A decreases. Postulate $R2$ makes the dual statement for agents that become less trusted. Postulate $R3$ states that changing the trust level associated with an agent A does not affect the trust level of any other agent.

We also introduce two postulates to ensure that $\star$ treats all agents equally. In other words, the magnitude of the trust change is equal for all agents:

R4. $T(A) - [T \star A(A)] = T(B) - [T \star B(B)]$.
R5. $[T \star \text{-}A](A) - T(A) = [T \star \text{-}B](B) - T(B)$.

Finally, the change in trust induced by $\star$ is the same for all trust states; the magnitude of trust change is determined by $\star$ and not by the initial trust state:

R6. $T(A) - [T \star A(A)] = T'(A) - [T' \star A(A)]$.
R7. $[T \star \text{-}A(A)] - T(A) = [T' \star \text{-}A(A)] - T'(A)$.

Taken together, these postulates define a class of basic trust change operators.

Definition 3. *A basic trust change operator T that satisfies postulates $R1 - R7$ is called a graded trust change operator.*

Some basic properties follow immediately.

Proposition 1. *Let T be a graded trust change operator. Then the following conditions hold: (1) If $A \in \alpha(T)$, then $A \in \alpha(T \star A)$ and (2) If $A \notin \alpha(T)$, then $A \notin \alpha(T \star \text{-}A)$.*

Hence, trusted agents remain trusted when we use $\star$ to increase trust. The reverse holds for untrusted agents that lose trust. These properties are immediate consequences of $R1$ and $R2$, respectively.

The following proposition states that an agent can always become trusted after a finite number of trust strengthenings, and they can always become untrusted after a finite number of weakenings.

Proposition 2. *If T is a graded trust change operator and $A \in \mathbf{A}$, then there is some n such that $A \in \alpha(T \star^n A)$. Similarly, there is some m such that $A \notin \alpha(T \star^m \text{-}A)$.*

This property is reminiscent of the key postulate for belief improvement operators [10]. This is not surprising, as graded trust change operators are also defined to induce incremental change.

3.3 Representation Result

We introduce a class of transformations on ranking functions over agents.

Definition 4. *Let $r : \mathbf{A} \rightarrow \mathbf{Z}$. If $n \in \mathbf{N}$, then define the ranking functions $r + (A, n)$ and $r - (A, n)$ as follows:*

$$[r + (A, n)](B) = \begin{cases} r(A) + n, & \text{if } A = B \\ r(A), & \text{otherwise} \end{cases}$$

$$[r - (A, n)](B) = \begin{cases} r(A) - n, & \text{if } A = B \\ r(A), & \text{otherwise} \end{cases}$$

Hence $r + (A, n)$ increases the ranking of A and $r - (A, n)$ decreases the ranking.

We can now give a representation result for graded trust change operators.

Proposition 3. *The function $\star$ is a graded trust change operator if and only if there exist positive integers s, w such that*

$$T \star L = \begin{cases} T - (A, s), & \text{if } L = A \text{ for some } A \in \mathbf{A} \\ T + (A, w), & \text{if } L = \text{-}A \text{ for some } A \in \mathbf{A} \end{cases}$$

Proof. Suppose $\star$ is a graded trust operator. Let T_0 be a trust state, and let A_0 be a particular agent. By $R1$, there is some s such that

$$[T_0 \star A_0](A_0) + s = T_0(A_0).$$

By $R4$, it follows that $[T_0 \star A](A) + s = T_0(A)$ for all agents A. By $R3$, we also know that $[T_0 \star B] = T_0(B)$ for all $B \neq A$. Moreover, by $R6$, we know that these equalities are actually true for all trust states T. Therefore $T \star A = T - (A, s)$; so the result holds for positive literals. By parallel reasoning, we can use propositions $R2$, $R3$, $R5$ and $R7$ to show that there is some w that validates the result for negative literals as well.

To prove the converse, suppose that we have two positive integers s, w that define $\star$ as in the definition. Then $R1$ holds because $[T \star A](A) + s = T(A)$ and $s > 0$. Similarly $R2$ holds for w. Postulate $R3$ holds from the definition of the $+$ and $-$ operators, which only increment the ranking for one agent at a time.

For any A, B and any T, T', we have the following equalities:

$$T(A) - [T \star A](A) = s = T(B) - [T \star B](B)$$
$$T(A) - [T \star A](A) = s = T'(A) - [T' \star A](A)$$

The first equality shows that $R4$ holds, and the second equality shows that $R6$ holds. We can prove $R5$ and $R7$ holds through similar equalities, using w as the middle value. Hence, $\star$ is a graded trust change operator.

So graded trust change operators can be fully characterized by two positive integers: the strengthening constant s and the weakening constant w.

There are some interesting variations that we can give to characterize a larger set of basic trust change operators. The following is one such instance.

Proposition 4. *A basic trust change operator $\star$ satisfies postulates $R1 - R3$ and $R6 - R7$ if and only if, for each agent A there is a pair of positive integers s_A, w_A such that:*

$$T \star L = \begin{cases} T - (A, s_A), & \text{if } L = A \text{ for some } A \in \mathbf{A} \\ T + (A, w_A), & \text{if } L = \text{-}A \text{ for some } A \in \mathbf{A} \end{cases} \tag{1}$$

We call such an operator a non-uniform *graded trust operator.*

This proposition states that, if we omit postulates $R4$ and $R5$, then we have a class of operators that is characterized by strengthening and weakening constants that could be distinct for each agent. The proof is similar to Proposition 3.

Additional operators can be defined by modifying postulates $R6$ and $R7$. For example, we could model situations where trust is resilient by making trust decreases very small for strongly trusted agents. We leave a full exploration of such variations for future work.

3.4 Iterated Trust Change

We have defined graded trust change operators for a single agent literal L. However, we will generally be interested in sequences of literals $\overline{L} = L_1, \ldots, L_n$. We will write $T \star \overline{L}$ as a shorthand for $T \star L_1 \star \cdots \star L_n$. Each literal L_i represents a single data point, indicating evidence that a particular agent should be more (or less) trusted. We adopt the following notation:

$$\overline{L}_a = |\{A \mid A \text{ in } \overline{L}\}|$$
$$\overline{L}_c = |\{A \mid \text{-}A \text{ in } \overline{L}\}|$$

Hence $\overline{L}_a$ is the number of positive literals in $\overline{L}$ and $\overline{L}_c$ is the number of negative literals in $\overline{L}$. The a stands for *agreement* while the c stands for *conflict*.

Proposition 5. *Let $\star$ be a graded trust operator, defined by s and w. Then*

$$[T \star \overline{L}](A) = T(A) - \overline{L}_a s + \overline{L}_c w.$$

This result follows directly from Proposition 3, since each increase or decrease is handled independently. So the iterated trust over a sequence of changes is just the aggregate of individual trust change operations. As a result, for any operator $\star$, any sequence $\overline{L}$, and any agent A we can define the following value:

$$\Delta(\star, \overline{L}, A) = [T \star \overline{L}](A) - T(A).$$

Hence, Δ represents the *change* in trust for agent A given the operator $\star$ and the sequence $\overline{L}$. The properties of this change value are given below.

Proposition 6. *Let $\star$ be a graded trust change operator. Then:*

1. *If $\overline{L}_a = \overline{L}_c = 0$, then $\Delta(\star, \overline{L}, A) = 0$.*
2. *If $\overline{L}_a = 0$ and $\overline{L}_c > 0$, then $\Delta(\star, \overline{L}, A) > 0$.*
3. *If $\overline{L}_c = 0$ and $\overline{L}_a > 0$, then $\Delta(\star, \overline{L}, A) < 0$.*
4. *If $\overline{L}_c = \overline{M}_c$ and $\overline{L}_a > \overline{M}_a$ then $\Delta(\star, \overline{L}, A) < \Delta(\star, \overline{M}, A)$.*
5. *If $\overline{L}_a = \overline{M}_a$ and $\overline{L}_c < \overline{M}_c$ then $\Delta(\star, \overline{L}, A) > \Delta(\star, \overline{M}, A)$.*

Item (1) asserts that trust in A does not change if A does not occur in $\overline{L}$. Item (2) says that A becomes less trusted if they only occur in *conflict* literals, while item (3) says the reverse for agents that occur only in *agreement* literals. Item (4) compares different sequences. It says that, if two sequences include the same number of conflict literals, then the one with more agreement literals will have a more positive impact on trust for A. Item (5) makes a similar statement for the case where the number of agreement literals is the same.

Proposition 6 summarizes the properties of aggregate trust change. However, since s and w are not constrained, we can not say anything specific about the aggregate change due to a sequence that includes both conflict and agreement.

Proposition 7. *Let T be a trust state, let $A \in \mathbf{A}$, and let $\overline{L}$ be any sequence of agent literals that contains at least one instance of A and at least one instance of -A. Then there are graded trust change operators $\star_1$, $\star_2$ and $\star_3$ such that*

$$\Delta(\star_1, \overline{L}, A) < 0 = \Delta(\star_2, \overline{L}, A) < \Delta(\star_3, \overline{L}, A).$$

Hence, in the general case, there is no way to determine if $\Delta(\star, \overline{L}, A)$ is positive or negative. This flexibility allows us to define graded trust change operators that handle agreement and conflict very differently. For example, a single conflict might increase the trust ranking as much as a million agreements. So if $\overline{L}$ contains both a strengthening and a weakening for A, then we can not say anything about whether or not A will be trusted unless we know the specific change operator.

4 Interacting Trust and Belief

4.1 Reports and Histories

We now move to the case involving both trust and belief. So we need a signature that includes both agents and properties of the world.

Definition 5. *A multi-agent signature is a pair* $\langle \mathbf{A}, \mathbf{V} \rangle$ *where* $\mathbf{A}$ *is a set of agents,* $\mathbf{V}$ *is a propositional signature.*

The important connection between agents and formulas is that agents provide *reports*, and the content of a report is a propositional formula.

Definition 6. *A* report *is a pair* (A, ϕ) *where* $A \in \mathbf{A}$ *and* ϕ *is a formula over* $\mathbf{V}$. *We write* $\overline{R} = (A_1, \phi_1), \ldots, (A_n, \phi_n)$ *for a finite sequence of reports.*

The problems that we address involve both reports and *observations*. We will normally be concerned with sequences of reports followed by an observation. This concept is formalized below.

Definition 7. *A history-sensitive observation (hs-observation) is a pair* $\langle \overline{R}, \phi \rangle$ *where* $\overline{R}$ *is a report history and* ϕ *is a formula.*

Defining belief change with respect to hs-observations allows us to consider how the observation ϕ informs the extent to which the reports in $\overline{R}$ should be incorporated. In order to represent an agent's beliefs along with their trust in other agents, we define the following notion of an *epistemic trust state*.

Definition 8. *An* epistemic trust state *is a pair* $\langle \mathbb{E}, T \rangle$ *where* $\mathbb{E}$ *is an epistemic state over* $\mathbf{V}$ *and* T *is a trust state over* $\mathbf{A}$.

Note that $\mathbb{E}$ and T are independent, but we will define change operators that impact them both at the same time.

4.2 A Family of Combined Change Operators

We now define combined change operators for trust and belief. The first step is to show how an hs-observation defines a sequence of trust change operations.

Definition 9. *Let* $\langle \overline{R}, \phi \rangle$ *be an hs-observation where* $\overline{R} = (A_1, \phi_1), \ldots, (A_n, \phi_n)$. *For each* $i \leq n$, *let:*

$$L_i = \begin{cases} A_i, & \text{if } \phi_i \not\models \phi \\ \text{-}A_i, & \text{if } \phi \models \phi. \end{cases}$$

Let $\tau(\langle \overline{R}, \phi \rangle) = L_1, \ldots, L_n$.

Hence, $\tau(\langle \overline{R}, \phi \rangle)$ is a sequence of literals. The literal in position i is A_i if the formula reported by A_i in position i is consistent with ϕ. The literal in position i is -A_i if the formula reported by A_i in position i is inconsistent with ϕ. Intuitively, this sequence encodes how our trust in each agent should change given the hs-observation $\langle \overline{R}, \phi \rangle$; the agent should be more trusted if they have provided reports consistent with ϕ and they should be less trusted if they have provided reports inconsistent with ϕ.

We use Definition 9 to overload the $\star$ operator, by allowing it to take an hs-observation as input. Specifically, we adopt the following notation::

$$T \star \langle \overline{R}, \phi \rangle = T \star \tau(\langle \overline{R}, \phi \rangle).$$

Hence, when we given an hs-observation as an input to $\star$, we simply pass to the sequence of agent literals $\tau(\langle \overline{R}, \phi \rangle)$. This sequence of literals captures the number of conflict and agreement reports that each agent has provided.

We need one more piece of notation. Given a report history $\overline{R}$ and a set of agents β, we write $\overline{R} \restriction_\beta$ as a short hand for the sequence of formulas ϕ_i where $A_i \in \beta$. So if β represents the set of trusted agents, then $\overline{R} \restriction_\beta$ represents the sequence of formulas reported by trusted agents.

We can now define an approach to combined change for trust and belief.

Definition 10. *Let $\langle \mathbb{E}, T \rangle$ be an epistemic trust state, let $*$ be a DP operator, and let $\star$ be a graded trust change operator. Then $\circ$ is defined as follows:*[1]

$$\langle \mathbb{E}, T \rangle \circ \langle \overline{R}, \phi \rangle = \langle E * \overline{R} \restriction_\beta * \phi, T \star \langle \overline{R}, \phi \rangle \rangle$$

where $\beta = \alpha(T \star \langle \overline{R}, \phi \rangle)$.

Hence, the new trust state is obtained by strengthening and weakening trust in agents, based on whether they have provided reports that are consistent with the observation ϕ. The new epistemic state is obtained by iteratively revising by all reports from trusted agents, and then revising by the observation ϕ. Note that the set of trusted agents used for this revision is determined *after* any trust changes resulting from the given sequence of reports. We illustrate by returning to our motivating example.

Example 1. We can further formalize our motivating example involving Juan (J) and Alma (A) at the crime scene. Let T be the trust state where $T(J) = -1$ and $T(A) = 1$, which reflects our assumption that Juan is initially trusted, while Alma is not. Suppose that $\star$ is the operator defined by the constant 2 for both strengthening and weakening. Recall that Alma reports $F \wedge C$, then Juan reports $\neg F \wedge \neg C$, then we observe C. So we need to calculate the following:

$$\langle \mathbb{E}, T \rangle \circ \langle (A, F \wedge C), (J, \neg F \wedge \neg C), C \rangle.$$

From Definition 10, our new trust state T' assigns $T'(J) = 1$ and $T'(A) = -1$. So after all events, only Alma will both be trusted. This also means that the final epistemic state will be $\mathbb{E} * (F \wedge C) * C$. Hence, we will not only believe the crowbar is in the house, but we will also believe the door was forced open. This is because Alma's report has been incorporated, since she is now trusted.

[1] Note that $\circ$ actually depends on $*$ and $\star$, so it would be more appropriate to write $\circ_{*,\star}$. But this notation is cumbersome, so we omit the subscripts unless they are required to reduce ambiguity.

Note that the we can get a different result, if we return to the example with one small tweak.

Example 2. Consider the same example, except that $T(J) = -3$ while $T(A) = 1$. In this case, the new trust state T' assigns $T'(J) = -1$ and $T'(A) = -1$. So, despite the fact that Juan has provided an erroneous report, he is still trusted. The intuition here is that Juan has built such a strong reputation that he will still be trusted after a single mistake. In this case, the final epistemic state will be $\mathbb{E}*(F \wedge C)*(\neg F \wedge \neg C)*C$. Following this sequence of revisions, we will believe the crowbar is in the house but we will not believe the door was forced open. This holds despite the fact that Alma has reported otherwise, because Juan is still a trusted source.

Many other small tweaks that could be made to get different results. For example, if the $\star$ operator only strengthens trust with a constant $s = 1$, then Alma will not be trusted despite the accuracy of her report. In all of these cases, the basic framework handles the subtle distinctions without any difficulty.

4.3 An Observation-Consistent Variation

A *report filter* is any function that maps an hs-observation $\langle \overline{R}, \phi \rangle$ to a new hs-observation including a subsequence of the reports from the original. A report filter is *observation-consistent* if every report in the output subsequence must be consistent with ϕ.

Proposition 8. *Let T be a trust state. The report filter $\upharpoonright_{\alpha(T)}$ is not guaranteed to be observation consistent.*

This result is important, because it means the $\circ$ operator is based on a filter that can include reports that are inconsistent with the observation.

We have seen this in Example 2, Where Juan's report influences the final epistemic state despite the fact that it is inconsistent with the observation. This seems reasonable in this case, because the report is a conjunction and we end up keeping only the "consistent part." However, in some applications, it would be preferable to discard all inconsistent reports regardless of how much the sender is trusted. We can enforce this condition by providing a modified definition of $\circ$.

Let $\langle \overline{R}, \phi \rangle$ be an hs-observation and let γ be the set of agents that have provided a report in $\overline{R}$ that is inconsistent with ϕ. The following is immediate.

Proposition 9. *Let T be a trust state. For every hs-observation, the report filter $\upharpoonright_{\alpha(T)\cup\gamma}$ is observation consistent.*

This simple change gives us a variation of $\circ$ that is based on an observation-consistent filter. Specifically, we can modify the definition of $\circ$ to define $\circ_{OC}$ as follows:

$$\langle \mathbb{E}, T \rangle \circ_{OC} \langle \overline{R}, \phi \rangle = \langle E * \overline{R} \upharpoonright_{\beta\cup\gamma} * \phi, T \star \langle \overline{R}, \phi \rangle \rangle$$

where β is defined as it was previously. Hence $\circ_{OC}$ is just like $\circ$, except that it filters out all reports inconsistent with ϕ before performing belief revision.

5 Discussion

5.1 Related Work

There has been previous work on the interaction between observations and reports in [7], where a notion of *conflict* is used to determine which reports should be ignored. However, the notion of conflict introduced is restricted, as it is based solely on counting inconsistent reports. More importantly, the framework does not include trust rankings or any model of trust change. There has also been previous work on trust revision, in the tradition started with [12]; however, this work does not consider any direct connection with belief change operators.

Perhaps the closest work in the literature to our approach is in [2], where the authors argue that trust change and belief change can not be separated. They introduce a new class of *information revision* operators that operate on a hybrid state which includes both beliefs and trust. While the motivation of this work is similar to ours, the framework is quite different. Whereas information revision operators are built from scratch to model a single change operation, we build our approach from independent change operators for beliefs and trust. Hence, we maintain that belief and trust change are distinct operations; however, they need to be combined to effectively incorporate reports and observations. In future work, we intend to explore the formal relationship between the two approaches, and the extent to which information revision can be embedded in our work.

5.2 Practical Application

One natural application for our framework is for reasoning about *reputation systems*, where we have a seller that has been rated based on a series of transactions. The information provided by ratings need not simply be judgments about the "goodness" of the seller; they might include information about the product, the promptness of delivery, or anything else about the transaction. All of this can be encoded in a suitable logical theory. When we read a series of reviews and then make a purchase, we are able to simultaneously update our beliefs and our trust in the ratings through the framework introduced in this paper. Automating this process, we can implement a simple reputation system that can maintain a sound trust ranking over all agents providing reviews.

The automation step here is actually not difficult. We have already implemented prototype software for solving report revision problems. The input is a text file that specifies the initial beliefs, a series of reports, and a formula for revision:

```
(Av!B)^(B^!C)
Jordan:AvB, Alma:A^B, obs:AV!B
```

The values for incrementing trust are in a configuration file that can be edited separately. This software will be described in a companion paper; for the moment, we simply point out that automating report revision is straightforward.

5.3 Conclusions

We introduced graded trust change operators, which let us incrementally change how much we trust information-providing agents. We then introduced a set of trust-change postulates, and proved that every operator satisfying the postulates is defined by two values: a strengthening constant and a weakening constant. Trust change operators can be combined with DP belief revision operators to define a new class of combined report-revision operators. These operators take a sequence of reports along with an observation as input, and they simultaneously revise the agent's beliefs and modify their trust in reporting agents. The result is a single operator that combines two rational change functions to ensure both beliefs and trust are changed appropriately.

There are many directions for future work. As noted, we are currently working on an efficient implementation of our approach. At a theoretical level, we remark that basic trust change operators are quite restrictive in that they can only take a literal as input. In future work, we would like to extend the vocabulary of "trust formulas" to permit updates by more complex trust statements. Another direction for future research is to axiomatize the interaction properties of report revision operators. Right now, we know that the revision part satisfies the DP postulates and the trust part satisfies our new trust change postulates. But it would be useful to provide a further set of interaction postulates to describe the properties that must hold when the operators are combined.

References

1. Alchourrón, C.E., Gärdenfors, P., Makinson, D.: On the logic of theory change: partial meet functions for contraction and revision. J. Symb. Log. **50**(2), 510–530 (1985)
2. Yasser, A., Ismail, H.O.: Trust is all you need: from belief revision to information revision. In: Faber, W., Friedrich, G., Gebser, M., Morak, M. (eds.) JELIA 2021. LNCS (LNAI), vol. 12678, pp. 50–65. Springer, Cham (2021). https://doi.org/10.1007/978-3-030-75775-5_5
3. Booth, R., Hunter, A.: Trust as a precursor to belief revision. J. Artif. Intell. Res. **61**, 699–722 (2018)
4. Darwiche, A., Pearl, J.: On the logic of iterated belief revision. Artif. Intell. **89**(1–2), 1–29 (1997)
5. Dong, X., et al.: Knowledge-based trust: estimating the trustworthiness of web sources. Proc. VLDB Endow. **8** (2015)
6. Govindaraj, R., Govindaraj, P., Chowdhury, S., Kim, D., Tran, D.T., Le, A.N.: A review on various applications of reputation based trust management. Int. J. Interact. Mobile Technol. **15**(10) (2021)
7. Hunter, A.: Reports, observations, and belief change. In: Proceedings of the 36th Australasian Joint Conference on Artificial Intelligence, pp. 54–65 (2023)
8. Huynh, T.D., Jennings, N.R., Shadbolt, N.R.: An integrated trust and reputation model for open multi-agent systems. Auton. Agent. Multi-Agent Syst. **13**(2), 119–154 (2006)
9. Katsuno, H., Mendelzon, A.: Propositional knowledge base revision and minimal change. Artif. Intell. **52**(2), 263–294 (1992)

10. Konieczny, S., Pérez, R.P.: Improvement operators. In: Eleventh International Conference on Principles of Knowledge Representation and Reasoning (KR'08), pp. 177–186 (2008)
11. Liu, F., Lorini, E.: Reasoning about belief, evidence and trust in a multi-agent setting. In: 20th International Conference on Principles and Practice of Multi-agent Systems, PRIMA 2017, vol. 10621, pp. 71–89 (2017)
12. Ma, J., Orgun, M.A.: Trust management and trust theory revision. IEEE Trans. Syst. Man Cybern. Part A Syst. Hum. **36**(3), 451–460 (2006)
13. Salehi-Abari, A., White, T.: Towards con-resistant trust models for distributed agent systems. In: Proceedings of the 21st International Joint Conference on Artificial Intelligence (IJCAI), pp. 272–277 (2009)
14. Wang, J., Yan, Z., Wang, H., Li, T., Pedrycz, W.: A survey on trust models in heterogeneous networks. IEEE Commun. Surv. Tut. **24**(4), 2127–2162 (2022)

Highlighting Case Studies in LLM Literature Review of Interdisciplinary System Science

Lachlan McGinness[1,2], Peter Baumgartner[1,2(✉)], Esther Onyango[3], and Zelalem Lema[4]

[1] Australian National University, Canberra, Australia
Lachlan.McGinness@anu.edu.au
[2] Data61—CSIRO, Canberra, Australia
Peter.Baumgartner@data61.csiro.au
[3] CSIRO, Agriculture and Food, Canberra, Australia
Esther.Onyango@csiro.au
[4] CSIRO, Agriculture and Food, Canberra, Australia
Zelalem.Moti@csiro.au

Abstract. Large Language Models (LLMs) were used to assist four Commonwealth Scientific and Industrial Research Organisation (CSIRO) researchers to perform systematic literature reviews (SLR). We evaluate the performance of LLMs for SLR tasks in these case studies. In each, we explore the impact of changing parameters on the accuracy of LLM responses. The LLM was tasked with extracting evidence from chosen academic papers to answer specific research questions. We evaluate the models' performance in faithfully reproducing quotes from the literature and subject experts were asked to assess the model performance in answering the research questions. We developed a semantic text highlighting tool to facilitate expert review of LLM responses.

We found that state of the art LLMs were able to reproduce quotes from texts with greater than 95% accuracy and answer research questions with an accuracy of approximately 83%. We use two methods to determine the correctness of LLM responses; expert review and the cosine similarity of transformer embeddings of LLM and expert answers. The correlation between these methods ranged from 0.48 to 0.77, providing evidence that the latter is a valid metric for measuring semantic similarity.

Keywords: Systematic Literature Review · Large Language Models · Highlighting

1 Introduction

The scientific community is currently full of hype and hope for the use of Artificial Intelligence (AI) to accelerate research [28]. Messeri and Crockett claim

M. Gong et al. (Eds.): AI 2024, LNAI 15442, pp. 29–43, 2025.
https://doi.org/10.1007/978-981-96-0348-0_3

that scientists are too trusting and lack awareness of the biases and errors of Large Language Models (LLMs) [26]. They present 'AI as Oracle' as a vision of the future where LLMs overcome the problem of too much literature to digest by efficiently searching and summarising information [26]. Many research groups are optimising and improving LLM tools for literature review [2,7,18] including the CSIRO's (Commonwealth Scientific and Industrial Research Organisation's) Science Digital program [10].

Many tools exist to enhance or automate literature review, including LitLLM [1], Scite [7], Elicit [18] and Scopus AI [2]. The techniques of these tools remain undisclosed as commercial secrets. However they all appear to use a combination of the same strategies: calling LLMs through APIs, prompt engineering (in-context learning), Retrieval Augmented Generation (RAG) and fine tuning [6]. The lack of transparency about these tools make it difficult to determine and accelerate best practice in AI systematic review methods.

Systematic Literature Review (SLR) was developed in the field of Evidence-Based Medicine as a method of reducing bias by sticking to strict protocols [6]. It has since been adopted in many disciplines including social science, education and environmental science [6]. SLR tasks include planning, searching, screening, extraction and synthesis [6]. In the extraction phase, desired information is extracted from a set of selected studies. Few studies have attempted to objectively measure the capability of LLMs for SLR [11,32,33,36,41] and even fewer for the extraction and screening phases. Previous studies found that LLMs are unreliable SLR tools as they 'hallucinate' references that do not exist [35].

In this paper we evaluate the performance of GPT-3.5 Turbo, and GPT-4 Turbo, one of the best models currently available, on SLR tasks. We will refer to these models as GPT3 and GPT4 respectively. We present four case studies where LLMs were used to assist CSIRO interdisciplinary systems science researchers, including the last two authors of this paper, in different stages of SLR. In each we systematically explore the impact of changing a parameter on the accuracy of LLM responses.

Although automatically checking LLM responses is highly desirable, currently there are no tools that can perfectly check the correctness or semantic similarity of texts. An important and tedious step of using LLMs in SLR is verifying their responses. To make this task easier we contribute a highlighting algorithm, in analogy to highlighting with a text marker on paper. It aims to provide a human reader with visual clues to quickly scan generated text. The algorithm is driven by a small set of user-supplied keywords provided by a domain expert. We describe the algorithm and report on experiments and experiences with application to our case studies.

The rest of this paper is structured as follows. Section 2 summarizes our methodology for application to the case studies chosen. This includes statistical evaluation methods and LLM techniques. Section 3 introduces an algorithm for explainable text relevance in terms of semantic similarity, and communicating it through text highlighting. Section 4 reports on experimental results, and Sect. 5 discusses these results and Sect. 6 summarises the conclusions.

2 Methodology

We explore four case studies with interdisciplinary system scientists. All four studies used Microsoft Azure endpoints to call either GPT3 and GPT4. The first case study focuses on the health impacts of agri-food transitions. In the second case study, the researcher had performed an SLR extracting the enablers and constraints from twelve papers on coordinated responses to crises. They were interested to know if they had missed any key points when completing the SLR. The third case study verifies an SLR involving sixty papers on a sustainable transitions SLR task [27]. The final case study focuses on screening papers for an SLR on the use of generative AI for marking student responses to exam questions. The studies investigate the impact on overall performance when using different models, asking LLMs to provide evidence for their answers, and splitting tasks into several calls.

An example research question from the first case study is "What are the health outcomes of an agri-food transition"? The LLM was tasked with answering the question and finding evidence from an academic paper to support its answer. Evidence would be a quote from the paper such as "An overabundance of food supply alone has been identified as a key cause of the obesity epidemic" and the LLM answer could be "Obesity".

2.1 Statistical and Evaluation Techniques

In this section we outline the statistical methods used to analyse the results. In cases where multiple similar data are available these results are summarised using mean (μ_x) and population standard deviation (σ_x) defined as usual.

In our analysis we use two methods to determine the correctness of LLM responses; expert review and automated similarity metrics of LLM and expert answers. The fist automated similarity metric is SpaCy Semantic Similarity [14]. This method compares two strings the average embedding vector of the tokens in each of the strings is calculated and the cosine similarity is taken. The second method is the cosine similarity between transformer-based embeddings [19,30,38], which we will refer to as 'transformer similarity'. Transformer based embeddings have the ability to take order of words into account.

To compare these two metrics we use Pearson correlation coefficient shown in Eq. 1.

$$\text{Correlation} = r = \frac{\sum_{i=1}^{n}(\mu_x - x_i)(\mu_y - y_i)}{\sigma_x \sigma_y} \tag{1}$$

Uncertainty in correlation values was calculated using the Fisher transformation of correlation [13] with a 95% confidence interval ($Z = 1.96$) using Eq. 2. In order to make these values more comparable to standard deviation they were divided by Z.

$$\text{Correlation Uncertainty} = \Delta r = \frac{1}{Z}\tanh\left(\text{arctanh}(r) \pm \frac{Z}{\sqrt{n-3}}\right) \tag{2}$$

One of the SLR tasks involves the using an LLM to screen papers and determine their relevance. If a paper is relevant we refer to this as positive. A true positive (TP) result is when a model classifies a relevant paper (as determined by experts) as relevant. A false positive (FP) occurs when an irrelevant paper is classified as relevant. True negatives (TN) and false negatives (FN) are defined similarly. False positive and False Negative rates in the normal way.

2.2 LLM Techniques

We converted the research papers from PDF to text with the `pdftotext` utility[1]. Instead of using standard RAG techniques, we provided the entire paper to the LLM as the context windows were large enough. The research scientists were concerned by the biases of LLMs [4,5,31]. Here we outline these and our approach to minimising them.

Selection bias: The LLM might favour, for example, well known authors or recent papers. To avoid this type of bias, the research scientist selected the papers for the SLRs.
Inability to understand nuance: There are many studies which note that Large Language Models are effective for general tasks but can struggle with domain specific knowledge and the nuance of specialised tasks [9,21,24]. We were concerned that LLMs may lack the nuance to determine whether information relates to the study in the paper or referenced studies. To overcome this issue, researchers removed specific pages based on the task that the LLM needed to perform. For example, when asking about the geographic location of the paper, only the abstract and introduction were provided. We also addressed this issue by using in-context learning to provide the model with specific definitions provided by domain experts.

Lack of domain specific knowledge: The domain or even subject specific vocabulary in scientific literature poses challenges for LLM-based analysis. LLMs are trained on vast corpuses of which the subject specific matter is only a small portion. In agri-food transitions and co-ordinated responses to crises research

[1] https://www.xpdfreader.com/pdftotext-man.html.

litterature terms have very specific meanings and LLMs would miss the nuances of these words. We mitigated this problem by providing the LLMs with definitions of subject specific vocabulary as part of the prompts.

Hallucinations: It is well known that LLMs can 'make up' information [17]. To manage this, tasks were broken into two steps. First, the Large Language Model was asked to find quotes that provide evidence of the desired information. Binary (yes/no) verification of the quotes failed because of trivial errors related to unicode characters. Therefore quotes were instead verified using fuzzy text matching using `thefuzz` implementation[2] of the Levenshtein distance metric [20]. The LLM was then given the quotes and asked for a final answer.

3 Semantic Text Highlighting

In this section we introduce a method for semantic text highlighting, or highlighting for short. The idea of semantic highlighting is not new and was originally formulated for information retrieval [16]. In our context we apply highlighting to LLM retrieved evidence. Because the amount of retrieved evidence can still be overwhelming, further support is needed to aid a human reviewer. This is where highlighting comes in.

Like with a text marker on paper, highlighting does not need to be perfect to be useful. However, the highlighted words should be semantically related to a given specific SLR research question. Our method requires a set of keywords representing the research question. Highlighting then boils down to determining which words in a given sentence are semantically related to the keywords and conveying the findings in a useful way. For that, we rely on readily available information-theoretic similarity measures and a carefully curated corpus of English, which requires no training and is explainable.

Our method works as follows. The SLR researcher provides sets of keywords comprising of entities E, relations R and properties P as nouns, (possibly transitive) verbs and adjectives or adverbs, respectively. For example, in our first SLR case study on the topic influences of agri-food transitions on health outcomes the following were used:

Entities E: health, disease, outcome, food, lifestyle
Relations R: explain, affect, improve, stimulate
Properties P: environmental

A word w in a given text is highlighted if it is deemed related to a keyword according to the procedure Similarity(w, C, t) in Algorithm 1, where $C = E \cup R \cup P$. The given text is first parsed with the part-of-speech (POS) parser SpaCy [15]. It assigns a grammatic role t to every word w for determining the best suited subset of C for similarity. The algorithm computes two similarity scores between

[2] https://github.com/seatgeek/thefuzz.

w and that subset in the range $[0, 1]$. One of them is word vector similarity, as readily provided by spaCy, and the other is Wu-Palmer similarity [34,40] based on hypernym-reachability in WordNet [12].

Using WordNet for word similarity is an established and well-researched topic [23]. In our algorithm, the search for similar words is broad and includes (one-step) synonyms, pertainyms, and related derived forms, weighted for each category. This was a design choice motivated by the case study in Sect. 4.2, where the researcher wants to ensure they did not miss any key points in their manual review. Similarity takes scores higher than given thresholds to determine whether w should be highlighted or not. We found that Wu-Palmer similarity often produces results more similar to what a human user expects from a highlighter. Hence, vector similarity acts only as a fall-back. Here are some examples for highlighted evidence text with the keywords above.

1. It is also likely that climate change will contribute to novel occurrences of disease emergence and transmission.
2. Foodborne illnesses significantly influence individuals nutritional status.
3. Changing lifestyles, mainly due to work commitment, have fuelled the increase in numbers eating out and the need for convenience foods.
4. Significant changes have occurred in food systems in the last decades that have contributed to widen such 'holes' in the barriers from phase to phase: agricultural intensification and industrialization causing major environmental deterioration, the increasing distance traveled by food in global markets, and the nutrition transition towards diets rich in ultra - processed food and animal protein are the three cornerstones of such changes.

Entities (nouns, noun chunks) are colored red and relations (verbs) are colored blue. Additional colors are used for supporting words according to their grammatical roles. Properties (adjectives) of colored entities are purple.

For each word, the algorithm can provide an explanation of why it is highlighted. These explanations are helpful for customising parameter settings; for example, we get:

2. Foodborne illnesses `(NCP(Foodborne illnesses, [SimilarTo('disease', 0.95, 'wup')]))` significantly influence `(SimilarTo('affect, 0.84, 'wup'))` individuals nutritional status `(NCP(nutritional status, [SimilarTo('food', 0.91, 'wup')]))`.

In these annotations, `NCP` means 'NounChunkPart', and the similarity of the highlighted word(s) to keyword(s) is indicated as in `SimilarTo(keyword, similarity, kind)`, where `'wup'` is Wu-Palmer similarity.

Highlighting can be useful beyond marking up text excerpts. In one of our case studies below we take the 'highlighting rate' as a statistical measure to assess the similarity between the LLM's response and the researcher's benchmark evaluation.

Algorithm 1 Similarity

Similarity(w, C, t)
Input: w word, C keywords in canonical form (lemmas), t type of w (noun, verb, adjective ...)
Output: Similarity score for w
P-Weight ← 0.95 RF-Weight ← 0.95 WUP-Threshold ← 0.8 VEC-Threshold ← 0.95
$\text{best}_{\text{wup}} \leftarrow \max_{c \in C}$ WUP-X(w, c, t) **or else** 0.0
$\text{best}_{\text{vec}} \leftarrow \max_{c \in C}$ VEC(w, c) **or else** 0.0
if $\text{best}_{\text{wup}} \geq$ WUP-Threshold **and** $\text{best}_{\text{wup}} \geq \text{best}_{\text{vec}}$ **then return** best_{wup}
elif $\text{best}_{\text{vec}} \geq$ VEC-Threshold **then return** best_{vec}
else return 0.0

WUP-X(w, c, t) // Extended Wu-Palmer similarity, considers reachable words from w and c
S_w = Extend(WN-Synsets(w, t)) // WN-Synsets returns synonyms of w
S_c = Extend(WN-Synsets(c, t))
return $\max_{(s_w \xrightarrow{\omega_w} r_w, s_c \xrightarrow{\omega_c} r_c) \in S_w \times S_c} \omega_w \cdot \omega_c \cdot$ WN-WUP(r_w, r_c, t) // Wu-Palmer from WordNet

VEC(w, c) // Vector similarity
if $w = c$ **then return** 1.0
elif both w and c have vector embeddings **then**
  return cosine-similarity of the embeddings of w and of c
else return 0.0

Extend(S, t)
Input: S a set of WordNet synsets
Output: Weighted extension of S by pertainyms and derivationally related forms
$R \leftarrow \emptyset$ // Result relation
for $s \in S$ **do**
  $R \leftarrow R \cup \{s \xrightarrow{1.0} s\}$ // R is reflexive
  for $l \in$ WN-Lemmas(s) **do**
    $R \leftarrow R \cup \{s \xrightarrow{\text{P-Weight}} \text{WN-SynSet}(v) \mid v \in \text{WN-Pertainyms}(l)\}$
    $R \leftarrow R \cup \{s \xrightarrow{\text{RF-Weight}} \text{WN-SynSet}(v) \mid v \in \text{WN-RelatedForms}(l)\}$
return R

4 Results

4.1 Case Study 1: Similarity Metrics in Agri-Food Transition SLR

In this first case study we determine the accuracy of the chosen metrics and explore if there is a significant difference in performance between GPT3 and GPT4. There were eight tasks (as shown in the first column of Table 1) and ten papers. GPT3 and GPT4 were given identical instructions to answer each task in separate calls. We compared the researcher's and models' answers.

The models were asked to record three quotes (evidences), then give a final answer in a second call. Cases where the average fuzzy string similarity was less than 90% were manually investigated. There were 25 and 38 cases where this occurred for GPT4 and GPT3 respectively, resulting in overall error rates of 2% and 5%.

The semantic similarity between LLM and expert answers was calculated using transformer similarity and SpaCy similarity. An example of an LLM/expert answer could be "The Global context is Africa" which has a word count of 5. The correlation between the Transformer similarity score and the expert's judgement of the model answers was 0.48 ± 0.09 and there was almost no correlation (-0.07 ± 0.08) between the SpaCy similarity and expert judgment.

Table 1. Case Study 1: Average and standard deviation for similarity scores, fuzzy matching scores, average word count and expert judged model accuracy are presented for GPT3 and GPT4.

Information	Complexity	Model	Quote Fuzzy Matching Score	Model Average Word Count	Expert Average Word Count	Transformer Similarity	SpaCy Similarity	Model Accuracy
Global Context	Low	GPT-3	97 ± 5	3.5	4.8	0.74 ± 0.25	0.58 ± 0.16	0.9
		GPT-4	98 ± 6	5.2	4.8	0.84 ± 0.06	0.57 ± 0.14	1.0
Associated Health Focus	Low	GPT-3	97 ± 5	5.6	1.6	0.82 ± 0.13	0.57 ± 0.16	0.75
		GPT-4	98 ± 4	7	1.6	0.85 ± 0.05	0.60 ± 0.14	0.94
Transition Pathway	Moderate	GPT-3	95 ± 10	13.4	5.4	0.81 ± 0.04	0.66 ± 0.08	0.65
		GPT-4	97 ± 8	23	5.4	0.85 ± 0.06	0.65 ± 0.21	1.0
Agri-food Boundary	Moderate	GPT-3	97 ± 4	32	17	0.83 ± 0.04	0.77 ± 0.14	0.5
		GPT-4	98 ± 6	50.6	17	0.87 ± 0.03	0.79 ± 0.12	0.85
Public Health Risk	Moderate	GPT-3	99 ± 7	8.8	6.5	0.85 ± 0.05	0.59 ± 0.16	0.7
		GPT-4	97 ± 6	20.5	6.5	0.87 ± 0.06	0.74 ± 0.17	0.95
Synergies	High	GPT-3	97 ± 5	31.3	26.4	0.83 ± 0.03	0.84 ± 0.07	0.25
		GPT-4	98 ± 5	58	26.4	0.81 ± 0.05	0.83 ± 0.07	0.1
Constraints	High	GPT-3	97 ± 5	35	18	0.82 ± 0.02	0.81 ± 0.08	0.44
		GPT-4	98 ± 5	59	18	0.84 ± 0.02	0.83 ± 0.07	1.0
Integrated Solutions	High	GPT-3	97 ± 4	28	30	0.89 ± 0.04	0.90 ± 0.07	0.88
		GPT-4	99 ± 3	50	30	0.89 ± 0.05	0.89 ± 0.07	1.0

4.2 Case Study 2: Impact of Evidence on Coordinated Response to Crisis SLR

In this case study we compare LLM output based on two methods. In the first method ('evidence') the LLM first obtains quotes to support its answer to the question and then writes it's answer. In the second method ('direct') the LLM writes an answer without searching for or providing evidence.

Providing evidence is a good way to increase the trustworthiness of LLM responses. However it will increase the number of completion tokens and therefore cost. We test our highlighting algorithm as an automated similarity metric by calculating the correlation between the highlighted fraction of each expert answer and model answers.

Table 2 shows that the evidence method results in slightly lower SpaCy Semantic Score, vector embedding cosine similarity, human judged accuracy and highlighting correlation. The highlighting correlation score changes by a more significant margin, but also has a greater uncertainty than the other measures.

Table 2. Case Study 2: Prompt tokens, completion tokens, SpaCy similarity, transformer similarity, human judged accuracy and Highlighting Correlation are presented for comparison of direct and evidence-based conditions.

Experimental Condition	Prompt Tokens ($\times 10^3$)	Completion Tokens	SpaCy Similarity	Transformer Similarity	Human Judged Accuracy	Highlighting Correlation
Evidence	20.2 ± 5.7	842 ± 397	0.85 ± 0.06	0.87 ± 0.06	69%	-0.18 ± 0.24
Direct	20.0 ± 5.7	213 ± 64	0.88 ± 0.06	0.90 ± 0.04	72%	0.13 ± 0.25

The number of prompt tokens is not significantly changed by asking the model for evidence, but the number of completion tokens increases fourfold. As the majority of tokens are prompt tokens, it might be expected that the number of completion tokens would have a small impact on the overall cost. However the computational cost of running a transformer in this case is proportional to the number of completion tokens. A technique to more effectively reduce the cost using an LLM for literature review is grouping multiple tasks into a single call as explored in the next case study.

4.3 Case Study 3: Impact of Combining Tasks on Sustainable Transitions SLR

In this case study, we compare conditions that we name 'separate' and 'together'. For the separate condition, GPT4 is called ten times. In each call the paper is provided and the LLM is asked to find an answer and evidence for a research question. In the together condition, the model is given all tasks in one call. The results for both experimental conditions are in Table 3. Note that papers were classified into three groups according to their geographical scale as requested by the researcher. There were twenty papers in each category.

Table 3. Case Study 3: Quote and final answer metrics for both conditions. The expert found significant errors in the first paper from the global scale responses for the together condition. They decided that the together method was not worth pursuing and did not evaluate the remaining together responses.

Experimental Condition	Scale of paper	Prompt Tokens $\times 10^5$	Completion Tokens $\times 10^3$	Fuzzy Text Matching	SpaCy Similarity	Transformer Similarity	Instances of failing to find quotes	Expert Judged Accuracy
Separate	Global	1.1 ± 0.2	1.5 ± 0.2	98.4 ± 2.5	0.84 ± 0.4	0.85 ± 0.01	0	82%
	International /national	4.7 ± 1.7	1.3 ± 0.1	95.3 ± 6.8	0.79 ± 0.12	0.86 ± 0.01	5	84%
	Subnational /Local	2.2 ± 0.7	1.4 ± 0.2	99.4 ± 0.6	0.83 ± 0.03	0.84 ± 0.06	2	75%
Together	Global	0.12 ± 0.02	0.97 ± 0.37	99.5 ± 0.4	0.79 ± 0.5	0.83 ± 0.03	0	N/A
	International /national	0.48 ± 0.16	1.3 ± 0.2	93.1 ± 7.2	0.77 ± 0.16	0.84 ± 0.01	9	N/A
	Subnational /Local	0.22 ± 0.07	0.8 ± 0.2	98.0 ± 1.0	0.81 ± 0.03	0.84 ± 0.01	1	N/A

The researcher did not like the results of the together condition because GPT4 jumbled its responses resulting in a frame-shift. Despite the near tenfold increase in prompt tokens and cost, they preferred the separate condition. In the separate condition the model had no awareness of its answers to the other questions resulting in overlap of the ideas presented for each sub-question.

4.4 Case Study 4: Prompt Variations for Automated Screening

In the final case study the researcher wanted to automatically extract information from papers for the purpose of screening. The researcher had manually reviewed 14 papers from the arXiv and 20 from Pubmed of which 12 and 3 were relevant respectively. These were used as datasets to measure the performance of LLM information extraction. These datasets provide opposite extremes, one where the LLM needs to accept nearly all of the papers as relevant and another where it needs to reject nearly all the papers.

This allowed for a study in simultaneously avoiding false positives and false negatives, see Table 4. Three prompts were used: 'prompt relevant' which indicated that the paper was likely to be relevant, 'prompt irrelevant', and a 'neutral prompt' which makes no indication of the paper's relevance.

Table 4. Case Study 4: SpaCy and transformer similarity, expert verified accuracy, false positive and false negative rates for three different prompts on both datasets. The correlation between expert accuracy and the similarities were calculated to measure the quality of these metrics.

Condition	Dataset	SpaCy Similarity	Transformer Similarity	Expert Accuracy	Spacy-Expert Correlation	Transformer-Expert Correlation	False Positive Rate	False Negative Rate
Prompt	arXiv	0.87 ± 0.08	0.94 ± 0.05	0.83	0.64 ± 0.08	0.77 ± 0.06	1.00	0.00
Relevant	PubMed	0.70 ± 0.17	0.90 ± 0.07	0.86	0.23 ± 0.28	0.69 ± 0.21	0.94	0.00
Neutral	arXiv	0.88 ± 0.7	0.94 ± 0.04	0.82	0.66 ± 0.09	0.66 ± 0.09	1.00	0.08
Prompt	PubMed	0.68 ± 0.17	0.90 ± 0.08	0.83	-0.10 ± 0.25	0.44 ± 0.27	0.88	0.00
Prompt	arXiv	0.87 ± 0.08	0.94 ± 0.04	0.85	0.55 ± 0.1	0.67 ± 0.08	0.00	0.00
Irrelevant	PubMed	0.70 ± 0.15	0.89 ± 0.08	0.83	-0.05 ± 0.26	0.44 ± 0.27	0.12	0.00

The false positive and false negative rates in Table 4 show the LLM was likely to state that a paper was relevant when in fact it was not. The correlations show transformer similarity correlated better with expert review than SpaCy similarity. It was found that SpaCy similary is heavily impacted by changes in capitalisation. In some cases changes in capitalisation alone reduced the similarity to 0.3.

5 Discussion

In the first case study, GPT4 made less mistakes than GPT3 when extracting exact quotes from documents. This may be because of the quality of the model or GPT3's shorter maximum context length. The longest context length available for GPT3 models was 16,000 tokens which was not always sufficient to read an entire paper. When the paper was broken into multiple sections the model was less likely to find relevant quotes. We noticed that GPT3 was more likely to select quotes from the beginning of the context window than GPT4, whose quotes were more evenly distributed from the entire paper.

Across all studies it was found that transformer similarity correlates more strongly with expert opinion than than SpaCy similarity. This indicates that transformer embeddings are a better metric. This is expected given that it is able to take the positions of words into account. The correlation between amounts of text highlighted increased with increased human judged accuracy, showing the correct trend as a measure of semantic similarity. However the uncertainty values were very high and more work would need to be performed to determine if this is a suitable metric.

The highlighting tool's primary purpose is to aid a researcher in sifting through evidence and other LLM response text. The researchers anecdotally confirmed its value.

We originally anticipated that SpaCy's similarity would be a better measure of semantic similarity than transformer similarity because the transformer similarity scores only ranged between 0.7 and 1, while SpaCy similarities scores had much larger ranges. Contrary to our original expectations, transformer similarity ubiquitously correlated more strongly with expert opinion. To make transformer similarity more interpretable for humans we recommend scaling these values before interpretation. This is because a transformer similarity of 0.8 is actually low, despite normal human expectations.

The second case study found that when a model was asked to provide evidence for its claims, it was slightly less accurate on all metrics including expert judgement. This is an unexpected result as normally 'chain of thought' or asking a model to explain its reasoning improves performance [25,39]. As trustworthiness is increased when the model provides verifiable evidence for its answers, this result indicates that there is an unfortunate trade off between accuracy and trustworthiness. The reason could be that the model is 'overloaded' when it needs to focus on multiple tasks at once. This was confirmed by the third case study (Sect. 4.3) where model performance also decreased when the number of tasks it was asked to complete in a single call increased.

In third third case study, as expected the number of prompt tokens is approximately ten times higher for the separate condition compared to the together condition. As far as the researcher was concerned the reduced number of errors in the results was worth the the extra cost. The major issue with the together condition was the possibility for tasks to be jumbled. One area for future work is to apply more advanced parsing techniques to avoid frameshift errors and therefore make the computationally cheaper technique more desirable for researchers. Another area for future work is to provide the model with specific keywords for each call to guide the LLM responses.

The fourth case study demonstrated that SpaCy semantic similarity can be heavily impacted by the capitalisation of words for medium sized models. This is not a desirable property for a system measuring semantic similarity; we argue that writing a sentence in all capital letters makes little change to the semantic meaning.

Readers may be tempted to think that scores such as SpaCy or transformer similarity could be used as better alternatives to subject expert review as they do not contain human biases. However, one needs to be careful in assuming that there is no bias when using metrics like these. There can be an illusion of objectivity, when in fact these models have been trained and validated on data which contains significant unknown biases. In this study, we highly value the opinion of experts who are part of the active research community and have observed that they demonstrate a strong awareness of their own biases.

False positive rates were consistently higher than false negative rates; GPT4 was more likely to think that an irrelevant paper was in fact relevant and most accurately screened papers when prompted to expect that papers may be irrelevant.

Overall GPT3 and GPT4 were able to find and correctly reproduce quotes from a text with 95% and 98% accuracy respectively. For low complexity tasks like finding the title or location of a paper, GPT4 performed with close to 100% accuracy, but accuracy was lower for more nuanced tasks such as identifying enablers in agri-food transitions. The overall approximate average accuracy of GPT4 in answering research questions was 83%.

Our highlighting workflow requires keyword calibration to determine a suitable set of keywords for each research question in an SLR. Currently, keyword calibration is semi-automated process with an expert in the loop. The expert proposes keywords, runs highlighting experiments on sample papers and adjusts the set of keywords according to their observations.

From our experiments we were able to derive some guidelines for calibration. A keywords set yielding a highlighting rate of around 0.4 ± 0.1 on evidence texts often seems to be a good compromise. If it is much higher, often too many irrelevant words are highlighted. If much lower, the domain has not been covered sufficiently and relevant keywords are missing. It is better to use evidence text than expert answers for keyword calibration as they are almost always proper English sentences. Expert assessments however can vary widely and sometimes are just lists of keywords. Hence the proportion of highlighted words is less reliable in this case, resulting in a less meaningful hit rate.

If the hit rate is unusually high this could be because of denser writing (less filler words) or because of denser information. The latter includes the possibility that surprising, additional insights have been unveiled. This helps to get a more complete picture of the problem. Conversely, we found that a much lower highlighting rate typically applies to irrelevant texts.

6 Conclusion

Large Language Models (LLMs) were used to assist interdisciplinary system scientists to conduct four Systematic Literature Reviews (SLR). The topics of the reviews were agri-food system transitions, coordinated responses to crises, sustainable transitions and automated marking. GPT-3.5 Turbo and GPT-4 Turbo had error rates of 5% and 2% when extracting exact quotes from research papers. Levenshtein distance accurately determined the faithfulness of quotes produced by LLMs and was robust to unexpected unicode characters and hyphenations.

When GPT-4 Turbo completed multiple tasks in a single prompt the number of prompt tokens decreased tenfold but with significant losses in accuracy in some cases due to frameshift errors. One area for future work would be to use more advanced parsing techniques to avoid frameshift or other 'jumbling' errors.

The accuracy of the models' answers was found to decrease with complexity of the task. For very simple tasks, expert rating of LLM answer correctness was close to 100% while for highly nuanced tasks it could be as low as 10%. On average it was found that GPT-4 Turbo was able to extract information from papers with approximately 83% accuracy. When screening papers it was found that GPT-4 Turbo was more likely to include irrelevant papers than exclude relevant papers. One area for future work is to provide the model with specific keywords to focus its answers, this may help a model focus on the desired ideas while trying to complete a nuanced task.

It was found that taking the cosine similarity of transformer embeddings of expert and LLM answers was a measure of accuracy that correlated more strongly with expert opinion than SpaCy's semantic similarity score. Nearly all of these transformer embedding cosine scores were in the range of 0.7 to 0.95. In order to make these cosine similarities more human interpretable, we recommend scaling them to take up the full range between 0 and 1. An area of future work would be to use cosine similarity of transformer embeddings for sentence-wise comparison of researcher and LLM answers in order to determine if any important pieces of information are missing.

Although highlighting is designed to assist researchers with manual checking of answers, correlation between amounts of highlighted text is showing some promise as an automated method for measuring the quality of LLM responses. Additional research would need to be conducted to see if this can be used as a valid similarity metric. Another idea for future work is to re-formulate the semantic similarity algorithm (Algorithm 1) with probabilistic logic programming. This would allow for a more flexible and expressive framework. In addition, weight parameters could be rephrased as probabilities and be learned from examples by maximum likelihood estimation.

Acknowledgements. This research was supported by funding from CSIRO Data61 and its Valuing Sustainability Future Science Platform initiative. We thank Enayat Moallemi for providing knowledge, expertise and data for the third case study. We thank Stephen Wan and Shima Khanehzar for helpful discussions.

References

1. Agarwal, S., Laradji, I.H., Charlin, L., Pal, C.: LitLLM: a toolkit for scientific literature review. (2024). https://arxiv.org/abs/2403.08399
2. Agüilera-Cora, E., Lopezosa, C., Fernández-Cavia, J., Codina, L.: Accelerating research processes with Scopus AI: a place branding case study. Rev. Panam. De Comun. **6**(1) (2024). https://doi.org/10.21555/rpc.v6i1.3088
3. Antu, S.A., Chen, H., Richards, C.K.: Using LLM (Large Language Model) to improve efficiency in literature review for undergraduate research. WS on Empowering Education with LLMs (2023)
4. Bender, E.M., Gebru, T., McMillan-Major, A., Shmitchell, S.: On the dangers of stochastic parrots: can language models be too big? In: Proceedings FAccT '21, ACM (2021). https://doi.org/10.1145/3442188.3445922
5. Blodgett, S.L., Barocas, S., Daumé III, H., Wallach, H.: Language (technology) is power: a critical survey of "Bias" in NLP. In: Proceedings 58th Annual Meeting of the ACL. ACL (2020). https://doi.org/10.18653/v1/2020.acl-main.485
6. Bolanos, F., Salatino, A., Osborne, F., Motta, E.: Artificial intelligence for literature reviews: opportunities and challenges. arXiv:2402.08565 [cs] (2024)
7. Brody, S.: Scite. J. Med. Libr. Assoc. **109**(4), 707–710 (2021). https://doi.org/10.5195/jmla.2021.1331

8. Brown, T., et al.: Language models are few-shot learners. In: Proceedings NeurIPS (2020)
9. Chang, Y., et al.: A survey on evaluation of large language models. ACM Trans. Intell. Syst. Technol. **15**(3) (2024) https://doi.org/10.1145/3641289
10. Trust in CINTEL. Collaborative Intelligence Future Science Platform (2024)
11. De Silva, A., Wijekoon, J.L., Liyanarachchi, R., Panchendrarajan, R., Rajapaksha, W.: AI insights: a case study on utilizing ChatGPT intelligence for research paper analysis. arXiv:2403.03293 [cs] (2024)
12. Fellbaum, C.: WordNet: An electronic Lexical Database. Bradford Books (1998)
13. Fisher, R.A.: Frequency distribution of the values of the correlation coefficient in samples from an indefinitely large population. Biometrika **10**(4), 507–521 (1915). https://doi.org/10.2307/2331838
14. Honnibal, M., Montani, I.: Linguistic Features · spaCy Usage Documentation. https://spacy.io/usage/linguistic-features
15. Honnibal, M., Montani, I.: spaCy 2: Natural language understanding with Bloom embeddings, convolutional neural networks and incremental parsing (2017)
16. Hussam, A., Ford, B., Hyde, J., Merayyan, A., Plummer, B., Anderson, T.: Semantic highlighting. In: CHI 98 Conference Summary on Human Factors in Computing Systems. CHI '98, ACM (1998). https://doi.org/10.1145/286498.286667
17. Ji, Z., et al.: Survey of hallucination in natural language generation. ACM Comput. Surv. **55**(12) (2023). https://doi.org/10.1145/3571730
18. Kung, J.: Elicit (product review). Journal of the Canadian Health Libraries Association / Journal de l'Association des bibliothèques de la santé du Canada 44(1) (2023). https://doi.org/10.29173/jchla29657
19. Laskar, M.T.R., Huang, J.X., Hoque, E.: Contextualized embeddings based transformer encoder for sentence similarity modeling in answer selection task. In: Proceedings Twelfth LREC. ELRA (2020)
20. Levenshtein, V.I.: Binary codes capable of correcting deletions, insertions and reversals. Sov. Phys. Dokl. **10**, 707 (1966)
21. Li, T.O., et al.: Nuances are the key: unlocking ChatGPT to find failure-inducing tests with differential prompting. In: 2023 38th IEEE/ACM International Conference on Automated Software Engineering (ASE), pp. 14–26. (2023). https://doi.org/10.1109/ASE56229.2023.00089
22. Li, Y., Chen, L., Liu, A., Yu, K., Wen, L.: ChatCite: LLM agent with human workflow guidance for comparative literature summary. arXiv:2403.02574 [cs] (2024)
23. Manna, S., Mendis, B.S.U.: Fuzzy word similarity: a semantic approach using wordnet. In: International Conference on Fuzzy Systems, pp. 1–8. IEEE, Barcelona, Spain (2010). https://doi.org/10.1109/FUZZY.2010.5584785
24. Meng, X., et al.: The application of large language models in medicine: a scoping review. iScience **27**(5) (2024). https://doi.org/10.1016/j.isci.2024.109713
25. McGinness, L., Baumgartner, P.: Automated theorem provers help improve large language model reasoning. In: Bjørner, N., Heule, M., Voronkov, A. (eds.). In: Proceedings of 25th Conference on Logic for Programming, Artificial Intelligence and Reasoning. EPiC Series in Computing, vol. 100, pp. 51–69. EasyChair (2024). https://doi.org/10.29007/2n9m
26. Messeri, L., Crockett, M.J.: Artificial intelligence and illusions of understanding in scientific research. Nature **627**(8002), 49–58 (2024). https://doi.org/10.1038/s41586-024-07146-0
27. Moallemi, E., et al.: Entry points for accelerating transitions towards a more sustainable future (2024). EarthArXiv pre-print, https://doi.org/10.31223/X5C68X

28. Editorial, N.: Why scientists trust AI too much - and what to do about it. Nature **627**(8003), 243–243 (2024). https://doi.org/10.1038/d41586-024-00639-y
29. OpenAI: GPT-4 Technical Report (2023)
30. Ormerod, M., Martínez del Rincón, J., Devereux, B.: Predicting semantic similarity between clinical sentence pairs using transformer models: evaluation and representational analysis. JMIR Med. Info. **9**(5), e23099 (2021)
31. Rogers, A.: Changing the world by changing the data. In: Proceedings 59th Annual Meeting of the ACL and the 11th IJCNLP. ACL (2021). https://doi.org/10.18653/v1/2021.acl-long.170
32. Sami, A.M., et al.: System for systematic literature review using multiple AI agents: concept and an empirical evaluation (2024) https://arxiv.org/abs/2403.08399
33. Shaib, C., Li, M., Joseph, S., Marshall, I., Li, J.J., Wallace, B.: Summarizing, simplifying, and synthesizing medical evidence using GPT-3 (with Varying Success). In: Proceedings 61st Annual Meeting of the ACL (Short Papers). ACL (2023)
34. Shenoy, M.: A New Similarity measure for taxonomy based on edge counting. Int. J. Web Semant. Technol. **3**(4), 23–30 (2012). https://doi.org/10.5121/ijwest.2012.3403
35. Smith, L.C.: Reviews and reviewing: approaches to research synthesis. An Annual Review of Information Science and Technology (ARIST) paper. J. ASIS&T **75**(3), 245–267 (2024). https://doi.org/10.1002/asi.24851
36. Spillias, S., et al.: Human-AI collaboration to identify literature for evidence synthesis. Research Square (2023). https://doi.org/10.21203/rs.3.rs-3099291/v1
37. de la Torre-López, J., Ramírez, A., Romero, J.R.: Artificial intelligence to automate the systematic review of scientific literature. Computing **105**(10), 2171–2194 (2023). https://doi.org/10.1007/s00607-023-01181-x
38. Turton, J., Smith, R.E., Vinson, D.: deriving contextualised semantic features from Bert (and other transformer model) embeddings. In: Proceedings RepL4NLP-2021. ACL (2021). https://doi.org/10.18653/v1/2021.repl4nlp-1.26
39. Wei, J., et al.: Chain-of-thought prompting elicits reasoning in large language models. In: Advances in Neural Information Processing Systems 35 (NeurIPS 2022) (2022)
40. Wu, Z., Palmer, M.: Verb semantics and lexical selection. In: 32nd Annual Meeting of the ACL. ACL (1994). https://doi.org/10.3115/981732.981751
41. Ye, A., Maiti, A., Schmidt, M., Pedersen, S.J.: A hybrid semi-automated workflow for systematic and literature review processes with large language model analysis. Future Internet **16**(5), 167 (2024). https://doi.org/10.3390/fi16050167
42. Zou, Y., et al.: Divide and Conquer: text semantic matching with disentangled keywords and intents. In: Findings of the ACL. ACL (2022)

Legal Judgment Prediction Through Argument Analysis

Azmi[1(✉)], Meladel Mistica[1,3], Inbar Levy[2], and Eduard Hovy[1,4]

[1] Computing and Information Systems, The University of Melbourne, Melbourne, Australia
muhaam@student.unimelb.edu.au, {misticam,eduard.hovy}@unimelb.edu.au
[2] Melbourne Law School, The University of Melbourne, Melbourne, Australia
inbar.levy@unimelb.edu.au
[3] Melbourne Data Analytics Platform, Melbourne, Australia
[4] Melbourne Connect, Melbourne, Australia

Abstract. Predicting the eventual judgement in legal cases requires consideration of more than just the facts since judges also consider the quality of the argumentation presented by the parties. Most prior NLP work focuses on just the facts and overlooks the essential element of legal argumentation within the court process. Working toward argument extraction and comparison technology, this paper describes the construction of a dataset comprising 8364 cases, including judgements. The arguments supporting each allegation have been identified and preprocessed. We show that there is no trivial solution to legal judgment prediction using simple correlation features. Facilitating a novel machine learning perspective on the Legal Judgment Prediction task, this is the first dataset of its kind; previous datasets do not provide this level of argumentation detail.

Keywords: legal judgement prediction · legal reasoning · AI and the law

1 Introduction

Legal Judgment Prediction (LJP) is the task of predicting case outcomes based on facts of a case [8]. However, LJP is very challenging due to the complexity of legal reasoning [14]. In court, judges do not make a decision by considering only the evidence, but they decide the outcome of a case by evaluating the legal arguments presented by the plaintiff and the defendant. Although these arguments are based on the facts of the case, predicting the outcome directly and only from the facts is very difficult since the model has to perform a series of complex legal reasoning steps to, in essence, reconstruct some of the argumentation indirectly.

Additionally, legal facts only include facts that are deemed to be relevant and reliable. This ambiguity indirectly influences machine learning models' prediction. While there is no complete separation between facts and legal arguments, as legal arguments apply the law to the facts of the case, the ambiguity in legal

M. Gong et al. (Eds.): AI 2024, LNAI 15442, pp. 44–58, 2025.
https://doi.org/10.1007/978-981-96-0348-0_4

arguments arises from the subjective interpretation of legal texts or principles by the plaintiff and defendant and not just from the courts.

Previous research indicates low agreement among legal experts on relevant facts, which highlights the subjectivity and complexity in LJP [35]. To facilitate agreement, we believe that AI models should be developed to extract the argumentation and help predict case outcomes. These models should analyze legal arguments instead of relying solely on factual information in the case. This approach aligns with the court's process, and the outcome of this alternative task can better aid legal practitioners in preparing their arguments.

The process of extracting and comparing arguments from legal case descriptions is complex. Prior approaches have viewed legal judgment merely as a court document classification task, failing to consider the critical aspects of legal argumentation, which are central to the court process [5,6,8,21,23,28,31,35]. The challenge is to identify and carefully format exactly the relevant components of arguments made by different parties in order to enable the application of machine learning and large language models (LLMs).

In this paper, we describe the construction of a dataset for the training of LJP using argumentation. Our contributions are:

- We develop a new dataset focused on legal arguments from court judgment documents. This dataset shifts the Legal Judgment Prediction task from evidence analysis to argument analysis. The dataset provides a valuable resource for legal professionals and researchers.
- We analyzed the complexity of Legal Judgment Prediction on our new argumentation focused dataset, ensuring it didn't permit trivial solutions through fortuitous correlations with judgments. Additionally, we established a baseline for future research to evaluate the effectiveness of new LJP models.

This work is part of an ongoing project to build LJP technology that can be used to assist legal scholars and practitioners in constructing and analysing arguments.

2 Related Work

Legal Judgment Predictions can be approached from various perspectives, such as recommending relevant articles, predicting charges and prison terms, predicting judgment from pleas, and generating court views [9]. These tasks are interrelated, with information from one task potentially benefiting another. Unfortunately, the available datasets contain distinct information tailored solely for each sub-task.

The article recommendation task includes two subtasks: predicting the article(s) relevant to an allegation and predicting the article(s) corresponding to the judgment violation [6]. For both tasks, the input is the factual summary of the case [2,4,13]. Similar to the article recommendation task, a factual description of a case is also used to predict charges or prison terms for the defendant [7,16,37,38].

Plea Judgment Prediction involves predicting the judgments from a case's facts. The difference between this task and the article recommendation task is that this task predicts only whether or not the defendant is guilty without predicting for which article they are guilty. Many datasets have been developed for this task that consider only a fact summary of the case [1,18,26,32]. However, a case fact summary is written after the court hearing, during which a significant part of the legal reasoning process has already been completed. In contrast, the plaintiff's plea is the original statement by the plaintiff in a case, which poses a more applicable (and more challenging) task. The Plea Judgment Prediction task that includes the plaintiff's plea as input is usually placed in one document along with case information, factual descriptions, and the court's assessment of the case [23,31]. Mai et al. [22], and Semo et al. [29] extract the plaintiff's plea to ensure the model only utilizes the plea as the input. While we use criminal law terminology, the same logic could be applied to civil law cases.

Recent research uses attention-based models for LJP and provides justification for the judgment by highlighting the facts that the model pays attention to [5,8,21,28,35]. However, there is a large misalignment between model outputs and expert rationales: as shown by Xu et al. [35], there is also a lack of consensus even among legal experts. Yu et al. [36] and Trautmann et al. [33] explore writing effective prompts for LLM multi-step reasoning, but they are still far from effective [3]. These results indicate that the facts of the case (as described in writing) are not enough information for machine learning models to decide the outcome of a case.

There is a need for datasets that include *all* aspects of a case: facts, plaintiff's claims, defendant's responses, and court assessments. Current datasets often miss this full picture, focusing mainly on fact summaries. In contrast, court processes involve detailed fact-finding through argumentation in a court hearing, which leads to court assessments before coming to a decision. Therefore, it is essential to develop datasets for Legal Judgment Prediction that analyze arguments thoroughly to allow models to predict judgments and demonstrate legal reasoning capability.

3 Dataset Creation

3.1 ECHR HUDOC Database

The European Court of Human Rights (ECHR) publishes the Human Rights Documentation (HUDOC) database, which provides access to case law documents. Many datasets have been developed using this database, including Legal Judgment Prediction and Argument Mining, which makes this database suitable as our main source of data. The ECHR judgment documents include the following sections [25]:

1. **Introduction**: Information about the case name and application number, and the stakeholders (Judge, Plaintiff, Defendant, Lawyers, etc.) and the court's procedure.

2. **The Facts**: All factual information related to the case is presented in chronological order.
3. **Proceeding Before The Commission**: The court's prior rulings of this case and the previous applicant's actions before the court.
4. **The Law**: The allegations made by the applicant and the court's deductive process to reach a conclusion.
5. **For The Reason (Conclusion)**: The reasons for a court decision that describes whether or not the defendant is guilty of violating the human rights convention.

Sections within this document do not hold equal relevance to the argumentative process. In this legal context, it is generally expected that the majority of the argumentation will be focused on the "The Law" section [25].

3.2 Argumentation Structuring

To construct the dataset, our primary objective is to extract arguments corresponding to each allegation made in a case. To achieve this, we focus specifically on "The Law" section of the ECHR judgment documents, which contains detailed argumentation from all parties involved in the court proceedings.

The structure of this section can vary significantly. Some judgment documents include general approaches or preliminary observations at the beginning, while others detail the scope of the court assessment of the evidence. However, a common thread across all variations is the inclusion of discussions about the alleged violation of an article of the convention, as well as the application of some articles to the case [25].

Since one case can involve multiple allegations under different articles, it is important to separate the arguments for each specific allegation. This way, we can clearly and accurately analyze the relevant arguments for each alleged violation. In order to extract arguments corresponding to each allegation made in a case, we do the following:

1. **Obtain alleged violated articles**. We extract articles for each alleged violation subsection using regular expressions. For instance, in the case of "Mahmudov v. Azerbaijan" [27], we obtain three allegedly violated articles: Article 5 Section 1 (5-1), Article 2 of Protocol No. 4 (p4-2), and Article 10 (10).
2. **Split law section into different allegations for each article**. We collect every argument under the law section except for arguments that are under different alleged violations articles.
3. **Find the verdict for each alleged violated article**. We extract the verdict label and separate allegations into two classes (violation, no-violation). In the case of "Mahmudov v. Azerbaijan" [27], the courts only consider allegations for Article 5 and Article 2 of Protocol No.4. In such cases we ignore the allegations of Article 10 obtained in the first step.

3.3 Author Classification

It is fortunate that the "The Law" section contains not only arguments from the plaintiff and the defendant but also the assessments by the court itself in which the final decision is explained, as in the following example (from "SMOKOVITIS AND OTHERS v. GREECE" for alleged violation of Article 6 Section 1):

> Paragraph 28 "Smokovitis And Others v. Greece" [11]:
> "In conclusion, the State infringed the applicants' rights under Article 6 §1 by intervening in a manner which was decisive to ensure that the outcome of proceedings in which it was a party was favourable to it. There has therefore been a violation of that Article."

This final paragraph is the court's conclusion, which provides the gold-standard answer for system evaluation. We therefore marked such assessment paragraphs in our dataset as not to be used as inputs since they are what we want to determine automatically and saved them separately for analysis.

To identify paragraphs authored by the court, we assign to each paragraph a label signifying its author. The five types of authors involved in the European Courts of Human Rights are [15]:

1. **ECHR**: All arguments presented by the court, including the court's assessment and potentially its decision.
2. **Applicant**: The individual or contracting Party claiming a violation of fundamental rights as outlined in the Convention.
3. **State**: The respondent accused by the applicant of being responsible for the alleged violation.
4. **Third Parties**: Other participants in the proceedings, like Amnesty International or Human Rights Watch.
5. **Commission/Chamber**: Includes arguments originating from the Commission (prior to its dissolution in 1998) or from a Chamber, especially in Grand Chamber decisions, which are then reiterated by the ECHR.

To identify the author(s) of each argument, we use Legal-BERT [15], which achieves a 0.918 F1-score on author classification. This model uses the standard BIO (Beginning, Inside, Outside) entity labelling paradigm. We consider each paragraph separately in the law section. To segment by the author label, we use the following rule:

- **Handle non-continuous span**. If there is a non-continuous span of the same entity class, convert it into a continuous span by using the "I-" prefix for all intermediate tokens. This rule ensures that an entity's span remains continuous without interruptions.
- **Handle Short Sequences**. If there are other classes with fewer than four tokens in a row[1], connect them to the previous class (except for Non-Argument). This rule is used to correct small, potentially incorrect entity spans.

[1] This number is determined manually from observation in the dataset [15], where the shortest argument clause is "The government disagrees." with four tokens.

Using these rules, we segment paragraphs into arguments and classify the author of each argument. The author classification evaluation can be seen in Table 1 below.

Table 1. Author Classification Evaluation

Author	Precision	Recall	F1-score
ECHR	94.6	96.2	95.4
Applicant	94.3	87.6	90.8
State	95.1	88.8	91.9
Third Parties	100.0	88.5	93.9
Commission/Chamber	94.7	100.0	97.3
Non-Argument	89.8	94.4	92.1
Macro-average	94.8	92.6	93.6

4 Comparison with Similar Datasets

Effective Legal Judgment Prediction systems require data that include detailed information about facts, legal arguments, and court assessments to evaluate the legal reasoning capability of a machine learning model. Current datasets often focus only on case summaries, missing the full scope of legal argumentation. However, we cannot use court documents in their entirety [23, 29], because they also contain the court view. As mentioned earlier, we carefully extract legal arguments for each allegation in the case.

A detailed comparison with similar existing datasets in legal judgment prediction is presented in Table 2.

Table 2. Dataset Comparison

Dataset	Source	Language	Input	Judgment
ECHR Task A & B [6]	ECHR	English	Facts	Multi-Label Article
CJO Dataset [34]	China Judgment Online	Chinese	Facts + Court View	Binary
Multilingual LJP [26]	Swiss Federal Supreme Court	German, French, Italian	Facts	Binary
Sulea et al. [32]	French Supreme Court	French	Facts	Binary
Lage-Freitas et al. [18]	Brazilian Higher Court	Portuguese	Facts	Yes, No, Partial
JUSTICE [1]	US Supreme Court	English	Facts	Binary
AutoJudge [20]	China Judgment Online	Chinese	Facts + Plaintiff Pleas	Binary
Class Action Prediction [29]	US Court	English	Facts + Plaintiff Pleas	Binary
ILDC [23]	Indian Court	English	Court Document	Binary
Strickson et al. [31]	UK Court	English	Court Document	Binary
Ours	ECHR	English	Facts + Plaintiff's & Defendant's Argument + Court View	Binary

To illustrate, Fig. 1 shows an example from our dataset, showing why analysing arguments for legal judgment prediction is important. If we look at the fact summary in Fig. 1, we can see that the legal issue is about the previous owner remaining registered as a permanent resident at the applicant's house due to a failure by local authorities. However, how can we determine the judgment without knowing the defendant's argument? The judge must evaluate the case through the arguments to decide how the law applies to the facts. These arguments help us understand each party's perspective and add context to the facts of the case. Furthermore, having the court's assessment address both parties' arguments gives context to the judgment label and explains the rationale behind the decision.

In contrast, LJP datasets for the ECHR do not have access to both parties' arguments and predict alleged violated articles (ECHR Task A or B [6]) only based on the fact summary, overlooking the broader argumentative context where legal reasoning takes place. Another difference is that our dataset predicts judgments only on articles considered by the court (this corresponds to ECHR Task A|B, which predicts judgments based on the allegations), making it a more meaningful task.

5 Dataset Statistics

This dataset includes a total of 8,364 cases. Each case can have multiple allegations, and the records of this dataset are based on allegations. There are 12,947 allegations documented within these cases, with 10,646 (82.23%) of these allegations having "violation" judgment and 2,301 (17.77%) having 'no-violation' judgment. A statistical overview of this dataset is summarized in Table 3.

Table 3. Overview

#cases	8,364
#allegations	12,947
#paragraphs	338,418
#tokens	35,547,426

Our dataset initially segments documents into paragraphs, then further divides paragraphs into one or more arguments based on the party to which each argument belongs. Table 4 presents the results obtained from this segmentation. Analyzing the table, we observe a slightly higher number of arguments from the applicant and state in cases where the judgment indicates "no-violation". In the following section, we critically evaluate this dataset to assess whether such distributional differences might influence the predictive accuracy of models in determining judgments.

Fact Summary

- On 18 August 1995 the applicant and her husband entered into a contract for the purchase of a family house and a plot of land from Mr and Mrs D.
- Mr D., who was registered as permanently resident (trvalý pobyt) at the house's address, undertook in the contract to secure his deregistration by 21 August 1995.
- The applicant and her husband then became the owners of the property and moved into the house. However, Mr D. failed to deregister himself from the relevant register, which was kept by the municipal office (Obecný úrad) of Nové Zámky
- . . . .
- As Mr D. had not applied for registration of a new permanent place of residence, the Nové Zámky municipal office had no legal authority under the existing legislation to cancel his previous registration at the applicant's home address.
- On 22 March 2001 the applicant and Mr D. made a joint written declaration to the effect that Mr D. was not resident at the house in question, ... Mr D. further declared that he had applied to be deregistered as permanently resident at the house, that his application had been dismissed and that he was homeless and staying in different locations.
- In a letter of 15 October 2001, in response to the applicant's enquiry, the Nové Zámky municipal office informed her that, in the circumstances, **it would not be possible to cancel Mr D.'s registration as permanently resident at her house**, even when the 1998 Registration of Citizens' Place of Residence Act came into force.
- . . . .

Applicant's Arguments

- The applicant complained that it was impossible for her to obtain cancellation of Mr D.'s registration as permanently re sident at her house, a fact which disturbed her and adversely affected the assessment of her situation in various cont exts, such as for the purpose of social contributions towards dwelling costs and the calculation of fees for the removal of household waste. She alleged a violation of her right to respect for her private life and home under Article 8 of the C onvention . . . .

State's Arguments

- The Government argued that the applicant had not exhausted domestic remedies as required by Article 35 § 1 of the Convention. They argued, firstly, that the applicant could have claimed redress in the ordinary courts under the Civil C ode by way of an action for the protection of her personal integrity, but had failed to do so. . . . .

Court's Asessment

- . . . .
- The Court observes that the principal thrust of the present complaint is the repercussions of Mr D.'s continued registra tion as residing permanently in her house on the applicant's rights as protected under Article 8 of the Convention and t he fact that, under domestic legislation as it currently stands, it is impossible to obtain cancellation of his registration.
- As to the remedies advanced by the Government, the Court observes that the mechanism under the Civil Code for th e protection of personal integrity allows for a request to be made for an order restraining unjustified interference with p ersonal integrity; . . . .

State's Arguments

- The Government accepted that, under the current legislation, it was impossible to deregister a citizen who was unable to be registered as permanently resident elsewhere. They pointed out that it was a condition of the contract of sale of 18 August 1995 that Mr D. would cancel his registration at the applicant's house, and maintained that the applicant should have sought enforcement of this contractual clause in the ordinary courts. . . . .

• • • • • • • •

Court's Asessment

- . . . .
- The Court finds that the impact on the applicant's Article 8 rights, resulting from the fact that Mr D. could not secure his deregistration, was sufficiently serious to amount to an interference with her right to respect for her private life and home. . . . .
- There has accordingly been a violation of Article 8 of the Convention on that account.

Fig. 1. Example from this dataset for the Case of BABYLONOVA v. SLOVAKIA.

Table 4. The non-zero average of arguments per agents

Agent	No Violation	Violation
Applicant	5.99	4.81
Commission/Chamber	2.81	2.58
ECHR	21.22	19.10
Non-Argument	5.09	4.51
State	5.36	4.10
Third Parties	3.30	4.68

6 Problem Complexity

Having extracted and structured the data, it is necessary to confirm that the LJP problem cannot be solved trivially. If some indicator were correlated highly enough with the eventual judgment to predict it reliably, the problem would be simple, and the dataset would have no substantial value.

We therefore investigate whether attributes like the number of paragraphs and the number of tokens in arguments can be used to predict judgment. Our investigation explores correlation using, first, traditional machine learning and, second, a generative LLM.

6.1 Confirming Problem Complexity by Machine Learning

To investigate the influence of document length on judgment prediction, we here evaluate how well judgments can be predicted using the count of paragraphs and tokens (words) for arguments presented by the state, the applicant, and third parties.

The Bag of Words (BoW) model is often used as a baseline in LJP tasks [2,4,8]. Here, we use BoW with n-grams of $n \in [1, 2, 3]$. The feature vectors are normalized to a unit vector with L1 normalization to remove the information about the document length. We remove stop words, lemmatize words, and replace names, dates, geopolitical entities (such as cities, countries, etc.) and nationalities (or religious or political groups) with the placeholders "NAME", "DATE", "PLACE", and "NORP" respectively[2]. The precision, recall and F1-score evaluations are presented in Table 5.

These results indicate that the number of paragraphs and words in the arguments of the plaintiff, defendant, and third parties are predictive features. This might be because the number of paragraphs and tokens in an argument reflects the quality and persuasiveness of the argument and the complexity of the case.

The Bag of Words (BoW) model achieves an F1-score of 68.1% and acts as the baseline for our dataset. This result is similar to the baseline of prior legal judgment prediction on ECHR, which also uses BoW, achieving an F1-score of

[2] implemented using spaCy.

Table 5. Evaluation (Precision/ Recall/ F-score) using length and BoW as features

	Length	BoW
Random	50.0/50.0/50.0	50.0/50.0/50.0
Gaussian Naive Bayes	60.0/21.4/31.6	71.4/58.1/64.0
KNN (k=5)	53.2/51.8/52.5	71.2/65.1/68.1
Logistic Regression	59.1/45.1/51.1	63.3/69.2/66.1
Decision Tree	57.8/59.8/58.8	64.6/53.1/58.3
Random Forest	59.2/55.8/57.4	70.0/65.6/67.7

70.9% [4]. Furthermore, this result confirms that predicting judgment from legal argumentation is as complex if not more complex than the previous facts-based legal judgment predictions on ECHR [4].

6.2 Confirming Problem Complexity by LLM

Our task can be seen as a rule-conclusion process, where the objective is to derive conclusions based on predefined rules. This involves tasks that require a machine learning model to determine the legal outcome of a given set of facts under a specified rule. The LegalBench benchmark [14], which evaluates the performance of language models in legal reasoning tasks, indicates that large language models (LLMs) can achieve decent performance in these rule-conclusion tasks. We want to test this potential on our dataset.

Table 6. Evaluation using Large Language Models

	Precision	Recall	F1-score	Accuracy
GPT 3.5	66.7	19.6	30.3	54.9
GPT 4	66.7	28.6	40.0	57.1

Table 6 shows that the LLM does worse than the Bag of Words model. This is because the LLM is very biased towards judging violations, as we can see from the confusion matrix in Table 7.

Table 7. GPT-4 Confusion Matrix

	Predicted violation	Predicted no-violation
Actual violation	0.86	0.14
Actual no-violation	0.71	0.29

7 Conclusion and Future Work

This study presents a novel approach to Legal Judgment Prediction (LJP) by emphasizing the importance of legal argumentation over mere factual analysis. The construction of a comprehensive dataset comprising 8364 cases, where arguments and their attributions are meticulously identified and preprocessed, lays the groundwork for developing AI models capable of mimicking judicial reasoning more closely. This dataset shifts the focus from case description analysis to a more nuanced argument-based analysis, reflecting the actual judicial process more accurately.

Our findings underscore the complexity of legal reasoning and the inadequacy of a simple end-to-end approach in predicting case outcomes. This reinforces the necessity for models that can extract, represent, and compare legal arguments effectively. The dataset present is not only a valuable resource for researchers but also a practical tool for legal professionals aiming to enhance their argumentation strategies[3]

Future research should aim to solve legal judgment prediction by focusing on deciding whose argument to accept and determining which party's argument is stronger and more persuasive. Additionally, a more formal approach to argument evaluation should be implemented, ensuring rigorous assessment aligned with legal standards and practices. Leveraging large language models (LLMs), which are good in legal understanding [14], can enhance this process by performing step-by-step argument analysis, thereby minimizing hallucinations and improving the accuracy and reliability of predictions [17]. By addressing these areas, future work can build on our findings and further advance AI's capabilities in legal contexts, making it a more applicable tool for legal professionals and researchers.

8 Ethics Consideration

There are several ethical concerns regarding AI applications in legal systems, including privacy violations, demographic bias, and the potential misuse of models, especially in crucial applications such as judgment prediction [19]. We acknowledge that these issues are also present in this dataset. However, our goal is to enhance the applicability of AI systems to ensure their responsible use in legal practice.

This research aims not to replace legal professionals but to support them with a system that has legal reasoning capability. To achieve this, detailed datasets that separate facts, arguments, and court assessments are essential. Distinguishing these elements helps in understanding how courts assess legal arguments, establish facts, and make judgments. This approach allows researchers to develop models that not only predict outcomes statistically but also follow human-like reasoning by identifying legal issues, considering formal legal rules, applying those rules, and evaluating arguments to reach a judgment [14,30].

[3] This dataset is available on https://github.com/azminajid/echr-args-dataset.

Although some might view the law as a set of rules that can be applied objectively to the facts of a case [10], bias can still occur due to the judge's subjective interpretation of those facts. This bias arises from how facts are selected, framed, and presented [24]. Consequently, we propose legal judgment prediction through argument analysis, as hearing arguments from both parties helps present a balanced view of the facts and enhances fairness in AI systems.

Legal documents often contain sensitive personal information. In the ECHR, details from written and oral proceedings, including identifiable information about applicants or third parties, are made public unless anonymity is requested [12]. Factors such as gender, race, nationality, and socio-economic status are included in the ECHR public database and, by extension, in our dataset. This may unintentionally introduce biases, resulting in unfair predictions and reinforcing systemic inequalities. Therefore, debiasing is crucial to ensure fairness when using this dataset.

References

1. Alali, M., Syed, S., Alsayed, M., Patel, S., Bodala, H.: JUSTICE: a benchmark dataset for supreme court's judgment prediction (2021)
2. Aletras, N., Tsarapatsanis, D., Preoţiuc-Pietro, D., Lampos, V.: Predicting judicial decisions of the European court of human rights: a natural language processing perspective. PeerJ Comput. Sci. **2**, e93 (2016). https://doi.org/10.7717/peerj-cs.93
3. Chalkidis, I.: ChatGPT may pass the bar exam soon, but has a long way to go for the LexGLUE benchmark, March 2023. arXiv arXiv:2304.12202 [cs]
4. Chalkidis, I., Androutsopoulos, I., Aletras, N.: Neural legal judgment prediction in English. In: Proceedings of the 57th Annual Meeting of the Association for Computational Linguistics. Association for Computational Linguistics (2019). https://doi.org/10.18653/v1/p19-1424
5. Chalkidis, I., Fergadiotis, M., Tsarapatsanis, D., Aletras, N., Androutsopoulos, I., Malakasiotis, P.: Paragraph-level rationale extraction through regularization: a case study on European court of human rights cases. In: Proceedings of the 2021 Conference of the North American Chapter of the Association for Computational Linguistics: Human Language Technologies, Online, pp. 226–241. Association for Computational Linguistics (2021). https://doi.org/10.18653/v1/2021.naacl-main.22, https://aclanthology.org/2021.naacl-main.22
6. Chalkidis, I., et al.: LexGLUE: a benchmark dataset for legal language understanding in English. In: Proceedings of the 60th Annual Meeting of the Association for Computational Linguistics (Volume 1: Long Papers), Dublin, Ireland, pp. 4310–4330. Association for Computational Linguistics (2022). https://doi.org/10.18653/v1/2022.acl-long.297, https://aclanthology.org/2022.acl-long.297
7. Chen, H., Cai, D., Dai, W., Dai, Z., Ding, Y.: Charge-based prison term prediction with deep gating network. In: Proceedings of the 2019 Conference on Empirical Methods in Natural Language Processing and the 9th International Joint Conference on Natural Language Processing (EMNLP-IJCNLP), Stroudsburg, PA, USA. Association for Computational Linguistics (2019)

8. Cui, J., Shen, X., Nie, F., Wang, Z., Wang, J., Chen, Y.: A survey on legal judgment prediction: datasets, metrics, models and challenges (2022). arXiv arXiv:2204.04859 [cs]
9. Cui, J., Shen, X., Wen, S.: A survey on legal judgment prediction: datasets, metrics, models and challenges. IEEE Access **11**, 102050–102071 (2023). https://doi.org/10.1109/ACCESS.2023.3317083
10. Dagistanli, S.: Legal Formalism (2017). https://doi.org/10.1002/9781118430873.est0205
11. ECtHR: European court of human rights. https://hudoc.echr.coe.int/
12. European Court of Human Rights: rules of court, March 2024. https://www.echr.coe.int/documents/d/echr/rules_court_eng. Accessed 23 Sep 2024
13. Ge, J., huang, Y., Shen, X., Li, C., Hu, W.: Learning fine-grained fact-article correspondence in legal cases (2021). https://doi.org/10.48550/ARXIV.2104.10726, https://arxiv.org/abs/2104.10726
14. Guha, N., et al.: LegalBench: a collaboratively built benchmark for measuring legal reasoning in large language models (2023). arXiv arXiv:2308.11462 [cs]
15. Habernal, I., et al.: Mining legal arguments in court decisions. Artificial Intelligence and Law (2023). https://doi.org/10.1007/s10506-023-09361-y, arXiv arXiv:2208.06178 [cs]
16. Hu, Z., Li, X., Tu, C., Liu, Z., Sun, M.: Few-shot charge prediction with discriminative legal attributes. In: Bender, E.M., Derczynski, L., Isabelle, P. (eds.). Proceedings of the 27th International Conference on Computational Linguistics, DSanta Fe, New Mexico, USA, August, pp. 487–498. Association for Computational Linguistics (2018)
17. Huang, L., et al.: A survey on hallucination in large language models: principles, taxonomy, challenges, and open questions (2023)
18. Lage-Freitas, A., Allende-Cid, H., Santana, O., Oliveira-Lage, L.: Predicting Brazilian court decisions. PeerJ Comput. Sci. **8**, e904 (2022)
19. Leins, K., Lau, J.H., Baldwin, T.: Give me convenience and give her death: who should decide what uses of NLP are appropriate, and on what basis? In: Proceedings of the 58th Annual Meeting of the Association for Computational Linguistics, Online, pp. 2908–2913. Association for Computational Linguistics (2020). https://doi.org/10.18653/v1/2020.acl-main.261, https://www.aclweb.org/anthology/2020.acl-main.261
20. Long, S., Tu, C., Liu, Z., Sun, M.: Automatic judgment prediction via legal reading comprehension. In: Sun, M., Huang, X., Ji, H., Liu, Z., Liu, Y. (eds.) CCL 2019. LNCS (LNAI), vol. 11856, pp. 558–572. Springer, Cham (2019). https://doi.org/10.1007/978-3-030-32381-3_45
21. Luo, B., Feng, Y., Xu, J., Zhang, X., Zhao, D.: Learning to predict charges for criminal cases with legal basis. In: Proceedings of the 2017 Conference on Empirical Methods in Natural Language Processing, Copenhagen, Denmark, pp. 2727–2736. Association for Computational Linguistics (2017). https://doi.org/10.18653/v1/D17-1289, http://aclweb.org/anthology/D17-1289
22. Ma, L., et al.: Legal judgment prediction with multi-stage case representation learning in the real court setting. In: Proceedings of the 44th International ACM SIGIR Conference on Research and Development in Information Retrieval, New York, NY, USA, July 2021. ACM (2021)

23. Malik, V., et al.: ILDC for CJPE: Indian legal documents corpus for court judgment prediction and explanation. In: Proceedings of the 59th Annual Meeting of the Association for Computational Linguistics and the 11th International Joint Conference on Natural Language Processing (Volume 1: Long Papers), Stroudsburg, PA, USA. Association for Computational Linguistics (2021)
24. Medvedeva, M., Mcbride, P.: Legal judgment prediction: if you are going to do it, do it right. In: Preotiuc-Pietro, D., Goanta, C., Chalkidis, I., Barrett, L., Spanakis, G., Aletras, N. (eds.) Proceedings of the Natural Legal Language Processing Workshop 2023, Singapore, December, pp. 73–84. Association for Computational Linguistics (2023). https://doi.org/10.18653/v1/2023.nllp-1.9, https://aclanthology.org/2023.nllp-1.9
25. Mochales, R., Moens, M.F.: Study on the Structure of Argumentation in Case Law, July 2008
26. Niklaus,J., Chalkidis, I., Stürmer, M.: Swiss-judgment-prediction: a multilingual legal judgment prediction benchmark. In: Proceedings of the Natural Legal Language Processing Workshop 2021, Stroudsburg, PA, USA. Association for Computational Linguistics (2021)
27. Quemy, A., Wrembel, R., Łopuszyńska, N., Papadakis, G., Delgado, A.D.: A large reproducible benchmark on text classification for the legal domain based on the ECHR-OD repository. Inf. Syst. **119**, 102258 (2023). https://doi.org/10.1016/j.is.2023.102258, https://linkinghub.elsevier.com/retrieve/pii/S0306437923000947
28. Santosh, T., Xu, S., Ichim, O., Grabmair, M.: Deconfounding legal judgment prediction for European court of human rights cases towards better alignment with experts. In: Proceedings of the 2022 Conference on Empirical Methods in Natural Language Processing, Abu Dhabi, United Arab Emirates, pp. 1120–1138. Association for Computational Linguistics (2022). https://doi.org/10.18653/v1/2022.emnlp-main.74, https://aclanthology.org/2022.emnlp-main.74
29. Semo, G., Bernsohn, D., Hagag, B., Hayat, G., Niklaus, J.: ClassActionPrediction: a challenging benchmark for legal judgment prediction of class action cases in the US. In: Proceedings of the Natural Legal Language Processing Workshop 2022, Stroudsburg, PA, USA. Association for Computational Linguistics (2022)
30. Stockmeyer, N.O.: Legal reasoning? It's all about IRAC (2021)
31. Strickson, B., De La Iglesia, B.: Legal judgement prediction for UK courts. In: Proceedings of the 2020 the 3rd International Conference on Information Science and System, New York, NY, USA, March 2020. ACM (2020)
32. Şulea, O.M., Zampieri, M., Vela, M., van Genabith, J.: Predicting the law area and decisions of French supreme court cases. In: Recent Advances in Natural Language Processing Meet Deep Learning, RANLP 2017, Shoumen, Bulgaria, November 2017. Incoma Ltd. (2017)
33. Trautmann, D., Petrova, A., Schilder, F.: Legal prompt engineering for multilingual legal judgement prediction (2022). arXiv arXiv:2212.02199 [cs]
34. Wu, Y., et al.: De-biased court's view generation with causality. In: Proceedings of the 2020 Conference on Empirical Methods in Natural Language Processing (EMNLP), Stroudsburg, PA, USA. Association for Computational Linguistics (2020)
35. Xu, S., Santosh, T.Y.S.S., Ichim, O., Risini, I., Plank, B., Grabmair, M.: From dissonance to insights: dissecting disagreements in rationale construction for case outcome classification, February 2024. arXiv arXiv:2310.11878 [cs]

36. Yu, F., Quartey, L., Schilder, F.: Exploring the effectiveness of prompt engineering for legal reasoning tasks. In: Findings of the Association for Computational Linguistics, ACL 2023,Toronto, Canada, pp. 13582–13596. Association for Computational Linguistics (2023). https://doi.org/10.18653/v1/2023.findings-acl.858, https://aclanthology.org/2023.findings-acl.858
37. Zhong, H., Guo, Z., Tu, C., Xiao, C., Liu, Z., Sun, M.: Legal judgment prediction via topological learning. In: Proceedings of the 2018 Conference on Empirical Methods in Natural Language Processing, Stroudsburg, PA, USA. Association for Computational Linguistics (2018)
38. Zhong, H., Wang, Y., Tu, C., Zhang, T., Liu, Z., Sun, M.: Iteratively questioning and answering for interpretable legal judgment prediction. Proc. Conf. AAAI Artif. Intell. **34**(01), 1250–1257 (2020)

Conditional Prototypical Optimal Transport for Enhanced Clue Identification in Multiple Choice Question Answering

Wangli Yang[1], Jie Yang[1(✉)], Wanqing Li[1], and Yi Guo[2]

[1] School of Computing and Information Technology, University of Wollongong, Wollongong, Australia
{wangli,jiey,wanqing}@uow.edu.au
[2] School of Computer, Data and Mathematical Sciences, Western Sydney University, Penrith, Australia
y.guo@westernsydney.edu.au

Abstract. This paper introduces the Conditional Prototypical Optimal Transport (CPOT) algorithm for clue identification in Multiple Choice Question Answering (MCQA) tasks. Existing clue-based methods suffer from inefficiencies, often relying on pseudo-labels or external resources, which introduce noise and additional computational demands. By contrast, the proposed CPOT method formulates clue identification as a sentence-oriented prototyping task, then further identifies sentences closest to prototype centroids as clues. Additionally, by leveraging the question and options as contextual guides, CPOT extends traditional Optimal Transport (OT) theory with constraints for unique assignment and uniform distribution across prototypes. This approach ensures semantically similar features converge within their prototypes and also maintains diversity among identified clues, enhancing answer accuracy. Empirical studies on several competitive benchmarks consistently demonstrate the superiority of our proposed method over different traditional approaches, with a substantial average improvement of 1.1—3.5 absolute percentage points in answering accuracy.

Keywords: Multiple Choice Question Answering · Conditional Prototypical Optimal Transport · Clue Identification

1 Introduction

Multiple Choice Question Answering (MCQA) is a crucial task within Machine Reading Comprehension (MRC). This task involves selecting the correct answer

The authors would like to thank anonymous reviewers for their valuable suggestions to improve the quality of the article. This work is partially supported by the Australian Research Council Discovery Project (DP210101426), the Australian Research Council Linkage Project (LP200201035), and Telstra-UOW Hub for AIOT Solutions Funding (2024–2025).

M. Gong et al. (Eds.): AI 2024, LNAI 15442, pp. 59–72, 2025.
https://doi.org/10.1007/978-981-96-0348-0_5

from a collection of alternatives based on a given passage and its corresponding question [9,16]. In recent years, the use of Pre-trained Language Models (PLMs), such as BERT [4], has garnered widespread attention and popularity within the field of Natural Language Processing (NLP), including MCQA. Consequently, a significant amount of research effort has been dedicated using PLMs [15,16, 18,19,21,22,24]. These methods typically unfold in three stages. Initially, PLMs serve as context encoders to extract latent representations for the input passage, question, and candidate options. This is followed by a semantic matching process to assess the coherence among these components. Finally, a decision-making layer, usually a Multi-Layer Perceptron (MLP) classifier, is employed to identify the option with the highest semantic coherence [22,24].

Based on the processed passage, existing methods can be further categorized into two main approaches: **holistic and clue-based** methods. Holistic methods rely on the entire passage to select the correct answer. While these methods can leverage the full context, they often struggle with long passages due to information dilution, the equal importance assigned to each sentence, and increased computational complexity. Clue-based methods, on the other hand, focus on identifying and leveraging specific clues (sentences or keywords) within the passage that directly relate to the question. This approach can be more efficient in pinpointing relevant information. As such, numerous studies have employed clue-based methods for MCQA [6,7,23]. However, existing clue-based methods have primarily depended on pseudo-labels for clues, which may introduce noise, or have necessitated external resources to identify clues, thus incurring additional computational costs and resource demands.

To address these limitations, we propose the **C**onditional **P**rototypical **O**ptimal **T**ransport (***CPOT***) algorithm for identifying relevant clues. Specifically, the clue identification problem is approached as a sentence-oriented prototyping task, where sentences nearest to prototype centroids are identified as clues due to their representativeness. Guided by the prior condition (*i.e.*, question and options), the method clusters sentences from a passage using an extended Optimal Transport (OT) theory, incorporating constraints to ensure each sentence belongs to only one prototype and is evenly distributed across prototypes. This approach ensures semantically similar features relevant to the prior condition converge within their prototypes, resulting in diverse and accurate clue identification. To the best of our knowledge, this study is the first to integrate a conditional-based prototyping approach in MCQA tasks. Our experiments across several competitive MCQA benchmarks demonstrate that our approach outperforms existing methods, showing a consistent improvement in accuracy. In summary, our major contributions are three-fold:

- We propose the **C**onditional **P**rototypical **O**ptimal **T**ransport (***CPOT***) algorithm, which neither relies on pseudo-labels nor external resources, to identify clues in the MCQA task;
- *CPOT* extends traditional OT theory by incorporating two constraints: uniform distribution and unique assignment, while also leveraging the question-option prior condition for prototyping;

- Empirically, the proposed method is validated on three highly competitive MCQA benchmarks. Our method consistently outperforms established models, achieving an improvement of 1.1-3.5 absolute points on average.

2 Related Work

The task of Multiple Choice Question Answering (MCQA) involves taking a background passage and a question as input, then identifying the correct answer from a collection of alternative options. MCQA plays a vital role in various downstream tasks, such as chatbots, educational assessments, and online recommendations [17]. A variety of methods have been proposed in the literature, which can be categorized into two main approaches: holistic and clue-based methods. Specifically, the former approach employs techniques, such as context understanding and semantic analysis, to infer the answer using the entire passage. In contrast, the clue-based approach relies on extracting and leveraging specific clues (few sentences or keywords) from the passage to find the correct answer. Hereafter, both holistic- and clue-based methods are briefly reviewed.

2.1 Holistic Methods

This line of research relies on assessing semantic relationships among the *triple* (*i.e.*, the passage, question, and answer option). Typically, Pre-trained Language Models (PLMs) are used to encode the context and form feature representations for the triple, followed by a semantic matching step to estimate the correlation based on their latent representations. Subsequently, a Multi-Layer Perceptron classifier selects the option with the highest semantic correlation as the final answer. For example, the **DCMN+** method utilizes unidirectional matching to analyze the relationships between the question-passage, question-answer, and passage-answer pairs individually, before aggregating them for final triple matching [21]. Zhang *et al.* [22] employ syntactic parse trees and propose **SG-Net** to capture syntactic relationships among tokens, leveraging these relationships to estimate semantic correlations. **DUMA** introduces multi-head attention mechanisms to identify the correct answer [24]. Similar work can be found in [19]. Recently, the **CTM** method has been proposed, which selects a single element from the triple (such as the passage) as the context to evaluate its semantic relationship with the other two components [18]. Additionally, **DIMN** explores global interactions among candidate triples through advanced embedding integration, leveraging co-attention and convolution mechanisms [16]. The work from [15] investigates the sensitivity of language models to the order of answer options in MCQA, and further proposes to enhance the model performance by calibrating top-2/3 choices among adjacent options.

Despite their efforts, the above holistic methods primarily depend on processing the entire passage, which can lead to high computational costs and difficulties in effectively managing the variations and complexities of language across diverse contexts. As a result, they may demonstrate suboptimal performance in scenarios that demand precise understanding and contextual awareness. To address these challenges, an increasing body of research has started exploring the use of key sentences and/or keywords from the passage.

2.2 Clue-Based Methods

This line of work involves either identifying evidential and question-relevant clues (answer-supporting sentences/keywords) from the passage or introducing clues from external sources to enhance the interpretation of questions and options.

In the early stages, supervised learning is applied when ground-truth clue labels are available [2,10]. Alternatively, without the presence of labels, clues can be selected based on pre-defined similarity measurements. For instance, **EAM** implements information retrieval methods to find clues based on their relevance to both questions and answers [12]. Additionally, **RekNet** introduces the *Reference Span* technique through estimating the significance of specific text segments, supported by associated confidence metrics [23]. **POE** [13] is a two-step scoring method that leverages the external knowledge of a large language model to evaluate questions and options comprehensively, first eliminating options with below-average scores and then selecting the highest-scoring correct answer from the remaining options.

Alternatively, some methods leverage additional knowledge bases for generating clues. For example, **JEEVES** uses an external corpus to train a combined retriever-reader model for effective clue retrieval [6], while **GenMC** utilizes generative models with external resources to synthesize clues [7]. **ExcMC** is proposed in [11] to generate clues based on the given question using a pre-trained encoder-decoder architecture. It then dynamically eliminates incorrect options to optimize the encoder for improving the answering accuracy. Additionally, Chen *et al.* propose **CASE** to identify keywords by assigning importance weights to individual words based on their semantic relations [3]. This dynamic weighting approach outperforms traditional methods by reducing noise from unimportant words and incorporating implicit commonsense knowledge.

Our method ($CPOT$) differs from existing clue-based methods in several key ways: (1) Unlike methods relying on keyword matching or semantic similarities, $CPOT$ creates prototypes to capture semantic correlations within the passage. (2) $CPOT$ uses the condition (question and options) to guide the clustering process, ensuring that the clues are highly relevant to the specific condition. (3) Compared to other clue generation methods, $CPOT$ relies less on external knowledge bases, making it more computationally efficient and affordable.

2.3 Optimal Transport

Originating from practical challenges in optimal transportation and resource allocation, the theory of Optimal Transport (OT) has been extensively applied across various NLP tasks. These include information retrieval [5], text style transfer [14], table ranking [20], among others.

From a mathematical perspective, Optimal Transport (OT) deals with the problem of moving resources from a set of suppliers $\boldsymbol{a} = \{a_i | i = 1, \ldots, |\boldsymbol{a}|\}$ to a set of customers $\boldsymbol{b} = \{b_j | j = 1, \ldots, |\boldsymbol{b}|\}$. In this context, a_i denotes the amount of resources available from the i-th supplier, and b_j indicates the demand of the j-th customer. Let $p_{ij} \geq 0$ represent the quantity allocated from supplier i to customer j, with $c_{ij} \geq 0$ being the associated transportation cost. The objective is to determine an optimal allocation plan $\mathbf{Z}^*$ that minimizes the total transportation cost while ensuring that the supply capacities and customer demands are fully satisfied:

$$\mathbf{Z}^* = \text{argmin}_{\mathbf{Z}} \sum_{i}^{|\boldsymbol{a}|} \sum_{j}^{|\boldsymbol{b}|} p_{ij} c_{ij}, \quad \text{s.t.} \quad \sum_{j=1}^{|\boldsymbol{b}|} p_{ij} = a_i, \quad \sum_{i=1}^{|\boldsymbol{a}|} p_{ij} = b_j. \tag{1}$$

Unlike traditional Optimal Transport (OT) methods, our method models the association between sentences and the question with candidate options as a conditional Optimal Transport problem guided by the prior condition. By utilizing entropy regularization and the Sinkhorn-Knopp algorithm, it optimizes the assignment of sentences to prototype under hard assignment and uniform distribution constraints. This approach effectively selects sentences that are highly relevant to the question and diverse across the passage, enhancing the performance of the MCQA system.

3 Proposed Method

This section introduces the **C**onditional **P**rototypical **O**ptimal **T**ransport (*CPOT*) algorithm (shown in Fig. 1). Specifically, *CPOT* effectively clusters sentences from the given passage, guided by the prior condition provided by the question and options. We then identify the prototypes pertinent to the prior condition to select the most representative sentences (those closest to the prototype centroids) as clues for answer inference.

3.1 Problem Formulation

Our approach formulates the clue identification in MCQA as a prototyping (clustering) task, where sentences close to the prototype centroids that serve as clues. That is, given a passage of N sentences $\boldsymbol{p} = [s_1, \ldots, s_i, \ldots, s_N]$, where s_i represents the i-th tokenized sentence, and a prior condition $\boldsymbol{z}_m$ comprising a question $\boldsymbol{q}$ and the m-th candidate option $\boldsymbol{o}_m$ (*i.e.*, $\boldsymbol{z}_m = \boldsymbol{q} \oplus \boldsymbol{o}_m$), with $m \in [1, M]$ (where M is the total number of candidate options), we prototype all N sentences from $\boldsymbol{p}$ into K clusters based on $\boldsymbol{z}_m$. This approach ensures that the identified clues are not only diverse across different prototypes but also highly relevant to the given question and options, thereby enhancing the answering accuracy.

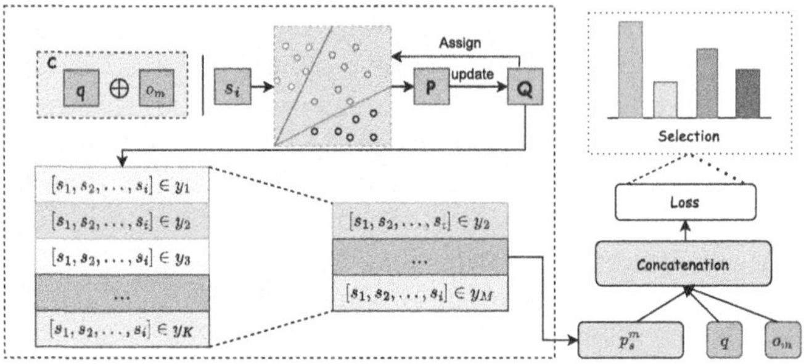

Fig. 1. The proposed **C**onditional **P**rototypical **O**ptimal **T**ransport (*CPOT*) algorithm for MCQA. The core idea is to prototype the passage sentences by modeling semantic information based on the question and options, and then identify the most representative sentences from the relevant prototypes as clues for question answering.

3.2 *CPOT*

A naive prototyping approach can be implemented by utilizing a similarity measure (such as the dot product or cosine function) to assign sentences to their closest clusters. However, this approach has two drawbacks. First, such a prototyping process lacks contextual guidance, as it does not consider a prior condition (*i.e.*, the question and/or option answers in the MCQA context). Second, the direct association of prototypes with instances often leads to an imbalance, where the majority of instances could be allocated to only a few prototypes [1,17]. To alleviate the aforementioned problems, we propose solving this prototyping issue as an Optimal Transport problem guided by the prior condition.

To illustrate, an encoder $Enc(\cdot)$ is leveraged to extract latent representations from the paragraph $\boldsymbol{p}$ and the condition $\boldsymbol{z}_m$ (*i.e.*, the concatenation of the question and option answer):

$$\Phi(s_i) = Enc(s_i), \quad \Phi(\boldsymbol{z}_m) = Enc(\boldsymbol{z}_m), \tag{2}$$

where $\Phi(s_i)$ and $\Phi(\boldsymbol{z}_m)$ denote the feature vectors for the i-th sentence (*i.e.*, s_i from $\boldsymbol{p}$) and $\boldsymbol{z}_m$, respectively. Subsequently, the problem is defined as assigning N individual sentences to K clusters according to $\boldsymbol{z}_m$. We further denote y as the category label, where each sentence will be assigned to a category $y \in \{1, 2, \dots, K\}$. The optimization objective is expressed as:

$$\mathcal{L}(p, q) = -\frac{1}{N}\sum_{i=1}^{N}\sum_{y=1}^{K} q(y|\Phi(s_i), \Phi(\boldsymbol{z}_m)) \log p(y|\Phi(s_i), \Phi(\boldsymbol{z}_m)), \tag{3}$$

where $p(y|\Phi(s_i), \Phi(\boldsymbol{z}_m))$ represents the model-predicted probability of the category y given both $\Phi(s_i)$ and $\Phi(\boldsymbol{z}_m)$. This probability can be estimated by applying a softmax function to the output of a model h (where h represents a Multi-Layer Perceptron layer to compute the logits for each category based on the features $\Phi(s_i)$ and $\Phi(\boldsymbol{z}_m)$):

$$p(y|\Phi(s_i), \Phi(\boldsymbol{z}_m)) = \text{softmax}_y(h(\Phi(s_i), \Phi(\boldsymbol{z}_m))). \tag{4}$$

Additionally, $q(y|\Phi(s_i), \Phi(\boldsymbol{z}_m))$ is a soft label, or the silver probability, of sentence s_i belonging to the y cluster. Hereafter, we omit the explicit notation on y and its conditions for simplicity, writing $q(\cdot)$ for $q(y|\Phi(s_i), \Phi(\boldsymbol{z}_m))$ and $p(\cdot)$ for $p(y|\Phi(s_i), \Phi(\boldsymbol{z}_m))$. By maximizing the probability $p(\cdot)$, *CPOT* ensures that the prototyping results are aligned with the contextual guidance provided by the prior condition ($\boldsymbol{z}_m$), thus improving the answering accuracy.

Furthermore, the proposed method encourages sentences to be uniformly distributed into K clusters to avoid cases where the majority of instances are grouped into only a few prototypes. To achieve this, two constraints are introduced to Eq. (3), using the principle of OT:

$$\min_{p,q} \mathcal{L}(p, q) \quad \text{s.t.} \quad \forall y : q(\cdot) \in \{0, 1\}, \quad \sum_{i=1}^{N} q(\cdot) = \frac{N}{K}, \tag{5}$$

where N is the total number of sentences in the passage, and K is the total number of prototypes. The first constraint ($q(\cdot) \in \{0, 1\}$) implies that each sentence can only be assigned to one prototype. That is, s_i is assigned to the y prototype, if and only if $q(\cdot) = 1$. Additionally, the second constraint ($\sum_{i=1}^{N} q(\cdot) = \frac{N}{K}$) ensures that each prototype is assigned an equal number of sentences, meaning each cluster contains $\frac{N}{K}$ sentences.

Notably, the optimization in Eq. (5) extends the traditional Optimal Transport problem we introduced in Eq. (1), but incorporating the prior condition of $\Phi(\boldsymbol{z}_m)$. To optimize this, we define Q as a $K \times N$ probability matrix where $Q_{yi} = \frac{1}{N} q(\cdot)$, representing the silver probability of assigning sentence s_i to cluster y. Similarly, P is the probability matrix predicted by the model, where $P_{yi} = \frac{1}{N} p(\cdot)$. The above constraint in Eq. (5) can be converted into a transportation polytope $U(r, c)$ defined as:

$$U(r, c) = \{Q \in \mathbb{R}_{+}^{K \times N} \mid Q\mathbb{1} = r, \quad Q^{\top}\mathbb{1} = c\}, \tag{6}$$

where $\mathbb{1}$ is a vector of all ones, and r and c are the marginal distributions of the matrix Q on its rows and columns, respectively. In this case, r represents the uniform distribution over prototypes ($r = \frac{1}{K} \cdot \mathbb{1}$), and c represents the uniform distribution over sentences (and $c = \frac{1}{N} \cdot \mathbb{1}$). Furthermore, existing solvers (such as the Sinkhorn-Knopp algorithm) can be used to solve Eq. (5) via:

$$\min_{Q \in U(r,c)} \langle Q, -\log P \rangle + \frac{1}{\lambda} \text{KL}(Q \| rc^{\top}), \tag{7}$$

where $\langle Q, -\log P \rangle$ denotes the Frobenius inner product, which encourages Q to align with the model's predictions P, $\text{KL}(\cdot)$ is the Kullback-Leibler divergence, and λ is a regularization parameter that adjusts the weight of entropy in the optimization process, balancing the uniformity of distribution and alignment with the prior condition. Specifically, with the Sinkhorn-Knopp algorithm, we can compute the optimal Q efficiently, the solution takes the form of:

$$Q = \mathrm{diag}(\alpha) P^{\lambda} \mathrm{diag}(\beta), \tag{8}$$

where P^{λ} is element-wise exponentiation of P raised to the power λ, α and β are two vectors of scaling coefficients chosen so that the resulting matrix Q is also a probability matrix. Algorithm 1 provides the pseudocode for the proposed *CPOT* to estimate the prototyping-probability matrix Q.

Algorithm 1: Pseudocode for Sentence Clustering using *CPOT*.

Input: Passage $\boldsymbol{p}$ with s_i as the i-th sentence, Question and option pair $\boldsymbol{z}_m$, Encoder Enc($\cdot$), Number of clusters K.
Output: Cluster assignments Q.
Initialize cluster assignments Q randomly;
Initialize model parameters θ for Enc($\cdot$);
Extract features $\Phi(s_i)$ and $\Phi(\boldsymbol{z}_m)$ using Enc($\cdot$);
while *max iterations not reached* **do**
    `// Compute probabilities for each cluster`
    for *each* $s_i \in \boldsymbol{p}$ **do**
        for *each* $y \in \{1, 2, \ldots, K\}$ **do**
            $P(y|\Phi(s_i), \Phi(\boldsymbol{z}_m)) \leftarrow \mathrm{softmax}(h(\Phi(s_i), \Phi(\boldsymbol{z}_m)))$;
        end
    end
    `// Update cluster assignments based on maximum probability`
    for *each* $s_i \in \boldsymbol{p}$ **do**
        $Q_{yi} \leftarrow \arg\max_y P(y|\Phi(s_i), \Phi(\boldsymbol{z}_m))$;
    end
    `// Update model parameters`
    Update θ by optimizing Eq. (7) using current P and Q;
end
return Q;

After obtaining the optimized assignment matrix Q, *CPOT* identifies relevant sentences based on $\boldsymbol{z}_m$ through a three-step process. First, a relevance score (RS) for each cluster y is calculated. This score is the sum of the cosine similarities between the feature vectors of each sentence $\Phi(s_i)$ (where s_i belongs to the y-th prototype) and $\Phi(\boldsymbol{z}_m)$. Second, the top K_{clu} prototypes with the highest RS are selected, as they contain sentences that are most relevant to $\Phi(\boldsymbol{z}_m)$. Third, we further select the top K_{sen} sentences from each individual prototype, quantified by their probabilities in the matrix Q:

$$K_{\mathrm{sen}} = \mathrm{argmax}_{s_i \in \mathrm{Cluster}\ y}(Q_{yi}). \tag{9}$$

The impact of K_{clu} and K_{sen} is analyzed in the ablation study.

3.3 Clues for Answering

Let the identified clues from *CPOT* be denoted as $\boldsymbol{p}_s^m$. We then combine $\boldsymbol{p}_s^m$ with the original passage $\boldsymbol{p}$, allowing the clues to emphasize key information while maintaining the entire context. The combined input is then to generate embeddings using the encoder Enc(·):

$$H_p^m = \text{Enc}(\boldsymbol{p} \oplus \boldsymbol{p}_s^m), \quad H_q = \text{Enc}(\boldsymbol{q}), \quad H_{o_m} = \text{Enc}(\boldsymbol{o}_m), \tag{10}$$

where $\oplus$ denotes concatenation. Existing approaches, such as DIMN [16], can be applied here to estimate the matching scores: $\text{score}_m = \text{Match}(H_p^m, H_q, H_{o_m})$. Finally, with M options, the model produces a vector of matching scores [score_1, score_2, ..., score_M]. Given the ground-truth label $\boldsymbol{a} = [a_1, a_2, \ldots, a_M]$ (where $a_m = 1$ if correct, otherwise 0), the following cross-entropy loss is leveraged to fine-tune the model:

$$\mathcal{L}_{overall} = -\sum_{m=1}^{M} a_m \log(\text{softmax}(\text{score}_m)). \tag{11}$$

Similarly, for inference, the final prediction is the option with the highest probability, $\hat{a} = \arg\max_m \text{softmax}(\text{score}_m)$.

4 Experiment

4.1 Datasets and Implementation

This study utilizes the competitive MCQA benchmarks from the RACE and DREAM datasets. The former includes two sections, *i.e.*, RACE-M and RACE-H, respectively, aimed at middle and high school levels, while the latter focuses on reading comprehension tests in conversational formats. Details of these benchmarks are listed in Table 1. We use the BERT-Base and BERT-Large models as encoders, setting a dropout rate of 0.1 and fine-tuning them with the Adam optimizer. Training involves a learning rate of $2e^{-5}$, a batch size of 4, and 3 entire epochs, with a maximum input sequence length of 320. In *CPOT*, only Top 25% prototypes and their 75% sentences are employed for clues (this selection's impact is examined in the ablation study). In addition, selected clues are combined with the original passage for semantic matching. The *Accuracy* performance is measured using the ratio of correctly answered questions to the total number of questions. All methods are run with a 24GB NVIDIA RTX3090 GPU.

4.2 Main Results

We employed several holistic (semantic-matching) methods including *Base*, *DIMN* [16], and *DUMA* [24], with *Base* using the `[CLS]` token's representation for input matching. Additionally, we utilized clue-based methods such as *EAM* [12], *GenMC* [7], and *RekNet* [23], all detailed in Sect. 2.

Table 1. Summary of employed datasets, including average option numbers (Opt.), and words per passage/question/answer (W/P, W/Q, W/A), respectively.

Dataset	Passages	Questions	Opt.	W/P	W/Q	W/A
RACE-M	7,139	28,293	4	231.1	9.0	3.9
RACE-H	20,794	69,394	4	353.1	10.4	5.8
DREAM	6,444	10,197	3	85.9	8.6	5.3

Table 2. Performance comparison between *CPOT* and existing work across three MCQA benchmarks. *CPOT* shows statistically significant improvements, marked with † for *p*-values below 0.01. The notation in brackets specifies the encoder type used, either BERT-Base (B) or BERT-Large (L).

	Database	Clue					Holistic			
		Base	EAM	GenMC	RekNet	*CPOT*	DIMN	+*CPOT*	DUMA	f+*CPOT*
(B)	RACE-M	71.1	73.5	72.0	71.7	**74.6**†	72.8	**74.8**†	73.3	**74.7**†
	RACE-H	62.3	65.1	63.4	65.2	**66.8**†	63.7	**67.2**†	64.1	**67.2**†
	DREAM	63.2	64.7	64.4	65.8	**66.5**†	65.2	**66.9**†	64.8	**66.8**†
(L)	RACE-M	76.6	78.4	76.8	77.9	**79.3**†	78.3	**80.9**†	78.5	**80.4**†
	RACE-H	70.1	73.7	71.5	73.6	**74.9**†	74.1	**76.3**†	74.6	**75.3**†
	DREAM	66.8	67.8	67.1	67.1	**69.0**†	68.4	**71.4**†	68.5	**70.2**†

We repeat our experiments five times using randomly initialized seeds and summarize the averaged results in Table 2. Key findings include: (1) BERT-Large models outperform BERT-Base models, suggesting benefits from larger-scale language models. (2) *CPOT* consistently enhances performance over traditional approaches across three benchmarks. For example, using BERT-Base, *CPOT* outperforms *EAM* by 1.1, 1.7, and 1.8 absolute points on RACE-M, RACE-H, and DREAM, respectively, demonstrating its effectiveness and stability. Additionally, the significance of these improvements is confirmed through one-sample T-tests, with *p*-values of 2.6×10^{-4}, 9.3×10^{-5}, and 6.2×10^{-5} for RACE-M, RACE-H, and DREAM respectively, affirming the robustness and efficacy of *CPOT*. (3) Integrating *CPOT* with existing matching methods like *DIMN* and *DUMA* results in significant performance improvements across all datasets, with gains of 2.0, 3.5, and 1.7 points for *DIMN*, and 1.4, 3.1, and 2.0 points for *DUMA* respectively. This shows that *CPOT* can effectively enhance existing semantic matching models.

5 Ablation Study

Robustness Across Different Encoders. In this experiment, the BERT-Base encoder is replaced by ALBERT-Base [8] with a learning rate of $1.5e^{-5}$, other configurations remain unchanged. Table 3 shows *CPOT* consistently achieving the highest performance across all datasets. However, when transitioning to different encoders, the RekNet method performs worse on RACE-H compared to using BERT-Base. These findings demonstrate the stability and robustness of our method with various encoders. Subsequent ablation studies are conducted using BERT-Base.

Table 3. Results on the encoder flexibility (replacing BERT with ALBERT). The result in brackets shows the combination of the original model and *CPOT*.

Dataset	Base	EAM	RekNet	*CPOT*	DIMN(*CPOT*)	DUMA(*CPOT*)
RACE-M	75.6	77.9	74.4	**78.5**	76.1(**78.9**)	76.3(**79.1**)
RACE-H	66.8	68.5	68.3	**69.6**	66.9(**70.1**)	67.2(**70.5**)
DREAM	64.5	65.2	66.0	**69.2**	68.5(**69.7**)	67.3(**69.1**)

On the Selection Size. In this experiment, we systematically examine the impact from the selected number of prototypes (K_{clu}) and the sentences (K_{sen}) on the overall performance. Specifically, we analyze the effects of selecting 10%, 25%, 50%, 75%, and 100% of the total available prototypes, along with an equivalent percentage of sentences from each selected prototypes (the selection is rounded up if necessary). The comparison across three datasets in Fig. 2 shows optimal performance with K_{clu} =25% and K_{sen} =75%. Notably, selecting too many prototypes and sentences simultaneously (*i.e.*, 75%) reduces performance due to the redundant details. Conversely, selecting a smaller K_{clu} and K_{sen}, such as 10%, results in insufficient information which also impacting performance. Accordingly, the following experiments adopt K_{clu} =25% and K_{sen} =75%.

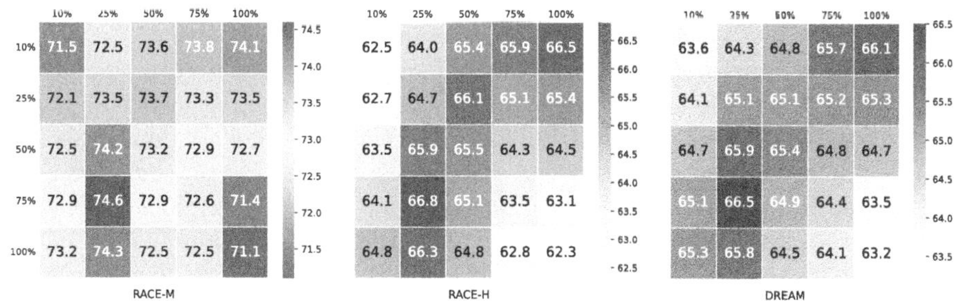

Fig. 2. Performance comparison as a function of the selection size. The horizontal axis represents the proportion of prototypes selected (K_{clu}), while the vertical axis represents the proportion of sentences (K_{sen}) from each selected prototype.

On the Prior Condition. *CPOT* utilizes the question and option answer as prior conditions for the clue extraction. We compare this strategy (called *Main*) with several alternatives, including:

- *OnlyQ*: using only the question as the condition, that is, $\boldsymbol{z}_m = \boldsymbol{q}$;
- *WithoutQO*: extracts clue without prior conditions;
- *Random*: randomly selecting sentences from the passage;

The experimental results in Fig. 3 yielding that firstly, *Main* consistently outperforms all other strategies, highlighting the necessity of employing conditions for gathering sufficient clues, while the *Random* performs the poorest, serving as the lower bound. Secondly, the *OnlyQ* strategy falls short compared to *Main*, indicating the importance of candidate options in identifying relevant sentences. Thirdly, *WithoutQO* results are inferior, as it navigates a large search space without guiding conditions, leading to irrelevant outcomes.

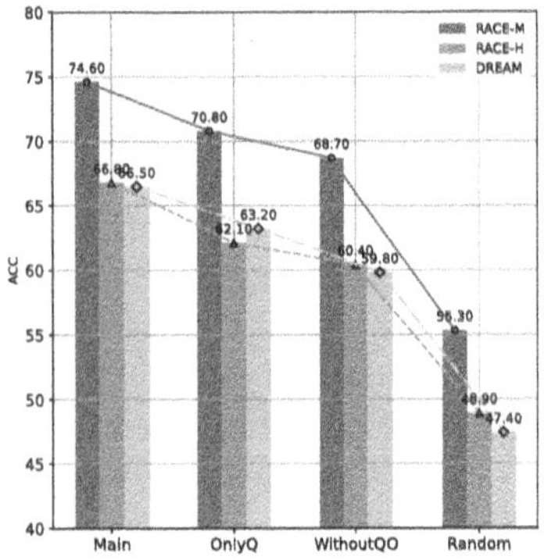

Fig. 3. Performance comparison as a function of the selection size across three benchmarks.

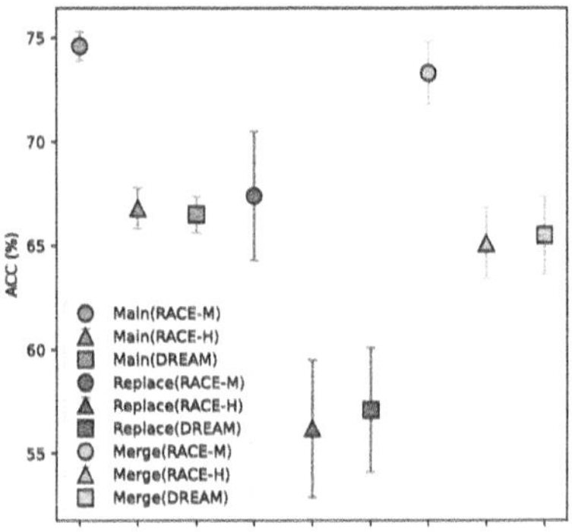

Fig. 4. Impact analysis of the proposed COT on the clue extraction against the other three strategies.

On the Combination Strategy. In *CPOT*, extracted clues are combined with the original passage for model fine-tuning. The following experiment compares two alternative strategies for utilizing extracted clues: (1) *Replace*: abandons the original passage and uses only the extracted clues; and (2) *Merge*: combines top-30% of clues from different options as the final input. As observed from the results in Fig. 4, the *Merge* variant achieves an accuracy of approximately 73.3, 65.1, 65.5 in RACE-M, RACE-H, and DREAM respectively, while *CPOT* achieves an averaged 0.9 absolute points over *Merge*. This is due to using diverse clues from various candidates, which may conflict, risking logical coherence when merged from incorrect options. On the other hand, *Replace* performs worst, even below the vanilla model, due to loss of contextual information from the original passage. By using extracted clues as additional passages combined with the original, our method retains the context and emphasizes key clues, enhancing overall model comprehension and confirming our method's superiority.

6 Conclusion

This paper proposes the **C**onditional **P**rototypical **O**ptimal **T**ransport (***CPOT***) algorithm for identifying relevant clues in Multiple Choice Question Answering (MCQA) tasks. Specifically, *CPOT* leverages the question and options as prior condition to cluster sentences from the background passage semantically. This is achieved by extending traditional Optimal Transport (OT) theory with constraints of unique assignment and uniform distribution across prototypes. We conduct extensive experiments using three MCQA benchmarks. When compared to existing work, our method consistently outperforms them, showcasing significant improvements in answering performance.

References

1. Asano, Y.M., Rupprecht, C., Vedaldi, A.: Self-labelling via simultaneous clustering and representation learning. arXiv preprint arXiv:1911.05371 (2019)
2. Berzak, Y., Malmaud, J., Levy, R.: STARC: Structured annotations for reading comprehension. In: Proceedings of the 58th Annual Meeting of the Association for Computational Linguistics, pp. 5726–5735 (2020)
3. Chen, W., Ravi, S., Shwartz, V.: CASE: commonsense-augmented score with an expanded answer space. In: Findings of the Association for Computational Linguistics: EMNLP 2023, pp. 2732–2744 (2023)
4. Devlin, J., Chang, M.W., Lee, K., Toutanova, K.: BERT: pre-training of deep bidirectional transformers for language understanding. In: Proceedings of the 2019 Conference of the North American Chapter of the Association for Computational Linguistics: Human Language Technologies, Volume 1 (Long and Short Papers), pp. 4171–4186 (2019)
5. Huang, Z., Yu, P., Allan, J.: Improving cross-lingual information retrieval on low-resource languages via optimal transport distillation. In: Proceedings of the Sixteenth ACM International Conference on Web Search and Data Mining, pp. 1048–1056 (2023)
6. Huang, Z., Wu, A., Shen, Y., Cheng, G., Qu, Y.: When retriever-reader meets scenario-based multiple-choice questions. In: Findings of the Association for Computational Linguistics: EMNLP 2021, pp. 985–994 (2021)
7. Huang, Z., Wu, A., Zhou, J., Gu, Y., Zhao, Y., Cheng, G.: Clues before answers: generation-enhanced multiple-choice QA. In: Proceedings of the 2022 Conference of the North American Chapter of the Association for Computational Linguistics: Human Language Technologies, pp. 3272–3287 (2022)
8. Lan, Z., Chen, M., Goodman, S., Gimpel, K., Sharma, P., Soricut, R.: Albert: a lite BERT for self-supervised learning of language representations. arXiv preprint arXiv:1909.11942 (2019)
9. Li, R., Jiang, Z., Wang, L., Lu, X., Zhao, M., Chen, D.: Enhancing transformer-based language models with commonsense representations for knowledge-driven machine comprehension. Knowl.-Based Syst. **220**, 106936 (2021)

10. Liu, C., Yang, J., Li, W.: Extractive question answering with contrastive puzzles and reweighted clues. In: Barney Smith, E.H., Liwicki, M., Peng, L. (eds.) Document Analysis and Recognition - ICDAR 2024: 18th International Conference, Athens, Greece, August 30–September 4, 2024, Proceedings, Part VI, pp. 97–112. Springer Nature Switzerland, Cham (2024). https://doi.org/10.1007/978-3-031-70552-6_6
11. Liu, X., Shi, Y., Liu, R., Bai, G., Chen, Y.: Narrow down before selection: a dynamic exclusion model for multiple-choice QA. In: ICASSP 2023 - 2023 IEEE International Conference on Acoustics, Speech and Signal Processing (ICASSP), pp. 1–5 (2023)
12. Luo, D., et al.: Evidence augment for multiple-choice machine reading comprehension by weak supervision. In: 30th International Conference on Artificial Neural Networks, pp. 357–368 (2021)
13. Ma, C., Du, X.: POE: process of elimination for multiple choice reasoning. In: Proceedings of the 2023 Conference on Empirical Methods in Natural Language Processing, pp. 4487–4496 (2023)
14. Nouri, N.: Text style transfer via optimal transport. In: Proceedings of the 2022 Conference of the North American Chapter of the Association for Computational Linguistics: Human Language Technologies, pp. 2532–2541 (2022)
15. Pezeshkpour, P., Hruschka, E.: Large language models sensitivity to the order of options in multiple-choice questions. In: Findings of the Association for Computational Linguistics: NAACL 2024, pp. 2006–2017 (2024)
16. Wei, Q., Ma, K., Liu, X., Ji, K., Yang, B., Abraham, A.: DIMN: dual integrated matching network for multi-choice reading comprehension. Eng. Appl. Artif. Intell. **130**, 107694 (2024)
17. Yang, J., Ma, J., Howard, S.K.: Usage profiling from mobile applications: a case study of online activity for Australian primary schools. Knowl.-Based Syst. **191**, 105214 (2020)
18. Yao, X., et al.: Context-guided triple matching for multiple choice question answering. In: 2023 IEEE Smart World Congress (SWC), pp. 1–8. IEEE (2023)
19. Yao, X., et al.: Killing many birds with one stone: single-source to multiple-target domain adaptation for extractive question answering. In: 2023 IEEE Smart World Congress (SWC), pp. 1–8 (2023)
20. Yao, X., Zhang, Z., Hu, X., Yang, J., Guo, Y., Zhu, D.D.: COTER: conditional optimal transport meets table retrieval. In: Proceedings of the 17th ACM International Conference on Web Search and Data Mining, pp. 911–919. WSDM '24, Association for Computing Machinery, New York, NY, USA (2024)
21. Zhang, S., Zhao, H., Wu, Y., Zhang, Z., Zhou, X., Zhou, X.: DCMN+: dual co-matching network for multi-choice reading comprehension. In: Proceedings of the AAAI Conference on Artificial Intelligence, vol. 34, pp. 9563–9570 (2020)
22. Zhang, Z., Wu, Y., Zhou, J., Duan, S., Zhao, H., Wang, R.: SG-Net: syntax-guided machine reading comprehension. In: Proceedings of the AAAI Conference on Artificial Intelligence, vol. 34, pp. 9636–9643 (2020)
23. Zhao, Y., Zhang, Z., Zhao, H.: Reference knowledgeable network for machine reading comprehension. IEEE/ACM Trans. Audio, Speech Lang. Process. **30**, 1461–1473 (2022)
24. Zhu, P., Zhang, Z., Zhao, H., Li, X.: DUMA: reading comprehension with transposition thinking. IEEE/ACM Trans. Audio, Speech Lang. Process. **30**, 269–279 (2022)

REFINE on Scarce Data: Retrieval Enhancement Through Fine-Tuning via Model Fusion of Embedding Models

Ambuje Gupta, Mrinal Rawat(✉), Andreas Stolcke, and Roberto Pieraccini

Uniphore, Palo Alto, CA 94304, USA
mrinal.rawat@uniphore.com

Abstract. Retrieval augmented generation (RAG) pipelines are commonly used in tasks such as question-answering (QA), relying on retrieving relevant documents from a vector store computed using a pretrained embedding model. However, if the retrieved context is inaccurate, the answers generated using the large language model (LLM) may contain errors or hallucinations. Although pretrained embedding models have advanced, adapting them to new domains remains challenging. Fine-tuning is a potential solution, but industry settings often lack the necessary fine-tuning data. To address these challenges, we propose REFINE, a novel technique that generates synthetic data from available documents and then uses a model fusion approach to fine-tune embeddings for improved retrieval performance in new domains, while preserving out-of-domain capability. We conducted experiments on the two public datasets: SQUAD and RAG-12000 and a proprietary TOURISM dataset. Results demonstrate that even the standard fine-tuning with the proposed data augmentation technique outperforms the vanilla pretrained model. Furthermore, when combined with model fusion, the proposed approach achieves superior performance, with a **5.76%** improvement in recall on the TOURISM dataset, and **6.58 %** and **0.32%** enhancement on SQUAD and RAG-12000 respectively.

Keywords: RAG · fine-tuning · LLM

1 Introduction

Question Answering (QA) systems for enterprise data have been a longstanding research focus [10]. Many efforts have centered on using BERT-based models that predict the start and end points of answer spans from the given context [12]. While initially promising, in practical business settings, there is an increasing need for abstractive QA that can provide more nuanced responses beyond extracting spans.

The rise of large language models (LLMs), such as ChatGPT [17] and Llama2 [23], has advanced abstractive QA capabilities substantially by enabling more

M. Gong et al. (Eds.): AI 2024, LNAI 15442, pp. 73–85, 2025.
https://doi.org/10.1007/978-981-96-0348-0_6

sophisticated response generation. However, a key challenge with LLMs is their lack of up-to-date, real-world knowledge, and tendency to hallucinate with confidence [28]. To alleviate this issue, retrieval-augmented generation (RAG) [14] has emerged as a promising technique that augments LLMs with relevant contextual information retrieved from databases. However, RAG's effectiveness depends on providing accurate context—if the retrieved context is flawed, errors can propagate into the generated responses. The retrieval process in RAG typically involves ingesting documents, segmenting them, and representing them using embedding models such as BGE [2]. These representations are then stored in vector databases. At run-time, the query embedding is computed and matched against the pre-computed document embeddings to retrieve the most relevant matches.

A key challenge is that general pretrained embedding models may not perform well for domain-specific enterprise data. Moreover, these models may fail to provide precise results for retrieval use cases as their training objective did not focus specifically on matching queries to relevant documents. One potential solution is to fine-tune embeddings using domain-specific dataset, but obtaining labeled data is difficult in many application scenarios. Although unlabeled documents are usually available, they can still be scarce in practical scenarios, complicating the situation further. Some approaches [9,11] have therefore employed continued pretraining on unlabeled documents. However, this method requires a substantial amount of data, which may not be available.

To address these problems, we propose REFINE—a novel approach to enhance retrieval performance using unlabeled documents, particularly when data is scarce. Our contributions can be summarized as follows:

- A novel approach to generate contrastive training datasets for fine-tuning embedding models suited for retrieval use-cases, using only available unlabeled documents, by leveraging an LLM.
- Second, we demonstrate how the generated dataset can enhance the embedding model through standard fine-tuning methods, without any additional techniques.
- Third, we introduce a model fusion technique during fine-tuning that incorporates both pretrained representations and the new data-specific learning, boosting performance further while retaining general capabilities for out-of-domain datasets.

2 Related Work

Document retrieval systems have revolutionized information access, allowing users to find relevant content from large databases quickly. These systems are particularly critical in question-answering (QA) tasks, where relevant documents provide the necessary context for generating accurate answers [24]. The emergence of large language models (LLMs) has increased the importance of document retrieval, especially in RAG applications. The LLM context is augmented with the right documents so that it generates the correct answers without hallucinations [14]. The effectiveness of such retrieval systems often depends upon the quality of the document embeddings that are used to represent and retrieve the relevant information. Initially, systems such as BM-25, which employ sparse retrievers, were common [20]. Although effective when exact word matches are present, they struggle with semantic understanding. To address this, the use of dense retrievers became prevalent. These retrievers, which are essentially embedding models, provide semantic text representations. DPR [13] demonstrates the practicality of using dense representations for retrieval with a dual encoder framework trained on a modest set of questions and passages.

A major breakthrough in embedding models occurred with the introduction of BERT-based models [4], which provide contextual representations through Masked Language Modeling (MLM), a self-supervised learning technique. Subsequently, considerable effort was made to create generalized embeddings, such as Sentence Transformers e.g. MPNet [19], OpenAI embeddings, and BGE embeddings [26]. These embeddings trained on a vast corpus using contrastive learning have demonstrated significant improvements. However, in many scenarios, they could still fail to correctly represent the text if the domain is very specialized. In these cases, continual pretraining and fine-tuning can be very helpful. Usually, if the domain-specific dataset is large, the model pretraining is first continued in a self-supervised fashion [9], which does not require labeled data. Then, the model is fine-tuned on the target task to further enhance performance for that specific domain and task. However, this fine-tuning requires labeled data, preventing its use in many data-starved application domains. Synthetic data generation has been proposed as a solution to the data scarcity problem. By generating artificial samples that mimic the characteristics of a target domain, the natural training set is augmented, allowing adaptation to new domains without extensive new annotation [22,25].

Furthermore, while fine-tuned models excel on the domains they are trained on, they can lead to a significant drop in performance on more general tasks, a phenomenon known as catastrophic forgetting [3,8]. To counteract this, various strategies have been proposed, including replay-based methods [21] and regularization-based methods [18]. More recently, the concept of model merging has been shown effective to harness the strengths of different models. This technique involves combining models that have been trained on different datasets or that employ distinct architectures [27]. However, this merging typically occurs after fine-tuning and without continued training. Our proposed REFINE method, by contrast, incorporates model merging during the training phase. This strategy allows for more effective representation learning as the model continues using knowledge from the pretrained model while adapting to the new domain.

3 Approach

In this work, we introduce a data augmentation technique employed to fine-tune the embedding model. This technique aims to enrich the representations and enhance the retrieval performance. Additionally, we propose a model fusion

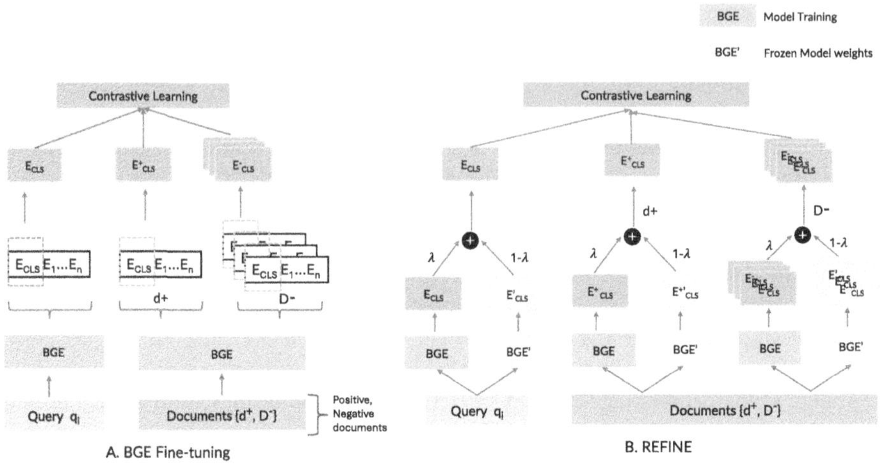

Fig. 1. (A) standard BGE process followed for fine-tuning. (B) training process for REFINE.

technique to further improve the fine-tuning process. In the subsequent subsections, we will elaborate on these concepts in detail.

3.1 Data Augmentation for Fine-Tuning

LLMs have been utilized for data augmentation or synthetic data creation, having shown good out-of-the-box performance. This process involves automating the laborious and time-consuming task of creating and annotating data by humans. With LLMs, we can generate synthetic data rapidly, and in some cases, even outperform human text annotation quality [7]. With the advent of larger and newer LLMs, the quality of LLM-generated annotations and data can be expected to improve further.

We utilize LLM to generate high-quality synthetic data that is used to fine-tune the embedding model, BGE in our case. We start with a set of given unlabeled documents D, and for each document, we prompt the LLM to generate a set of derived queries $Q = \{q_1, q_2, \ldots, q_k\}$. The prompt we used for generating queries is:

```
You are a query generator bot. Generate 10 distinct queries
from the document D
```

In our experiments, we set k to 10. Now, for each generated query $q_i \in Q$, since we know from which document the question was generated, its corresponding source document becomes the positive sample passage d_i^+. For contrastive fine-tuning of the embedding model, we also require negative sample passages, so that the model learns to discriminate between positive and negative matches between queries and passages.

Algorithm 1: Pseudo-code for data augmentation

Input: Documents D, pretrained BGE model M

$D_A \leftarrow \emptyset$; `// Augmented Dataset`

$DB \leftarrow create\_vectordb(D, M)$

foreach $d_i^+ \in D$ **do**

  $Q \leftarrow generate\_queries(d_i^+, k = 10)$ `// Generates 10 queries using LLM from the document`

  foreach $q_i \in Q$ **do**

    $R \leftarrow retrieve\_docs(q_i, DB)$

    $D^- \leftarrow select\_neg\_docs(R)$ `// pick hard negative documents with similarity scores obtained from retriever between 0.5 and 0.7, excluding those in the top 5.`

    $D_A \leftarrow D_A \cup \{(q_i, d_i^+, D^-)\}$

We aim to find "hard negative" training samples to enhance the model's discriminative power. To that end, we first vectorize and ingest all documents to create a vector store, using a pretrained embedding model. Now, for each query $q_i \in Q$, we use the retriever to locate the top 50 most similar documents. Next, we filter the documents, selecting those that have similarity scores between 50% and 70% and are not in the top $k = 5$, to keep near misses while ruling out accidental matches. We found these hyperparameters empirically by running trials on a held-out 15% of the training set. In this way, we obtain $D^- = \{d_1^-, d_2^-, \ldots, d_m^-\}$ as negative training samples (cf. Algorithm 1).

3.2 Fine-Tuning via Model Fusion

Contrastive learning on diverse data has demonstrated good performance in similarity learning based on embedding models, as it learns to push negative samples away from the query, while pulling positive samples closer to it. Although refining the model through additional fine-tuning can improve its effectiveness within a particular domain [26], it is important to note that this specialization might deteriorate performance outside of the fine-tuning domain.

In this work, we also utilize contrastive learning to fine-tune the model using a model fusion approach (Fig. 1); specifically the `BAAI/bge-large-en-v1.5` model, is fine-tuned on a domain-specific dataset obtained using the technique discussed in Sect. 3.1. We first initialize two models, BGE and BGE', with the pretrained BGE weights, freezing the weights of BGE' while allowing only BGE to undergo fine-tuning [26]. Given a query q_i, we pass it through both BGE and BGE' to obtain the representations of the `[CLS]` token, denoted as E_{CLS} and E'_{CLS}, respectively. Next, we interpolate these two representations to create a final representation E_{CLS}:

$$E_{CLS} = \lambda \cdot E_{CLS} + (1 - \lambda) \cdot E'_{CLS} \quad (1)$$

where λ is a hyperparameter. Increasing its value will place greater emphasis on the new domain.

Next, we apply the same technique to the positive passage d_i^+ and the negative passages D_i^-, yielding E_{CLS}^+ and $\{E_{CLS_1}^-, E_{CLS_2}^-, \ldots, E_{CLS_m}^-\}$, respectively. We then fine-tune the model using the objective

$$\max \sum \log \frac{e^{sim(E_{CLS}, E_{CLS}^+)/\tau}}{e^{sim(E_{CLS}, E_{CLS}^+)/\tau} + \sum_{i=1}^{m} e^{sim(E_{CLS}, E_{CLS_i}^-)/\tau}} \tag{2}$$

where τ is a temperature hyperparameter and $sim()$ is the cosine similarity.

4 Experiments

4.1 Datasets

We conducted experiments on two public and one propriety datasets:

SQUAD[1] is a question-answering dataset along with the provided context. To simulate limited dataset conditions, we utilized a subset comprising 300 randomly selected instances from the publicly available dataset.

RAG 12000[2] is a collection of triples featuring `context`, `question`, and `answer` fields, designed for building models optimized for RAG. Similar to SQUAD dataset, we took a subset of 100 randomly selected instances from the dataset. We will refer to it as the RAG dataset for brevity.

TOURISM is a proprietary dataset focusing on the tourism domain. It contains both unstructured text and structured, tabular data.

To evaluate the retriever, we require both the question and the supporting document (evidence). With the SQUAD & RAG dataset, this is straightforward due to the inherent mapping present within the raw dataset, allowing us to create a dataset of question-document (context) pairs for evaluation. However, for the TOURISM dataset, our annotation team manually curated a question from each document. For fine-tuning, we generated a dataset exclusively from documents and synthetic questions. To perform hyperparameter tuning, we set aside 15% of the generated data as a validation set for each of the datasets. Table 1 summarizes the sizes of these datasets.

4.2 Baselines

We present a direct comparison of our approach with the following baseline methods:

- **Vanilla BGE:** Here, we compute the metrics by simply utilizing the pre-trained BGE[3] embedding model on the two evaluation sets.

[1] rajpurkar/squad.
[2] RAG 12000 Dataset.
[3] BAAI/bge-large-en-v1.5.

Table 1. Statistics of the two datasets, with the number of documents and questions for evaluation, as well as training and validation (tuning) samples generated through our data augmentation method.

Dataset	# Docs	# Ques.	# Train	#Val	Avg Words (Docs)
SQUAD	300	300	2550	450	165
RAG 12000	100	100	850	150	190
TOURISM	58	58	493	87	138

- **Vanilla snowflake-arctic-embed-l:**[4] We use an embedding model that attained state-of-the-art performance on the MTEB leaderboard [15].
- **Vanilla text-embedding-3-large:** Metrics are computed using a state-of-the-art off-the-shelf text embedding model [16].
- **Fine-tuned BGE:** We fine-tune the BGE model according to the method outlined by [26] on the dataset generated synthetically with our approach, and subsequently compute the metrics on the evaluation data. Figure 1(A) illustrates this technique.
- **LM Cocktail:** We performed model merging after training [27], combining vanilla BGE with the fine-tuned BGE model. We applied the same weighting as in our REFINE model, i.e., 0.35 for BGE and 0.65 for fine-tuned BGE.

4.3 Evaluation Metrics

We use the following commonly used metrics for evaluating the retriever:

- **Mean average precision (MAP)** measures the average precision across multiple queries. Higher values indicate better performance, rewarding systems that return relevant documents earlier in the ranked list.
- **Normalized discounted cumulative gain (NDCG)** assesses ranking quality by considering relevance and position in the ranked list. Higher scores are given to highly relevant documents at the top.
- **Mean reciprocal rank (MRR)** evaluates systems producing ranked lists by considering the rank of the first relevant document. Higher values indicate the relevant document is ranked higher.
- **Recall** measures the fraction of relevant documents successfully retrieved by the system. Higher values indicate better performance in retrieving all relevant document

4.4 Evaluation Setup

To evaluate the retriever's performance, we initially ingest documents from each dataset into a FAISS vector store [5], separately. Then, for every query in the question set, we derived its representation using the model fine-tuned through

[4] Snowflake/snowflake-arctic-embed-l.

our proposed method. Subsequently, we used this representation to fetch relevant documents from the vector store, utilizing the retriever from the LangChain library. For most experiments, and unless indicated otherwise, we fixed the top_k parameter at 3 (retrieving the three most relevant documents). (The number of retrieved documents is commonly notated as "@k" following the metric, e.g., "Recall@3."). We computed the evaluation metrics for all the baseline approaches described in Sect. 4.2. We want to highlight that we did not concentrate on hyperparameter tuning. However, to account for the variability in results due to different random initializations, we report the average performance across five experimental runs with different random seeds for the models that involved training.

4.5 Training Details

In this work, we leveraged the `BAAI/bge-large-en-v1.5` as the backbone model. To train both the BGE way and REFINE, identical hyperparameters were used. The learning rate was set to 10^{-5}, and the batch size was adjusted according to GPU availability, with 4 gradient accumulation steps implemented for both approaches. Notably, in the case of REFINE, we froze the layers of BGE' while allowing gradient updates for BGE. In the REFINE approach, a crucial parameter is the value of λ, which we set at 0.35, chosen empirically. This implies that the frozen model (BGE') contributed 0.35 to the overall output, while the fine-tuned model (BGE) contributed 0.65. All experiments were conducted using g5.12xlarge GPUs.

5 Results and Discussion

Our evaluation is based on the metrics defined in Sect. 4.3. We present our findings in two parts: 1) The enhancement in retrieval performance achieved through the generation of synthetic data and the model fusion approach. 2) We discuss how the REFINE approach not only outperforms other baseline methods but also enables the model to maintain its general capability, thereby mitigating catastrophic forgetting.

The key findings from the part 1 experiment, where results were obtained on the TOURISM (private), SQUAD (public) and RAG (public) datasets, are presented in Table 2. Our approach of using synthetic data combined with model fusion boosted the performance of the retriever significantly, leading to improvements of **5.79%**, **6.58%** and **0.32%** in Recall@3 for the TOURISM, SQUAD and RAG datasets, respectively compared to the baseline Vanilla BGE model. The gain in recall did not come at the expense of precision (MAP), which also improved. The substantial improvement observed on the TOURISM dataset highlights the importance of fine-tuning in specific domain settings. Furthermore, even the standard BGE fine-tuning on synthetic data led to improved metrics across both datasets, except for recall on the RAG dataset. An important observation was that the TOURISM dataset was comprised mostly of structured tabular data. Upon further analysis, we observed that the vanilla BGE failed to correctly identify documents for the queries, especially when the queries involved specific details within tables. For instance, when querying"Which hotel's phone number is xxxx?" where xxxx refers to a value within a table cell, the vanilla BGE model struggled to provide accurate representations, resulting in unreliable cosine similarity. However, with augmented data used for fine-tuning, performance improved notably, particularly in the case of structured data retrieval.

Table 2. Results on TOURISM, RAG, and SQUAD datasets using the baselines defined in the paper and REFINE (our approach). For all evaluation metrics, higher values indicate better performance.

Model	TOURISM Dataset				RAG Dataset				SQUAD			
	MAP @3	NDCG @3	MRR @3	Recall @3	MAP @3	NDCG @3	MRR @3	Recall @3	MAP @3	NDCG @3	MRR @3	Recall @3
Vanilla BGE	0.781	0.800	0.789	0.884	0.863	0.858	0.904	0.937	0.763	0.789	0.763	0.866
snowflake-arctic-embed-m	0.794	0.817	0.796	0.888	0.874	0.857	**0.921**	0.916	0.708	0.728	0.707	0.786
text-embedding-3-large	0.847	0.857	0.850	0.915	0.878	0.866	0.920	0.935	0.808	0.828	0.808	0.883
Fine-tuned BGE	0.870	0.883	0.873	0.929	0.876	0.865	0.915	0.929	0.834	0.853	0.834	0.907
LM-cocktail (BGE+FT-BGE)	0.862	0.881	0.869	0.936	0.870	0.861	0.906	0.934	0.829	0.852	0.829	0.920
REFINE (ours)	**0.873**	**0.885**	**0.877**	**0.937**	**0.881**	**0.867**	0.919	**0.940**	**0.846**	**0.866**	**0.846**	**0.923**

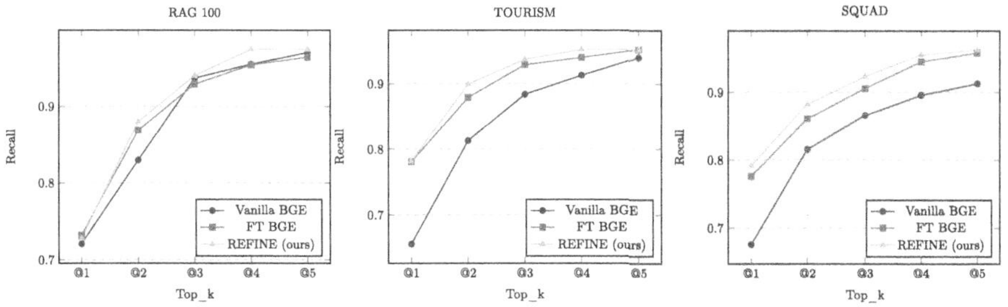

Fig. 2. Recall at different top_k values for vanilla (pretrained) BGE, fine-tuned (FT) BGE, and REFINE.

Figure 2 plots the performance of the three models when different values of top_k were used in the retriever. Recall was measured with top_k $= 1, 2, 3, 4, 5$. The figure shows that REFINE outperforms the other baselines across the different top_k settings on both datasets. A significant improvement can be observed in the TOURISM and SQUAD dataset, even for top_k $= 1$ w.r.t vanilla model. This demonstrates the superior performance of REFINE when retrieving only one or few relevant documents. Refer Table 4 in Supplementary material for the detailed results. We also want to highlight that in preliminary experiments, we attempted unsupervised fine-tuning on the unlabeled training data. However, this approach led to a degradation in performance. The reason for this is that the objective of pre-training is to reconstruct the text, and the pre-trained model cannot be directly used for similarity computation. Instead, it requires fine-tuning to achieve that capability [6,26].

Cross-Dataset Generalization We further conducted experiments to evaluate the impact of fine-tuning on the generalizability of a model trained on domain-specific data. While fine-tuning generally improves performance within the targeted domain (in-domain), it often results in a loss of generalization across other domains (out-of-domain), a phenomenon known as catastrophic forgetting [1]. Our findings, presented in Table 3, show the performance of a model initially trained on the SQUAD dataset and then evaluated on the RAG dataset, serving as an out-of-domain test case (OOD). We also compared our results with the LM-cocktail method, a popular model-merging technique. Our approach improved performance, achieving a recall of **0.938** compared to baseline methods. Interestingly, even fine-tuned BGE exhibited improved rather than degraded performance. We want to highlight that our primary focus was on enhancing the retriever for the specific dataset, we aimed to examine the effects of catastrophic forgetting and acknowledge the need for further exploration in future studies. Therefore, we limited our analysis to the comparison between the SQuAD and RAG datasets, rather than examining all possible combinations. Additionally, when we reversed the training and test datasets (using RAG as the training data and SQUAD as the test data), a similar trend was observed. Fine-tuned BGE, LM-cocktail showed performance improvements, with REFINE outperforming others achieving a recall of **0.896**, highlighting the generalizability of our approach.

Table 3. The results shown represent the model's performance when it was trained on a specific dataset and evaluated on an out-of-domain (OOD) dataset. The row with a "-" indicates the metrics for the vanilla (pretrained) BGE model, which did not undergo fine-tuning.

Train Dataset	OOD Dataset	Model	MAP @3	NDCG @3	MRR @3	Recall @3
-	RAG	Vanilla BGE	0.863	0.858	0.904	0.937
SQUAD		FT BGE	0.879	0.866	0.916	0.932
		LM-cocktail	0.878	**0.866**	**0.917**	0.934
		REFINE (ours)	**0.881**	0.865	0.915	**0.938**
-	SQUAD	Vanilla BGE	0.708	0.728	0.707	0.786
RAG		FT BGE	0.743	0.771	0.743	0.852
		LM-cocktail	0.769	0.794	0.769	0.866
		REFINE (ours)	**0.801**	**0.825**	**0.801**	**0.896**

6 Conclusions and Future Work

In this work, we introduced a novel approach to improve retrieval models used in RAG. First, we generated synthetic contrastive data to fine-tune the embedding models. Then, we introduced a model fusion approach, which combines the pretrained (vanilla) model with the newly trained model during the training phase. This fusion method led to significant performance gains across various metrics and improved the model's ability to generalize to unseen data. Although this study focused on scarce data to simulate real-life scenarios, we believe our approach can be extended to larger datasets, potentially achieving even greater performance improvements. In the future, we plan to apply our technique to more (larger) datasets and further investigate catastrophic forgetting in retrieval embedding models. Additionally, we aim to explore multi-hop retrieval fine-tuning, which is a challenging task.

A Detailed Results

The detailed recall results at different top_k values for both datasets are presented in Table 4. Our approach significantly outperforms all the baselines except in two cases. For the TOURISM dataset, LM-cocktail performed slightly better than REFINE for top_k = 5, while `snowflake-arctic-embed-l` performed better for top_k = 1 in case of RAG. Notably, on the TOURISM & SQUAD dataset, OpenAI's `text-embedding-3-large` embedding model demonstrated significantly better performance than other embedding models used in a vanilla (without fine-tuning) setting.

Table 4. Recall at different top_k values for the baselines defined in the paper and REFINE (our approach).

Dataset	Model	Recall				
		@1	@2	@3	@4	@5
SQUAD	Vanilla BGE	0.676	0.816	0.866	0.896	0.913
	snowflake-arctic-embed-l	0.643	0.743	0.786	0.823	0.850
	text-embedding-3-large	0.743	0.856	0.883	0.903	0.916
	Fine-tuned BGE	0.776	0.861	0.906	0.945	0.958
	LM-cocktail	0.781	0.862	0.908	0.946	0.950
	REFINE (ours)	**0.791**	**0.882**	**0.923**	**0.955**	**0.962**
TOURISM	Vanilla BGE	0.656	0.813	0.884	0.913	0.940
	snowflake-arctic-embed-l	0.665	0.840	0.888	0.9	0.9
	text-embedding-3-large	0.738	0.895	0.915	0.936	0.954
	Fine-tuned BGE	0.781	0.879	0.929	0.94	0.952
	LM-cocktail	0.762	0.872	0.936	0.946	**0.955**
	REFINE (ours)	**0.782**	**0.899**	**0.937**	**0.952**	0.952
RAG 100	Vanilla BGE	0.721	0.83	0.937	0.955	0.97
	snowflake-arctic-embed-l	**0.739**	0.865	0.916	0.954	0.954
	text-embedding-3-large	0.734	0.865	0.935	0.959	0.964
	Fine-tuned BGE	0.732	0.869	0.929	0.954	0.964
	LM-cocktail	0.720	0.859	0.934	0.962	0.963
	REFINE (ours)	0.728	**0.88**	**0.94**	**0.974**	**0.974**

References

1. Aleixo, E.L., Colonna, J.G., Cristo, M., Fernandes, E.: Catastrophic forgetting in deep learning: a comprehensive taxonomy. arXiv:2312.10549 (2023)
2. Chen, J., Xiao, S., Zhang, P., Luo, K., Lian, D., Liu, Z.: BGE M3-Embedding: multi-lingual, multi-functionality, multi-granularity text embeddings through self-knowledge distillation. arXiv:2309.07597 (2023)
3. Chen, S., Hou, Y., Cui, Y., Che, W., Liu, T., Yu, X.: Recall and Learn: fine-tuning deep pretrained language models with less forgetting. In: Proceedings of Conference on Empirical Methods in Natural Language Processing (EMNLP), pp. 7870–7881 (2020)
4. Devlin, J., Chang, M.W., Lee, K., Toutanova, K.: BERT: pre-training of deep bidirectional transformers for language understanding. In: Burstein, J., Doran, C., Solorio, T. (eds.) Proceedings of the 2019 Conference of the North American Chapter of the Association for Computational Linguistics: Human Language Technologies, Volume 1 (Long and Short Papers), pp. 4171–4186. Minneapolis, Minnesota (2019)
5. Douze, M., et al.: The Faiss library. arXiv:2401.08281 (2024)
6. FlagOpen: FlagEmbedding: retrieval and retrieval-augmented LLMs. https://github.com/FlagOpen/FlagEmbedding (2024), gitHub repository
7. Gilardi, F., Alizadeh, M., Kubli, M.: ChatGPT outperforms crowd workers for text-annotation tasks. In: Proceedings of National Academy of Sciences, vol. 120, issue 30, e2305016120 (2023)
8. Goodfellow, I.J., Mirza, M., Xiao, D., Courville, A., Bengio, Y.: An empirical investigation of catastrophic forgetting in gradient-based neural networks. arXiv:1312.6211 (2015)
9. Gururangan, S., et al.: Don't stop pretraining: adapt language models to domains and tasks. In: Jurafsky, D., Chai, J., Schluter, N., Tetreault, J. (eds.) Proceedings of 58th Annual Meeting of the Association for Computational Linguistics, pp. 8342–8360 (2020)
10. Hirschman, L., Gaizauskas, R.: Natural language question answering: the view from here. Natl. Lang. Eng. **7**, 275–300 (2001)
11. Huang, Y., Wang, K., Dutta, S., Patel, R., Glavaš, G., Gurevych, I.: AdaSent: efficient domain-adapted sentence embeddings for few-shot classification. In: Bouamor, H., Pino, J., Bali, K. (eds.) Proceedings of Conference on Empirical Methods in Natural Language Processing, pp. 3420–3434 (2023)
12. Joshi, M., Chen, D., Liu, Y., Weld, D.S., Zettlemoyer, L., Levy, O.: SpanBERT: improving pre-training by representing and predicting spans. arXiv:1907.10529 (2020)
13. Karpukhin, V., Oguz, B., Min, S., Lewis, P., Wu, L., Edunov, S., Chen, D., Yih, W.t.: Dense passage retrieval for open-domain question answering. In: Webber, B., Cohn, T., He, Y., Liu, Y. (eds.) Proceedings of Conference on Empirical Methods in Natural Language Processing (EMNLP), pp. 6769–6781 (2020)
14. Lewis, P., et al.: Retrieval-augmented generation for knowledge-intensive NLP tasks. In: Proceedings of 34th International Conference on Neural Information Processing Systems. NIPS '20 (2020)
15. Merrick, L., Xu, D., Nuti, G., Campos, D.: Arctic-Embed: scalable, efficient, and accurate text embedding models. arXiv:2405.05374 (2024)
16. OpenAI: New embedding models and API updates (2024). https://openai.com/index/new-embedding-models-and-api-updates/

17. Ouyang, L., et al.: Training language models to follow instructions with human feedback (2022). arXiv:2203.02155
18. Rebuffi, S.A., Kolesnikov, A., Sperl, G., Lampert, C.H.: iCaRL: incremental classifier and representation learning (2017). arXiv:1611.07725
19. Reimers, N., Gurevych, I.: Sentence-BERT: sentence embeddings using Siamese BERT-networks. In: Proceedings of Conference on Empirical Methods in Natural Language Processing (2019)
20. Robertson, S., Zaragoza, H.: The probabilistic relevance framework: BM25 and beyond. Found. Trends Inf. Retr. **3**(4), 333–389 (2009)
21. Rolnick, D., Ahuja, A., Schwarz, J., Lillicrap, T.P., Wayne, G.: Experience replay for continual learning. arXiv:1811.11682 (2019)
22. Thakur, N., Reimers, N., Daxenberger, J., Gurevych, I.: Augmented SBERT: data augmentation method for improving bi-encoders for pairwise sentence scoring tasks. In: Proceedings of Conference of the North American Chapter of the Association for Computational Linguistics: Human Language Technologies, pp. 296–310 (2021)
23. Touvron, H., et al.: LLaMA 2: open foundation and fine-tuned chat models. arXiv:2307.09288 (2023)
24. Voorhees, E.M., Tice, D.M.: The TREC-8 question answering track. In: Gavrilidou, M., Carayannis, G., Markantonatou, S., Piperidis, S., Stainhauer, G. (eds.) Proceedings of 2nd International Conference on Language Resources and Evaluation (LREC'00). European Language Resources Association (ELRA), Athens, Greece (2000)
25. Wu, M., Cao, S.: LLM-augmented retrieval: enhancing retrieval models through language models and doc-level embedding. arXiv:2404.05825 (2024)
26. Xiao, S., Liu, Z., Zhang, P., Muennighoff, N.: C-Pack: packaged resources to advance general Chinese embedding. arXiv:2309.07597 (2024)
27. Xiao, S., Liu, Z., Zhang, P., Xing, X.: LM-Cocktail: resilient tuning of language models via model merging. arXiv:2311.13534 (2023)
28. Xu, Z., Jain, S., Kankanhalli, M.: Hallucination is inevitable: an innate limitation of large language models. arXiv:2401.11817 (2024)

Leveraging LLM in Genetic Programming Hyper-heuristics for Dynamic Microservice Deployment

Zhengxin Fang[1(✉)], Hui Ma[1], Gang Chen[1], Sven Hartmann[2], and Chen Wang[3]

[1] Centre for Data Science and Artificial Intelligence and School of Engineering and Computer Science, Victoria University of Wellington, Wellington, New Zealand
{zhengxin.fang,hui.ma,aaron.chen}@ecs.vuw.ac.nz

[2] Department of Informatics, Clausthal University of Technology, Clausthal-Zellerfeld, Germany
sven.hartmann@tu-clausthal.de

[3] National Institute of Water and Atmospheric Research, Wellington, New Zealand
chen.wang@niwa.co.nz

Abstract. Microservice deployment in cloud computing is a challenging combinatorial optimization problem due to the complex dependencies among microservices and the intricate trade-offs among different QoS requirements, e.g., minimizing Energy Consumption (EC) vs. minimizing Communication Overhead (CO). Recently, some hyper-heuristics methods, particularly Genetic Programming Hyper-Heuristics (GPHH), have been proposed to automatically generate heuristics for solving dynamic microservice deployment problems. Meanwhile, Large Language Models (LLMs) are becoming popular for solving various domain-specific problems thanks to their strong ability to learn problem-related knowledge. However, hybridizing GPHH with LLM by combining their abilities in solving complex optimization problems remains unexplored. In this paper, we propose an LLM-enhanced Genetic Programming Hyper-Heuristic (LLM-GPHH) algorithm to evolve heuristics for the dynamic deployment of applications composed of microservices, to jointly optimize EC and CO. Our experiments on real-world datasets demonstrate the effectiveness of the newly proposed LLM-GPHH.

Keywords: LLM · genetic programming · hyper-heuristics · dynamic microservice deployment · cloud computing

1 Introduction

Recent years have witnessed an increasing number of applications being developed by composing microservices due to their advantages in terms of scalability, maintainability and resilience [14]. A microservices application comprises multiple independent microservices, each responsible for a specific function and communicating with others via data transmission. Microservice applications deployed in container-based clouds become popular thanks to containers'

M. Gong et al. (Eds.): AI 2024, LNAI 15442, pp. 86–97, 2025.
https://doi.org/10.1007/978-981-96-0348-0_7

lightweight nature and scalability [14]. The deployment of microservice applications in container-based clouds requires deploying microservices to containers, which are then allocated to virtual machines (VMs). Subsequently, VMs are allocated to Physical Machines (PMs). The process of deploying microservices applications in container-based clouds gives rise to the problem of **M**icroservice **D**eployment in **C**ontainer-based clouds (called the MDC problem), which is NP-hard [4,16,17].

In real-world practices, microservice applications arrive dynamically in cloud data centers in real-time. As a result, a *VM selection heuristic* is required to either select existing VMs or create new VM instances for real-time microservice applications deployment. Similarly, a suitable PM should be selected for newly created VM instances to achieve more effective resource utilization through a *PM selection heuristic*. Improper *VM selection heuristic* and *PM selection heuristic* causes poor CPU and memory allocation, increasing the Energy Consumption (EC) in cloud data centers [2]. In addition, improper deployment of microservices increases the communication data volumes between different PMs through the physical network, resulting in a large Communication Overhead (CO) in cloud data centers, which affects the performance of microservice applications [1]. Therefore, in this paper, we aim to propose an effective method to automatically generate *VM selection heuristic* and *PM selection heuristic* to minimize both EC and CO in cloud data centers [4] for dynamic MDC problems.

Various heuristics have been proposed in the literature to solve MDC problems [7,8,11]. However, manually designed heuristics can lose effectiveness due to the dynamic workload arriving in clouds [20]. Hyper-heuristics (HHs), in particular Genetic Programming Hyper-Heuristics (GPHHs), have been proposed to learn heuristics automatically for many complex dynamic combinatorial optimization problems [16,17,20]. Nevertheless, the randomness of genetic operators (i.e., crossover and mutation) of GP do not explicitly exploit the patterns among high-fitness individuals [9], leading to the high possibility of producing bad individuals during evolution. In recent years, the rapid rise of Large Language Models (LLMs) has introduced new possibilities for the exploitation of GP. LLMs can take a sequence of tokens as input and generate a new sequence of tokens, leveraging a large amount of semantic knowledge during training. This means that LLMs have the potential to capture the complex patterns among tokens (i.e. nodes of the GP tree) in high-fitness GP individuals, then output new promising GP individuals by generating new tokens.

Based on the above analysis, this paper proposes a LLM-enhance GPHH (LLM-GPHH) for generating *VM selection heuristics* and *PM selection heuristics* in dynamic MDC problems to jointly optimize CO and EC. The contributions of this paper are as follows.

- A novel *LLM evolution* is proposed in LLM-GPHH to capture the complex relationship among tokens in high-fitness individuals of GPHH and generate new promising individuals to improve the performance of GPHH in terms of jointly optimizing CO and EC for dynamic MEC problems.

- We conduct an experimental evaluation of LLM-GPHH based real-world cloud dataset. Our experiments demonstrate that the proposed LLM-GPHH can effectively generate better heuristics than existing algorithms in most scenarios.
- To the best of our knowledge, this is the first approach that combines LLM with GPHH to automatically generate heuristics for dynamic MDC problems. Ablation studies are also conducted to show the effectiveness of *LLM evolution* in some scenarios.

2 Related Work

Various heuristics have been proposed to solve microservice deployment problems as they are easy to understand and apply. For example, the Best-Fit [11] heuristic deploys applications with the most suitable resource (i.e., VM or PM), which has enough capacity (CPU and memory) to host the application. To minimize the CO among PMs, the *Min-cut* [21] mechanism has been applied in [7] to partition an application into several groups of microservices, which can then be allocated to the same PMs. Resource utilization is enhanced by allocating groups of microservices to the most-loaded machine. Similar to [7], groups of containers are allocated in [8] using a concurrent container scheduling algorithm. Specifically, the process of microservice deployment in [8] is modelled as the minimum cost flow problem (MCFP) in a weighted network, on which containers with dependencies will be merged into an aggregator node to be subsequently allocated to VMs. Although the approaches in [7,8] can optimize both CO and EC, they assume that workloads are known in advance. As a result, their methods can lose effectiveness for handling MDC problems where microservice deployment requests and workloads are arriving dynamically [17].

To automatically learn effective heuristics for dynamic microservice deployment, a hybrid GPHH based approach, named Hybrid-Evo, is proposed in [17]. This approach learns VM selection heuristics while combining the Best-Fit heuristic with its learned PM selection heuristics. Specifically, Hybrid-Evo applies Best-Fit to generate new PM instances. These two GPHH based approaches [16,17] only consider EC, without paying attention to microservices and inter-service communication. However, existing GPHH methods do not explicitly exploit the patterns among good individuals. LLM is a potential solution to enhance the exploitation ability of GPHH. For example, Romera et al. [10] applied LLM for mathematical discovery by capturing the complex patterns in the equations.

As summarized, GPHH is particularly suitable for handling dynamic MDC problems, and LLMs can be used to enhance the exploitation ability of GPHH due to LLM's proficiency in capturing complex but promising patterns and semantic relationships in good heuristics. In this paper, we will propose a novel GPHH algorithm (i.e., LLM-GPHH) that applies LLM to enhance the abilities of GPHH to automatically generate heuristics that can effectively solve dynamic MDC problems.

3 Problem Definition

Microservice applications can be represented as Directed Acyclic Graphs (DAGs) with weighted edges [13]. Specifically, an application is denoted by $G(M, E)$ with M as a set of nodes where each $m_i \in M$ represents a microservice. Meanwhile, E stands for a set of directed edges with weights where each $e_{i,j} \in E$ represents direct data flow between m_i and m_j with its weight quantifying the corresponding communication data volume.

In line with existing research [4,5,15], each microservice can only run on a single container at any time. We assume that we are given a list C of containers, a set V of VM types, and a set P of PM types in a container-based cloud where containers are assigned to VM instances and VMs are assigned to PMs. We represent a container c_i by a pair $(\zeta^{cpu}(c_i), \zeta^{mem}(c_i))$ where the first denotes the CPU requirement, and the second denotes the memory requirement of the container. We represent a VM type γ_t by a tuple $= (\Omega^{cpu}(\gamma_t), \Omega^{mem}(\gamma_t), \pi^{cpu}(\gamma_t), \pi^{mem}(\gamma_t))$, where the first two denote the CPU and memory capacity, and the last two give the CPU and memory overhead of the VM type. Every VM instance v_i belongs to one VM type γ_t. We represent a PM type τ_t by a tuple $(\Omega^{cpu}(\tau_t), \Omega^{mem}(\tau_t), E^{idle}(\tau_t), E^{full}(\tau_t))$, where $\Omega^{cpu}(\tau_t)$, $\Omega^{mem}(\tau_t)$, $E^{idle}(\tau_t)$, and $E^{full}(\tau_t)$ denote the CPU capacity, the memory capacity, the EC per time unit for the idle state, and the EC per time unit for fully loaded state, respectively. Every PM instance p_i belongs to one PM type τ_t.

Following the non-linear energy model in [3], we use Eq. (1) to express the EC of a PM instance p_i.

$$E(p_i) = E^{idle}(p_i) + (E^{full}(p_i) - E^{idle}(p_i)) \times (2\mu^{cpu}(p_i) - (\mu^{cpu}(p_i))^{1.4}) \quad (1)$$

where $E^{idle}(p_i)$ and $E^{full}(p_i)$ are the EC of the PM instance p_i when it is idle and fully loaded, respectively. $\mu^{cpu}(p_i)$ is the CPU utilization level [3] of the PM instance p_i.

Assuming that n_t PM instances are active at timestamp t, the EC at timestamp t is calculated as follows:

$$EC(t) = \sum_{i=1}^{n_t} E(p_i) \quad (2)$$

To determine CO, we use a matrix to represent the deployment of microservices to PMs. Let X denote a binary matrix, in which $x_{i,j}$ is 1 if microservice m_i is deployed on PM instance p_j, and is 0 otherwise. Another matrix A is used to quantify the communication data volume between each pair of microservices, i.e., in A the value of $a_{i,j}$ indicates the communication data volume between microservice m_i and microservice m_j.

Assuming that n_t denotes the number of active PMs at timestamp t, and m_t stands for the number of microservices deployed at timestamp t, the CO at timestamp t can be calculated through Eq. (3):

$$CO(t) = \sum_{i=1}^{n_t}\sum_{j=1}^{n_t} a_{i,j} \cdot (1 - \sum_{k=1}^{m} x_{i,k} \cdot x_{j,k}) \tag{3}$$

Note that the term $(1-\sum_{k=1}^{m} x_{i,k} \cdot x_{j,k})$ in Eq. (3) equals 1 if the microservices m_i and m_j are on different PMs, and otherwise it is 0.

In this paper, for the joint optimization of CO and EC in the dynamic MDC problem, the optimization objective (J) is defined over a time period T in Eq. (4):

$$J = \omega \cdot \int_0^T Nor(CO(t))dt + (1-\omega) \cdot \int_0^T Nor(EC(t))dt \tag{4}$$

where $0 < \omega < 1$ controls the importance of the CO, while $1 - \omega$ controls the importance of the EC. Since CO and EC are equally important, ω is set to 0.5 in this paper. $Nor(CO(t))$ and $Nor(EC(t))$ are normalized $CO(t)$ and $EC(t)$ by using the common min-max normalization technique [6].

4 The Proposed Approach: LLM-GPHH

4.1 Overview

The flowchart of LLM-GPHH is shown in Fig. 1. LLM-GPHH starts with creating an *initial population* that consists of randomly generated individuals. Each individual includes a *VM selection heuristic* and a *PM selection heuristic*. Both heuristics are represented as trees. Individuals in every generation are evaluated by the optimization objective.

There are two ways to generate new individuals for the next generation population. On the one hand, selection is used to select a subset of individuals, and traditional evolutionary operators, i.e., crossover, mutation and reproduction, are applied to this subset of individuals to generate new individuals. On the other hand, in our newly proposed *LLM evolution*, high-fitness individuals will be stored in an *archive* based on an *archive* update strategy. Subsequently, LLM is leveraged to capture the implicit knowledge among high-fitness individuals in the *archive*, to generate improved new *VM selection heuristics* and *PM selection heuristics*. The details of the LLM-GPHH are described as follows. In Fig. 1, the main contributions of this paper are highlighted in blue.

4.2 Representation

LLM-GPHH adopts a tree structure to represent the *VM selection heuristic* and the *PM selection heuristic*, as commonly used in GPHH [18]. Examples of the *VM selection heuristic* and *PM selection heuristic* are shown in Fig. 2. Each tree in Fig. 2 has one root, multiple non-leaf nodes and leaf nodes. The non-leaf nodes include elementary functions, e.g., $+, /, -, \times$. Each leaf node, named terminal, represents one of the problem-dependent features of the MDC problem.

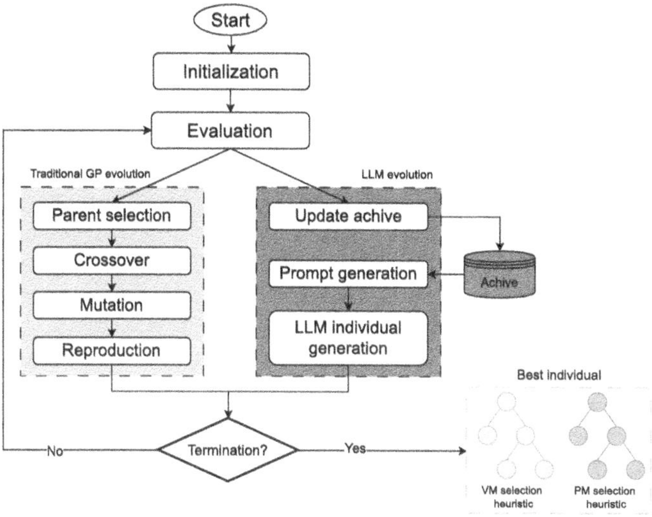

Fig. 1. The overall process of LLM-GPHH

To capture the features of dynamic MDC problems, we elaborately design a set of terminals, which are shown in Table 1. Specifically, the *VM selection heuristic* uses VM-specific terminals. The *PM selection heuristic* uses PM-specific terminals. Besides the terminals already used in existing research [16,17], we designed four new terminals (highlighted in red). The main idea of these newly designed terminals is to capture the features of 1) communication data volume within VMs, or PMs (i.e., VIC or PIC) and 2) external communication data volume of VMs or PMs (i.e., VOC or POC), which are important for optimizing the CO during the evolution.

4.3 Fitness Evaluation

To evaluate the performance of individuals, the *VM selection heuristic* and *PM selection heuristic* of each individual are used to select VMs and PMs for microservice applications. Firstly, the min-cut heuristic [4,7] is utilized to divide the microservice application into different partitions, such that the communication data volumes across each partitions are minimal. Then, microservices in the same partitions are deployed together to the same VM.

Figure 2 illustrates how to use examples of a *VM selection heuristic* (i.e., $\frac{CC}{LVC} - (LVM + COV)$) and a *PM selection heuristic* (i.e., $(VMC + VMM) + PMC$) for MDC problems. According to the table in Fig. 2, the priority values of VM_1, VM_2 and VM_3 in Fig. 2 are calculated by the *VM selection heuristic*, i.e., $\frac{21}{10} - (0.7 + 1.2) = 0.8$, $\frac{14}{11} - (0.9 + 0.2) = 0.17$ and $\frac{15}{1.8} - (3.1 + 4.4) = 0.83$. Therefore, VM_3 is selected for the microservices deployment as it has the highest priority value. Similarly, for each newly created VM instance, PM_1 is selected based on the *PM selection heuristic* shown in Fig. 2. After the deployment of

Table 1. Terminals used by LLM-GPHH. The newly designed terminals are highlighted in blue

VM-related	Description
CC	CPU requirement of a container
CM	Memory requirement of a container
LVC	Lefting CPU of a VM instance
LVM	Lefting memory of a VM instance
COV	The CPU overhead of a VM instance
MOV	The Memory overhead of a VM instance
VOC	communication data volume transfer externally by the PM that a VM is located
VIC	communication data volume within the PM that a VM is located
PM-related	**Description**
VMC	The CPU capacity of a VM instance
VMM	The memory capacity of a VM instance
LPC	Lefting CPU of a PM
LPM	Lefting memory of a PM
PMC	The CPU capacity of a PM
PMM	The memory capacity of a PM
PCore	The number of cores of a PM
POC	communication data volume transfers externally of a PM
POC	communication data volume within a PM

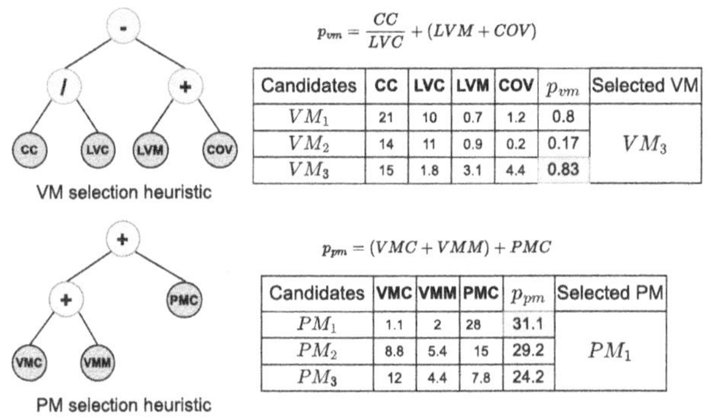
-
/
+
CC
LVC
LVM
COV
VM selection heuristic
$p_{vm} = \frac{CC}{LVC} + (LVM + COV)$
Candidates | CC | LVC | LVM | COV | p_{vm} | Selected VM
VM_1 | 21 | 10 | 0.7 | 1.2 | 0.8
VM_2 | 14 | 11 | 0.9 | 0.2 | 0.17 | VM_3
VM_3 | 15 | 1.8 | 3.1 | 4.4 | 0.83
+
+
PMC
VMC
VMM
PM selection heuristic
$p_{pm} = (VMC + VMM) + PMC$
Candidates | VMC | VMM | PMC | p_{pm} | Selected PM
PM_1 | 1.1 | 2 | 28 | 31.1
PM_2 | 8.8 | 5.4 | 15 | 29.2 | PM_1
PM_3 | 12 | 4.4 | 7.8 | 24.2

Fig. 2. Example of using *VM selection heuristic* and *PM selection heuristic* for the microservice deployment

newly arrived microservices and newly created VMs into the selected VMs and PMs, the fitness value of an individual is calculated by Eq. (4).

4.4 Traditional GP Evolution

The traditional evolutionary operators of LLM-GPHH include crossover, mutation, and reproduction. Crossover exchanges a sub-tree of one heuristic (i.e., *VM selection heuristic* or *PM selection heuristic*) in the parent individual with a sub-tree of the same heuristic type in the second parent individual. The mutation is applied to an individual by replacing a sub-tree of a heuristic with a randomly generated sub-tree. The process of reproduction is straightforward and directly preserves selected individuals to the next generation.

4.5 LLM Evolution

The process of *LLM evolution* is shown in Algorithm 1. The inputs of *LLM evolution* include an *archive*, the best individual in current generation ind_{best}, the number of heuristics chosen for prompt *prompt_size*, and the number of heuristics generated by LLM *num*. Then *LLM evolution* outputs new individuals.

Firstly, an *archive* update strategy is designed in this paper (line 1). Specifically, when there is no individual in the *archive* that is identical to the ind_{best} in the tree structure, heuristics in ind_{best} will be stored in the archive; otherwise, the ind_{best} will not be stored to maintain the diversity in the *archive* [18]. When the *archive* reaches a fixed capacity (e.g., 5), the earliest stored individual within the *archive* is removed due to the earliest individual lacks several generations of evolution, which is most likely to perform poorly. Then, a prompt is generated by sampling promising individuals in the *archive* (lines 3–4). An example of a generated prompt is shown in Fig. 3, which includes three parts: *task description*, *promising heuristics* and *output format*. To be specific, the *task description* part gives details of the problem to be solved (i.e., resource allocation problem in container-based cloud for microservice applications). Meanwhile, the format of promising heuristics and the expected outputs from LLMs are detailed and described in the *task description*. In the *promising heuristics* part, *prompt_size* promising individuals from the *achive* are sampled with an equal probability for each individual (line 3).

The response of LLM includes *num* new heuristics (Line 5). A response with three new heuristics is presented in Fig. 3 (b). Then, new heuristics will be checked for their validity (Line 6). For example, $LVM - COV+$ is not a correct heuristic since the absence of the second operand of +. Correct heuristics will be converted into tree-represented individuals (Line 7). Individuals generated by *LLM evolution* are combined with the individuals generated by traditional GP evolution as a new population for the next generation.

5 Experimental Evaluation

We compare LLM-GPHH with two heuristic algorithms, i.e., KP_HP [7], and ECSched [8], as well as one state-of-the-art hyper-heuristic algorithm, Hybrid-Evo [17]. The average performance results obtained under 30 independent runs are verified and compared based on the Wilcoxon rank-sum test with a significance level of 0.05.

Simulation. We simulate the state of a cloud data center by using a real-world dataset [12,17] that records the arrivals of a large number of applications in cloud data centers during a long time period. Each application belongs to one of the popular application structures studied in [13]. Six scenarios are used to

Algorithm 1. The process of *LLM evolution*

Input: An *archive*; the best individual in current generation: ind_{best}; the number of heuristics chosen for prompt: *prompt_size*; the number of heuristics generated by LLM: *num*;

Output: new individuals generated by LLM;

```
archive ← update_archive(ind_best)
new_indivduals ← []
heuristics ← sample(archive, prompt_size)
prompt ← generate_prompt(prompt_size, num, heuristics)
new_heuristics ← LLM(prompt)
valid_heuristics ← Check(new_heuristics)
new_individuals.add(Generate(valid_heuristics))
```

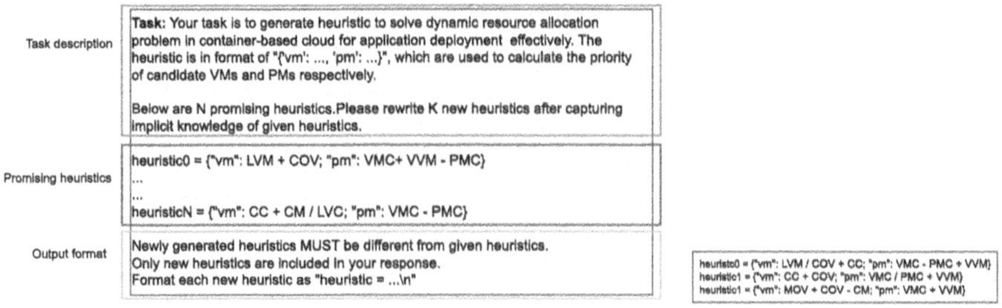

(a) Example of prompt (b) Example of response

Fig. 3. Examples of prompt and the response of LLM

simulate different workload patterns of new arrival applications and the number of different OS types, following [17].

Parameter Settings. The parameter settings of GPHH are based on the existing state-of-the-art GPHH methods [16,17]. Specifically, the population size is 512. The tournament selection size is 7. The rates of crossover, mutation and reproduction are 0.8, 0.1 and 0.1, respectively. The archive size of LLM-GPHH is 10, while the prompt_size is 5. The number of LLM generated individuals is 1. GPT-3.5-turbo-0613[1] API is used in this paper under the consideration of cost and performance.

Experiment Results. Table 2 compares the test performance results across all scenarios. We can observe from Table 2 that LLM-GPHH significantly outperforms the other three algorithms (i.e., KP-HP, ECSched and Hybrid-Evo) in terms of CO in all six scenarios. As for EC, LLM-GPHH consumes less EC in most scenarios when compared to KP-HP, ECSched and Hybrid-Evo.

Number of Active VMs and PMs. We see from Fig. 4 that LLM-GPHH requires more active VMs compared to other algorithms, but it minimizes the

[1] https://openai.com/pricing.

Table 2. The comparison results of test instances in 6 scenarios in terms of mean and standard deviation of EC (kWh) and CO (Mbps). (*Notice that smaller values in the table mean better performance; "+", "-" and "=" indicate significantly better, significantly worse and no significant difference respectively*)

Scenarios	Objectives	KP-HP [7]	ECSched [8]	Hybrid-Evo [17]	LLM-GPHH
W1OS3	EC	86123.45 ± 352.58	119882.72 ± 423.55	87849.80 ± 8610.68	83392.60 ± 3443.20 (+)(+)(+)
	CO	152102.27 ± 2302.83	181091.05 ± 3993.89	895782.48 ± 132158.76	84528.65 ± 9254.59 (+)(+)(+)
W1OS4	EC	79900.73 ± 536.62	119128.81 ± 384.50	85797.29 ± 9164.63	80519.16 ± 9230.01 (=)(+)(+)
	CO	140231.02 ± 2487.36	161973.62 ± 3545.41	688004.95 ± 235592.43	73013.34 ± 8491.053 (+)(+)(+)
W1OS5	EC	74794.61 ± 1140.74	113037.15 ± 343.56	78659.62 ± 7192.01	74105.31 ± 3926.38 (=)(+)(=)
	CO	127200.59 ± 2544.78	155613.84 ± 3663.14	664116.27 ± 225881.42	74340.36 ± 11513.97 (+)(+)(+)
W2OS3	EC	77425.84 ± 768.97	132472.29 ± 528.86	90405.21 ± 12136.97	79318.25 ± 13292.66 (+)(+)(+)
	CO	130645.07 ± 1926.53	171362.58 ± 3361.51	668220.39 ± 204766.39	84928.44 ± 6608.78 (+)(+)(+)
W2OS4	EC	75623.89 ± 863.04	125253.16 ± 550.92	85107.43 ± 48059.55	77736.15 ± 22240.93 (+)(=)(+)
	CO	121785.46 ± 2618.48	159210.25 ± 4696.93	929877.51 ± 140681.28	78142.90 ± 9769.10 (+)(+)(+)
W2OS5	EC	66919.23 ± 1006.71	117324.06 ± 452.58	69191.39 ± 8164.00	73121.75 ± 10912.93 (-)(+)(=)
	CO	108456.06 ± 2543.06	146288.07 ± 3167.12	760304.42 ± 229171.43	76007.46 ± 2219.03 (+)(+)(+)

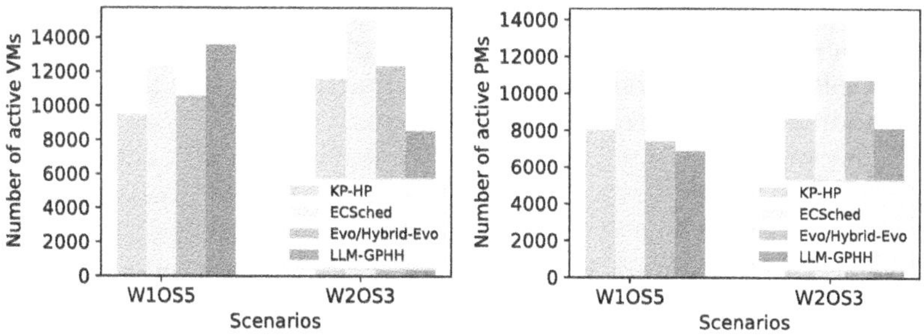

(a) Number of VMs

(b) Number of PMs

Fig. 4. The number of active VMs and PMs

Table 3. The comparison results of ablation studies. (*Notice that smaller values in the table mean better performance; "+", "-" and "=" indicate significantly better, significantly worse and no significant difference respectively*)

Scenarios	Objectives	GPHH-noLLM	LLM-GPHH
W1OS3	**EC**	88235.35 ± 9292.04	83392.60 ± 3443.20 (+)
	CO	80215.42 ± 10028.89	84528.65 ± 9254.59 (=)
W2OS3	**EC**	85933.33 ± 10005.24	79318.25 ± 13292.66 (+)
	CO	88168.45 ± 3303.04	84928.44 ± 6608.78 (+)

number of active PMs in W1OS5. Meanwhile, both the active VMs and PMs of LLM-GPHH are minimized in W2OS3. As a result, LLM-GPHH can save computational resources by reducing the overhead of VMs and PMs [19].

Effectiveness of LLM Evolution. We compare LLM-GPHH with the GPHH algorithm designed in this paper but without the assistance of LLM (GPHH-noLLM) in two scenarios. The results in Table 3 show the effectiveness of *LLM* *evolution* as LLM-GPHH outperforms GPHH-noLLM in terms of EC in both W1OS3 and W2OS3. Meanwhile, LLM-GPHH consumes less CO than GPHH-noLLM in W2OS3, while has no significant difference with GPHH-noLLM in W1OS3. The results show the effectiveness of *LLM evolution* in some scenarios. We also test different settings of *LLM evolution*, e.g., the number of promising heuristics in the prompt and the number of response heuristics, the most effective settings are described in the parameters setting part.

6 Conclusions and Future Work

In this paper, to solve the dynamic MDC problems, we proposed *LLM evolution* to enhance the ability of GPHH to generate effective heuristics. Experimental results showed that LLM-GPHH outperforms existing heuristics and hyper-heuristics methods in terms of minimizing energy consumption and communication overhead. Meanwhile, ablation studies also showed the effectiveness of *LLM evolution*. This suggests that LLM indeed has the potential to enhance GPHH's capabilities in searching effective heuristics. In the future, we will continue to explore the ability of LLM and try to further improve the capability of LLM in addressing complex real-world problems.

References

1. Alzahrani, A., Tang, M.: A microservice-based SaaS deployment in a data center considering computational server and network energy consumption. In: 2023 IEEE 16th International Conference on Cloud Computing (CLOUD), pp. 505–515. IEEE (2023)
2. Buyya, R., Ilager, S., Arroba, P.: Energy-efficiency and sustainability in new generation cloud computing: a vision and directions for integrated management of data centre resources and workloads. Softw. Pract. Exper. **54**(1), 24–38 (2024)
3. Dayarathna, M., Wen, Y., Fan, R.: Data center energy consumption modeling: a survey. IEEE Commun. Surv. Tutorials **18**(1), 732–794 (2015)
4. Fang, Z., Ma, H., Chen, G., Hartmann, S.: Energy-efficient and communication-aware resource allocation in container-based cloud with group genetic algorithm. In: Monti, F., Rinderle-Ma, S., Ruiz Cortés, A., Zheng, Z., Mecella, M. (eds.) Service-Oriented Computing: 21st International Conference, ICSOC 2023, Rome, Italy, November 28 – December 1, 2023, Proceedings, Part I, pp. 212–226. Springer Nature Switzerland, Cham (2023). https://doi.org/10.1007/978-3-031-48421-6_15
5. Fang, Z., Ma, H., Chen, G., Hartmann, S.: A group genetic algorithm for energy-efficient resource allocation in container-based clouds with heterogeneous physical machines. In: Liu, T., Webb, G., Yue, L., Wang, D. (eds.) AI 2023: Advances in Artificial Intelligence: 36th Australasian Joint Conference on Artificial Intelligence, AI 2023, Brisbane, QLD, Australia, November 28–December 1, 2023, Proceedings, Part II, pp. 453–465. Springer Nature Singapore, Singapore (2024). https://doi.org/10.1007/978-981-99-8391-9_36
6. Gajera, V., Gupta, R., Jana, P.K., et al.: An effective multi-objective task scheduling algorithm using min-max normalization in cloud computing. In: 2016 2nd International Conference on Applied and Theoretical Computing and Communication Technology (iCATccT), pp. 812–816. IEEE (2016)

7. Hu, Y., de Laat, C., Zhao, Z.: Optimizing service placement for microservice architecture in clouds. Appl. Sci. **9**(21), 4663 (2019)
8. Hu, Y., Zhou, H., de Laat, C., Zhao, Z.: Concurrent container scheduling on heterogeneous clusters with multi-resource constraints. Futur. Gener. Comput. Syst. **102**, 562–573 (2020)
9. Meyerson, E., et al.: Language model crossover: variation through few-shot prompting. arXiv preprint arXiv:2302.12170 (2023)
10. Romera-Paredes, B., et al.: Mathematical discoveries from program search with large language models. Nature **625**(7995), 468–475 (2024)
11. Sharma, H.C., Bisht, M.: Best fit resource allocation in cloud computing. International J. Comput. Sci. Eng., E-ISSN, 2347–2693 (2019)
12. Shen, S., Van Beek, V., Iosup, A.: Statistical characterization of business-critical workloads hosted in cloud datacenters. In: 15th IEEE/ACM International Symposium on Cluster, Cloud and Grid Computing, pp. 465–474. IEEE (2015)
13. Shi, T., Ma, H., Chen, G., Hartmann, S.: Location-aware and budget-constrained application replication and deployment in multi-cloud environment. In: 2020 IEEE International Conference on Web Services (ICWS), pp. 110–117. IEEE (2020)
14. Sorgalla, J., Sachweh, S., Zündorf, A.: Exploring the microservice development process in small and medium-sized organizations. In: Morisio, M., Torchiano, M., Jedlitschka, A. (eds.) PROFES 2020. LNCS, vol. 12562, pp. 453–460. Springer, Cham (2020). https://doi.org/10.1007/978-3-030-64148-1_28
15. Tan, B., Ma, H., Mei, Y.: A NSGA-II-based approach for multi-objective microservice allocation in container-based clouds. In: 2020 20th IEEE/ACM International Symposium on Cluster, Cloud and Internet Computing (CCGrid), pp. 282–289. IEEE (2020)
16. Tan, B., Ma, H., Mei, Y., Zhang, M.: A cooperative coevolution genetic programming hyper-heuristics approach for on-line resource allocation in container-based clouds. IEEE Trans. Cloud Comput. **10**(3), 1500–1514 (2020)
17. Wang, C., Ma, H., Chen, G., Huang, V., Yu, Y., Christopher, K.: Energy-aware dynamic resource allocation in container-based clouds via cooperative coevolution genetic programming. In: Correia, J., Smith, S., Qaddoura, R. (eds.) Applications of Evolutionary Computation: 26th European Conference, EvoApplications 2023, Held as Part of EvoStar 2023, Brno, Czech Republic, April 12–14, 2023, Proceedings, pp. 539–555. Springer Nature Switzerland, Cham (2023). https://doi.org/10.1007/978-3-031-30229-9_35
18. Wang, S., Mei, Y., Zhang, M.: Explaining genetic programming-evolved routing policies for uncertain capacitated arc routing problems. IEEE Trans. Evol. Comput. (2023)
19. Xu, F., Liu, F., Jin, H., Vasilakos, A.V.: Managing performance overhead of virtual machines in cloud computing: a survey, state of the art, and future directions. Proc. IEEE **102**(1), 11–31 (2013)
20. Yang, Y., Chen, G., Ma, H., Zhang, M.: Dual-tree genetic programming for deadline-constrained dynamic workflow scheduling in cloud. In: Troya, J., Medjahed, B., Piattini, M., Yao, L., Fernández, P., Ruiz-Cortés, A. (eds.) Service-Oriented Computing: 20th International Conference, ICSOC 2022, Seville, Spain, November 29 – December 2, 2022, Proceedings, pp. 433–448. Springer Nature Switzerland, Cham (2022). https://doi.org/10.1007/978-3-031-20984-0_31
21. Yuan, J., Bae, E., Tai, X.C.: A study on continuous max-flow and min-cut approaches. In: 2010 IEEE Computer Society Conference on Computer Vision and Pattern Recognition, pp. 2217–2224. IEEE (2010)

Bidirectional Dependency Representation Disentanglement for Time Series Classification

Tianren Zhao, Hua Zuo(✉), and Guangquan Zhang

Australian Artificial Intelligence Institute, Faculty of Engineering and IT, University of Technology Sydney, Sydney, NSW, Australia
tianren.zhao@student.uts.edu.au, {hua.zuo,guangquan.zhang}@uts.edu.au

Abstract. Time series classification an important and challenging real-world problem and has been extensively studied by deep learning methods. However, the non-stationary property of time series data hinders further improvement. Thus domain generalization technique has been introduced due to such property, and existed methods adopted domain generalization have overlooked the potential mutual dependency between domain-invariant and domain-specific representations. In response, we present bidirectional dependency representation disentanglement for generalization (BiRep), a novel approach to time series classification that leverages the power of transfer learning through domain generalization to enhance model generalizability by disentangling domain-specific and domain-invariant representations using a single GPT-2 backbone with prompt tuning, thus improving classification accuracy. Extensive experiments on various gesture recognition and human activity datasets demonstrate the superior performance of BiRep compared to other related methods, highlighting its robustness and effectiveness in handling diverse and unseen data distributions.

Keywords: Time Series Classification · Domain Generalization

1 Introduction

With the increase in availability of time series data, such as bioelectric and human activity data, time series classification has become an increasingly important yet challenging task in the realm of data mining and machine learning [8]. The dynamic nature of time series data, characterized by its temporally non-stationary property, presents significant challenges. There have been significant efforts in time series classification includes the RNN-based methods [7], and recently popular transformer-based methods [9]. However, these methods struggle to maintain performance when confronting data from unseen distributions.

M. Gong et al. (Eds.): AI 2024, LNAI 15442, pp. 98–110, 2025.
https://doi.org/10.1007/978-981-96-0348-0_8

To address this issue, transfer learning was introduced to time series classification: specifically, domain generalization (DG). DG is designed to extract features that generalize across multiple related source distributions, enabling the model to perform effectively on previously unseen target distributions during test time [21]. Some previous research has tried to disentangle relevant domain-invariant representations from domain-specific representations to perform the classification task [14,22]. However, previous works overlook the mutual dependency between domain-invariant and domain-specific features, which could significantly benefit disentanglement and improve classification accuracy.

In this paper, we propose a novel time series classification algorithm, bidirectional dependency representation disentanglement for generalization (BiRep). The core concept of BiRep is to characterize the latent domains, then bridge the gap between them by disentangling domain-specific representations and domain-invariant representations, thus enhancing the performance of the subsequent classification task on unseen target domains. BiRep consists of bidirectional dependent adversarial representation disentanglement with dynamic characterization of latent domains: we employ the latent domain classifier and the classification classifier to act as each other's discriminators to facilitate adversarial learning under dynamic latent domain label assignment. It is important to note that representation disentanglement presents challenges, as it requires the model to simultaneously extract domain-invariant and domain-specific representations. To serve the dual purpose, we leverage the pretrained language model GPT-2 [15] as the backbone for representation extraction and disentanglement empowered by prompt tuning, which is a unique technique to language models. It enables the backbone to capture either domain-invariant or domain-specific features by adding tunable suffixes to the data, enabling a dual purpose without extensive parameter tuning. We provide extensive experimental results across diverse datasets to show the effectiveness of BiRep, as well as demonstrating robustness in a more difficult scenario.

Our contributions can be summarized as the following:

- **A novel framework**: We propose BiRep, a novel framework for performing bidirectional dependent adversarial representation disentanglement, assisted by prompt tuning.
- **Superior performance**: Comprehensive experimental results on gesture recognition and human activity recognition datasets with various comparison methods to demonstrate the effectiveness of BiRep, as well as abundant analysis showing the robustness of BiRep under a more difficult scenario.

2 Related Work

2.1 Time Series Classification(TSC)

Extensive research effort has been devoted to the time series classification task, as it is a challenging problem. Progress was made when deep learning was introduced to times series classification. Recurrent Neural Network (RNN) [7] can effectively capture temporal dependency of data. However, RNNs often encounter the vanishing gradient problem when trained on long time series data, it hinders further development on the TSC task. Since the tremendous success of transformer architecture [24] in the natural language processing field, such architecture is applied in many different fields. After [11] proposed an efficient method to seamlessly integrate time series forecast task with transformers, transformer was later introduced to TSC. Soon, transformer variations like [23] shows superior performance on TSC task at its time. Although these models showcased state-of-the-art (SOTA) performance during their respective periods, their assumption of stationary distribution inhibited further progress. Until transfer learning techniques was introduced to TSC and made further progress.

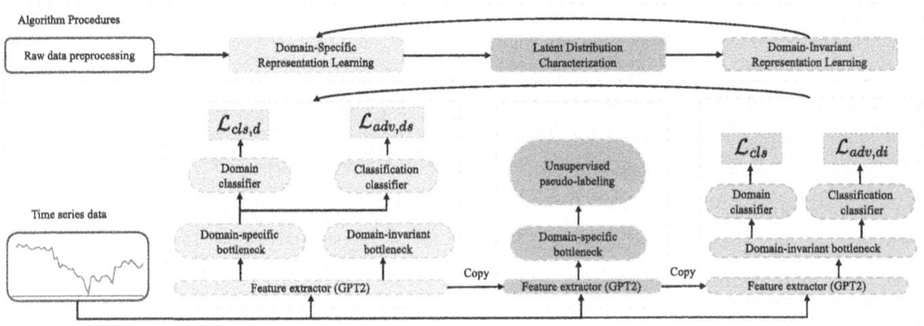

Fig. 1. The framework for BiRep (Color figure online)

2.2 Domain Generalization

Domain generalization is a transfer learning technique, that aims to achieve generalization of unseen distributions using only given distributions. Existing approaches including data manipulation technique, enhancing generalizability by training model with randomized and augmented data. On the other hand, rather than training the model to project the entire data sample into a domain-invariant feature, the representation disentanglement technique allows the model to decompose relevant domain-invariant representations from distracting domain-specific representations for the subsequent classification task [21]. [2] proposed a masking technique to filter out domain-specific representations from image data, while [13] not only extracted domain-invariant representation, but also removed domain-specific representations upon it. However, most existing methods are limited to the image data type.

Domain generalization is later introduced to time series classification task due to the non-stationary property of time series data, but soon confronting another challenge. Most of existed methods require explicit domain labels in datasets, while domain generalization with latent domain labels are not as extensively studied, despite such case is more common in time series classification. To address such challenge, AdaRNN [3] and Diversify [22] creatively merged the distribution perspective into the time series classification task, leveraging domain generalization techniques to obtain the new SOTA. They also integrate existing methods by introducing characterization of latent domain, reducing the reliance of domain labels. BiRep further develops these approaches with advancements in effectiveness.

3 Methodology

3.1 Problem Definition

Take a time series training dataset $\mathcal{D}^{tr} = \{(x_i, y_i)\}_{i=1}^{n}$, where $x \in \mathbb{R}^D$ denotes data samples with dimension D and its label space $y_i \in \mathcal{Y} = \{1, \ldots, C\}$. Whereas an unseen target dataset serves as the test set $\mathcal{D}^{te}$ sharing the same input and label spaces, it is inaccessible during training. The goal is to train a model on the $\mathcal{D}^{tr}$ to minimize the prediction error on $\mathcal{D}^{te}$ while the training set joint distribution $\mathbb{P}^{tr}(x, y)$ differs from the test set joint distribution $\mathbb{P}^{te}(x, y)$.

We formulate latent domain characterization as follows. Due to the non-stationary property of time series data, a dataset may consist of K latent domains. We formulate it as $\mathbb{P}^{tr}(x, y) = \sum_{i=1}^{K} a_i \mathbb{P}^i(x, y)$, where $\mathbb{P}^i(x, y)$ stands for the joint distribution of the i-th latent domain and $\sum_{i=1}^{K} a_i = 1$.

3.2 BiRep

In this paper, we propose BiRep to perform a time series classification task. BiRep using an iterative process shown in Fig. 1 consists four steps. From left to right, raw time series undergoes preprocessing in the first step. Then the preprocessed data is used for domain-specific representation learning, where domain classifiers and classification classifiers are employed to extract domain-specific features using a GPT-2 feature extractor. Furthermore, latent distribution characterization is performed through unsupervised pseudo-labeling technique. Lastly, domain-invariant representation learning is conducted, utilizing a similar structure to extract domain-invariant features leveraging feature extractor. This process then back to the second step (green in Fig. 1), iterating until satisfactory performance. We introduce technical details of steps after raw data preprocessing in following sections.

Domain-Specific Representation Learning. We now introduce the second step of representation disentanglement (green in Fig. 1): extracting the domain-specific representation. The goal is to train an accurate domain classifier that can later function as a vital component for subsequent domain-invariant representation generalization. We propose a training scheme adopting adversarial training

$$\begin{aligned}\mathcal{L}_{cls,d} + \mathcal{L}_{adv,ds} &= \mathbb{E}_{(x,y)\sim\mathbb{P}^{tr}} \ell\left(h_{c,ds}\left(h_{b,ds}\left(h_f(x_{ds})\right)\right), d'\right) \\ &\quad + \lambda_{ds} \ell\left(h_{c,di}\left(R_{\lambda_r}\left(h_{b,ds}\left(h_f(x_{ds})\right)\right)\right), y\right) \qquad (1)\end{aligned}$$

where $x_{ds} = concat[x, p_{ds}]$ stands for the data x concatenated with a domain-specific tunable prompt p_{ds}. $\mathcal{L}_{cls,d}$ denote the latent domain classification loss using cross-entropy loss associating with h_f, $h_{b,ds}$, $h_{c,ds}$ which are the feature extractor, bottleneck for extracting domain-specific representation and domain classifier respectively. Whereas $\mathcal{L}_{adv,ds}$ denote the adversarial loss in the domain-specific representation learning step, the discrimination function is supported by the classification classifier $h_{c,di}$; it serves a dual purpose by helping to distinguish domain-invariant features. Also, R_{λ_2} is the gradient reverse layer with hyperparameter λ_r. Domain labels $d' \in \{0, \ldots, K-1\}$ are initialized into $d' = 0$ in the first iteration and will be updated in the following step. The discriminator's feedback ensures the latent domain classifier focuses on representations that effectively differentiate between latent domains, while maintaining the capacity to generalize across them for effective domain characterization.

Latent Distribution Characterization. This step primarily characterizes latent domains in the training set (blue in Fig. 1). In our datasets, distribution shifts are not specified, indicating that domain labels are not provided. We utilize an unsupervised pseudo-labeling strategy proposed by Diversify [22] to obtain domain labels.

Firstly, the domain-specific feature centroid weighted is obtained by logit softmax output:

$$\tilde{\mu}_k = \frac{\sum_{x_{i,ds} \in \mathcal{X}^{\text{tr}}} \delta_k \left(h_{c,ds}\left(h_{b,ds}\left(h_f(x_{i,ds})\right)\right)\right) h_{b,ds}\left(h_f(x_{i,ds})\right)}{\sum_{x_{i,ds} \in \mathcal{X}^{\text{tr}}} \delta_k \left(h_{c,ds}\left(h_{b,ds}\left(h_f(x_{i,ds})\right)\right)\right) h_f(x_{i,ds})}, \tag{2}$$

where $\tilde{\mu}_k$ is the initial centroid of the kth latent domain, and δ_k is the kth element of the logit softmax output. In BiRep, latent domain number K is a hyperparameter. We then assign pseudo domain labels for each sample to the nearest k^{th} centroid using Euclidean distance D:

$$\tilde{d}'_i = \arg\min_k D\left(h_{b,ds}\left(h_f(x_{i,ds})\right), \tilde{\mu}_k\right). \tag{3}$$

We then update the centroid and the pseudo domain labels:

$$\mu_k = \frac{\sum_{x_{i,ds} \in \mathcal{X}^{\text{tr}}} \mathbb{I}(\tilde{d}'_i = k) h_{b,ds}\left(h_f(x_{i,ds})\right)}{\sum_{x_{i,ds} \in \mathcal{X}^{\text{tr}}} \mathbb{I}(\tilde{d}'_i = k)}, \quad \tilde{d}'_i = \arg\min_k D\left(h_{b,ds}\left(h_f(x_{i,ds})\right), \mu_k\right), \tag{4}$$

where $\mathbb{I}(a) = 1$ when a is true, otherwise $\mathbb{I}(a) = 0$. This means that the kth centroid is updated by finding the mean of features from samples already assigned in the the kth domain.

Domain-Invariant Representation Learning. Classification task is performed in this step via domain-invariant representation learning(pink in Fig. 1). We propose an adversarial training scheme for generalization in this step.

$$\begin{aligned}\mathcal{L}_{cls} + \mathcal{L}_{adv,di} = \mathbb{E}_{(x_{di},y)\sim\mathbb{P}^{\text{tr}}} \ell\left(h_{c,di}\left(h_{b,di}\left(h_f(x_{di})\right)\right), y\right) \\ + \lambda_{di} \ell\left(h_{c,ds}\left(R_{\lambda_2}\left(h_{b,di}\left(h_f(x_{di})\right)\right)\right), d'\right)\end{aligned} \tag{5}$$

where $x_{di} = concat[x, p_{ds}]$ stands for the data x concatenated with a domain-invariant promptp_{ds}, and $h_{b,di}$, $h_{c,di}$ are bottleneck and classifier for domain-invariant representations, respectively. $\mathcal{L}_{adv,di}$ denote the adversarial loss in domain-invariant representation learning step, and the discrimination function is supported by the domain classifier $h_{c,ds}$, which serves a dual purpose by helping to distinguish domain-specific features. λ_{di} is a hyperparameter for adversarial loss.

Training Details. During training, the aforementioned steps are repeated in orders until convergence. In addition, we use six layers of GPT-2 as the featurizer while only parameters in layer normalization layers are trainable and other parameters are frozen. In addition, to fit time series data into the transformer-based backbone, we slide data into patches following [11]. Furthermore, during training of domain-specific representation learning, only associated layers like h_f,$h_{b,ds}$ and $h_{b,ds}$ are trainable while others are frozen. A similar setting is deployed in the domain-invariant representation learning. To achieve better discrimination performance, classifiers require a pretraining on their primary function until performance is satisfactory.

4 Experiment

We show extensive experimental results to evaluate BiRep on gesture recognition and sensor-based activity recognition datasets. Specifically, the experiment setting aims to train a generalized model from a group of individuals, and the model is expected to perform well on unseen individual data samples.

Table 1. Domain splits of each dataset

Dataset	0	1	2	3	4
EMG	0–8	9–17	18–26	27–35	-
DSADS	0,1	2,3	4,5	6,7	-
USCHAD	0,1,2,11	3,5,6,9	7,8,10,13	4,12	-
PAMAP	2,3,8	1,5	0,7	4,6	-
UCIHAR	0–5	6–11	12–17	18–23	24–29

4.1 Experiment Setting and Data Preprocessing

Our experiments are conducted in a leave-one-domain-out setting in each dataset. Given that all datasets are collected from individuals, we divide the data of individuals into several groups, shown in Table 1, where each group is regarded as a domain. In each dataset, one domain is left unseen as the target domain while other domains are considered as source domains. Domains are

Table 2. Final dimension and sliding window settings for all datasets

Dataset	Dimensions	Window size	Step size
EMG	$8 \times 1 \times 200$	200	100
DSADS	$45 \times 1 \times 125$	-	-
USCHAD	$6 \times 1 \times 200$	200	100
PAMAP	$27 \times 1 \times 200$	200	100
UCIHAR	$6 \times 1 \times 128$	-	-

treated as the target domains in turn to evaluate the overall performance of the model on each dataset.

For data preprocessing, we introduce how data cleaning is conducted and how the sliding window technique, a common technique in a time-series classification task, is implemented in each dataset. Data cleaning includes eliminating redundant and saturated data in each dataset. More details will be introduced in the following sections. The purpose of the sliding window technique is to obtain time series data segmentation. It is specified by window size and step size; the specifications differ in each dataset and describe the length of each segment and distance by which it slides forward each time, respectively. Final dimensions of the processed datasets are shown in Table 2. To construct the cross-person generalization experiment scenario, we evenly divided individuals in each dataset into four domains, shown in Table 1.

4.2 Comparison Methods

Given the limited research on time series domain generalization, we implement several domain generalization algorithms from DomainBed [5] as comparison methods: Emperical Risk Minimization(ERM), DANN [4], CORAL [19], Mixup [25], GroupDRO [18], RSC [6] and ANDMask [12]. We also use three strong time series domain generalization comparison methods: GILE [14], AdaRNN [3] and Diversify [22]. All DomainBed [5] algorithms mentioned above follow the settings of [20], and are deployed with a feature network consisting of two sequential blocks, each comprising a convolutional layer, a pooling layer, and a batch normalization layer. The time series domain generalization comparison methods follow their original network settings.

4.3 Experiment Results

Gesture Recognition Task. The Gesture Recognition Task Electromyography (EMG) data for the gestures dataset [10] contains hand gesture data collected from 36 individuals and 7 classes in total. Six classes are used in our experiments, while the class labeled as the gesture transitions is eliminated. We normalize each segment with $\tilde{x} = \frac{x - \min X}{\max X - \min X}$, where X contains all data. Temporal distribution shifts frequently occur in EMG data, meaning that the same gesture from

Table 3. Cross-people generalization results on the EMG dataset. The domain treated as the unseen test domain is noted in the Target row. The best performance is shown in bold.

Target	0	1	2	3	avg
ERM	59.8	67.0	68.9	69.6	66.3
DANN	60.5	63.8	66.4	58.4	62.3
CORAL	60.0	71.9	69.5	66.4	67.0
Mixup	69.7	79.8	77.8	71.9	74.8
GroupDRO	68.8	79.6	77.6	76.1	75.6
RSC	65.7	73.3	74.3	75.8	72.3
ANDMask	70.0	75.9	76.1	74.2	74.0
GILE	63.1	68.8	71.0	65.0	67.0
AdaRNN	68.1	83.6	78.5	70.7	75.2
Diversify	71.7	82.4	76.9	77.3	77.1
BiRep	**74.8**	**84.6**	**79.1**	**80.2**	**79.7**

the same person may perform differently each time. Also, the EMG is sensitive to environmental noise since they are bioelectric signals. Evaluation of BiRep on the EMG dataset is shown in Table 3. BiRep achieves the best performance, outperforming all the strong time series domain generalization comparison methods, with 2.6% better performance than the second best method.

Sensor-Based Activity Recognition Task. Sensor-Based Activity Recognition Task. We evaluate BiRep on four sensor-based activity recognition datasets. Daily and Sports Activities Data Set (DSADS) [1] contains data from 19 daily activities collected from 8 individuals (4 females and 4 males between the ages 20 and 30). Data are collected by wearable sensors including a triaxial accelerometer, triaxial gyroscope, and triaxial magnetometer on five body parts: torso, right arm, left arm, right leg, and left leg. Data is segmented, and no sliding window is needed. The University of Southern California Human Activity Dataset (USCHAD) [26] collected data from 12 activities via 14 individuals (7 males and 7 females). Motion sensors including triaxial accelerometer and triaxial gyroscope are placed on subjects' hips. The Physical Activity Monitoring (PAMAP) dataset [16] contains data from nine individuals wearing three inertial measurement units (IMUs), where each IMU consists of two 3-axis accelerometers, one 3-axis magnetometer, and one 3-axis gyroscope. Note that data collected from the 3-axis accelerometers with upper and lower bounds of $\pm 6m/s^2$ is discarded due to data saturation. We selected 8 common activities out of 18 in total for our experiments. Human Activity Recognition Using Smartphones (UCIHAR) [17] consists of data collected from 30 volunteers within an age range of 19–48 years. Volunteers performed eight activities while wearing a smartphone on the waist

Table 4. Performance comparison across DSADS, USCHAD, PAMAP, and UCI-HAR datasets with ALL AVG representing the overall average. The following method names are abbreviated for presentation: GDRO(GroupDRO), ANDM(ANDMask), ARNN(AdaRNN), DIV(Diversify). The best performance is shown in bold.

Target	ERM	DANN	CORAL	Mixup	GDRO	RSC	ANDM	GILE	ARNN	DIV	BiRep
DSADS											
0	81.4	78.6	83.8	87.7	87.0	84.2	82.8	77.3	88.7	90.4	**92.8**
1	77.9	76.3	75.2	82.8	81.3	81.0	79.6	73.9	82.5	**86.5**	83.1
2	89.2	82.2	87.4	86.9	84.3	84.3	84.9	80.4	84.2	86.1	**91.2**
3	81.3	67.1	82.1	83.1	80.6	78.9	78.9	74.6	87.9	86.1	**90.0**
avg	82.4	76.0	82.1	85.0	83.5	82.6	81.5	76.5	85.2	88.2	**89.5**
USCHAD											
0	75.6	75.9	75.2	74.3	75.4	74.8	77.8	67.0	70.1	**82.6**	79.6
1	75.3	77.9	74.9	76.8	78.4	78.9	74.9	73.9	67.2	63.5	**81.5**
2	66.3	65.7	67.4	67.4	65.8	67.1	66.8	66.1	70.2	**78.7**	78.1
3	80.4	77.3	77.8	75.8	71.6	71.8	78.7	69.8	70.4	71.3	**81.6**
avg	74.4	74.2	73.8	73.6	74.0	74.4	74.5	69.9	70.2	74.0	**80.2**
PAMAP											
0	89.9	88.4	90.0	90.3	88.2	92.1	90.6	87.3	88.7	91.0	**93.4**
1	77.2	73.5	75.2	77.5	76.0	75.5	75.5	71.1	71.2	**84.3**	82.4
2	47.5	57.5	46.8	58.2	50.2	59.6	51.4	47.4	43.3	**60.5**	59.9
3	84.1	86.7	86.1	87.1	87.1	87.6	87.6	80.0	78.6	87.7	**89.4**
avg	74.6	76.6	74.5	78.3	75.5	77.8	76.3	72.5	70.5	80.8	**81.3**
UCIHAR											
0	92.8	93.4	93.3	92.6	91.9	93.0	93.7	90.8	89.5	93.4	**94.6**
1	90.1	87.0	89.5	89.3	90.3	89.0	89.0	85.1	88.1	88.1	**91.4**
2	89.2	86.4	90.1	89.6	87.9	88.8	89.7	88.0	88.1	88.1	**92.4**
3	98.1	97.4	98.1	95.7	95.6	96.4	96.6	94.1	95.7	97.9	**98.8**
4	93.6	92.3	92.9	94.7	91.7	94.6	94.2	92.1	93.0	92.1	**95.8**
avg	92.8	91.3	92.9	92.6	91.7	93.3	92.2	90.9	92.4	93.0	**94.6**

to collect data via a triaxial accelerometer and triaxial gyroscope. The data is segmented, and no sliding window is needed.

All results are present in Table 4. BiRep outperformed all comparison methods on all datasets, and outperformed the second best comparison method by 1.3%, 5.7%, 0.5% and 1.3% respectively. These results demonstrate the robustness and effectiveness of BiRep in enhancing classification accuracy across diverse time series datasets compare to other methods.

4.4 Analysis

In the following section, we present ablation study and evaluation of BiRep on a more challenging scenario.

Table 5. Ablation study

Datasets	EMG	DSADS	USCHAD	PAMAP	UCIHAR	avg
B	74.3	83.7	79.2	76.6	91.3	81.0
B+RD	36.9	**89.8**	78.8	**82.1**	94.4	76.4
B+RD+LDC	78.6	88.7	79.8	80.4	94.6	84.4
B+RD+BA	78.6	88.4	78.2	80.9	94.4	84.1
BiRep	**79.7**	89.5	**80.2**	81.3	**94.6**	**85.1**

Table 6. Cross-position generalization results of DSADS dataset.

Target	0	1	2	3	4	avg
GILE	35.1	30.6	37.5	29.8	24.6	31.5
AdaRNN	37.7	22.3	29.6	24.1	23.6	27.5
Diversify	47.7	32.9	**44.5**	**31.6**	**30.4**	37.4
BiRep	**49.7**	**36.4**	42.3	30.7	29.4	**37.7**

Ablation Study. We present the ablation study to investigate the contribution of each essential part of BiRep. We decompose BiRep into 4 parts, the backbone(B), prompt tuning enabled representation disentanglement(RD), latent distribution characterization(LDC) and bidirectional adversarial learning(BA). The Table 5 have shown different combination of 4 parts and evaluation of overall performance. Note that when LDC is disabled, we present random assignment to domain labels to enable domain classification. We have observed that only B and RD can cause catastrophic failure on EMG dataset despite better performance on DSADS and PAMAP dataset. Adding either LDC or BA can avoid failure on the EMD dataset but cannot reach the best performance. Despite BiRep have shown the best overall performance, removing some parts of BiRep, even under random domain label assignment, can benefit performance on some specific datasets. These findings provide valuable insights into potential future directions for BiRep. We aim to build on this research to achieve further improvements

Performance on a More Challenging Scenario. In the following analyses, we present strong comparison methods that are specifically designed for time series domain generalization so that BiRep is assessed against the more relevant benchmarks. We evaluate BiRep in a more challenging scenario where domains are split in a more diverse manner. As mentioned above, DSADS dataset consists of data from five different body parts; we leverage this propensity and design cross-position generalization. Instead of splitting domains by individuals, we train the model on data from four body parts while the data from one body part serves as unseen target data. We expect BiRep to handle both individual diversity and body part diversity. The results are shown in Table 6. Note that we only compare BiRep to algorithms specifically designed for time series classification. BiRep achieves the best average performance.

5 Conclusion

In this study, we propose the Bidirectional Dependency Representation Disentanglement for Generalization (BiRep) method to address the time series classification task. By utilizing GPT-2 for representation disentanglement and leveraging prompt tuning, BiRep simultaneously captures domain-invariant and domain-specific features using bidirectional dependent adversarial learning. Thus, it enhances performance on unseen data. Our extensive evaluations across multiple datasets, including gesture and human activity recognition, show that BiRep outperforms existing methods, achieving notable improvements in classification accuracy and generalization ability.

In the future work, we aim to enhance BiRep following the insight in the ablation study section. We aim to investigate how random domain label assignment can be helpful in the latent domain characterization, thereby improving the model performance to a higher level.

Acknowledgment. This work is supported by the Australian Research Council under Discovery Early Career Researcher Award DE220101075.

References

1. Barshan, B., Yüksek, M.C.: Recognizing daily and sports activities in two open source machine learning environments using body-worn sensor units. Comput. J. (2014)
2. Chattopadhyay, P., Balaji, Y., Hoffman, J.: Learning to balance specificity and invariance for in and out of domain generalization. In: Vedaldi, A., Bischof, H., Brox, T., Frahm, J.-M. (eds.) Computer Vision – ECCV 2020: 16th European Conference, Glasgow, UK, August 23–28, 2020, Proceedings, Part IX, pp. 301–318. Springer International Publishing, Cham (2020). https://doi.org/10.1007/978-3-030-58545-7_18
3. Du, Y., et al.: AdaRNN: adaptive learning and forecasting of time series. In: Proceedings of the 30th ACM International Conference on Information & Knowledge Management, pp. 402–411 (2021)
4. Ganin, Y., et al.: Domain-adversarial training of neural networks. J. Mach. Learn. Res. (2016)
5. Gulrajani, I., Lopez-Paz, D.: In search of lost domain generalization. In: International Conference on Learning Representations (2021)
6. Huang, Z., Wang, H., Xing, E.P., Huang, D.: Self-challenging improves cross-domain generalization. In: Vedaldi, A., Bischof, H., Brox, T., Frahm, J.-M. (eds.) Computer Vision – ECCV 2020: 16th European Conference, Glasgow, UK, August 23–28, 2020, Proceedings, Part II, pp. 124–140. Springer International Publishing, Cham (2020). https://doi.org/10.1007/978-3-030-58536-5_8

7. Hüsken, M., Stagge, P.: Recurrent neural networks for time series classification. Neurocomputing **50**, 223–235 (2003)
8. Ismail Fawaz, H., Forestier, G., Weber, J., Idoumghar, L., Muller, P.A.: Deep learning for time series classification: a review. Data Min. Knowl. Disc. **33**(4), 917–963 (2019)
9. Liang, Y., et al.: TrajFormer: efficient trajectory classification with transformers. In: Proceedings of the 31st ACM International Conference on Information & Knowledge Management, pp. 1229–1237. CIKM '22, Association for Computing Machinery (2022)
10. Lobov, S., Krilova, N., Kastalskiy, I., Kazantsev, V., Makarov, V.A.: Latent factors limiting the performance of sEMG-interfaces. Sensors (2018)
11. Nie, Y., H. Nguyen, N., Sinthong, P., Kalagnanam, J.: A time series is worth 64 words: long-term forecasting with transformers. In: International Conference on Learning Representations (2023)
12. Parascandolo, G., Neitz, A., Orvieto, A., Gresele, L., Schölkopf, B.: Learning explanations that are hard to vary. In: International Conference on Learning Representations (2021)
13. Peng, X., Huang, Z., Sun, X., Saenko, K.: Domain agnostic learning with disentangled representations. In: International Conference on Machine Learning, pp. 5102–5112. PMLR (2019)
14. Qian, H., Pan, S.J., Miao, C.: Latent independent excitation for generalizable sensor-based cross-person activity recognition. In: Proceedings of the AAAI Conference on Artificial Intelligence, vol. 35, pp. 11921–11929 (2021)
15. Radford, A., Wu, J., Child, R., Luan, D., Amodei, D., Sutskever, I.: Language models are unsupervised multitask learners. OpenAI (2019)
16. Reiss, A.: PAMAP2 Physical Activity Monitoring. UCI Machine Learning Repository (2012)
17. Reyes-Ortiz, J.L., Anguita, D., Ghio, A. Oneto, L. Parra, X.: Human Activity Recognition Using Smartphones. UCI Machine Learning Repository (2012)
18. Sagawa, S., Koh, P.W., Hashimoto, T.B., Liang, P.: Distributionally robust neural networks for group shifts: on the importance of regularization for worst-case generalization. In: International Conference on Learning Representations (2019)
19. Sun, B., Saenko, K.: Deep CORAL: correlation alignment for deep domain adaptation. In: Hua, G., Jégou, H. (eds.) Computer Vision – ECCV 2016 Workshops: Amsterdam, The Netherlands, October 8-10 and 15-16, 2016, Proceedings, Part III, pp. 443–450. Springer International Publishing, Cham (2016). https://doi.org/10.1007/978-3-319-49409-8_35
20. Wang, J., Chen, Y., Hao, S., Peng, X., Hu, L.: Deep learning for sensor-based activity recognition: a survey. Pattern Recogn. Lett. **119**, 3–11 (2019)
21. Wang, J., et al.: Generalizing to unseen domains: a survey on domain generalization. IEEE Trans. Knowl. Data Eng. **35**(8), 8052–8072 (2022)
22. Wang, L., et al.: Diversify: a general framework for time series out-of-distribution detection and generalization. arXiv (2023)
23. Wu, H., Hu, T., Liu, Y., Zhou, H., Wang, J., Long, M.: TimesNet: temporal 2D-variation modeling for general time series analysis. arXiv preprint arXiv:2210.02186 (2022)
24. Zerveas, G., Jayaraman, S., Patel, D., Bhamidipaty, A., Eickhoff, C.: A transformer-based framework for multivariate time series representation learning. In: Proceedings of the 27th ACM SIGKDD Conference on Knowledge Discovery & Data Mining, pp. 2114–2124 (2021)

25. Zhang, H., Cisse, M., Dauphin, Y.N., Lopez-Paz, D.: mixup: Beyond empirical risk minimization. In: International Conference on Learning Representations (2017)
26. Zhang, M., Sawchuk, A.A.: USC-HAD: a daily activity dataset for ubiquitous activity recognition using wearable sensors. In: ACM International Conference on Ubiquitous Computing (Ubicomp) Workshop on Situation, Activity and Goal Awareness (SAGAware) (2012)

SCODA - Framework for Software Capability Representation and Inspection

Hsu Myat Win[1(✉)], Sebastian Rodriguez[1], John Thangarajah[1], and Andrew Warhurst[2]

[1] Royal Melbourne Institute of Technology, Melbourne, Australia
{hsu.myat.win,sebastian.rodriguez,john.thangarajah}@rmit.edu.au
[2] Defence Science and Technology Group, Canberra, Australia
andrew.warhurst1@defence.gov.au

Abstract. Software composition remains a significant challenge in today's technology landscape, especially when dynamically integrating and collaborating among different systems. Before software composition, analyzing software capabilities (SC) in a system is a crucial and challenging step. In this work, we introduce an ontology-based framework for SC representation and inspection. Our novelty lies in our model called the Software Capabilities for Open and Dynamic Architectures(SCODA), which represents software capability in the system by providing essential data for effective software inspection. Unlike previous approaches that primarily focused on software quality or reusability, SCODA emphasizes a structured representation of capabilities by capturing dynamic factors such as preconditions and effects. We used the Multi-Agent Programming Contest (MAC) to develop the model and the Drone Courier System (DCS) to illustrate our approach, demonstrating its utility in real-world scenarios. We have implemented a tool called QueryCap designed for inspecting software capabilities from the dataset (i.e., a list of SCs found in the system) for a given task.

Keywords: Software capability · Ontology

1 Introduction

Software composition presents one of the most challenging issues in today's rapidly evolving technological landscape, particularly when integrating multiple dynamic software systems. Before engaging in software composition, it is crucial to inspect the functionalities of existing software. This step is also challenging because misrepresenting software functional capabilities may lead to deviations from the task's goal or provide inaccurate information. For example, one of the faults in integrating a Web API like Adyen's payment system is incomplete or insufficient information on how to implement specific API features or functionalities, which can hinder proper integration [21]. Under ISO 9000:2015(en), capability refers to "the ability of an object to realize an output that will fulfil the

M. Gong et al. (Eds.): AI 2024, LNAI 15442, pp. 111–123, 2025.
https://doi.org/10.1007/978-981-96-0348-0_9

requirements for that output". In the area of system engineering, capability refers to "the ability of a system to execute a particular course of action or achieve a desired effect, under a specified set of conditions " [22]. In software engineering, the capability maturity model (CMM) has introduced processes as capabilities, defining how to assess specific qualities of software engineering processes, and offering an understanding of the current state of software systems [13]. We follow the definition from system engineering perspective because in our project, we specifically focus on functional capability in the software rather than the software engineering process or business requirement.

Prior research has primarily focused on assessing and evaluating software capabilities within specific requirements, catering solely to organizational needs [8]. Similarly, some work typically focuses on software quality to determine if it meets business requirements [9]. However, this approach often neglects the exploration of inherent abilities within the existing software. On the other hand, some researchers have introduced feature extractions from systems to compose software artifacts written in different languages [10,11]. Additionally, software product lines have aimed to identify reusable functionalities from previous software versions [1,12]. Since they focus on software composition, there is still a need to address the investigation of software abilities across different systems or software before composition. Similarly, some researchers explore assessing the reusability of software components through software capability profiling [2–4]. However, there remains a gap in addressing effects that can impact the features of the entity and influence the expected output. For instance, available charge in a drone's battery could affect its ability to complete its task, yet existing studies often overlook such critical factors.

In this project, we introduce an ontology-based framework for representing software capability, which we have termed the SCODA model. The SCODA model includes six main concepts: entity, action, effect, condition, feature, and resource. We chose the standard ontology because it provides us with a consistent framework for representing and organizing knowledge, promoting reusability across different applications and projects. It also enables seamless integration with other existing ontologies, facilitating data interoperability and reducing the effort required to combine different data sources. We used the Multi-Agent Programming Contest (MAC)[1] as our initial study for developing the model. MAC is an annual event that allows participants to showcase their innovations in the field of multi-agent systems and artificial intelligence (AI). This competition involves developing multi-agent systems to solve cooperative tasks in a dynamically changing environment. Agents are autonomous entities that collaborate to solve tasks. An agent possesses the ability to learn and make autonomous decisions, leveraging interactions with neighboring agents or the environment to acquire new knowledge and execute actions to accomplish their designated tasks within the system [20]. We chose MAC for analyzing software capability because (1)in MAC, agents (e.g., a drone or a truck) work dynamically to achieve specific goals, aligning with SCODA's emphasis on open and dynamic architecture,

[1] https://multiagentcontest.org/2018/.

and (2) our research group has participated in this competition in the past, giving us expertise within that domain. For illustration, we developed a Drone Courier System (DCS) as a second case study due to its simplicity for the paper. Specifically, we created instances of the SCODA model for the DCS. We then developed a mechanism to inspect and assess the software capabilities within the dataset for a given task (e.g., do we have a software capability that can deliver a package from a warehouse to a customer?). We choose SPARQL Protocol and RDF Query Language (SPARQL) for developing software capability inspection mechanism because it allows for powerful querying and manipulation of data within Resource Description Framework (RDF) graphs, enabling complex data retrieval and integration in ontology-based systems. Overall, our aim is to provide insights into the software capabilities of existing software for inspection and our contributions are as follows:

- developing a model for software capability representation;
- providing a query language to inspect the software capability; and
- presenting a case study that illustrates the use of our model.

2 Background

In this section, we provide background information on the relevant technologies for the project: ontology engineering, utilized for modeling, and SPARQL, employed for SC inspection. **Ontology engineering** involves creating knowledge maps (ontologies) to efficiently organize information. An ontology represents structured knowledge within a specific domain, encompassing concepts, and relationships [16]. Ontologies serve as specifications for sharing and reusing knowledge across applications [15]. **Web Ontology Language (OWL)** is a standard endorsed by the W3C for constructing OWL knowledge models [14]. It is a semantic web language designed to model intricate knowledge about entities, groups, and their relationships. OWL has found extensive use in modeling concepts such as access control policies [18] and privacy in medical data [17]. Knowledge expressed in OWL can be utilized by computer programs to verify consistency or make implicit knowledge explicit. Hence, our approach is designed based on ontology engineering principles. The ontology is developed using information gathered from human sources (such as domain experts), structured sources (like databases), or unstructured sources (such as books). This information is then assigned to the ontology in the form of concepts, relationships, and definitions. The proposed ontology, referred to as SCODA, was constructed using Protégé 5.0. Protégé is a freely available ontology editor and information management framework developed by the Biomedical Informatics Research Center at Stanford University. **SPARQL** is a standardized query language and protocol for querying RDF data. It allows users to retrieve and manipulate data in RDF format, and supports the principles of the Semantic Web by enabling the integration and querying of data with rich semantics. It also allows applications to understand and reason about the relationships between entities, making it valuable for knowledge representation and discovery.

Some studies emphasize examining software capabilities and resource awareness on available hardware capacity such as CPU and memory [5], neglecting other required resource types (e.g., battery power). Proposals for defining capability include preconditions, effects, input/output, hardware, and parameters during domain planning [6,7]. However, these often focus on post-conditions rather than dynamic factors influencing process outcomes. In our approach, we consider both effects and preconditions as integral parts of software capability, as they significantly influence software functionality. For example, when a drone moves from one location to another, its battery decreases (we call this an effect), which could lead to the drone not reaching its destination. Therefore, having enough battery should be a precondition for a drone to perform certain tasks and provide specific output. Solely observing the output of software functions is insufficient for defining software capability. Instead, analyzing the preconditions and effects that influence the outcome is crucial. Therefore, our approach aims to address the shortcomings of previous research by considering dynamic factors that influence software capability outcomes, thus providing a more robust model for representing software capabilities (SC).

3 Approach

As shown in Fig. 1, our ontology-based approach begins with domain knowledge elicitation followed by software capability (SC) representation modeling (i.e., SCODA model). We also developed a querying mechanism for SC inspection. The objective of the domain knowledge elicitation step is to gather and extract relevant information within the domain (e.g., Multi-Agent City contest (MAC)). This process involves comprehending the context, and identifying their functionalities and essential elements. We begin by reviewing documents and conducting simulations to verify functionalities. For instance, in the context of MAC, we validated that a drone can perform functions such as loading, moving, and charging. Subsequently, we analyze the necessary elements and conditions required to support these functionalities. E.g., for the moving function, the drone needs a battery and knows the destination.

3.1 SCODA Model

Inspired by Jabardi and Hadi [19], who utilized ontological engineering to detect and classify fake accounts on Twitter, our ontology development comprises four steps: (1) defining the domain and scope of the ontology (specifically focusing on software capability), (2) establishing classes and their hierarchical structure, (3) specifying class properties, and (4) assigning values or assertions to these properties (referred to as "slot facts"). Specifically, we explored candidate components to represent Software Capability (SC), considering fundamental elements in systems or software. We defined a component as an asset representing a complete aspect of software capability. We then examined which components could influence the expected output of a given task and their characteristics. We identified

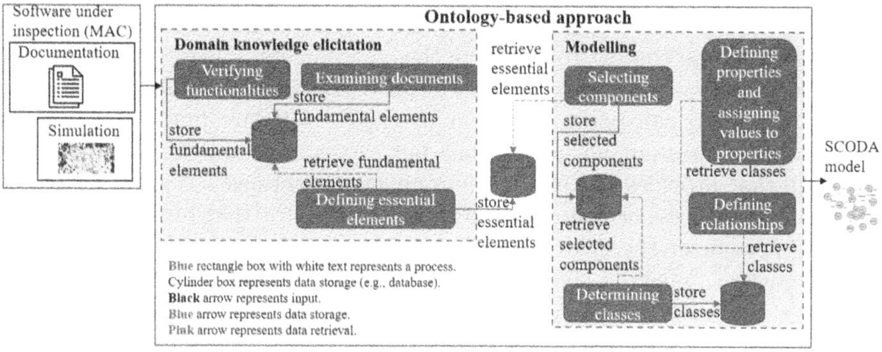

Fig. 1. The framework of the proposed method.

the essential components required to represent SC, seeking a balance between generality and specificity. Additionally, we refined the definition and naming of each SC component, clarifying terms like *resource*. We built our model using Protégé, where classes represent components in SC. Subsequently, we investigated connections among these components (i.e., relationships between classes) and brainstormed class properties while assigning values to these properties.

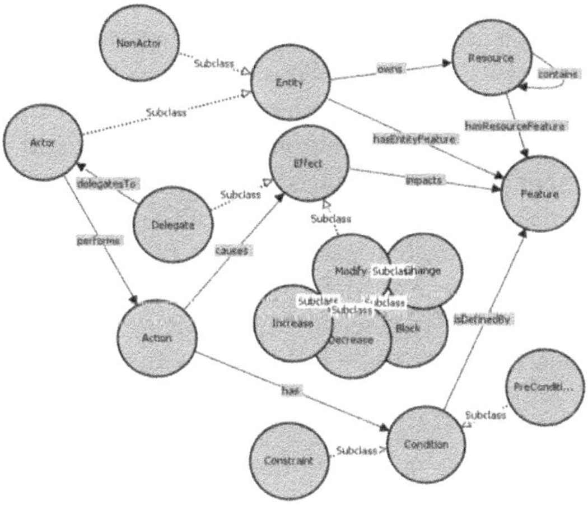

Fig. 2. SCODA model.

Figure 2 illustrates the SCODA model, comprising six key concepts:

1. **Entity**: It is a complete and functional part of a system, capable of existing independently as either a software component, hardware component, or a combination of both. For example, consider a drone that needs to go to a warehouse. In this context, both the drone and the warehouse would be entities;

however, the drone can perform the action of going, while the warehouse simply provides the context to complete the environment. Therefore, we created two subclasses under entity, which we called **Actor** and **Non-actor**. Specifically, an actor can perform actions or trigger software functions, whereas a non-actor does not initiate actions but serves as a functional part of the system that provides context or environment for actions.

2. **Action**: A process or operation executed by an actor that results in changes to the system or the environment. An action typically involves manipulating resources, affecting conditions, and changing properties. For example, a drone's movement action could change its location.
3. **Resource**: A resource represents a physical or informational element/component that an entity possesses, which can be used, accessed, or depended on by actors to perform tasks. It does not function independently but supports the functioning of the system, being utilized, consumed, or accessed during system operations. For instance, a drone's battery.
4. **Condition**: It refers to any requirement or circumstance that influences an action. For example, before going to a warehouse, a drone must have enough battery. In this context, we introduced **Precondition** as a subclass of condition. Besides precondition, we may need some limitations, such as "Drone must reach a warehouse within 1 min". Therefore, we introduced **Constraint** as a subclass as well. Specifically, a precondition refers to the condition that must be satisfied for an action to occur, while a constraint denotes any limitation or restriction that governs the action.
5. **Effect**: A measurable or observable change or impact when an action is triggered. It describes the impact of actions on entities and resources. For example, upon going to a warehouse, a drone's battery would experience a negative impact. Conversely, upon charging, a drone's battery would experience a positive impact. Therefore, we introduced the **Modify** concept as a subclass of **Effect**. Specifically, the **Modify** effect can alter the feature of an entity or resource by *increasing*, *decreasing*, *changing*, or *blocking* it. In addition to altering features, an effect can delegate a task to an actor to perform. For example, in MAC, an initiator agent responsible for evaluating and allocating all types of jobs can delegate the selected job to an agent who can perform the action such as "building a well". Thus, we introduced the **Delegate** concept as a subclass of **Effect**.
6. **Feature**: An attribute or property of an entity or a resource. For instance, the location of an agent.

Relationships between concepts. After defining each concept in our model, we now explore how these concepts relate to each other, revealing the connections and interactions among them. We have provided relationships between concepts in Fig. 1. Dotted lines represent subclass relationships, while solid lines represent relationships between the concepts. Following are the relationships between the concepts -

performsAction represents a relationship between an actor and an action. Specifically, an actor can perform an action.

causalEffect represents the relationship between an action and an effect on resources or entities. It helps us understand how a specific action leads to observable changes or effects on resources or entities.
impactsFeature represents the relationship to illustrate how an effect impacts and changes the specific feature of a resource or entity. It helps us to understand the direct consequences or changes impacted by a particular effect on the relevant features.
hasResourceFeature represents the relationship between a resource and its associated attributes, represented by features.
hasEntityFeature represents the relationship between an entity and its associated attributes, represented by features.
ownsResource represents a relationship between an entity and a resource. Specifically, an entity can own a resource.
contains represents a relationship between a resource and its own resource. Specifically, a resource can contain resources.
hasCondition represents a relationship between an action and conditions. Before triggering an action, conditions need to be satisfied.
isDefinedBy represents a relationship between a condition and features. Specifically, the condition is defined by features.
delegatesTo represents a relationship between a delegate effect and an actor.

Insights. We derived the concept of an entity from MAC, where multiple agents (i.e., drones, trucks, or cars) can perform actions, while warehouses and charging stations support the context of MAC. We then applied this theory to generalize real-world systems and arrived at the definition above. Some researchers have suggested that a resource is a type of entity [20]. However, in our model, we distinguish resources from entities because resources have distinct features, such as supporting functions without being functional parts of a system (its features determine the expected output), and they exist dependently on entities. Our analysis revealed that rather than direct impacts (i.e., a direct relationship) from operations (referred to as *Actions*) to resources, we need a conceptual representation (an abstract representation) between them, which we term *Effect*. This possesses characteristics distinct from both actions and resources and encompasses more than just a relationship between them. Consequently, we introduced *Effect* as a concept in representing software capability and defined it as such if it can alter the current state of a resource and it is triggered by an action. Since resources can have varied states, we introduced the *Feature* concept (e.g., battery can have not only *current_charge* but also *current_temperature*) and an effect could impact these features rather than the resource itself. Specifically, we define the state as a feature if it can be changed, increased, decreased, or blocked by an effect. Finally, our study revealed that certain actions need to trigger the main action, leading us to introduce the concept of the *Delegate* effect.

3.2 SC Inspection Mechanism

The objective of the SC inspection mechanism is to facilitate the analysis and understanding of software capabilities. We used SPARQL-based query because,

Algorithm 1: Inspecting SC

```
Input: Feature
Procedure InspectingSC (Feature)
    effectsTypes ← QueryEffectsTypes(Feature);
    DisplayListofEffectsTypes(effectsTypes);
    selectedEffectType ← GetUserSelection(effectsTypes);
    suggestedSC ← RetrieveSC(selectedEffectType);
    DisplaySCDetails(suggestedSC);
Procedure QueryEffectsTypes(Feature)
    /* Return effect types based on the provided feature. */
Procedure DisplayListofEffectsTypes(effectsTypes)
    /* Display the effect types list (effectsTypes) to select. */
Procedure GetUserSelection(effectsTypes)
    /* Return the selected effect type chosen by the user */
Procedure RetrieveSC(selectedEffectType)
    /* Return the SC details based on the provided effectType */
Procedure DisplaySCDetails(suggestedSC)
    /* Display suggested SC with corresponding actor, action, effect,
       feature, and entity. */
Main:;
Feature ← Input("Select the feature: ");
InspectingSC(Feature);
```

ideally, we aimed for it to be as close to natural language as possible. This would require Natural Language Processing (NLP); which we aim for future work. As shown in Algorithm 1, we begin with an input parameter *Feature*, which represents the specific feature of the entity or resource. Initially, the procedure *InspectingSC* queries the ontology using *QueryEffectsTypes(Feature)* to retrieve a list of effect types that impact the specified feature. These effect types are then displayed to the user through *DisplayListofEffectsTypes(effectsTypes)*, allowing for user selection via *GetUserSelection(effectsTypes)*. Once the user selects an effect type, *RetrieveSC(selectedEffectType)* fetches details of the corresponding Software Capability (SC). Finally, *DisplaySCDetails(suggestedSC)* presents the suggested SC along with its associated actor, action, effect, feature, and entity details to the user.

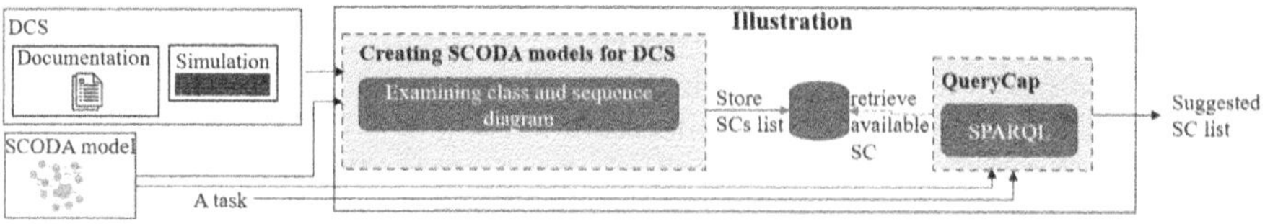

Fig. 3. The process for illustrating our model.

Implementation. We implemented in a tool named QueryCap, designed for querying software capabilities from the dataset (i.e., a list of SCODA models of a system). We use RDF for dataset creation because RDF is a standard data model in semantic web technologies, enabling structured description and exchange of data on the web. QueryCap is a standalone Java program and requires JDK 17.0.9, JavaFX SDK 21.0.1, and Apache Jena 4.9.0. It is a Windows application and comprises three components: the User Interface (UI), Data Extraction, and SPARQL scripts. When passing parameters across the class, data is stored in the object model.

4 Case Studies

To illustrate our approach, we present a much simpler case study (i.e., Drone Courier System (DCS)) because Multi-Agent Programming Contest(MAC), which we used for model development, requires a lot of domain knowledge to understand. Therefore, for simplicity in illustration purposes, we developed DCS example as a second case study to present our work here. DCS case study involves delivering packages to customers, where the vehicle scheduler coordinates with the warehouse to prepare the package, and a drone is deployed to transport and deliver the package to the customer's address.

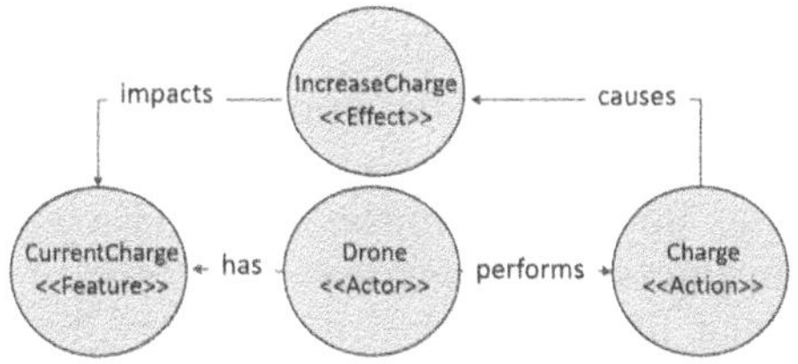

Fig. 4. Software capability that can increase the Drone's battery.

The process of our illustration is shown in Fig. 3. First, we created SCODA models using our ontology-based model, DCS's documentation and DCS's simulation. Next, we stored the dataset (i.e., list of SCs) on the local machine, and then the SC inspection mechanism implemented as QueryCap extracts the relevant software capabilities based on the given task.

Creating SCODA Models. Using SCODA model and DCS's UML class diagram and sequence diagram, we created SCODA models for DCS. We take a class as an actor if it can trigger actions/operations while we define it as a resource if it can be consumed/changed and we define it as a non-actor if neither of that. We identified attributes of each class as features if they possess the distance property (e.g., drone's *currentLoad*). These attributes are further classified based on whether they can be changed, increased, decreased, or blocked, thus determining their effects and types. We classified operations of a class as actions if they impact any of the defined features. Because of page limitation, we illustrate SC which can charge Drone's battery. As shown in Fig. 4, when *Drone* triggers *Charge* action, it causes a positive effect on the battery's current charge.

```
SELECT ?feature
WHERE {
  ?entity hasEntityFeature ?feature .
  FILTER(str(?entity) = str(Drone))
}
```

(a) SPARQL

	feature
1	BatteryCapacity
2	DroneStorageCapacity
3	CurrentCharge
4	DroneLocation
5	CurrentLoad

(b) Result

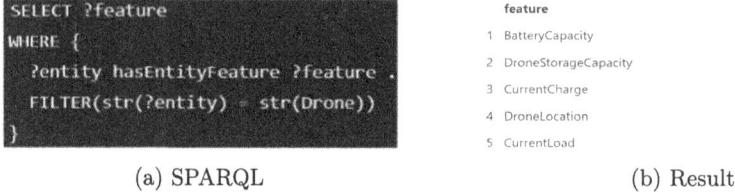

Fig. 5. Finding features for Drone.

4.1 Inspecting Software Capability

We use SPARQL to extract SC from the dataset. We provided a sample query in Fig. 5. In the **WHERE** clause, **?entity hasEntityFeature ?feature .** is a triple pattern to retrieve all features (**?feature**) belonging to entities (**?entity**). **FILTER(str(?entity) = str(Drone))** is a filter clause to check if the **?entity** is **Drone**. Specifically, this query retrieves all features of Drone, using RDF triples and a filter condition in SPARQL.

Based on algorithm 1, to illustrate our approach, we investigate SCs that can alter a drone's battery. The query begins by selecting types of effects that impact a drone's battery charge. Specifically, in the **WHERE** clause, the query defines conditions **?effect ?impactsObj ?feature .** and **?entity ?hasObj ?feature .** to select all effects that impact the feature of entity, filtered with conditions **str(?feature)=str(CurrentCharge)** and **str(?entity)=str(Drone)** to verify if the feature is *CurrentCharge* of *Drone*. This query specifically retrieves all effects (i.e., An effect that can increase the battery and An effect that can decrease battery) that impact the Drone's battery. We assume the user chooses "An effect which can increase battery", our inspection continues to extract actions which cause that effect and actors who perform those actions. Specifically, in the **WHERE** clause, the query defines conditions **?action ?causes ?effect .** and **?actor ?performs ?action .** to extract all actions and corresponding actors. In the filter clause, we have **str(?effect)= str(Increase)** to verify if the effect is *Increase*. This query specifically retrieves all actions that can increase Drone's battery and actors that can perform those actions. The result of that query is shown in Fig. 6.

5 Discussion and Conclusion

In this project, we developed a model for representing software capabilities which we termed the SCODA model - an ontology-based framework to represent

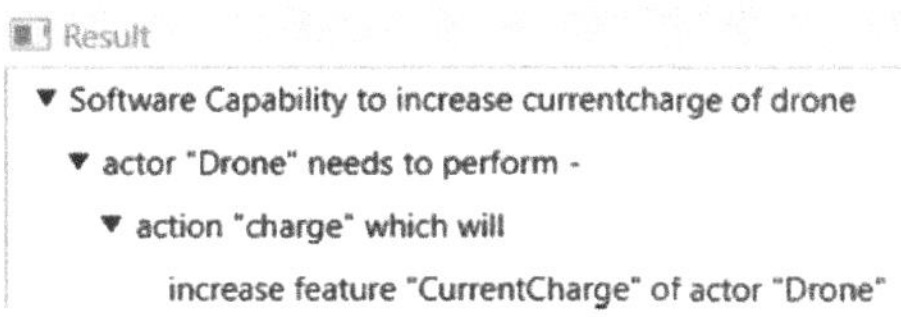

Fig. 6. SC which can able to increase Drone's battery.

and inspect software capabilities. We used Multi-Agent Programming Contest (MAC) for developing the model and the Drone Courier System (DCS) to illustrate our approach. We also developed an inspection mechanism for software capabilities with the help of SCODA model. For illustration purposes, we implemented a tool named QueryCap[2] for querying capabilities from the dataset and our demonstration shows that our approach can able to extract the relevant software capabilities for a given task. These insights enable us to make informed decisions when dynamically integrating and collaborating among different systems.

Although this approach provides a robust framework for conceptualizing software capabilities, it lacks real-time simulation capabilities and instance-level modeling. Without dynamic analysis and runtime implementation, potential runtime issues such as concurrency problems remain unaddressed, providing scope for future work. However, our current approach lays the foundation for software capability inspection using SCODA models and, in the future, we aim to integrate dynamic analysis into our models to address runtime issues such as concurrency problems.

Acknowledgments. This research is supported by the Commonwealth of Australia as represented by the Defence Science and Technology Group.

References

1. Debbiche, J., Lignell, O., Krüger, J., Berger, T.: Migrating Java-based apo-games into a composition-based software product line. In: Proceedings of the 23rd International Systems and Software Product Line Conference-Volume A, pp. 98–102 (2019)
2. Belfadel, A., Laval, J., Cherifi, C.B., Moalla, N.: Toward service orchestration through software capability profile. In: Popplewell, K., Thoben, K.-D., Knothe, T., Poler, R. (eds.) Enterprise Interoperability VIII: Smart Services and Business Impact of Enterprise Interoperability, pp. 385–395. Springer International Publishing, Cham (2019). https://doi.org/10.1007/978-3-030-13693-2_32
3. Belfadel, A., Laval, J., Bonner Cherifi, C., Moalla, N.: Semantic software capability profile based on enterprise architecture for software reuse. In: Ben Sassi, S., Ducasse, S., Mili, H. (eds.) ICSR 2020. LNCS, vol. 12541, pp. 3–18. Springer, Cham (2020). https://doi.org/10.1007/978-3-030-64694-3_1

[2] https://github.com/hmteams/scoda.

4. Belfadel, A., Amdouni, E., Laval, J., Cherifi, C.B., Moalla, N.: Towards software reuse through an enterprise architecture-based software capability profile. Enterprise Inf. Syst. **16**(1), 29–70 (2022). https://doi.org/10.1080/17517575.2020.1843076
5. Rolia, J., Cherkasova, L., Arlitt, M., Andrzejak, A.: A capacity management service for resource pools. In: Proceedings of the 5th International Workshop on Software and Performance, pp. 229–237 (2005)
6. Hendler, J., Wu, D., Sirin, E., Nau, D., Parsia, B.: Automatic web services composition using Shop2. In: Proceedings of The Second International Semantic Web Conference (ISWC) (2003)
7. Buehler, J., Pagnucco, M.: A framework for task planning in heterogeneous multi robot systems based on robot capabilities. In: Proceedings of the AAAI Conference on Artificial Intelligence, vol. 28, no. 1 (2014)
8. Paulk, M.C., Curtis, B., Chrissis, M.B., Weber, C.V.: Capability maturity model, version 1.1. IEEE Softw. **10**(4), 18–27 (1993). https://doi.org/10.1109/52.219617
9. Bollinger, T., McGowan, C.: A critical look at software capability evaluations: an update. IEEE Softw. **26**(5), 80–83 (2009). https://doi.org/10.1109/MS.2009.119
10. Apel, S., Lengauer, C.: Superimposition: a language-independent approach to software composition. In: Pautasso, C., Tanter, É. (eds.) Software Composition, pp. 20–35. Springer, Berlin, Heidelberg (2008). https://doi.org/10.1007/978-3-540-78789-1_2
11. Apel, S., Kastner, C., Lengauer, C.: Language-independent and automated software composition: the feature house experience. IEEE Trans. Softw. Eng. **39**(1), 63–79 (2013). https://doi.org/10.1109/TSE.2011.120
12. Linsbauer, L., et al.: Systematic software reuse with automated extraction and composition for clone-and-own. In: Lopez-Herrejon, R.E., Martinez, J., Guez Assunção, W.K., Ziadi, T., Acher, M., Vergilio, S. (eds.) Handbook of Re-Engineering Software Intensive Systems into Software Product Lines, pp. 379–404. Springer International Publishing, Cham (2023). https://doi.org/10.1007/978-3-031-11686-5_15
13. Chrissis, M.B., Konrad, M., Shrum, S.: CMMI for development: guidelines for process integration and product improvement. Pearson Education (2011)
14. W3C. Retrieved July 9, 2024. https://www.w3.org/OWL/
15. Neches, R., Fikes, R.E., Finin, T., Gruber, T., Patil, R., Senator, T., Swartout, W.R.: Enabling technology for knowledge sharing. AI Mag. **12**(3), 36–36 (1991)
16. Beimel, D., Peleg, M.: Using OWL and SWRL to represent and reason with situation-based access control policies. Data Knowl. Eng. **70**(6), 596–615 (2011). https://doi.org/10.1016/j.datak.2011.03.006
17. Rahmouni, H.B., Solomonides, T., Casassa Mont, M., Shiu, S.: Modelling and enforcing privacy for medical data disclosure across Europe. In: Medical Informatics in a United and Healthy Europe, pp. 695–699 (2009). IOS Press
18. Kayes, A.S.M., Rahayu, W., Dillon, T., Chang, E.: Accessing data from multiple sources through context-aware access control. In: 2018 17th IEEE International Conference On Trust, Security And Privacy In Computing And Communications/12th IEEE International Conference On Big Data Science And Engineering (TrustCom/BigDataSE), pp. 551–559 (2018). IEEE
19. Jabardi, M., Hadi, A.S.: Twitter fake account detection and classification using ontological engineering and semantic web rule language. Karbala Int. J. Mod. Sci. **6**(4), 8 (2020)
20. Dorri, A., Kanhere, S.S., Jurdak, R., Gauravaram, P.: Multi-agent systems: a survey. IEEE Access **6**, 28573–28593 (2018)

21. Aué, J., Aniche, M., Lobbezoo, M., van Deursen, A.: An exploratory study on faults in web API integration in a large-scale payment company. In: Proceedings of the 40th International Conference on Software Engineering: Software Engineering in Practice, pp. 13–22 (2018)
22. SEBoK Editorial Board. 2024. The Guide to the Systems Engineering Body of Knowledge (SEBoK), v. 2.10, N. Hutchison (Editor in Chief). Hoboken, NJ: The Trustees of the Stevens Institute of Technology. Accessed 15 July 2024. www.sebokwiki.org. BKCASE is managed and maintained by the Stevens Institute of Technology Systems Engineering Research Center, the International Council on Systems Engineering, and the Institute of Electrical and Electronics Engineers Systems Council

Some Considerations for the Preservation of Endangered Languages Using Low-Resource Machine Translation

Alastair Kho(✉), Duc-Son Pham, Susannah Soon, and Kit Yan Chan

School of Electrical Engineering, Computing and Mathematical Sciences, Curtin University, Bentley, Western Australia
alastair.kho@student.curtin.edu.au, dspham@ieee.org, {susannah.soon,kit.chan}@curtin.edu.au

Abstract. While widespread languages remain actively prevalent in digital mediums, endangered languages such as Indigenous Australian languages, are often scarce in textual resources and lack a substantial digital presence. This diminishes the survivability of their language and cultural heritages, making language preservation initiatives important. Neural Machine Translation (NMT) efforts for low-resource languages can help accelerate the digitisation and preservation of such languages by further enabling translation, information access, and second-language acquisition. This study explores the challenges and considerations with low-resource machine translation (MT), specifically focusing on primarily oral Indigenous Australian languages with minimally available textual resources. Additionally, we explore the existing challenges in the search for quality data and ethical research considerations in approaching Indigenous Cultural Intellectual Property (ICIP). As NMT performance often scales with the quality and quantity of multilingual corpora, we explore promising alternatives such as leveraging large language models (LLMs) to tackle severely low-resource MT as a few-shot prompting translation task. By employing a data imputation approach inspired by Continuous Bag-of-Words (CBOW) to strengthen a prompt's contextual relevancy, we enhance translations generated by LLMs, achieving a chrF score of 37.3 on imputed data, compared to a baseline of 31.6 with GPT-3.5, and 39.3 compared to a baseline of 38.3 on GPT-4. Through our work, we hope to establish a foundation for future efforts in preserving Indigenous Australian languages.

Keywords: Low-Resource Languages · Language Preservation · In-Context Learning · Machine Translation

1 Introduction

Modern translation systems heavily depend on abundant, quality data paired with state-of-the-art Neural Machine Translation (NMT) algorithms. The advent of Transformers with enhanced context-awareness has significantly bolstered

M. Gong et al. (Eds.): AI 2024, LNAI 15442, pp. 124–135, 2025.
https://doi.org/10.1007/978-981-96-0348-0_10

translation qualities, capturing subtle lexical relationships and contextual information from extensive corpora [18]. As a result, NMT algorithms have emerged as the dominant approach in modelling widespread languages with substantial digitised and written texts.

While the primary goal of language translation focuses on cross-cultural communication and enabling information access, NMT additionally supplements language and cultural preservation initiatives. By accurately capturing the essence of a language in a NMT model, cultural values and social attitudes can often be preserved within translations. For instance, languages akin to Japanese and Korean incorporate honorifics within communications as sociolinguistic customs, resulting in syntactic alterations in the language to denote respect and politeness [9]. This intrinsic interconnection across languages and cultures aligns NMT for language preservation with supporting cultural conservation efforts. Additionally, MT is also able to support second-language acquisition for language preservation, as translation models can support a future generation of speakers to carry forward languages.

The fluency and authenticity of translations generated by NMT models typically require extensive training on datasets with substantial sentence pairs from both source and target languages. This highlights a significant co-dependence between dataset sizes and NMT algorithms for quality translations. While Large Language Models (LLM) and multi-lingual models such as No Language Left Behind (NLLB-200) [15] have demonstrated improving performances in replicating linguistic characteristics across multiple languages, such capabilities are often attributed with the availability of substantial, web-extracted, language corpora.

However, many languages remain undigitized and lack sufficient parallel translation data, particularly Indigenous Australian languages spoken by smaller communities. Languages that are severely low in textual resources are often digitally underrepresented and insufficient for fine-tuning with large NMT models, as these models typically perform better with extensive bilingual sentence pairs. Consequently, the lack of available datasets presents significant hurdles in efforts for language and cultural preservation with MT.

We present an exploration into the challenges and considerations for the task of MT on severely low-resource languages, specifically targeting the preservation of digitally underrepresented Indigenous Australian languages. Our work is uniquely motivated by the dual objectives of language retention and cultural preservation. We delve into the ethical considerations of low-resource language preservation efforts, providing a comprehensive discussion that could serve as a guiding framework for future research in this area. Our primary technical contribution lies in our data imputation efforts with in-context learning MT, such as few-shot LLM-prompting for translations [13] [17]. We demonstrate that our imputation approach yields promising results, enhancing translation performances by improving the correspondence across supplementary examples within the LLM prompt context. Our findings presents new insights and opportunities for the application of MT in the context of low-resource languages.

2 Challenges and Considerations

2.1 Scarcity of Bilingual Translation Pairs in Indigenous Languages

Indigenous Australian languages pose a unique challenge for a MT task, primarily due to the endangered-classified majority with significant scarcity in written mediums. As a Western Desert language (WDL) with few speakers[1], Wangkajunga and similar dialects are profoundly rich in culture, with social systems deeply embedded within the language. For instance, Wangkajunga and other neighbouring Indigenous Australian cultures often unify familial terms; *mama* is used to refer to a father and or a fathers brothers, while *jurtu* is used to refer to female siblings and or female cousin relationships [8]. Such use of kinship terms within the language provides insights into the social systems of broader familial respect valued by their culture. The preservation of these languages, interwoven with their unique cultural attitudes, is significant in supporting the continuity of their cultural heritage and customs.

Other languages within the WDL groups, such as Martu Wangka, also epitomize a severely low-resource MT challenge. Spoken by approximately 791 individuals (see Footnote 1), Martu Wangka is primarily oral and has significantly limited availability of digital textual datasets. In contrast, other works such as NLLB-200 may have more favourable dataset sizes of low-resource languages, relying on available web-extracted data such as Common Crawl [15].

Prominent written resources in Martu Wangka mainly comprise hard-copy only materials, such as English-Martu dictionaries consisting of Martu words with English definitions and limited bilingual sentence samples [4]. Across our consultations with experienced and active linguists working with the Martu language, this dictionary has been noted as the most reliable in accuracy. The most extensive Martu Wangka text corpus found is a partially translated Christian Bible[2]. While this resource appears to contain the most abundant Martu texts that can be matched with an English equivalent, we note several limitations while considering the Bible as a resource candidate for a low-resource MT task:

- **Mismatched sentence counts across translations.** In our findings, we observed that among the 188 translated equivalent chapters between the Good News Bible (GNB) and the Martu Wangka Bible (MPJ), only 52 chapters contained identical number of verses (each often comprising multiple sentences). This disparity often varied between one to two verses, with few cases extending up to seven verses. We further investigated this issue with other English Bible versions such as RSV, KJV and ESV, uncovering similar inconsistencies in sentence counts when compared with MPJ. Our exchanges with a linguist-translator that facilitated the Martu Wangka Christian Bible indicates that this was due to the grammatical nuances of the language. This discrepancy

[1] The 2021 ABS census reports a total of six speakers for Wangkajunga and 791 for Martu Wangka. We note that these totals are applied with a slight random adjustment for confidentiality. Census data is accessible **here**.

[2] Online versions can be found via https://ebible.org/find/show.php?id=mpj.

in sentence alignment pose additional challenges in determining the correct corresponding sentence pairs.

- **Limited domain diversity and historic considerations.** Our observations also noted a limited domain diversity in the text for a MT task (also noticed in NLLB [15]), with Biblical texts largely comprising of theological narratives and concepts. Furthermore, Biblical texts frequently incorporate figurative and idiomatic language deeply rooted in various socio-cultural contexts, such as the Ancient Near Eastern, Greco-Roman, and Judeo-Christian traditions. This presents additional challenges for a MT task motivated by cultural preservation, as cultural semantics embedded within translations may diverge away from the intended cultural context. Additionally, the historic associations on colonialism with religious texts may need to be considered, particularly within the local context of Indigenous communities [6].
- **Translations employ a meaning-based approach.** Our exchanges with a Martu Bible linguist-translator indicate that translations can prioritize conveying theological meanings effectively in the receptor language, i.e. a meaning-based translation, rather than adhering strictly to a literal translation, i.e. word-for-word translations. While recognising the motivation behind meaning-based translations, MT tasks often favors literal translations rather than implicit meanings between bilingual pairs to ensure consistency and reducing ambiguity for NMT models.

Additional resources include studies on the language authored by linguists working together with local communities and limited samples of Martu Wangka narratives and learning materials[3]. This brings the total number of immediately available translated sentences to a range within the low hundreds. Consequently, the challenge of developing a performant translation model is significantly increased, due to the limited samples that can be maximised.

2.2 Morphological Considerations

Morphological and syntactic intricacies within Indigenous Australian languages present additional hurdles. While the arrangement of words within languages such as English typically follows a strict Subject-Verb-Object (SVO) order, Wangkajunga and similar WDL dialects in neighbouring Indigenous communities can employ a flexible word order requiring determination by pragmatics instead of syntax [8]. These languages may also adopt a complex case-marking system, affixes, and word compounding within grammatical structures, presenting additional challenges for sequence-based predictions. These linguistic intricacies necessitate extensive and typologically diverse training data to enable NMT models to capture language patterns effectively [2].

2.3 Ethical Considerations with Low-Resource Languages

While initially appearing intuitive that data mining and collection efforts must be vigorously and immediately endorsed, engaging with the origin community

[3] Materials can be find via SIL, https://www.sil.org/resources/search/language/mpj.

of the low-resource language involves significant ethical implications that must be addressed. Although widespread languages are often considered universally unowned, within the context of Indigenous Australian languages in particular, it is crucial to recognise the Indigenous community as the rightful language custodians. Given that language and cultural preservation efforts are intended to benefit their communities, it is imperative that such efforts are not decontextualised with a results-oriented approach [5]. Notably, we must ensure that the values and needs of language custodians are first acknowledged. In doing so, we acknowledge their central role in preserving and transmitting cultural heritage.

Works involving Indigenous Cultural Intellectual Property (ICIP), including languages, must be safeguarded from misuse, misappropriation and exploitation [7]. Consequently, research efforts with ICIP must be approached carefully and conducted with custodian representatives, through co-design and co-implementation. This raises challenges in acquiring language data, as establishing such relationships with communities cannot be immediately attained—it is earned with time and effective communication. Preservation initiatives with ICIP and custodians must inherently focus on fostering mutual trust, cooperation and, in accordance with Indigenous (data) sovereignty and governance principles [11].

We position this paper to lay the groundwork for facilitating future low-resource translation efforts for underrepresented Indigenous Australian languages and share our own experiences. Given the immediate complexities in navigating the ethical landscape when using existing ICIP materials, such as the English-Martu dictionary, acquiring quality datasets suitable for a MT task is increasingly challenging. In light of these constraints on MT for Indigenous Australian languages, we propose to use a low-resource but publicly available language (in this case, Kalamang) to mimic the translation to English. We hypothesise that our work is transferable towards future endeavors in applying MT for Indigenous Australian languages. Subsequently, we draw inspiration from similar works on severely low-resource languages to establish new findings. Notable recent works include the Machine Translation from One Book (MTOB) benchmark, which focuses on translating between English and Kalamang [17], a Trans-New Guinea language. The benchmark utilises a licensed Kalamang grammar book for the dataset [19] with 500 sentence pairs and 2,531 bilingual word lists as the dataset. The data availability constraints of the Kalamang language closely mirrors those of Martu Wangka, as both have limited digital presence, are primarily oral and have intricate morphologies to consider.

3 Machine Translation with LLM Prompting

Existing approaches to low-resource MT primarily involve leveraging models pre-trained on higher resource languages, such as NLLB-200's Mixture of Experts [15] to enhance the translation of low-resource languages. This approach transfers the learned parameters from a high-resource (parent) model to a low-resource (child) language pair model [21]. While transfer learning on bilingual or multilingual pre-trained models present overall improved performances, this method still requires a moderately sufficient amount of quality fine-tuning data for the low-resource language.

LLM-prompting MT involves maximising in-context learning with effective prompt engineering techniques, where supplementary information is injected into the context window of the LLM alongside a task prompt [3]. As shown in Fig. 1, in-context learning for a translation task involves incorporating language information such as dictionary entries, sentence examples and linguistic studies of the language, wrapped in a prompt to translate the source sentence.

Translate the following Kalamang sentence to English:
Kalamang sentence: [Source Sentence]
To help with the translation, here is the wordlists with the equivalent English:
[Kalamang Word 1] : [part of speech, English definition]
[Kalamang Word 2] : [part of speech, English definition]
...
To help with the translation, here are supplementary materials:
[Retrieved Sentence Pairs or Linguistic Studies Samples]
...
Output the translation only. Translation:

Fig. 1. LLM translation prompt example with additional context. This structure is akin to prompting approaches in MTOB [17].

Subsequently, the LLM decodes the patterns within the context window, generating translation predictions from the supplied few-shot prompt with contextual examples. Fine-tuning with MTOB has demonstrated weaker translation performances in comparison to the proposed LLM-prompting MT approach [17]. This is perhaps due to the limited number of training sentence pair samples that can be capitalised from. Such methods can be extended with data retrieval approaches, such as Retrieval Augmented Generation (RAG), where data samples are embedded into vector databases and retrieved based on vector similarities to the prompt [10]. In the context of translation, chunks of linguistic studies such as grammar books are retrieved and injected into an LLMs context, supplementing the original prompt with contextual information [17] [13]. Within this process, it is imperative that the test set remains unknown to the LLM, as using language or data samples an LLM has been exposed to is akin to information leakage, resulting in an overestimation of capabilities on unseen data[4].

The translation qualities of LLM prompting MT also appears to scale with parameter and context window size, presenting a dependence on large parameter models such as GPT-4 and Gemini 1.5 Pro [16]. These models with significantly

[4] In all figures, we omit use of Kalamang words within MTOB's train and test sets to reduce the likelihood of contamination into LLM training data.

larger context window sizes enable more contextual information to be injected for an in-context learning approach, yielding higher translation qualities compared to smaller models. While related works in LLM-prompting MT with languages such as Kalamang [17] present baseline performances that lag slightly behind human levels, the results appear promising for severely low-resource MT tasks with potential for further improvements.

4 MTOB Data Imputation

The authors of MTOB demonstrated higher performances when a combination of retrieved word lists with English meanings (**W**), bilingual sentence pairs (**S**), and retrieved grammar book (**G**) were injected into the prompt context as examples [17]. Specifically, we observed stronger quantitative improvements in chrF scores within the baseline results of MTOB where **W** or **S** were included within the LLM context [17]. This potentially indicates a higher dependence on **W** and **S** with LLMs for enhancing translation qualities.

Upon inspection of the Kalamang to English prompts and qualitative results, we observe that the presence of Kalamang words in the word list definitions **W** that match with words in the Kalamang source sentence achieves more favourable chrF scores. Notably, the relevance of the retrieved context **W**, **S** and **G** appears to increase the likelihood of LLMs predicting relevant keywords that are reflected within the ground truth translation. This contributes to higher chrF scores, as the LLM is more likely to generate translation sentences that closely match with ground truth from a more relevant prompt context.

In the MTOB benchmark dataset, we identified a significant number of instances where Kalamang words within the source sentences (**S**) of the train set had no known definition and were missing in the supplied word lists (**W**). While these Kalamang words lacked direct English equivalents in the word list, their meanings could be implicitly inferred from the context of the bilingual translation pairs as shown in Fig. 2.

We propose imputing the word lists **W** to include missing words and infer explicit meanings from the context of the source sentences **S**, thereby enhancing relevance in retrieved **W** with **S** when attempting translations with the LLM. Our method draws inspiration from Continuous Bag of Words (CBOW) as seen in Word2Vec [14], aiming to "fill in the unknown" word meanings by inferring them from sentence translation pairs and known adjacent Kalamang words.

This approach also requires handling of subwords and affixes. We employ nave whole word matching to retrieve known words. To tackle segmentation on words with known substrings and affixes, we adapt Word Break, a known algorithm following a Dynamic Programming approach [1], maximising for the longest substring of known Kalamang words (with English definitions) and appending these to a set (see Fig. 3).

We hypothesise that an increased prompt contextual clarity from word lists may result in enhanced LLM-driven MT performance. Specifically, the imputations enable a stronger correspondence between retrieved **W** and **S**. As imputa-

---

*Kalamang Sentence from **S**:* $[Kal_A]$ $[Kal_B]$ $[Kal_C]$ $[Kal_D]$
*English Translation from **S**:* $[Eng_A]$ $[Eng_B]$ $[Eng_C]$ $[Eng_D]$

*Retrieved known words in word list **W**:*

- $[Kal_A]$: $[English\ Definition\ of\ Kal_A]$
- $[Kal_C]$: $[English\ Definition\ of\ Kal_C]$
- $[Kal_D]$: $[English\ Definition\ of\ Kal_D]$

*Unknown Words in word list **W**:*

- $[Kal_B]$: $[Unknown]$

*Number of Kalamang known words (**k**): 3*
*Total number of words in Kalamang sentence (**n**): 4*
*The ratio (α) between **k** and **n** = 3/4 = 0.75*

Fig. 2. Example of missing word meanings. While Kal_B is an unknown word, the definition can be inferred from the current known definitions and the context of the English sentence. The feasibility to infer will depend on the ratio between **k** and **n**.

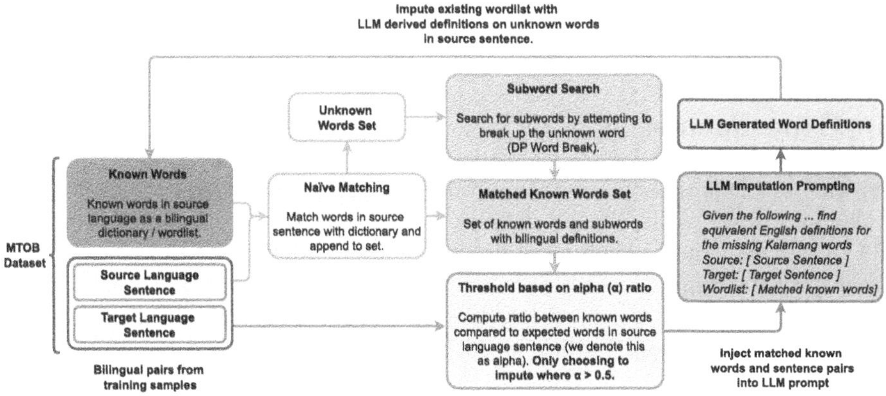

Fig. 3. Imputation Pipeline aims to extend word list with LLM derived definitions.

tions may be subject to LLM hallucinations, derived definitions are not intended to act as additions or modifications to the language.

Within our imputation process, we compute an α value to determine the ratio of known words within the Kalamang source sentences. Specifically, we define α as the ratio between the number of retrieved known words and the total number of words in the source Kalamang sentence (see Fig. 2). This ratio enables us to empirically determine the proportion of known words within the source Kalamang sentence. We only impute where the α ratio exceeds a certain threshold, i.e. the retrieved known words is at least some suitable proportion of the sentence length. This is to reduce the likelihood of hallucinated imputations on samples where a significant number of words in the source sentence are unknown. By prompting with GPT-4o, we aim to infer unknown word meanings on training samples using the prompt structure in Fig. 4.

Given the following sentence in Kalamang and the English translation, find the missing Kalamang words in the dictionary list and the equivalent English definitions.

Kalamang sentence: [Source Sentence]
English translation: [Ground Truth Translation]
word list: [Word set with English definitions]

Output only the missing Kalamang words in the sentence with the format 'Kalamang word: [Part of Speech, English definition]' as the value in a JSON structure:

Fig. 4. LLM imputation prompt example. Known word lists with English definitions are retrieved by whole-word matching or segmenting unknown words into known subwords and affixes. The output is parsed and appended to the current word list.

5 Experiments

Within our experiments, our imputation pipeline infers an additional 317 Kalamang words with English definitions. Appending these imputed words to MTOB's Kalamang-English word list yields a new total of 2,848 known words. We inspect the imputed words and note that the imputation includes new words, compounds of known words with affixes, and names. Consequently, the LLM is able to maximise these explicit word definitions in the word list to enhance translation qualities.

We observed higher performances with imputations where $\alpha > 0.5$, i.e. samples considered for imputation are where the known words comprise more than 50% of all words within the source Kalamang sentence. Increasing this α threshold reduced the imputations, while decreasing resulted in more hallucinated inferred definition. Additionally, we also found it most performant to maintain a temperature of 0.05 for all LLMs to reduce stochastic variability in results—this was also used in the reported baseline results of MTOB [17]. For consistency, we use the default longest common substring retrieval approaches in MTOB (i.e. $\mathbf{G}^s$ for the grammar book with a default 512 token-chunk size).

Our experiments focus on the word list context $\mathbf{W}$ and supplementing this with sentence pairs $\mathbf{S}$ and the grammar book extracts $\mathbf{G}^s$. We find improvements in select models as observed in Table 1, presenting promising results with the enhanced word list. Notably, we observe improved chrF scores in GPT-3.5 and GPT-4 in comparison to baseline results when the word list is imputed ($\mathbf{W}'$). This improvement is most prominent when the imputed word list is supplemented by the sentence and grammar book, i.e. $\mathbf{W}' + \mathbf{S}$ and $\mathbf{W}' + \mathbf{S} + \mathbf{G}^s$.

While Claude 2 presents an increase to 34.1 chrF from 29.7 chrF in using $\mathbf{W}'$, the chrF scores of $\mathbf{W}' + \mathbf{S}$ and $\mathbf{W}' + \mathbf{S} + \mathbf{G}^s$ are close to MTOB's baselines.

Table 1. chrF scores for Kalamang to English on MTOB. $\mathbf{W}'$ denotes our imputation attempts on the sentence pairs and word list of Kalamang. We compare the original baselines from MTOB [17] **(W, W + S, and W + S + G^s)** with our imputed wordlist alternatives **(W′, W′ + S, W′ + S + G^s)**. Bold indicates the highest value between baseline results and imputed word list results.

	W	W + S	W + S + G^s	W'	W' + S	W' + S + G^s
gpt-3.5-turbo	24.9	31.1	31.6	**28.3**	**32.2**	**37.3**
gpt-4	33.3	34.9	38.4	**34.4**	**37.4**	**39.3**
Claude 2	29.7	**43.6**	**42.9**	**34.1**	43.5	42.6

Our findings highlight a distinct advantage in improving alignments between word list **W** and sentences **S**. Qualitatively, the LLM-derived translations were still subject to some level of hallucinated generations, which may be attributed to model capabilities, insufficient contextual information, or hallucinated imputations. We expect that refining both **W** and **S** may enhance LLM-prompted translations. In this effort, the involvement of Indigenous community members will be instrumental in enhancing and quality checking of **W** and **S** [12].

6 Future Work

There remains significant potential for further improving performance in LLM-prompted MT through various research directions. One future area is refining word search and subword segmentation techniques that better consider the morphology (such as stem words and affixes) of the language for enhanced relevancy in **W**. Additionally, an alternative approach for retrieving sentence examples **S** could involve directly linking each sentence with words in **W** (similar to how a dictionary's example sentences are linked to a word). This may lead to tighter alignments between retrieved **W** and **S** when compared with the current longest-common substring approach employed in MTOB [17].

Some samples with alpha (α) below the 0.5 threshold have been observed to exceed this threshold after a prior sample imputes a common word. An enhancement to the imputation process may include sequencing the imputations, by starting from the highest α proceeding downwards to the threshold. This ensures that the most informational samples, i.e., higher α with an increased likelihood of accurate imputations, contribute to the imputation of samples with lower α to reduce hallucinations. Another promising avenue to explore is Chain-of-Thought (CoT) prompting with LLMs, which could further improve translation accuracy. CoT prompting enables reasoning abilities with LLMs through explicitly breaking down the thought processes as intermediate steps [20]. Since the translation and imputation prompts resemble reasoning tasks, CoT prompting may be a promising approach to improve the quality of the translations and imputations.

7 Conclusions

Modern machine translation (MT) techniques for low-resource languages, such as Indigenous Australian languages, supports the preservation of these languages and reducing the risk of cultural heritage losses. However, these efforts must be approached with careful consideration. In this paper, we explore the key challenges in applying MT to Indigenous languages, focusing on data availability constraints and ethical implications. We demonstrate clear improvements in LLM-prompting MT when the prompt context is more informational, as demonstrated by our imputed word list. Our results present enhanced performances where there exists a stronger correspondence between word lists and sentence examples. Despite these observations, our experiments also note that translation qualities across multiple test instances still lacks coherence and can often deviate from the semantics of ground truth. This highlights potential areas for future work in LLM-prompting MT. We aim to enable the transferability of our findings on the challenges of MT in severely low-resource languages and the potential solutions with LLM-prompting for MT to aid future preservation endeavors in Indigenous Australian languages. This is especially relevant when resources are limited to dictionaries and a small set of bilingual sentence samples. The implementation of our work is available at https://github.com/alastairrr/impute_mtob.

Acknowledgement. We thank the original authors of MTOB [17] for granting permission to use the Kalamang dataset. We are grateful to Professor Gareth Baynam and other experts for their valuable guidance on low-resource machine translation and ethical research practices. We also acknowledge the Martu Bible linguist-translator Ken Hansen for his insights into the Martu Wangka Biblical translation, and thank the many people for their valuable discussions.

Ethics Statement. We adhere to ethical principles to respect the languages and communities involved, and emphasize obtaining permissions and engaging with stakeholders. We obtained permission to use the Kalamang dataset and acknowledge Tanzer et al. Our exploration of WDL like Martu Wangka and Wangkajunga uses publicly available materials, and we engaged with multiple stakeholders to ensure respectful handling. We acknowledge that while focusing on technical challenges in MT for low-resource languages, future research should be guided by community needs and priorities.

References

1. Bellman, R.: Dynamic Programming. Princeton University Press, Princeton (1957)
2. Bisazza, A., Üstün, A., Sportel, S.: On the difficulty of translating free-order case-marking languages. Proc. TACL **9**, 1233–1248 (2021)
3. Brown, T., et al.: Language models are few-shot learners. In: Advances in Neural Information Processing Systems, vol. 33, pp. 1877–1901 (2020)
4. Burgman, A.: Martu Wangka Dictionary and Topical Finder Lists 2005: Draft 1. Wangka Maya Pilbara Aboriginal Language Centre (2005)

5. Cooper, N., Heldreth, C., Hutchinson, B.: It's how you do things that matters: attending to process to better serve indigenous communities with language technologies. In: Proceedings of EACL, pp. 204–211. St. Julian's, Malta (2024)
6. Hutchinson, B.: Modeling the sacred: considerations when using religious texts in natural language processing. In: Proceedings of NAACL (2024). ACL, Mexico City, Mexico (2024)
7. Janke, T.: True Tracks: Respecting Indigenous Knowledge and Culture. NewSouth Publishing, Sydney (2022)
8. Jones, B., et al.: A grammar of Wangkajunga: a language of the great sandy desert of north Western Australia. Research School of Pacific and Asian Studies, The Australian National University, Pacific Linguistics (2011)
9. Yoon, K.-J.: Not just words: Korean social models and the use of honorifics. Intercult. Pragmat.**1**(2), 189–210 (2004)
10. Lewis, P., et al.: Retrieval-augmented generation for knowledge-intensive NLP tasks. In: Proceedings of NeurIPS, vol. 33, pp. 9459–9474 (2020)
11. Lowitja Institute: Indigenous data governance and sovereignty (2021)
12. Mager, M., Mager, E., Kann, K., Vu, N.T.: Ethical considerations for machine translation of indigenous languages: giving a voice to the speakers. In: Proceedings of ACL (2023). Toronto, Canada (2023)
13. Merx, R., Mahmudi, A., Langford, K., de Araujo, L.A., Vylomova, E.: Low-resource machine translation through retrieval-augmented LLM prompting: a study on the Mambai language. In: Proceedings of EURALI. ELRA and ICCL, Torino, Italia (2024)
14. Mikolov, T., Chen, K., Corrado, G., Dean, J.: Efficient estimation of word representations in vector space. In: Proceedings of ICLR (2013)
15. NLLB Team: Scaling neural machine translation to 200 languages. Nature (2024)
16. Reid, M., et al.: Gemini 1.5: unlocking multimodal understanding across millions of tokens of context. arXiv preprint arXiv:2403.05530 (2024)
17. Tanzer, G., Suzgun, M., Visser, E., Jurafsky, D., Melas-Kyriazi, L.: A benchmark for learning to translate a new language from one grammar book. In: Proceedings of ICLR (2024)
18. Vaswani, A., et al.: Attention is all you need. In: Proceedings of NeurIPS (2017)
19. Visser, E.: A grammar of Kalamang. No. 4 in Comprehensive Grammar Library, Language Science Press, Berlin (2022)
20. Wei, J., et al..: Chain-of-thought prompting elicits reasoning in large language models. In: Proceedings of NeurIPS. Curran Associates Inc., Red Hook, NY, USA (2024)
21. Zoph, B., Yuret, D., May, J., Knight, K.: Transfer learning for low-resource neural machine translation. In: Proceedings of EMNLP (2016)

Trustworthy and Explainable AI

Improving Intersectional Group Fairness Using Conditional Generative Adversarial Network and Transfer Learning

David Quashigah Dzakpasu(✉), Jixue Liu, Jiuyong Li, and Lin Liu

University of South Australia, Adelaide, Australia
David.Dzakpasu@mymail.unisa.edu.au,
{Jixue.Liu,Jiuyong.Li,Lin.Liu}@unisa.edu.au

Abstract. Classification models often exhibit intersectional group unfairness due to the underrepresentation of minority intersectional groups, posing a significant challenge in ensuring fair classification for all groups. Intersectional group fairness considers fairness across multiple dimensions of identity, such as race, gender, and socioeconomic status, recognizing that individuals may experience adverse forms of discrimination or disadvantage based on the intersections of these identities. This paper proposes the Intersectional Fair Transfer Learning (IFTL) method to address this issue by introducing a two-phase strategy utilizing balanced synthetic data and transfer learning. In the first phase, a model is trained on a balanced synthetic dataset w.r.t intersectional groups and label values to establish a fair decision boundary. Subsequently, parameter sharing is employed in the second phase to adapt the pre-trained model to real-world data, aiming to improve classification accuracy while maintaining fairness. Experimental results on two depression datasets demonstrate the efficacy of the proposed IFTL method.

Keywords: Intersectional group fairness · Underrepresentation · Transfer learning · Conditional generative adversarial network

1 Introduction

With the growing influence of machine learning (ML) applications in critical domains, such as education, healthcare, and criminal justice, concerns about their potential to produce unintentionally biased decisions against marginalised groups or individuals have increased [3,4]. One prominent example of such bias includes the COMPAS recidivism tool's disproportionate labelling of blacks as high-risk criminals compared to whites with similar offences [4].

To assess the fairness of machine learning applications, various fairness metrics have been proposed across individual, group, and intersectional group

Supplementary Information The online version contains supplementary material available at https://doi.org/10.1007/978-981-96-0348-0_11.

M. Gong et al. (Eds.): AI 2024, LNAI 15442, pp. 139–153, 2025.
https://doi.org/10.1007/978-981-96-0348-0_11

domains [6]. Individual fairness metrics evaluate whether similar individuals are treated similarly. In contrast, group fairness metrics focus on the equal treatment of different groups defined by a *single* protected attribute. Intersectional group fairness takes this further by considering *multiple* protected attributes, assessing whether groups formed at the intersections of *all* considered protected attributes are treated equally. Research indicates that as the number of protected attributes increases, the fairness observed among intersectional groups tends to decrease [14]. This leads to the understanding that fairness algorithms achieving group fairness may not necessarily ensure intersectional group fairness, which is why there is a growing interest in studying intersectional group fairness [7,10,14,15].

Several methods have been developed to mitigate intersectional group unfairness [8–12]. The multiaccuracy boost method by Kim et al. [12] audits classifiers for intersectional groups fairness and post-process the prediction if the prediction result is unfair. The GerryFair method [8,11] prevents "fairness gerrymandering" among intersectional groups using a zero-sum game between a learner and an auditor (the learner optimises for accuracy while the auditor checks for intersectional group fairness violations). The empirical differential fairness method by Fould et al. [10] restricts the ratio of predictions among intersectional groups to a threshold. Also, the infofair method [7] revises prediction results to reduce the mutual information between multiple protected attributes and predicted labels.

One limitation of prior methods is their limited effectiveness in mitigating unfairness for underrepresented intersectional groups. In [8,11], group sizes are used as weights in fairness constraints, with underrepresented groups assigned smaller weights, leading to a reduced influence of unfairness on these groups in the fairness solution. In [10], a user-defined parameter is used to amplify the influence of fairness among underrepresented groups. However, determining the parameters in this complex optimization and balance issue is challenging. The primary reason for this limitation is the underrepresentation of small groups in these methods, which hinders successful unfairness mitigation. This underrepresentation arises because as the number of protected attribute dimensions increases, the size of each intersectional group rapidly diminishes, exacerbating the challenge of mitigating unfairness due to the curse of dimensionality.

In the context of machine learning, underrepresentation presents a formidable obstacle with far-reaching consequences. Models trained on imbalanced datasets, where certain groups are disproportionately represented, often produce biased and unreliable outcomes [20]. This imbalance leads to a skewed understanding of the data distribution, resulting in subpair performance, particularly for underrepresented groups. In critical domains like healthcare, finance, and criminal justice, biased models can perpetuate systemic inequities, exacerbating societal disparities. Addressing underrepresentation is therefore essential for ensuring fairness, equity, and effectiveness in algorithmic decision-making processes.

In this paper, we introduce a novel strategy centered on data augmentation and transfer learning. By employing a conditional generative adversarial network (GAN) model, we generate a perfectly balanced synthetic dataset w.r.t intersectional groups and labels. This initial step is crucial in rectifying the underrepresentation issue faced by minority intersectional groups, ensuring each group receives sufficient representation. Subsequently, we pre-train a classifier using this balanced synthetic dataset, thereby guaranteeing an *equal opportunity* for each intersectional group to be accurately classified. During the adaptation phase, we use the *parameter sharing* approach from transfer learning to fine-tune the pre-trained classifier using real-world data. The adaptation phase helps to improve the classification accuracy of the pre-trained model while maintaining the fairness attained in the pre-training phase.

The contributions of our work are summarized as follows: (1) We introduce a new Intersectional Fair Transfer Learning (IFTL) method which combines conditional GAN and transfer learning to enhance intersectional group fairness in classification models. (2) We address the underrepresentation of minority intersectional groups by creating a balanced synthetic dataset that encompasses all intersectional groups. (3) We validate the usefulness of IFTL method using real-world depression datasets.

The rest of the paper is organized as follows: We outline our proposed method in Sect. 2. Section 3 details the experiments, results and analysis. Section 4 reviews the related works. We conclude the paper in Sect. 5.

2 The Proposed IFTL Method

2.1 Problem Definition

Consider a dataset $\mathcal{D} = \{A, X, Y\}$, where $A = \{A_1, A_2, \cdots, A_p\}$ represent a set of binary protected attributes. $A_i = 0$ and $A_i = 1$ represent the protected and unprotected group's attribute value respectively. We use X to denote all other non-protected attributes. The binary outcome attribute is denoted by Y. Let $f(\cdot)$ represent a predictive model or classifier that takes X and A as input and output prediction $\hat{Y}$. The set of all intersectional groups is denoted by $\mathcal{G} = \{A_1 \times A_2 \times \cdots \times A_p\}$. Given the prediction, $\hat{Y}$ and an intersectional group $g \in \mathcal{G}$, the function $\pi_g(\hat{Y})$ returns the predictive outcomes for the intersectional group g.

Given a fairness metric F, the problem of this research is to build a model $f(\cdot)$ that produces *intersectional group fair decisions*. That is, for any two intersectional groups g_i and g_j in $\mathcal{G}$, $|F(\pi_{g_i}(\hat{Y})) - F(\pi_{g_j}(\hat{Y}))| \leq \alpha$, where α is a very small value, normally close to zero.

2.2 Overview of IFTL Method

As illustrated in Fig. 1, the architecture of IFTL consists of two primary components: synthetic data generation and intersectional group fair classifier construction, represented by the upper and lower rectangular boxes, respectively.

The synthetic data generation uses the CTAB-GAN model [23] to generate a source dataset $\mathcal{D}_S$ where sample sizes w.r.t label distribution in all intersectional groups are consistent. Also, $\mathcal{D}_S$ should have similar statistical properties as the

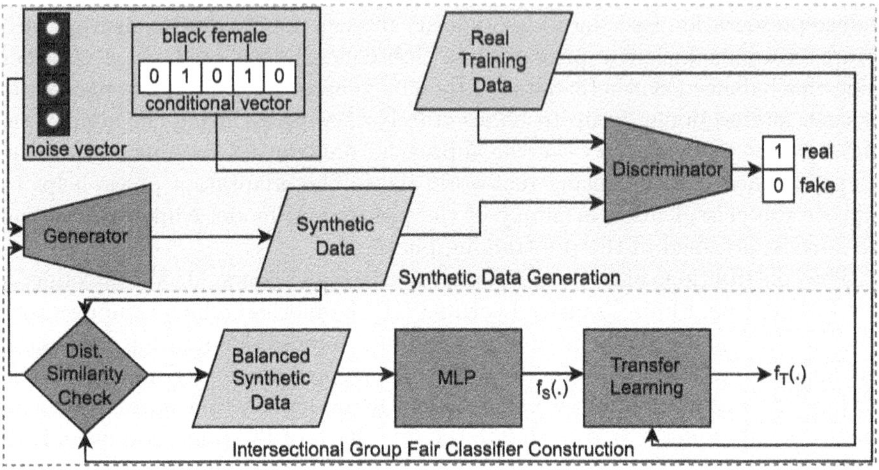

Fig. 1. The architecture of IFTL. The upper and lower rectangular boxes display synthetic data generation and intersectional group fair classifier construction, respectively.

real dataset $\mathcal{D}_T$. The intersectional group fair classifier construction is further divided into two phases: pre-training and adaptation. In the pre-training phase, we train model $f_S(\cdot)$ on $\mathcal{D}_S$ with the focus on learning a fair decision boundary. In the adaptation phase, we adapt $f_S(\cdot)$ to $f_T(\cdot)$ on $\mathcal{D}_T$, through transfer learning to improve classification accuracy.

In our problem setting, we operate under a transfer learning scenario where $P(X_S, A_S) \neq P(X_T, A_T)$ and $f_S(\cdot) \neq f_T(\cdot)$. $P(X_S, A_S) \neq P(X_T, A_T)$ implies a difference in the marginal distribution of the feature space between the source and target domains. Similarly, $f_S(\cdot) \neq f_T(\cdot)$ indicates a dissimilarity in $P(Y_S|X_S, A_S)$ and $P(Y_T|X_T, A_T)$, signifying a variance in the distribution of depression labels between the source and target domains.

2.3 Synthetic Data Generation

Zhao et al. [23] propose the CTAB-GAN model to tackle the challenge of dataset imbalance. Their approach integrates a conditional GAN with a training-by-sampling method. The key concept is to use a conditional vector to represent classes of underrepresented categorical attributes for augmentation. By resampling the conditional vector for the underrepresented classes and combining it with a random noise vector, the generator is trained to give these classes higher chances of adequate representation in the generated data.

We adopt the CTAB-GAN model to address the underrepresentation problem owing to its advantageous properties. Notably, it effectively balances the representation of underrepresented classes and maintains classification accuracy comparable to real data.

Recall 4 the synthetic dataset must fulfil two crucial requirements: first, it should exhibit statistical similarity to the real data; second, each intersectional group must be well represented. To ensure statistical similarity value of at least 0.1, we utilise metrics such as Wasserstein Distance [5] for continuous attributes (ranging from 0 to ∞) and Jensen-Shannon Divergence [2] for categorical attributes (ranging from 0 to 1). Lower values indicate a greater resemblance between synthetic and real data. To guarantee adequate representation of each intersectional group, the conditional vector is constructed to include the under-represented categories in each protected attribute of interest that we want to augment.

2.4 Construction of Intersectional Group Fair Classifier

Phase 1: Pre-training for Intersectional Group Fairness. In this phase, our objective is to learn a fair decision boundary that ensures fairness across intersectional groups using the classifier $f_S(\cdot)$, parameterised by θ_S. To achieve this, we train classifier $f_S(\cdot)$ using a perfectly balanced synthetic dataset.

Although the synthetic dataset contains well-augmented samples w.r.t each intersectional group, we further streamline the synthetic data to a balanced synthetic data $\mathcal{D}_S$ to give each intersectional group equal representation in the training dataset. In this balanced dataset, every intersectional group maintains an equal sample size concerning the label Y. Specifically, we balance $\mathcal{D}_S$ such that $|\{X \mid \mathcal{G} = g, Y = y\}| = c$, $\forall g \in \mathcal{G}$, where c denotes the group size per label value. The value of c is determined using the most represented intersectional group's size w.r.t the label in the real dataset $\mathcal{D}_T$ as highlighted in Table 3.

We design $f_S(\cdot)$ using a Multilayer Perceptron (MLP) classifier with three fully connected (FC) layers and ReLU activation functions. The output is a scalar that is squashed by the sigmoid activation function between 0 and 1 for our binary classification task. Training $f_S(\cdot)$ on $\mathcal{D}_S$ is performed using binary cross-entropy loss as in Eq. 1:

$$\mathcal{L}_S = -\frac{1}{N}\sum_{i=1}^{N}[y_i * log(\tilde{y}_i) + (1 - y_i) * log(1 - \tilde{y}_i)], \tag{1}$$

where, N is the number of samples, y_i is the actual label and $\tilde{y}_i$ is the predicted probability that i belongs to a particular label class.

Training $f_S(\cdot)$ on the balanced synthetic data $\mathcal{D}_S$ enables the learning algorithm to discern a decision boundary conducive to fair classification among the intersectional groups. The parameters θ_S of $f_S(\cdot)$ are saved for our transfer learning task in phase 2.

Phase 2: Adaptation for Improved Classification Accuracy. Although the pre-trained classifier $f_S(\cdot)$, parameterized by θ_S, appears to be synthetically intersectional group fair, its classification accuracy may not match that of a model trained directly on real data due to a lack of exposure to real-world data. We address this problem by improving the classification accuracy on real data $\mathcal{D}_T$ while retaining the intersectional group fairness attained in phase 1.

In adapting $f_S(\cdot)$ to $f_T(\cdot)$ on $\mathcal{D}_T$, we leverage the *parameter sharing* technique from transfer learning. This technique involves freezing or sharing most of the parameters in the pre-trained model and fine-tuning only a subset of parameters, typically the final layers, to suit the target task. Specifically, we extend the architecture of $f_S(\cdot)$ by adding a fully connected layer before the output layer. After this structural modification, we fine-tune $f_S(\cdot)$ using proportions of $\mathcal{D}_T$ to obtain the model $f_T(\cdot)$ while keeping parameters θ_S frozen. We incrementally vary the proportion of $\mathcal{D}_T$ by 25% each time to fine-tune the model in order to monitor the proportion of $\mathcal{D}_T$ at which we achieve desirable improvement in the classification accuracy without much degradation in fairness. The freezing of θ_S is crucial to preserving fair parameters attained in phase 1. The adapted model, denoted as $f_T(\cdot)$, is trained to minimize the loss on $\mathcal{D}_T$ using Eq. 2:

$$\theta_T = \underset{\theta}{\text{argmin}} \sum_{(x,a,y)\in\mathcal{D}_T} \mathcal{L}(f_\theta(x, a; \theta_S), y), \tag{2}$$

where, θ_T represents the adapted parameters for the target domain, θ_S denotes the parameters of the source domain model used for parameter sharing, and $\mathcal{L}$ denotes the binary cross-entropy loss function as defined in Eq. 1.

2.5 Training Intersectional Fair Transfer Learning (IFTL)

In this section, we provide the learning algorithm of IFTL in Algorithm 1. Also, we detail the architecture and hyper-parameters of IFTL in Table 1.

Algorithm 1. Intersectional Fair Transfer Learning (IFTL) Algorithm

1: **Input** Dataset $\{(x_i, a_i, r_i, y_i)\}_i^n$ where $\mathbf{x}$ is the input feature vector, $\mathbf{a}$ is the protected attribute vector, $\mathbf{r}$ is the conditional vector and $\mathbf{y}$ is the target label.
2: **Output** Intersectional Group Fair Classifier ($f_T(\cdot)$).
3: Initialize D, G, C, $f_S(\cdot)$, $f_T(\cdot)$ randomly.
4: **while** not converge **do**
5: 1. Update generator G to minimize $\mathcal{L}^G$ (cf. subsubsection 2.5)
6: 2. Update discriminator D to minimize $\mathcal{L}^D_{orig}$ (cf. subsubsection 2.5)
7: 3. Update classifier C to minimize $\mathcal{L}^C_{class}$ (cf. subsubsection 2.5)
8: **end while**
9: **Pre-training phase** Train $f_S(\cdot)$ on balanced synthetic data $\mathcal{D}_S$ from G.
10: **Adaptation phase** Adapt $f_S(\cdot)$ on real data $\mathcal{D}_T$ to obtain $f_T(\cdot)$.

Training for Synthetic Data Generation. The CTAB-GAN [23] model uses four loss terms to train the generator G: $\mathcal{L}^G = \mathcal{L}^G_{orig} + \mathcal{L}^G_{info} + \mathcal{L}^G_{class} + \mathcal{L}^G_{gen}$. Here, $\mathcal{L}^G_{orig}$ mirrors the original generator loss in the GAN model [17], $\mathcal{L}^G_{info} = ||\mathbb{E}[f_x]_{x\sim p(x)} - \mathbb{E}[f_{G(z)}]_{z\sim p(z)}||_2 + ||\mathbb{SD}[f_x]_{x\sim p(x)} - \mathbb{SD}[f_{G(z)}]_{z\sim p(z)}||_2$ matches distributions between real and synthetic samples, $\mathcal{L}^G_{class} = \mathbb{E}[|l(G_z) -$

$\mathcal{C}(fe(G(z)))||_{z\sim p(z)}$, where $l(\cdot)$ returns the target label and $fe(\cdot)$ returns the input features of a given row. $\mathcal{L}^G_{class}$ aligns the target label with the generated input features. $\mathcal{L}^G_{gen} = H(r_i, \hat{r}_i)$ guarantees the synthetic sample is generated according to the pre-specified conditions in the conditional vector r. The discriminator D loss $\mathcal{L}^D_{orig}$ remains consistent with [17], while the classifier C loss $\mathcal{L}^C_{class} = \mathbb{E}[|l(x) - \mathcal{C}(fe(G(z)))||_{x\sim p(x)}$ ensures similarity in classification accuracy between synthetic and real data. Each time the synthetic data similarity values are greater than 0.1, we tweak the hyper-parameters and re-train the model until the model generate samples with similarity value less than 0.1. We follow the training and hyper-parameter guide in [23].

Phase 1 Training: Pre-training for Intersectional Group Fairness. The network architecture comprises three fully connected (FC) layers, each employing ReLU activation, followed by a sigmoid activation for the output layer. Binary cross-entropy loss and Adam optimizer are used to train $f_S(\cdot)$ on $\mathcal{D}_S$ until convergence.

Phase 2 Training: Adaptation for Improved Classification Accuracy. Model $f_T(\cdot)$ is derived from pre-trained model $f_S(\cdot)$ using θ_S parameters. An additional FC layer with ReLU activation is introduced before the output layer. We incrementally increase proportions of $\mathcal{D}_T$ each time by 25% and train only the newly added layers while freezing the parameters θ_S of $f_S(\cdot)$. We use binary cross-entropy loss and Adam optimizer to train $f_S(\cdot)$ on $\mathcal{D}_T$ until convergence.

Table 1. Hyper-parameter for NHANES and COVID-19 Datasets.

Hyper-parameter	NHANES		COVID-19	
	Pre-training	Adaptation	Pre-training	Adaptation
Optimiser	Adam	Adam	Adam	Adam
Epoch	500	400	500	400
Learning rate	0.01	0.001	0.01	0.001
Weight decay	0.00001	0.00001	0.00001	0.00001
Hidden Layer 1	80	80	80	80
Hidden Layer 2	60	60	60	60
Hidden Layer 3	30	30	30	30
Hidden Layer 4	–	10	–	10
Output Layer	1	1	1	1

3 Experiments

3.1 Experiment Setup

Datasets. We evaluate IFTL on two benchmark depression datasets as summarized in Table 2. We treat sex and race as binary protected attributes in both datasets.

Table 2. Real Dataset Quantitative Statistics

Dataset	Sample #	Attribute #	Protected Attr. 1	Protected Attr. 2
NHANES	68,249	42	Sex	Race
COVID-19	19,525	149	Sex	Race

National Health and Nutrition Examination Survey (NHANES) [29]. This survey dataset covers the years 2005 to 2018 and includes a diverse range of demographic information, medical histories, lifestyle characteristics, prescription medication records, and socioeconomic data. In line with [25], depression status is determined using the Patient Health Questionnaire-9 (PHQ-9) [24]. The PHQ-9 has nine items, each scored from 0 to 3. Participants are assigned a major depressive disorder (MDD) status if their total score $\geq$ 10. Accordingly, a threshold of 10 is applied to designate individuals as depressed. There are 4,325 individuals labelled as depressed out of 68,249 participants.

COVID-19. The COVID-19 dataset obtained from the Data Foundation [30] offers insights into the real-time impact of the COVID-19 pandemic spanning April to June 2020. It encompasses a wide range of data on social dynamics, economic well-being, and both mental and physical health indicators. Depression labels are stratified into four groups: not at all or less than 1 day, 1–2 days, 3–4 days, and 5–7 days. Participants falling into the not at all or less than 1 day category are labelled as non-depressed, while those in the 3–4 days and 5–7 days categories are labelled as depressed. Individuals with 1–2 days are excluded as borderline cases. Out of 19,525 participants, there are 4,019 identified as non-depressed.

Each real dataset is randomly split into three distinct sets: training, validation, and testing, with proportions of 65%, 15%, and 20%, respectively. In the pre-training phase, we utilized a synthetic dataset where each intersectional group had a uniform sample size with respect to the label value. For the NHANES dataset, this resulted in 32,882 samples per group, totaling 263,056 observations. Similarly, for the COVID-19 dataset, each intersectional group contained 5,738 samples, amounting to 45,904 total observations. The synthetic dataset's intersectional group sizes were calibrated to match the size of the most represented intersectional group in the corresponding real dataset, as illustrated in Table 3. This approach ensures balanced representation across all intersectional groups while maintaining consistency with the scale of the real data.

Table 3. Maximum intersectional group size in real datasets vs. uniform group size in balanced synthetic datasets

Dataset	Real	Synthetic
NHANES	32,882	32,882
COVID-19	5,738	5,738

Baseline Methods. We compare IFTL against the following baselines: (i) Fairness Unaware (FU) shares the same architecture as our pre-trained classifier $f_S(\cdot)$ but it is trained on only the real dataset. (ii) Correlation Remover (CO) [27] prevents disparate impact by interpolating the original data distribution with an unbiased distribution. (iii) We extend Random Over-sampling (ROS) [1] to balance minority intersectional groups with the largest majority group, then train using the FU model structure. (iv) Empirical Differential Fairness (EDF) [10] ensures that the pairwise acceptance rates among intersectional groups do not exceed a pre-defined threshold, such as the 80% rule [26]. (v) InfoFair (IF) [7] ensures independence between multiple protected attributes and predicted labels using a mutual information minimization approach. (vi) GerryFair (GF) [8] ensures fairness for cost-sensitive classification through zero-sum game.

Evaluation Metrics. We describe evaluation metrics used in our analysis below.

Accuracy. We use Balanced Accuracy (BAcc) to assess classification accuracy due to the imbalanced distribution of class labels. BAcc is calculated as the average of the true positive rate (TPR) and the true negative rate (TNR). It ranges from 0 to 1.

Fairness. We adopt Equal Opportunity (EO) [28] to measure intersectional group fairness. EO requires the true positive rate (TPR) to be similar for all groups. EO is computed as:

$$P(\hat{y} = 1|y = 1, \mathcal{G} = g_i) = P(\hat{y} = 1|y = 1, \mathcal{G} = g_j), i \neq j \tag{3}$$

However, achieving EO directly is mostly difficult, instead, we adopt Equal Opportunity Difference (EOD). The value of α ranges from 0 to 1. The smaller the value of α, the fairer the model. EOD is computed as follows;

$$|P(\hat{y} = 1|y = 1, \mathcal{G} = g_i) - P(\hat{y} = 1|y = 1, \mathcal{G} = g_j)| \leq \alpha, i \neq j \tag{4}$$

Accuracy-Fairness Trade-Off. We compare accuracy and fairness by calculating the difference between the relative changes in both concerning the fairness-unaware (fu) model. We refer to methods with fairness intervention as fair-aware (fa) models. The trade-off value is the difference between the relative balanced accuracy and equal opportunity difference. The trade-off is calculated as follows:

$$\left[\frac{EOD_{fu} - EOD_{fa}}{EOD_{fu}}\right] - \left[\frac{BAcc_{fu} - BAcc_{fa}}{BAcc_{fu}}\right] \quad (5)$$

The first part of the equation quantifies the improvement or deterioration in the fairness of the model after applying fairness interventions. The second part assesses how much the accuracy of the model changes with fairness interventions. By subtracting the relative change in fairness from the relative change in accuracy, we obtain a single value representing the trade-off between accuracy and fairness. A higher value indicates a better trade-off between the two metrics. This trade-off value enables us to evaluate the effectiveness of fairness interventions while considering their impact on classification accuracy.

Wasserstein Distance (WD). This metric evaluates how closely the distributions of continuous attributes in the synthetic data resemble those in the real data. WD ranges from 0 to ∞, where lower values indicate higher similarity between the distributions.

Jensen-Shannon Divergence (JSD). JSD measures the divergence in probability mass distribution for each categorical attribute between the synthetic and real datasets. Values range from 0 to 1, with lower values indicating greater similarity between the distributions.

3.2 Results and Analysis

In this section, we assess IFTL's overall performance in terms of accuracy and fairness in Table 4. We analyze the trade-off between accuracy and fairness and show our results in Fig. 2. Our result also covers the impact of real dataset proportions on accuracy and fairness in Fig. 3. We also show the result on impact of balanced and unbalanced data on intersectional group fairness in Fig. 4. Finally, we report on the quality of the synthetic dataset. All results presented are averaged across 10 runs.

Overall Performance. Results in Table 4 illustrates that, IFTL, significantly improves intersectional group fairness compared to all baselines, while also maintaining comparable accuracy levels. In the NHANES dataset, we observe high unfairness in the FU model, indicating significant disparities among intersectional groups. Modest improvements in fairness are observed with various baselines, but the IFTL achieves the best intersectional group fairness.

Similar trends are observed in the COVID-19 dataset, with the FU method being the most unfair. IFTL outperforms baselines in improving intersectional group fairness, followed by the IF, ROS, CO, EDF, and GF.

Accuracy-Fairness Trade-Off. In Fig. 2, we elucidate the trade-off between classification accuracy and intersectional group fairness, showcasing the extent of

Table 4. Debiasing result of IFTL against baselines on NHANES and COVID-19 datasets. Bold for the best. ↑ / ↓ means the higher/lower, the better

Method	NHANES		COVID-19	
	BAcc ↑	EOD ↓	BAcc ↑	EOD ↓
FU	0.749 ± 0.003	0.188 ± 0.018	$\mathbf{0.774 \pm 0.012}$	0.048 ± 0.020
EDF	$\mathbf{0.777 \pm 0.008}$	0.123 ± 0.017	0.730 ± 0.011	0.042 ± 0.022
GF	0.718 ± 0.063	0.157 ± 0.072	0.767 ± 0.014	0.049 ± 0.035
IF	0.735 ± 0.028	0.109 ± 0.071	0.708 ± 0.138	0.034 ± 0.036
CO	0.757 ± 0.005	0.132 ± 0.016	0.764 ± 0.025	0.036 ± 0.013
ROS	0.752 ± 0.004	0.151 ± 0.015	0.757 ± 0.008	0.036 ± 0.016
IFTL (ours)	0.721 ± 0.025	$\mathbf{0.073 \pm 0.045}$	0.742 ± 0.038	$\mathbf{0.026 \pm 0.011}$

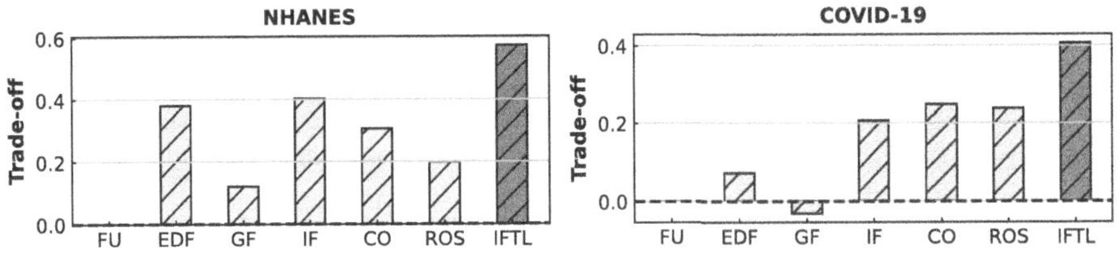

Fig. 2. Accuracy-Fairness trade-off of IFTL (blue bar) against baselines. The higher the bar, the better. (Color figure online)

accuracy sacrificed to enhance fairness relative to the fairness-unaware model. In the NHANES dataset, our IFTL approach outperforms baselines by sacrificing less accuracy to achieve significant improvement in fairness, followed by the IF, EDF, CO, ROS, and GF methods. Similarly, in the COVID-19 dataset, IFTL demonstrates superior trade-off compared to baselines, with CO, ROS, IF, EDF, and GF following in the ranking.

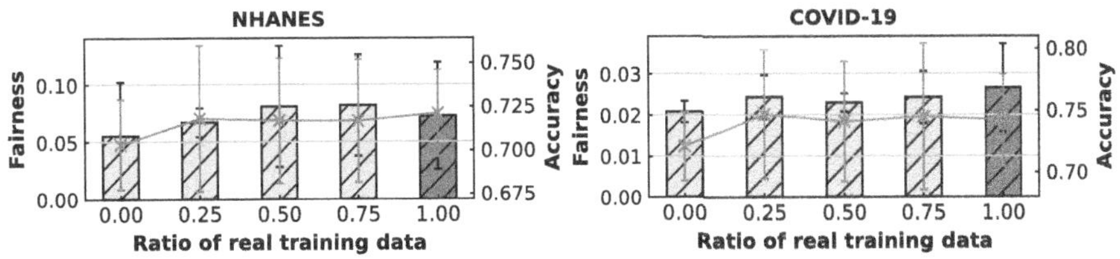

Fig. 3. Impact of proportions of real data on accuracy and fairness. The lower the bar, the better the fairness. The higher the line, the better the accuracy.

Impact of Proportions of Real Data on Accuracy and Fairness. We implement transfer learning with a 25% increment of the real data and analyze its impact on both classification accuracy and fairness. This analysis, as summarized in Fig. 3, aims to identify the proportion of real data at which we achieve the best classification accuracy and fairness.

In the NHANES dataset, we observe that when we initially train the model on only balanced synthetic data, fairness improves, but accuracy is low. However, as we fine-tune the model with incremental proportions of real data, accuracy progressively improves. The best fairness and accuracy are attained when we use 100% real data to fine-tune the model. Similarly, in the COVID-19 dataset, we observe identical trends. Fairness depreciates as we increase the amount of real data, while the opposite trend is observed for accuracy.

Impact of Balanced and Unbalanced Data on Intersectional Group Fairness. This analysis aims to determine whether augmenting datasets without considering equal representation of intersectional groups affects fairness. We utilize both unbalanced and balanced synthetic data to pre-train $f_S(\cdot)$, followed by adaptation of $f_S(\cdot)$ on 100% real data through transfer learning to obtain $f_T(\cdot)$ model. The decision to use 100% real data is based on our initial analysis, which demonstrated the best results. Our results in Fig. 4 confirm that maintaining equal representation among intersectional groups leads to fairer models.

In the NHANES dataset, we observe that the fairness of the $f_S(\cdot)$ model pre-trained on balanced synthetic data surpasses that of the model pre-trained using unbalanced synthetic data. Subsequently, when we adapt the two $f_S(\cdot)$ models on 100% real data to $f_T(\cdot)$, the fairness of the model $f_S(\cdot)$ initially pre-trained on unbalanced synthetic data deteriorates compared to the model pre-trained on balanced synthetic data, albeit with some slight degradation in fairness. This behavior may be due to pre-training the $f_S(\cdot)$ model on unbalanced synthetic data and then adapting it on unbalanced real data.

In the COVID-19 dataset, a similar trend is observed, with the fairness of the $f_S(\cdot)$ model pre-trained on balanced synthetic data outperforming that of the model trained on unbalanced synthetic data. Likewise, the model initially pre-trained on balanced synthetic data exhibits better fairness when adapted on 100% real data compared to the one pre-trained on unbalanced synthetic data. Disparities in the intersectional groups' size in the real data could be a contributing factor.

Evaluation of Quality of Synthetic Dataset. In the NHANES dataset, the average JSD for categorical features is 0.0343, indicating a low degree of dissimilarity between synthetic and real data distributions. For continuous features, the WD value is 0.0144, suggesting a close similarity between synthetic and real data distributions.

Similarly, in the COVID dataset, we observe comparable trends, with JSD values of 0.0339 for categorical attributes and a WD value of 0.0408 for con-

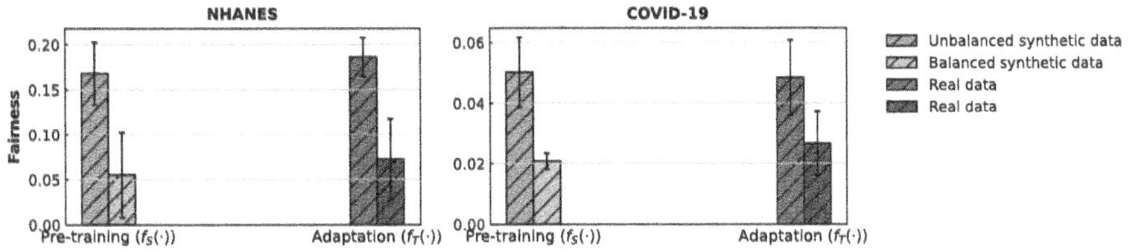

Fig. 4. Impact of balanced and unbalanced data on intersectional group fairness. The lower the bar, the better the fairness.

tinuous attributes. These results demonstrate that our synthetic data closely approximates the statistical properties of the real data.

4 Related Works

Group fairness has been extensively examined across multiple disciplines, such as education [6], criminal justice [4], and healthcare [25]. Strategies to achieve group fairness vary, with some scholars proposing data transformation techniques [6,19, 25], while others advocate for fairness regularizers within learning algorithms [13] or post-processing techniques [28]. However, these group-based fairness methods may not mitigate unfairness among intersectional groups [15,20].

Various approaches [8–12] have been proposed to address intersectional group unfairness. The GerryFair methods [8,11] introduce a zero-sum game to prevent fairness gerrymandering across multiple structured intersectional groups, but their use of group sizes as weights limits the contribution of underrepresented groups. Foulds et al. [10] developed an in-processing technique to reduce acceptance rate gaps among intersectional groups based on disparate impact law [26], using user-defined parameters to amplify the influence of unfairness in underrepresented groups. However, determining these parameters in this complex optimization remains challenging. Different from these methods, we use conditional GAN and transfer learning to solve the intersectional group fairness problem.

Some methods [18,21,22] employ Generative Adversarial Networks (GANs) to address fairness issues. Xu et al. [22] proposed FairGAN to generate fair synthetic datasets, promoting fairness in classifier training. Similarly, Rajabi et al. [21] introduced a two-phase training approach: the first phase optimizes for accuracy, while the second emphasizes generating data with specific fairness properties. While effective at mitigating group unfairness, these methods are limited to groups defined by single protected attributes. Our method differs by targeting both underrepresentation and intersectional group unfairness through the use of conditional GANs and transfer learning.

5 Conclusions

In this paper, we introduced the IFTL method to address intersectional group unfairness in classification models, particularly focusing on the underrepresentation of minority intersectional groups. IFTL employs a two-step strategy combining balanced synthetic data and transfer learning. Firstly, pre-training a classifier on a balanced synthetic dataset, ensuring sufficient representation of each intersectional group with respect to label values. This step aims to establish a fair decision boundary. Then, adapting the pre-trained model on real-world data to refine accuracy while preserving the fair decision boundary learned in the pre-training phase.

Empirical evaluation on two real-world depression datasets demonstrates that IFTL significantly improves intersectional group fairness with minimal sacrifice in classification accuracy.

Future research directions include: Exploring methods to adjust distributions of relevant non-protected attributes alongside protected attributes to further enhance intersectional group fairness. Also, investigating the performance of our method on single protected attribute group fairness.

Acknowledgements. The work has been supported by the Australian Research Council (DP200101210) and Postgraduate Research Scholarship of University of South Australia.

References

1. Sanja, R., Sandro, R., Boris, D.: Investigating oversampling techniques for fair machine learning models. In: ICDSST, pp. 110–123 (2021)
2. Lin, J.: Divergence measures based on the Shannon entropy. IEEE Trans. Inf. Theory **37**(1), 145–151 (1991)
3. Noble, S.U: Algorithms of Oppression: How Search Engines Reinforce Racism. New York University Press, New York (2018)
4. Angwin, J., Larson, J.: Bias in criminal risk scores is mathematically inevitable, researchers say. In: Ethics of Data and Analytics, pp. 265–267 (2016)
5. Kantorovich, L.V.: On the translocation of masses. J. Math. Sci. **133**(4), 1381–1382 (2006)
6. Mehrabi, N., et al.: A survey on bias and fairness in machine learning. ACM Comput. Surv. (CSUR) **54**, 1–35 (2021)
7. Kang, J., Xie, T., Wu, X., Maciejewski, R., Tong, H.: Infofair: information-theoretic intersectional fairness. In: BigData, pp. 1455–1464. IEEE (2022)
8. Kearns, M., Neel, S., Roth, A., Wu, Z.: Preventing fairness gerrymandering: auditing and learning for subgroup fairness. In: ICML, pp. 2564–2572 (2018)
9. Foulds, J., Islam, R., Keya, K., Pan, S.: Bayesian modeling of intersectional fairness: the variance of bias?. In: SDM, pp. 424–432 (2020)
10. Foulds, J., Islam, R., Keya, K., Pan, S.: An intersectional definition of fairness. In: IEEE 36th ICDE, pp. 1918–1921 (2020)
11. Kearns, M., Neel, S., Roth, A., Wu, Z.: An empirical study of rich subgroup fairness for machine learning. In: ACM FAccT, pp. 100–109 (2019)
12. Kim, M., Ghorbani, A., Zou, J.: Multiaccuracy: black-box post-processing for fairness in classification. In: AAAI/ACM AIES, pp. 247–254 (2019)
13. Zhang, B., Lemoine, B., Mitchell, M.: Mitigating unwanted biases with adversarial learning. In: AAAI/ACM AIES, pp. 335–340 (2018)

14. Buolamwini, J., Gebru, T.: Gender shades: intersectional accuracy disparities in commercial gender classification. In: ACM FAccT, pp. 77–91 (2018)
15. Ghosh, A., Genuit, L., Reagan, M.: Characterizing intersectional group fairness with worst-case comparisons. In: PMLR AIDBEI, pp. 22–34 (2021)
16. Morina, G., Oliinyk, V., Waton, J., Marusic, I., Georgatzis, K.: Auditing and achieving intersectional fairness in classification problems. arXiv preprint arXiv:1911.01468 (2019)
17. Ian, G. et al.: Generative adversarial nets. In: Advances in Neural Information Processing Systems, vol. 27 (2014)
18. Rajabi, A., Garibay, O. O.: Distance correlation GAN: fair tabular data generation with generative adversarial networks. In: HCII, pp. 431–445 (2023)
19. Rich, Z., Yu, W., Kevin, S., Toni, P., Cynthia, D.: Learning fair representations. In: ICML, pp. 325–333 (2013)
20. Wang, A., Ramaswamy, V. V., Russakovsky, O.: Towards intersectionality in machine learning: including more identities, handling underrepresentation, and performing evaluation. In: ACM FAccT, pp. 336–349 (2022)
21. Rajabi, A., Garibay, O.: TabfairGAN: fair tabular data generation with generative adversarial networks. MAKE **4**, 488–501 (2022)
22. Xu, D., Yuan, S., Zhang, L., Wu, X.: FairGAN: fairness-aware generative adversarial networks, In: BigData, pp. 570–575. IEEE (2018)
23. Zhao, Z., Kunar, A., Birke, R., Chen, L.: CTAB-GAN: effective table data synthesizing. In: ACML, pp. 97–112 (2021)
24. Kroenke, K., Spitzer, R., Williams, J.: The PHQ-9: validity of a brief depression severity measure. J. Gen. Intern. Med. **16**, 606–613 (2001)
25. Dang, V.N., et al.: Fairness and bias correction in machine learning for depression prediction across four study populations. Sci. Rep. **14**, 7848 (2024)
26. Peck, C.J.: Equal employment opportunity commission: developments in the administrative process 1965–1975. Wash. L. Rev. **51**, 831 HeinOnline (1975)
27. Kamiran, F., Calders, T.: Data preprocessing techniques for classification without discrimination. Knowl. Inf. Syst. **33**(1), 1–33 (2012)
28. Hardt, M., Price, E., Srebro, N.: Equality of opportunity in supervised learning. In: NIPS, pp. 3315–3323 (2016)
29. Centers for Disease Control and Prevention (CDC). National Center for Health Statistics (NCHS). National Health and Nutrition Examination Survey Data. U.S. Department of Health and Human Services, Centers for Disease Control and Prevention
30. Wozniak, A., Willey, J., Benz, J., Hart, N.: COVID Impact Survey. National Opinion Research Center, Chicago, IL (2020)

GPT-4 Attempting to Attack AI-Text Detectors

Alshehri Nojoud(✉) and Lin Yuhao

University of Adelaide, Adelaide, SA 5005, Australia
nojoudmushababh.alshehri@student.adelaide.edu.au,
yuhao.lin01@adelaide.edu.au

Abstract. Recent large language models (LLMs) generate machine content across a wide range of channels, including news, social media, and educational frameworks. The significant challenge of differentiating between AI-generated content and the content written by humans raised the potential misuse of LLMs. Academic integrity risks have become a growing concern due to the potential utilisation of these models in completing assignments and writing essays. Therefore, many detection tools have been developed to identify AI-generated and human-generated texts. The effectiveness of these tools against attack strategies and adversarial perturbations has not been adequately validated, specifically in the context of student essay writing. In this work, we aim to utilize GPT-4 model to apply a series of perturbations to an essay generated originally by GPT-4 in order to confuse three AI detectors: GPTZero, DetectGPT, and ZeroGPT. The proposed attack technique produces a text as an adversarial sample used to examine the effect on the detection accuracy of AI detectors. The results demonstrate that utilizing GPT-4 to rephrase and apply perturbation at the sentence and word level is able to confuse the detection models and reduce their prediction probabilities. Moreover, the final essay, after applying the series of perturbations, maintains a reasonable amount of both writing quality and semantic similarity with the original GPT-generated essay. This project will provide insights for further improvements to increase the robustness of AI detectors and future AI-generated text classification studies.

Keywords: LLM · AI-generated text · AI-text detectors

1 Introduction

Large Language Models (LLMs) have demonstrated exceptional proficiency in text generation tasks. Neural language representation models, like BERT, that have been trained on extensive collections of text are capable of accurately extracting rich semantic patterns from text and are able to further optimize to consistently enhance the performance of a wide range of natural language processing tasks [1]. The Generative Pre-Trained Transformer (GPT) operates on the principle of utilizing a large set of publicly available digital content data to process and generate human-like text [2]. This text exhibits creativity in the process of writing convincingly on a wide range of topics. Since LLMs are trained on texts written by humans, it might be challenging to differentiate between the content produced by these tools and human-written content [3]. Consequently, this

M. Gong et al. (Eds.): AI 2024, LNAI 15442, pp. 154–170, 2025.
https://doi.org/10.1007/978-981-96-0348-0_12

raised the potential misuse of LLMs, such as the generation of disinformation [4]. Yang and Menczer [5] stated that ChatGPT is employed to generate content that promotes suspicious websites and spreads harmful or negative social media comments. Despite its seamless potential benefits, ChatGPT has considerable limitations in the academic field. Qasem [6] conducted qualitative interviews consisting of seven experts in the field of AI and they reported that excessive use of ChatGPT can lead to less control over academic writing ethics, increasing the likelihood of plagiarism outcomes. The growing concerns about the risks to academic integrity led to a significant push to develop software capable of detecting AI-produced content efficiently [7, 8]. Therefore, many detection tools have been developed to identify AI-generated and human-generated texts.

The existing approaches for detecting machine-written content can be classified into watermarking [9], zero-shot detection [10–13], and fine-tuning based detection techniques [14]. Solaiman et al. [10] purported that simple classifiers can exhibit correct results with an accuracy of 97%. The authors' approach was initiated as follows: First, a simple logistic regression model based on GLTR [15]. Then, a pre-trained deep classifier model based on RoBERTa [16] for single text and QA detection was initiated. A watermark is an alteration to the produced text that is undetectable to human readers but detectable post-hoc by an algorithm. Recently, Kirchenbauer et al. [9] introduced a straightforward approach for watermarking by making small adjustments to the probability distribution during text generation. In DetectGPT by Mitchell et al. [13], the log probabilities were used to distinguish between the human-generated text and the GPT-2 generated text. The authors demonstrated that the text generated by the AI occupies the area of negative curvature regions of the model's log probability function. Building on the foundation laid by DetectGPT, Bao et al. [17] introduce Fast-DetectGPT, an optimized zero-shot detector that utilizes conditional probability curvature to distinguish between machine-generated and human-authored text more efficiently. Fast-DetectGPT achieves superior performance to DetectGPT while significantly reducing computational costs. In a separate study, Mitrović, Andreoletti, and Ayoub [14] employed a framework named DistilBERT, which is a simpler model trained on BERT and then fine-tuned using the transformer-based model. Venkatraman et al. [18] introduced GPT-who, a statistical detector inspired by the Uniform Information Density (UID) principle, which aims to differentiate between LLM-generated and human-generated texts.

In addition to the available online tools to detect the AI-generated text, such as GPTZero[1], ZeroGPT[2], Copyleaks[3], Originality.ai[4], Writer AI Content Detector[5], neither of these studies can be considered a benchmark for detecting the text generated by GPT-4. Weichert and Dimobi [19] demonstrated the potential for bypassing detectors by paraphrasing AI-generated texts. Therefore, in this work, we aim to utilize GPT-4 model to apply a series of perturbations to a text generated originally by GPT-4 in order to confuse three AI detectors: GPTZero, DetectGPT, and ZeroGPT. The proposed attack technique produces a text as an adversarial sample used to examine the effect on the

[1] https://gptzero.me/.
[2] https://zerogpt.com/.
[3] https://copyleaks.com/ai-content-detector.
[4] https://originality.ai/.
[5] https://writer.com/ai-content-detector/.

detection accuracy of AI detectors. As an investigation, we formulate the following research question: Is our attack sample able to confuse the selected AI detectors and reduce its accuracy? This project will provide insights for further improvements to increase the robustness of AI detectors and future AI-generated text classification studies.

2 Related Work

Since the initiation of approaches and tools to detect the text generated by LLMs, many studies seek to evaluate its prediction accuracy [7, 20, 21] and explore methods to bypass the detector [22]. However, most AI detectors are not fully accurate to detect AI-generated content convincingly in different contexts. Liang, Guerrero, and Alsmadi [23] proposed the use of mutation-based attack samples; one of them involved replacing a specified word with an empty string. These samples have the potential to confuse machine learning classifiers and reduce their prediction accuracy. They evaluated three detection models; one of them is RoBERTa, which was originally released by OpenAI and was extremely vulnerable to simple adversarial attacks. Furthermore, Lu et al. [24] developed an innovative in-context learning methodology intended to assist LLMs in producing text that successfully evaded six detectors, resulting in an average AUC reduction of 0.5. The authors in Cai and Cui [22] demonstrate that adding one space before a random comma in the given text is an attack strategy to evade detection. The robustness of detection algorithms like DetectGPT for paraphrased AI-generated text is called into question by Krishna et al. [25]. Additionally, ChatGPT can be instructed in an advanced way to answer or rephrase the text in a human way as an attempt to evade detection [26]. The authors employed a combination of features for the classification of human and AI-generated texts, such as perplexity features, semantic features, and error-based features. In the domain of essay generation, employing a range of text perturbation methods, such as replacements on sentence and word level, avoids the detection without compromising the quality of the generated essays [27]. In addition, Weichert and Dimobi [19] claimed that they effectively bypassed the detectors by focusing on prompt engineering to induce ChatGPT to paraphrase AI-generated essays to avoid detection. In this work, we will take advantage of the results that performed specifically well in this related work. However, GPTZero announced in August 2023 an improvement in their detection model by implementing a multilayered approach and a fine-tuned deep learning model, thereby increasing the prediction accuracy [28]. This continuous enhancement of detection tools has led to the need for a re-assessment of the performance and reliability of GPTZero in light of previous limitations.

3 Methodology

There are three main phases to our proposed approach, as represented in Fig. 1. First, we utilize GPT-4 to generate the GPT essay corresponding to the student essays. Then, we run a series of perturbations on the GPT essay. Lastly, we enter the perturbated essay to three AI detectors: GPTZero, DetectGPT, and ZeroGPT, in order to assess their robustness. The next methodology sections explain each phase in detail. Code and data of this work are released for public use[6].

[6] https://github.com/NojoudAlshehri/LLMs-Attack-GPTZero.

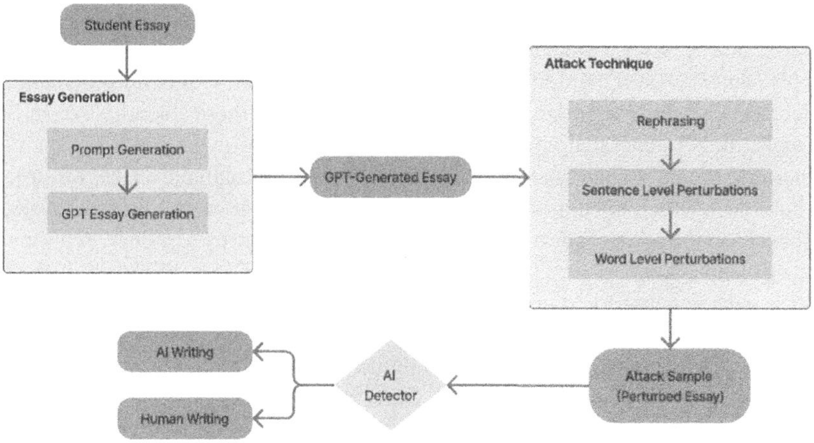

Fig. 1. An overview of the proposed approach

Notably, we decide to use GPT-4 as a main generator model for most of the tasks. GPT-4 is widely used for text generation. Due to its intensive training on large data sets, it is currently the most superior performing system.

3.1 Dataset Construction

The dataset used in this project is the British Academic Written English (BAWE) corpus [29], which consists of essays written by university students in a variety of subjects. We choose to focus on the academic context for two reasons: First, the concerns about the misuse of LLMs in text generation by students. Second, some AI detectors tend to serve educators in terms of providing an accurate AI detection tool, as their model is fine-tuned for student writing and academic settings. For sample of 20 student essays, we instruct GPT to generate a prompt with: `Generate a prompt corresponding to the student essay here: {student_essay}`

Then, we enter the generated prompt to GPT to generate a corresponding essay. We will call the output text: GPT-generated essay. The reason behind this step is that we need to assume that student essay and GPT essay is written as responses to the same question or prompt. As a result, we have two sets: 20 student essays and 20 GPT-generated essays.

3.2 Attack Technique

This part will introduce the method that we follow to produce an attack sample that will be used to confuse the AI detector. Mainly, our attack approach depends on applying a multilayer of manipulations to each GPT-generated essay. Here is an explanation of each step:

1. Instruct GPT to rewrite or rephrase the GPT-generated essay, following the style and structure of the provided student-written essay. The essay resulted from this step called: GPT-rephrased essay. Prompt: `{student_essay} Follow the`

`above text to rewrite and improve the following text below, generate your response to evade detection. {gpt_essay}`

2. Apply sentence-level manipulations to GPT-rephrased text. In this step, we randomly select 45% of the sentences and replace them with the mask token. Next, in order to make the AI-generated writings more diverse, we employ FLAN-T5 as a smaller generative model to generate replacements for the masked segments. Here we simulate a situation in which students work with generative models to manually rewrite specific sentences of the text produced by AI. An observation made by Mindner, Schlippe and Schaaff [26] is that AI-generated text often repeats keywords prompted in the title, which indicates that LLMs tend to overuse topical words in their generated output. This overutilization creates the potential for avoiding detection by replacing these keywords with synonyms, as we will introduce in the next step.
3. Apply word-level manipulations to the essay resulting from step 2. We adopted the algorithm introduced by Peng et al. [27]. It relies on finding high-frequency words in the essay and replacing it with mask. Next, leveraging WordNet, a lexical database of English words, to construct a synonym set for each masked word. We fixed the number of words to be replaced to 152. In Peng et al. [27], they sampled the high frequency words from both the essay and the prompt asked to generate the essay, as it includes the key word that LLMs tend to repeat in the text. However, we only extract the high frequency words from the essay text itself. Additionally, Peng et al. [27] utilized BERT-base model to predict the most suitable synonym for each word. If WordNet does not provide any candidates for the masked word, they select the word with the highest prediction score based on BERT. In this work, we randomly select one synonym for each masked word.

As a result of this proposed attack technique, an adversarial sample will be formed, which is the essays after all the perturbations have been applied. Finally, three AI detectors will be fed by this sample to explore to what extent we are able to reduce its accuracy.

3.3 Detectors

As mentioned in the introduction, we target attacking three AI detectors: GPTZero, DetectGPT, and ZeroGPT. This part introduces an overview of each detector.

GPTZero GPTZero is an advanced, closed-source AI detection tool developed at Princeton University following the release of ChatGPT. They assert a classification accuracy of 99% for human-written articles and 85% for AI-generated content [30]. GPTZero utilizes the measures of burstiness and perplexity to determine whether a text is human-generated or AI-generated. Since perplexity refers to a text's randomness, the higher the perplexity in a text, the more likely it is to be written by a human, and vice versa [21]. The term burstiness is related to the complexity of sentences in a text and the variation in sentence usage [31]. The higher scores for burstiness and perplexity indicate that the text is human-generated, whereas lower scores for these measures suggest that the text is AI-generated [30]. Some prior works indicated that GPTZero sometimes misclassifies portions of the text as either human-generated or AI-generated and, therefore, cannot be considered 100% accurate [32]. GPTZero is trained to detect

AI models such as ChatGPT, GPT4, Bard, and LLaMa. In this project, we utilize the GPTZero version tailored for GPT-4.

ZeroGPT ZeroGPT is an online AI text detector that returns an "AI GPT" percentage, and a classification f text as "Human written" or "AI/GPT Generated". They claimed that the accuracy rate of text detection is up to 98% [33]. ZeroGPT detection model consists of many components that analyse text in order to identify its source. It employs a multi-stage process that is specifically designed to maximise accuracy while minimising both false positives and false negatives [33]. The previous studies revealed varying capabilities and inconsistent assessments of ZeroGPT. Yeadon et al. [34] assessed the quality of AI and human authorship in academic writing by analyzing physics essays. Blinded evaluations of essays authored by humans and those generated by OpenAI's GPT-4 revealed no statistically significant differences in scores. ZeroGPT emerged as the most accurate software tool for identifying essay authorship, achieving a 98% accuracy rate and a precision score of 1.0. Meanwhile, Weichert and Dimobi [19] addressed concerns about distinguishing between human and AI text, particularly in academic settings. The study evaluated the effectiveness of AI text detectors, including ZeroGPT, which exhibited high false positive and false negative rates, highlighting the challenges in accurately detecting AI-generated content.

DetectGPT DetectGPT is an open-source, zero-shot AI detection tool introduced by Mitchell et al. [13] that leverages the negative curvature regions of an LLM's log probability function to identify text generated by the model. It produces random perturbations of the text and subsequently evaluates the log probability of these modifications using an internal language model, comparing them to the original text. Ultimately, the text is deemed to be AI-generated if the log probability of the original input text is notably greater than the log probability of any alterations made to it.

3.4 Semantic Similarity

Throughout the processing of the proposed approach, there will be five forms of essays: student essays, GPT-generated essays, GPT-rephrased essays, manipulated essays on sentence-level, and manipulated essays on word-level (final essays). In the dataset construction, we use the student essay to generate the corresponding question or prompt for that essay, which is then used to generate the GPT essay. As a result, student essays and GPT essays are now assumed to be generated as answers to the same prompt. Additionally, we reuse the student essay as a reference to follow during the rephrasing phase. Therefore, in order to assess the influence of human writing on the GPT writing style and prove that it is possible to maintain control over elements like phrasing, structure, and length in the GPT output, we incorporate computing the SimHash similarity [35] between the student essay and the GPT-rephrased essay. Then, we will be able to compare the similarities between the student essay and the original GPT-generated essay. As a high similarity value between two texts should indicate that the texts are semantically similar, it is expected to find a similarity value between the student essay and the GPT-rephrased essay relatively higher than the student essay and the original GPT-generated essay. In addition, we compute the SimHash similarity between the GPT-rephrased essay, manipulated essay on sentence-level, and manipulated essay on word-level (final essay) in order to measure, after applying the proposed perturbations, the extent to which they maintain roughly the same semantic content as the original AI-generated essay.

3.5 Essay Quality Scorer

According to Peng et al. [27], substituting the frequently used key words with synonyms increases the possibility of evading detection while preserving the high writing quality of the generated essays. In line with [19, 27], it is necessary to assess the text quality of the generated essays after the perturbations. Mainly, we utilize GPT-4 as a custom-powered automatic essay evaluator to assign a quality score to each generated essay after applying the perturbations. Quality scores will be on a scale of 1 to 10, where 1 represents poor quality and 10 represents the best possible quality.

3.6 Metrics

The metric used to evaluate our attack approach is prediction accuracy (ACC). We calculate the ACC for each detector as follows:

- GPTZero: "AI Probability" is given by the GPTZero API, which returns the probability of being written by an AI model; the returned result includes the probability for the entire essay $0 \geq$ AI Probability ≤ 1. Following the documentation of GPTZero, we assume a human classification as a text receiving an "AI Probability" of less than 0.5.
- DetectGPT: Negation of Human Probability, i.e., 1 – Human Probability, as we follow this open-source implementation[7] of DetectGPT.
- ZeroGPT: "Is GPT generated Probability" scores given by the API of ZeroGPT, $0 \geq$ GPT Probability ≤ 1.

4 Results and Discussion

In this section, you will find the results of the evaluation of attack attempt, generated essay quality, and semantic robustness.

4.1 Attack Method Evaluation

The detection accuracy of essays generated during the attack process under various detectors is summarized in Table 1. As shown in Table 1, GPT-generated essays without perturbations can be easily detected by GPTZero and ZeroGPT with 0.97 and 0.94 ACC values, respectively, while they are poorly detected by DetectGPT with an ACC of 0.38. Notably, Peng et al. [27] demonstrated the effectiveness of paraphrasing in attacking five different detectors, whereas the rephrasing step in our work has a slight effect on the detection accuracy of all three detectors that we tested. In addition, applying the sentence-level substitutions to the GPT-rephrased essay reflects a relatively impact on the

[7] https://github.com/BurhanUlTayyab/DetectGPT/tree/main.

ACC, as it slightly decreased from 0.80 to 0.71 in ZeroGPT while surprisingly increasing from 0.97 to 0.99 in GPTZero. Meanwhile, DetectGPT reserves approximately a performance around 0.31 compared to the previous ACC of the GPT-rephrased essay. However, in terms of detection effectiveness, word-level perturbations demonstrate better attack performance, as shown in the table in bold. The main conclusion drawn is that the application of perturbed methods to GPT-generated text effectively reduces the possibility of detection.

Table 1. Evaluation results of Three AI detectors on the four forms of essay

	Detection Accuracy (ACC)		
	GPTZero	ZeroGPT	DetectGPT
	Avg. "AI Prob."	Avg. "AI Prob."	Avg. "AI Prob."
GPT-generated essay	0.97	0.94	0.38
GPT-rephrased essay	0.97	0.80	0.32
Essay after Sentence-level perturbations	0.99	0.71	0.31
Essay after Word-level perturbations	**0.46**	**0.11**	**0.12**

The top-performing detector, GPTZero, has over 95% detection accuracy for GPT-generated, GPT-rephrased, and sentence-level perturbation essays. This conclusion comes in line with the work by Weichert and Dimobi [19], as they acknowledged that they succeeded in lowering the GPTZero "AI probability" of the AI-generated essay, while they failed in making GPTZero misclassify the AI text as human text. Therefore, the proposed attack method in this work indicates success in confusing GPTZero and reducing its detection accuracy from 0.97 to 0.46, which is the ACC after applying all perturbations. Indeed, the overall reduction in detection accuracy among the three detectors, particularly after the third phase of perturbation (word-level perturbations), demonstrates that implementing more layers of perturbations causes more lowering in the probability of detection. This validates the findings of Peng et al. [27], which demonstrate a negative correlation between the depth of perturbation and the chance of detection. As the number of perturbed words increases, the likelihood of detection by the detector decreases.

4.2 Essay Quality Scorer

Table 2 summarizes the average quality evaluation results of each type of essay. Clearly, the original GPT-generated essay and the GPT-rephrased essay share approximately the same average score, confirming that using LLMs such as GPT-4 for rewriting does not affect the quality of the writing. However, after applying sentence-level substitutions, the average quality score of the essay decreased slightly to 7.81. Moreover, the table represents a dramatic drop in the average essay quality score to be 4.81 for the final essay after applying the word-level substitutions, as shown in the table in bold. In fact, this number is lower than the average quality of a student-written essay. It is worth noting that the quality of the perturbed content in Peng et al. [27] additionally experiences a slight decrease.

Table 2. Quality evaluation results of each type of essay in a score range of 1 to 10, where 1 represents poor quality and 10 represents the best possible excellent quality.

	Avg. Quality Score (1–10)
Student essay	6.95
GPT-generated essay	8.43
GPT-rephrased essay	8.38
Essay after Sentence-level perturbations	7.81
Essay after Word-level perturbations	**4.81**

4.3 Semantic Similarity

The summary of the results of the semantic analysis between the five forms of essay used in this work is shown in Table 3. As mentioned in the methodology, a high similarity value between two texts should indicate that the texts are semantically similar. The average SimHash similarity between the student essay and the corresponding GPT-generated essay is 18.381. If we compare this number with 18.667, which is the average SimHash similarity between the student essay and the GPT rephrased essay, it is considered approximately the same. Obviously, because GPT-4 utilized the student essay to generate the prompt that was then used to generate the corresponding GPT essay. Regarding the similarity between the student essay and the GPT rephrased essay, the result was expected since we used the student essay as a reference to rephrase the original GPT-written essay. Surprisingly, the average SimHash similarity of the GPT-generated essay and the GPT-rephrased essay dropped to 7.714, even though they both share approximately the same similarity with the student essay. Then, the semantic similarity between the original GPT-generated essay and the word-substituted essay (final essay) increased, reaching an average SimHash similarity of 9.905. Appendix includes one sample essay with its five versions: student essay, corresponding GPT-generated essay, GPT-rephrased essay, sentence-level perturbated essay, and word-level perturbated essay.

Table 3. Average SimHash similarity

	Avg. Sim
Student essay vs. GPT-generated essay	18.381
Student essay vs. GPT-rephrased essay	18.667
GPT-generated essay vs. GPT-rephrased essay	7.714
GPT-generated essay vs. sentence substituted essay	7.81
GPT-generated essay vs. word substituted essay	**9.905**

5 Conclusion

The experimental results demonstrate that utilizing LLMs such as GPT-4 to rephrase and apply perturbation at the sentence and word level is able to confuse the GPTZero, Detect-GPT, and ZeroGPT detection models and reduce their prediction probabilities. Moreover, the final essay, after applying the series of perturbations, maintains a reasonable amount of both writing quality and semantic similarity with the original GPT-generated essay. The findings illuminate the weaknesses of existing detection approaches, demonstrating that they can be effortlessly bypassed by a straightforward automated adversarial attack. More precise and reliable detection approaches that are specific to the difficulties of AI-generated student essays are urgently required. As a future work, and according to Krishna et al. [25], employing a specialized smaller model such as DIPPER to perform the rephrasing step is effective in terms of paraphrase quality and diversity control.

Disclosure of Interests. The authors have no competing interests to declare that are relevant to the content of this article.

Appendix

Example of the Generated Essay

Student Essay

A mutation is a change in the genome of an organism. Most commonly, mutations are defined as changes in DNA structure that occur due to disruption of the sequence of genes or the emergence of new genes. In this paper, I would like to talk about spontaneous mutations. Spontaneous mutations occur by chance – for example if a sheep is suddenly born in a flock of sheep with shorter legs; such mutations can also occur in humans. The ADRB1 gene, for example, is responsible for regulating the amount of sleep a person needs each night to function properly. Most humans need eight hours of sleep or more, but about 5% of the population enjoys a slightly different mutant version. Tests done on a mother and daughter who had the mutation revealed the ability to sleep only 4 to 6 h each night (Servick, 2019). This mutation refers to a type of genomic mutation. The protein encoded by the mutation is a receptor for norepinephrine.

This mutation is dominant, autosomal, and beneficial in its effect on the viability of individuals. Changes in the ADRB1 gene occur primarily in humans. However, they can also be in animals, although no studies have yet been conducted. Such a mutation does not affect the human population in the general world. This is mainly due to the fact that it is not dangerous and is not life-threatening to its carriers. Thus, such a mutation is only a positive addition to the life abilities of a human carrier. This mutation is sporadic and is not a dominant trend in population genome changes despite its dominance. I find this example interesting because natural changes in the human genome open up new possibilities for the population. Thus, studying such changes is essential for human development and a deeper study of human capabilities.

GPT-Generated Essay

Spontaneous genetic mutations have been a driving force in human evolution and development, shaping our species in profound ways. These mutations, which occur randomly during DNA replication or due to environmental factors, can introduce new genetic variations into a population. While many mutations are neutral or even harmful, some can confer advantages that enhance survival and reproduction, contributing to the evolutionary success of humans. The ADRB1 gene mutation provides a compelling case study to illustrate the complex role and impact of these genetic changes.

Nature and Occurrence of Spontaneous Mutations

Spontaneous mutations are changes in the DNA sequence that occur without external influence. They can result from errors in DNA replication, spontaneous lesions, or the action of mobile genetic elements. The rate of mutation is relatively low, estimated at about 100 to 200 new mutations per human genome per generation. Despite their rarity, the cumulative effect of spontaneous mutations over generations can significantly impact genetic diversity and evolutionary trajectories.

ADRB1 Gene Mutation: A Case Study

The ADRB1 gene encodes the beta-1 adrenergic receptor, which plays a crucial role in the cardiovascular system. A specific mutation in this gene, identified in a small population, has been linked to shorter sleep duration without apparent negative effects on health. This mutation causes a change in the receptor's activity, leading to altered sleep patterns. This case exemplifies how a spontaneous mutation can introduce a new phenotype (shorter sleep duration) into a population.

Potential Benefits of Spontaneous Mutations

Spontaneous mutations can be beneficial by introducing genetic variations that allow populations to adapt to changing environments. For instance, the ADRB1 mutation might confer advantages in environments where shorter sleep is beneficial, perhaps by allowing more time for food gathering, social interactions, or vigilance against predators. Over time, such beneficial mutations can spread through a population, contributing to evolutionary adaptations.

Drawbacks of Spontaneous Mutations

However, not all spontaneous mutations are beneficial. Many mutations are neutral, having no apparent effect on the organism. Others can be deleterious, leading to diseases or reduced fitness. For example, mutations in the BRCA1 and BRCA2 genes significantly increase the risk of breast and ovarian cancer. The balance between beneficial and harmful mutations is a critical factor in the evolutionary process, with natural selection tending to eliminate harmful mutations over time.

Impact on Human Evolution and Development

The role of spontaneous mutations in human evolution is evident in the genetic diversity and adaptability of our species. Mutations like the one in the ADRB1 gene contribute to the rich tapestry of human phenotypes, influencing traits such as sleep patterns, metabolism, and disease susceptibility. Over time, the accumulation of beneficial mutations has enabled humans to colonize a wide range of environments and develop complex societies.

Conclusion

Spontaneous genetic mutations are a double-edged sword, capable of driving evolution and adaptation but also posing risks of deleterious effects. The ADRB1 gene mutation serves as a fascinating example of how a single genetic change can have a significant impact on human traits and potentially on evolutionary fitness. Understanding the role of spontaneous mutations in human evolution and development offers insights into our past and clues about our future as a species.

GPT-Rephrased Essay

The Role of Spontaneous Genetic Mutations in Human Evolution

Spontaneous genetic mutations are pivotal in shaping human evolution, introducing new genetic variations that have profoundly influenced our species' development. These mutations occur randomly, either during DNA replication or due to environmental factors, and their effects can range from neutral to highly impactful.

Understanding Spontaneous Mutations

Spontaneous mutations arise from errors in DNA replication, spontaneous lesions, or the activity of mobile genetic elements, without any external triggers. Although the mutation rate is relatively low, with about 100 to 200 new mutations per human genome per generation, their accumulated effects over multiple generations can significantly alter genetic diversity and influence evolutionary paths.

Case Study: The ADRB1 Gene Mutation

The ADRB1 gene, responsible for encoding the beta-1 adrenergic receptor crucial to the cardiovascular system, provides an insightful case study. A particular mutation in this gene has been observed in a small population, leading to shorter sleep durations without negatively impacting health. This mutation modifies the receptor's activity, showcasing how spontaneous genetic changes can introduce new traits to a population.

Benefits of Spontaneous Mutations

Spontaneous mutations can offer evolutionary advantages by allowing populations to adapt to environmental changes. For example, the ADRB1 mutation may offer benefits in scenarios where reduced sleep could enhance survival, such as extended opportunities for food gathering or increased vigilance. Over time, such advantageous mutations can proliferate through natural selection, fostering evolutionary adaptations.

Challenges Posed by Spontaneous Mutations

However, not all spontaneous mutations are beneficial. Many are neutral, exerting no significant effect on the organism, while others can be detrimental, leading to diseases or decreased fitness. Mutations in genes like BRCA1 and BRCA2, for instance, greatly increase the risk of breast and ovarian cancers. The evolutionary process is influenced by the balance between these beneficial and harmful mutations, with natural selection generally favoring the elimination of detrimental mutations.

Impact on Human Evolution

Spontaneous mutations have been instrumental in the diversification and adaptability of humans. Mutations such as those in the ADRB1 gene contribute to the variety of human phenotypes, affecting traits like sleep patterns, metabolism, and susceptibility to diseases. This genetic variability has enabled humans to thrive in diverse environments and develop complex societies.

Conclusion

Spontaneous genetic mutations are a fundamental aspect of human evolution, acting as agents of change that can both drive adaptation and pose significant health risks. The ADRB1 gene mutation exemplifies how a single genetic alteration can significantly influence human traits and potentially affect evolutionary fitness. By studying these mutations, we gain valuable insights into human evolutionary history and potential future developments.

Sentence-Level Perturbations

Many are neutral, exerting no significant effect on the organism, while others can be detrimental, leading to diseases or decreased fitness. These mutations occur randomly, either during DNA replication or due to environmental factors, and their effects can range from neutral to highly impactful.

Understanding Spontaneous Mutations

Spontaneous mutations arise from errors in DNA replication, spontaneous lesions, or the activity of mobile genetic elements, without any external triggers. Although the mutation rate is relatively low, with about 100 to 200 new mutations per human genome per generation, their accumulated effects over multiple generations can significantly alter genetic diversity and influence evolutionary paths. The ADRB1 gene mutation exemplifies how a single genetic alteration can significantly influence human traits and potentially affect evolutionary fitness. For example, the ADRB1 mutation may offer benefits in scenarios where reduced sleep could enhance survival, such as extended opportunities for food gathering or increased vigilance. This mutation modifies the receptor's activity, showcasing how spontaneous genetic changes can introduce new traits to a population.

Benefits of Spontaneous Mutations

Spontaneous mutations can offer evolutionary advantages by allowing populations to adapt to environmental changes. By studying these mutations, we gain valuable insights into human evolutionary history and potential future developments. Over time, such advantageous mutations can proliferate through natural selection, fostering evolutionary adaptations. A particular mutation in this gene has been observed in a small population, leading to shorter sleep durations without negatively impacting health. ### Case Study: The ADRB1 Gene Mutation the ADRB1 gene, responsible for encoding the beta-1 adrenergic receptor crucial to the cardiovascular system, provides an insightful case study. Mutations in genes like BRCA1 and BRCA2, for instance, greatly increase the risk of breast and ovarian cancers. The evolutionary process is influenced by the balance between these beneficial and harmful mutations, with natural selection generally favoring the elimination of detrimental mutations.

Impact on Human Evolution

Spontaneous mutations have been instrumental in the diversification and adaptability of humans. Mutations such as those in the ADRB1 gene contribute to the variety of human phenotypes, affecting traits like sleep patterns, metabolism, and susceptibility to diseases. This genetic variability has enabled humans to thrive in diverse environments and develop complex societies.

Conclusion

Spontaneous genetic mutations are a fundamental aspect of human evolution, acting as agents of change that can both drive adaptation and pose significant health risks. ### The Role of Spontaneous Genetic Mutations in Human Evolution Spontaneous genetic mutations are pivotal in shaping human evolution, introducing new genetic variations that have profoundly influenced our species' development. ### Challenges Posed by Spontaneous Mutations However, not all spontaneous mutations are beneficial.

Word-Level Perturbations

Many are impersonal, maintain no significant impression on the organism, while others can be detrimental, leadership to disease or decrease seaworthiness. These variation come_about every_which_way, either during DNA reverberation or due to environmental constituent, and their effects can mountain_range from neutral to extremely impactful.

empathise unwritten mutant

Spontaneous mutations go_up from mistake in DNA replication, ad-lib lesion, or the lifewish_action of Mobile factortic component, without any external induction. Although the mutation place is comparatively low, with about hundred to 200 new mutations per man genome per contemporaries, their accumulated effects over multiple propagation can significantly interpolate genetic variety and influence phylocistronisary itinerary. The ADRB1 gene mutation case how a unity genetic modification can significantly influence human trait and potential_differencely affect evolutionary fitness. For model, the ADRB1 mutation May fling profit in scenario where melt_off eternal_sleep could raise survival, such as prolonged chance for intellectual_nourishment get_together or increment alertness. This mutation qualify the sensory_receptor's activity, showcasing how spontaneous genetic switch can preface new traits to a population.

profit of Spontaneous sports

Spontaneous mutations can offer evolutionary reward by grant population to accommodate to environmental changes. By analyze these mutations, we derive worthful insight into human evolutionary account and potential next ontogeny. Over clip, such advantageous mutations can proliferate through natural option, foster evolutionary version. A especial mutation in this gene has been mention in a modest population, leading to unretentive sleep duration without negatively touch_on wellness. ### face work: The ADRB1 cistron Mutation The ADRB1 gene, creditworthy for encoding the beta-1 adrenergic receptor crucial to the cardiovascular system, cater an insightful font consider. Mutations in genes like BRCA1 and BRCA2, for instance, greatly increase the jeopardy of knocker and ovarian Cancer. The evolutionary process is tempt by the equipoise between these beneficial and harmful mutations, with natural selection generally favoring the elimination of detrimental mutations.

Impact on human development

Spontaneous mutations have been instrumental in the diversification and adaptability of mankind. Mutations such as those in the ADRB1 gene contribute to the variety of human phenotypes, affecting traits like sleep patterns, metabolism, and susceptibility to diseases. This genetic variability has enabled humans to thrive in diverse environments and develop complex societies.

Conclusion

Spontaneous genetic mutations are a fundamental aspect of human evolution, acting as agents of change that can both drive adaptation and pose significant health risks. ### The Role of Spontaneous Genetic Mutations in Human Evolution Spontaneous genetic mutations are pivotal in shaping human evolution, introducing new genetic variations that have profoundly influenced our species' development. ### Challenges Posed by Spontaneous Mutations However, not all spontaneous mutations are beneficial.

References

1. Yen-Chun, C., Gan, Z., Cheng, Y., Liu, J., Liu, J.: Distilling Knowledge Learned in BERT for Text Generation (2020). https://doi.org/10.48550/arxiv.1911.03829
2. Grassini, S.: Shaping the future of education: exploring the potential and consequences of AI and ChatGPT in educational settings. Educ. Sci. **13**(7), 692– (2023). https://doi.org/10.3390/educsci13070692
3. Weber-Wulff, D., Anohina-Naumeca, A., Bjelobaba, S., Foltýnek, T., Guerrero-Dib, J., Popoola, O.: Testing of detection tools for AI-generated text. Int. J. Educ. Integr. **19**(1), 26–39 (2023). https://doi.org/10.48550/arXiv.2306.15666
4. Stiff, H., Johansson, F.: Detecting computer-generated disinformation. Int. J. Data Sci. Anal. **13**(4), 363–383 (2022). https://doi.org/10.1007/s41060-021-00299-5
5. Yang, K.-C., Menczer, F.: Anatomy of an AI-powered malicious social botnet (2023). https://doi.org/10.48550/arxiv.2307.16336
6. Qasem, F.: ChatGPT in scientific and academic research: future fears and reassurances. Libr. Hi Tech News. **40**(3), 30–32 (2023). https://doi.org/10.1108/lhtn-03-2023-0043
7. Orenstrakh, M.S., Karnalim, O., Suarez, C.A., Liut, M.: Detecting LLM-generated text in computing education: a comparative study for ChatGPT cases. arXiv (Cornell University) (2023). https://doi.org/10.48550/arxiv.2307.07411
8. Verma, V., Fleisig, E., Tomlin, N., Klein, D.: Ghostbuster: detecting text ghost written by large language models. arXiv (Cornell University) (2023). https://doi.org/10.48550/arxiv.2305.15047
9. Kirchenbauer, J., Geiping, J., Wen, Y., Katz, J., Miers, I., Goldstein, T.: A watermark for large language models. arXiv (Cornell University) (2023). https://doi.org/10.48550/arxiv.2301.10226
10. Solaiman, I., et al.: Release strategies and the social impacts of language models. arXiv (Cornell University) (2019). https://doi.org/10.48550/arxiv.1908.09203
11. Zellers, R., et al.: Defending against neural fake news. arXiv (Cornell University) (2019). https://doi.org/10.48550/arxiv.1905.12616
12. Kushnareva, L., et al.: Artificial text detection via examining the topology of attention maps. In: Proceedings of the 2021 Conference on Empirical Methods in Natural Language Processing (2021). https://doi.org/10.18653/v1/2021.emnlp-main.50

13. Mitchell, E., Lee, Y., Khazatsky, A., Manning, C.D., Finn, C.: DetectGPT: zero-shot machine-generated text detection using probability curvature. arXiv (Cornell University) (2023). https://doi.org/10.48550/arxiv.2301.11305
14. Mitrović, S., Andreoletti, D., Ayoub, O.: ChatGPT or human? Detect and explain. Explaining decisions of machine learning model for detecting short ChatGPT-generated text. arXiv (Cornell University) (2023). https://doi.org/10.48550/arxiv.2301.13852
15. Gehrmann, S., Strobelt, H., Rush, A.M.: GLTR: statistical detection and visualization of generated text. arXiv (Cornell University) (2019). https://doi.org/10.48550/arxiv.1906.04043
16. Liu, Y., et al.: RoBERTa: a robustly optimized BERT pretraining approach. arXiv (Cornell University) (2019). https://doi.org/10.48550/arxiv.1907.11692
17. Bao, G., Zhao, Y., Teng, Z., Yang, L., Zhang, Y.: Fast-DetectGPT: efficient Zero-Shot detection of machine-Generated text via conditional probability curvature. arXiv (Cornell University) (2023). https://doi.org/10.48550/arxiv.2310.05130
18. Venkatraman, S., Uchendu, A., Lee, D.: GPT-WHO: an information density-based machine-generated text detector. arXiv (Cornell University) (2023). https://doi.org/10.48550/arxiv.2310.06202
19. Weichert, J., Dimobi, C.: DUPE: detection undermining via prompt engineering for deepfake text. arXiv (Cornell University) (2024). https://doi.org/10.48550/arxiv.2404.11408
20. Elkhatat, A.M., Elsaid, K., Almeer, S.: Evaluating the efficacy of AI content detection tools in differentiating between human and AI-generated text. Int. J. Educ. Integr. **19**(1), 1–16 (2023). https://doi.org/10.1007/s40979-023-00140-5
21. Chaka, C.: Detecting AI content in responses generated by ChatGPT, YouChat, and Chatsonic: the case of five AI content detection tools. J. Appl. Learn. Teach. **6**, 1–11 (2023). https://doi.org/10.37074/jalt.2023.6.2.12
22. Cai, S., Cui, W.: Evade ChatGPT detectors via a single space. arXiv (Cornell University) (2023). https://doi.org/10.48550/arxiv.2307.02599
23. Liang, G., Guerrero, J., Alsmadi, I.: Mutation-based adversarial attacks on neural text detectors. arXiv (Cornell University) (2023). https://doi.org/10.48550/arxiv.2302.05794
24. Lu, N., Liu, S., He, R., Wang, Q., Tang, K.: Large language models can be guided to evade AI-generated text detection. arXiv (Cornell University) (2023). https://doi.org/10.48550/arxiv.2305.10847
25. Krishna, K., Song, Y., Karpinska, M., Wieting, J., Iyyer, M.: Paraphrasing evades detectors of AI-generated text, but retrieval is an effective defense. arXiv (Cornell University) (2023). https://doi.org/10.48550/arxiv.2303.13408
26. Mindner, L., Schlippe, T., Schaaff, K.: Classification of human- and AI-generated texts: investigating features for ChatGPT. In: Lecture Notes on Data Engineering and Communications Technologies, pp. 152–170 (2023)
27. Peng, X., Zhou, Y., He, B., Sun, L., Sun, Y.: Hidding the Ghostwriters: an adversarial evaluation of AI-generated student essay detection. arXiv (Cornell University) (2024). https://doi.org/10.48550/arxiv.2402.00412
28. GPTZero Improves with Diverse Data, Surpasses Competitor AI Detectors. https://gptzero.me/news/gptzero-surpasses-competitors-in-accuracies. Accessed 20 Sept 2023
29. Nesi, H., et al.: (BAWE) British academic written English Corpus (2023)
30. GPTZero Homepage. https://gptzero.me/. Accessed 02 June 2024
31. Chaka, C.: Generative AI Chatbots - ChatGPT versus YouChat versus Chatsonic: use cases of selected areas of applied English language studies. Int. J. Learn. Teach. Educ. Res./Int. J. Learn. Teach. Educ. Res. **22**, 1–19 (2023). https://doi.org/10.26803/ijlter.22.6.1
32. Chaka, C.: Stylised-facts view of fourth industrial revolution technologies impacting digital learning and workplace environments: ChatGPT and critical reflections. Front. Educ. **8** (2023). https://doi.org/10.3389/feduc.2023.1150499

33. ZeroGPT Homepage. https://www.zerogpt.com. Accessed 06 May 2024
34. Yeadon, W., Agra, E., Inyang, O.-O., Mackay, P., Mizouri, A.: Evaluating AI and human authorship quality in academic writing through physics essays. arXiv (Cornell University) (2024). https://doi.org/10.48550/arxiv.2403.05458
35. Charikar, M.S.: Similarity estimation techniques from rounding algorithms. In: Annual ACM Symposium on Theory of Computing (2002)

Charting a Fair Path: FaGGM Fairness-Aware Generative Graphical Models

Vivian Wei Jiang(✉), Gustavo Batista, and Michael Bain

University of New South Wales, Sydney, Australia
{vivian.jiang1,g.batista,m.bain}@unsw.edu.au

Abstract. Machine learning (ML) models behind the decisions that increasingly affect our lives may rely on historical data fraught with biases, which may lead to discrimination. These biases can manifest in two significant ways: data may reflect historical prejudices, leading to models that perpetuate these issues, or there may be insufficient data for certain groups, hindering the models' ability to identify fair and accurate patterns. This paper introduces FaGGM, a fairness-aware generative graphical model designed to tackle these challenges. FaGGM incorporates a fairness regularization term into its graph structure learning algorithm, making the models based on these structures fairer. Additionally, it acts as a fair data generator and improves data representation for underrepresented groups. FaGGM is flexible and is compatible with most fairness definitions and score-based structure learning algorithms. Our experiments demonstrate that FaGGM mitigates bias and generates high-quality synthetic data, setting it apart from existing bias mitigation methods. ML models trained on this data show considerably increased fairness scores and smaller reductions in accuracy relative to comparable approaches.

Keywords: ML Fairness · Bias Mitigation · Graphical Models · Generative Models

1 Introduction

Machine Learning (ML) and Artificial Intelligence (AI) have become integral parts of everyday business applications across diverse industries [7], with their adoption driven by the ability to process vast data volumes and execute complex computations efficiently. However, despite the efficiency and potential for objective decision-making, there is a growing awareness of inherent biases in AI systems. These biases, often unintentional, can stem from the data used in AI algorithms [20], leading to ethical concerns and potential harm in real-life applications.

Numerous strategies exist for bias mitigation (for a recent review see [15]). Pre-processing strategies adjust biased data before training ML models through

Supplementary Information The online version contains supplementary material available at https://doi.org/10.1007/978-981-96-0348-0_13.

M. Gong et al. (Eds.): AI 2024, LNAI 15442, pp. 171–185, 2025.
https://doi.org/10.1007/978-981-96-0348-0_13

techniques like relabeling, perturbation, sampling, and representation learning. These techniques modify existing data labels, features, and distributions to align with fairness criteria. This paper focuses on a relatively under-explored approach: generating new fair synthetic data from fair graphical models.

Goal. From a biased dataset D, we are interested in learning a graph structure G that minimises biased pathways[1] while preserving variable relationships. This enhanced graph G can support reasoning and modelling tasks with greater fairness considerations. Furthermore, it can serve as a mechanism for data debiasing and generating a less biased synthetic dataset D' that can be used to train ML models. Consequently, the resulting ML models are fairer without significant compromise in model performance.

Solution. We approach this from a graph structure learning perspective, introducing a method we refer to as Fairness-aware Generative Graphical Models (FaGGM) (Fig. 1). Our method incorporates a fairness regularization term into the objective function to prevent unfair relationships. This integration is non-trivial because score-based structure learning algorithms perform incremental local searches, whereas fairness considerations typically require a global perspective. We calculate a fairness score for each potential relationship and integrate it into the structural scores, encouraging fairer pathways. Furthermore, we introduce a fairness regularization parameter to fine-tune and achieve a better balance between fairness and accuracy in the graph. The code for this project is available at https://github.com/VivianJiang-GingerRose/FaGGM.

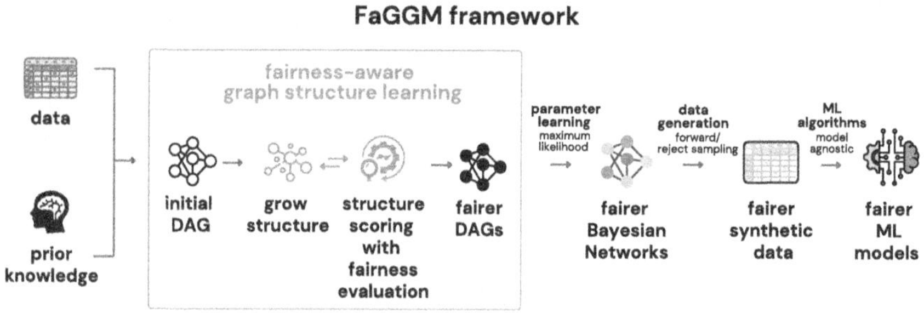

Fig. 1. Solution Overview.

Contributions. We introduce a framework focused on acquiring fair graph representations to facilitate the generation of fair synthetic data. We make four main contributions: (1) We introduce a fairness-aware, score-based graph structural learning algorithm. Our modification to a standard Hill-Climbing search algorithm incorporates a *fairness regularization* term to guide the algorithm towards discovering more equitable graphs. (2) We demonstrate that our fairness-enhanced algorithm can learn Directed Acyclic Graphs (DAGs) with less biased

[1] By "pathway" we mean a finite directed path in a directed acyclic graph [21].

pathways. (3) We show that synthetic data generated from Bayesian Networks based on fairer DAGs is itself fairer. This fairness is preserved across samples with different numbers of instances and class distributions, allowing the generation of data samples that better fit the needs of the downstream ML algorithm. (4) We highlight that ML models trained on synthetic data generated by FaGGM achieve increased fairness without significant classification performance loss.

The well-known fairness-accuracy trade-off [2,23] results from the fact that adjusting the values of fairness-related attributes in a dataset can reduce predictive accuracy. For example, a recent study found that accuracy decreased by between 42–66% for more than half of the bias mitigation techniques evaluated [8]. We note that in this paper our approach manages to achieve accuracies only 4–8% lower compared to the best (i.e., most accurate) alternative method we tested, with fairness scores that are 80–97% higher compared to the best (i.e., most fair) alternative method.

2 Related Work

In the machine learning literature on fairness, *bias* typically refers to disparate outcome predictions for individuals or groups identifiable by specific *protected attributes*, such as gender, race, socioeconomic status, or age [20]. Considerable research has focused on identifying and reducing such biases. In this paper we concentrate on studies related to pre-processing and graph-based approaches to mitigate bias, as these areas are most relevant to our research objectives.

Mitigating Bias Through Data Pre-processing. Bias mitigation methods in ML can be divided into pre-processing, in-processing, and post-processing, depending on when they are applied in the ML workflow [15]. Pre-processing methods adjust the training data to reduce bias before training. Notable pre-processing approaches include *Fairness through Unawareness* (FTU) [18], which eliminates protected attributes from the dataset; *Reweighing* (RW) [5], which adjusts the weights for different groups and labels; *Learning Fair Representations* (LFR) [28], which conceals protected attribute information to achieve fair representation; and *Disparate Impact Remover* (DIR) [12], which alters feature values to enhance fairness, maintaining the rank order within groups. Pre-processing methods like DIR and LFR alter data to improve fairness, potentially complicating the interpretability of subsequent machine learning models. Additionally, none of these methods can generate new, fairer data points.

Graph-Based Bias Mitigation. One area of research on bias mitigation is on creating unbiased synthetic data using causal graphical models. This approach constructs graphical models from real data, applies algorithms to minimise bias and then generates new, fairer synthetic data. In [29] causal graphs are used to discover and remove path-specific discrimination, adjusting Conditional Probability Tables (CPTs) by solving a quadratic programming problem. Building on this work, [30] simplifies the CPT modification process to ensure the sub-graph adheres to a non-discrimination criterion. Additionally, Generative Adversarial Network (GAN)-based methods, like FairGAN [26], CFGAN [25], DECAF

[4] and GOGGLE [19], have been developed towards ensuring that synthetic data both well-approximates the underlying distribution and is free from bias. CFGAN and DECAF use causal graphs to further target discrimination.

DAGs are essential in these models for mapping causal relationships and identifying which connections to remove to reduce bias. The success of bias correction depends on the accuracy of the causal graph, yet current research often assumes the graph's accuracy without directly assessing its fairness. For example, in approaches like DECAF the DAG must be provided [4]. Some methods [4,29,30] use DAGs with known biases to show how well they can correct these issues. Furthermore, strategies like those in DECAF, which eliminate biased connections, may compromise the graph's ability to reflect true causal relationships. This raises concerns about the trade-off between achieving fairness and accurately representing real-world dynamics.

3 Fairness-Aware Graph Structure Learning

There are two main types of Bayesian network structure learning from data: constraint-based or score-based learning [17].

Constraint-based algorithms like PC and FCI start with a fully connected graph and use conditional independence tests to remove or orient edges. In contrast, score-based algorithms like Hill-Climbing, TABU, SaiyanH, and Greedy Equivalence Search (GES) start from an empty graph or a simple graph and iteratively modify it to increase the objective score [17].

The Hill-Climbing search algorithm (HC) is a well-known score-based method for extracting graph structure from data [3,14]. HC is based on local search. It starts from an initial valid graph (usually a simple tree-structured DAG) and modifies it iteratively. HC uses three basic operations on pairs of nodes: adding an edge, removing it, or reversing its direction. A structure score is calculated for every operation, reflecting how effectively the operation models the dataset. The Bayesian Information Criterion (BIC), Akaike Information Criterion (AIC) and K2 score are commonly used scores [17].

A key characteristic of HC search is the decomposability of its scores. This means that when assessing a new node pair, it is only necessary to calculate the incremental change in score. This change, or delta structure score, informs the selection of the optimal operation for the node pair, or opting for no action when it does not improve the current DAG's score.

The Fairness-aware Hill-Climbing search algorithm developed in this work incorporates a fairness regularization term into its structural score calculations. The primary goal of this adaptation extends beyond simply uncovering structural relationships in the data; it aims to ensure these relationships align with fairness principles. Consequently, the algorithm identifies edges that balance accuracy with fair graph representation. This approach actively contributes to promoting fairness in the graphs it generates. Furthermore, our approach can be extended to other score-based structure search algorithms such as TABU, SaiyanH, Model-Averaging Hill-Climbing (MAHC) and GES [17].

3.1 Initialisation

The Fairness-aware Hill-Climbing search algorithm starts with an initial DAG that can be obtained in several ways: through a tree-based method like Chow-Liu [9], from domain expert input, or a mix of both. The initial graph establishes a baseline for fairness and accuracy, and serves as the starting point in the DAG search space for structure learning operations.

In the context of fairness, a DAG shows how a protected attribute relates to the outcome node. Protected attributes, also referred to as sensitive attributes, are attributes that require equitable treatment. This includes privileged groups, which receive favourable outcomes, and under-privileged groups, which experience less favourable outcomes. The outcome node in a DAG corresponds to the class attribute in the ML model. We assume that a domain expert defines the protected attributes and outcome node based on knowledge of the data.

3.2 Growing the Graph Structure

Next, the goal is to learn the graph structure from the data. Hill-Climbing search iterates over the possible node pairs and graph operations, selecting the combination with the highest improvement in the structure score until no further enhancements are possible. We extend the standard structure score by introducing a new fairness regularization term to incorporate fairness considerations into the structure learning process.

For graph structure G, dataset D of size N, variable X, and parent node set $\mathbf{U}$, the decomposed structure score with a fairness regularization term is represented as:

$$FS(G; D) = F(G; D) + \sum_{X\mathbf{U}} \text{Score}(X, \mathbf{U}; D) \tag{1}$$

where,

$$F(G; D) = -N \cdot \lambda_{\text{fairness}} \cdot F_D(G) \tag{2}$$

and

$$\text{Score}(X, \mathbf{U}; D) = -N \cdot H_D(X|\mathbf{U}) - \psi(N) \cdot \|X\mathbf{U}\| \tag{3}$$

Equation 2 evaluates the fairness of graph G during structure learning and has a similar mathematical form to the standard local score function in Eq. 3. $F_D(G)$ represents the fairness score for graph G in a dataset D. Algorithm 1 outlines the calculation process of $F(G; D)$.

The fairness parameter $\lambda_{\text{fairness}}$ in Eq. 2 controls how strongly fairness is enforced and can be fine-tuned to align with different objectives. For instance, if accurately modelling the dataset's relationships is more important than fairness, a lower $\lambda_{\text{fairness}}$ value lessens the fairness emphasis. Conversely, if addressing bias and enhancing fairness is the priority, a higher $\lambda_{\text{fairness}}$ encourages the algorithm to favour fairer edges.

Equation 3 is the standard local structure score, with the first component a penalty term favouring simpler structures and the second component being the log-likelihood. The conditional entropy $H_D(X|\mathbf{U}) = -P_D(x,u) \cdot log(P_D(x|u))$ measures how much information is needed to describe the target variable X given its parents $\mathbf{U}$, indicating the structure's effectiveness in capturing data dependencies. The term $||X\mathbf{U}||$ represents network complexity by evaluating the dimension of X's parent node set $\mathbf{U}$. When $\psi(N) = 1$, Eq. 3 denotes the AIC. When $\psi(N) = \frac{\log_2 N}{2}$, Eq. 3 is referred to as the BIC. In our experiments, we selected the BIC score for evaluating structure, although the algorithm is also compatible with AIC.

Algorithm 1 details the computation of $F(G;D)$ while Algorithm 2 presents the Fairness-aware Hill-Climbing search algorithm. In Algorithm 1, common fairness statistics, such as Statistical Parity Difference (SPD), Disparate Impact (DI), and Equalized Odds (EO), are all suitable for computing $F_D(G)$ to assess the fairness of the graph structure G. We used the SPD in our experimentation.

Algorithm 1: $F(G;D)$

Input : DAG G, dataset D, protected node P, outcome node Y, fairness measure *FairnessStatistic*, and a machine learning classifier C
Output: Fairness evaluation $F(G;D)$

$BN \leftarrow$ learn_parameters(G, D) `// Learn Bayes Net parameters from data`
$D_{\text{new}} \leftarrow$ simulate(BN) `// Forward or rejection sampling`
$M \leftarrow$ train(C, D_{new}) `// Train classifier with sampled data`
$\mathbf{y}' \leftarrow$ predict(M, D_{new}) `// Predict outcome label with classifier`
$\mathbf{y} \leftarrow D_{\text{new}}[Y]$ `// Ground truth outcome label`
$F_D \leftarrow$ *FairnessStatistic*$(P, \mathbf{y}, \mathbf{y}')$ `// Compute fairness statistic (e.g., SPD, DI, EO; see Section 4.3)`
return $-N \cdot \lambda_{\text{fairness}} \cdot \log F_D$

4 Experiments

This section assesses FaGGM's effectiveness using three well-known datasets that illustrate gender and race biases [12,30]: the Dutch Census dataset [10], the Adult dataset [1] and the Compas dataset [22]. Additionally, we use two datasets with synthetically added bias: Credit Card Clients [27] and Law School Admission [24]. Details of these datasets are provided in Appendix A.1.

Algorithm 2: Fairness-aware Hill-Climbing Search algorithm. Additional pseudo-code that accounts for fairness is highlighted in blue.

Input : dataset D, max in-degree M, initial DAG G_{initial}, fairness score function $F(G; D)$, structure score function $\text{Score}(X, \mathbf{U}; D)$

Output: DAG $G_{\max}$ that maximises structure score S_{G} and maximises fair edges in the search space of possible graphs

$G_{\text{bsf}} \leftarrow G_{\text{initial}}$ `// Best-so-far graph starts with the initial graph`
for *every node X with parents $\mathbf{U}$ in G_{bsf}* **do**
    $S[X] \leftarrow \text{Score}(X, \mathbf{U}; D)$
end
$S_{\text{bsf}} \leftarrow \sum_X S[X]$
$FS_{\text{bsf}} \leftarrow F(G_{\text{bsf}}; D) + S_{\text{bsf}}$
while *FS_{bsf} increases* **do**
    for *every edge addition, removal and reversal that modifies parent set U of a child node X* **do**
        if $G_{\text{new}} \models DAG, M$ **then**
            `// New graph is a DAG satisfying` M
            $S_{\text{new}} \leftarrow S_{\text{bsf}} - S[X] + \text{Score}(X, \mathbf{U}; D)$
            $FS_{\text{new}} \leftarrow F(G_{\text{new}}; D) + S_{\text{new}}$
            if $FS_{\text{new}} > FS_{\text{bsf}}$ **then**
                $G_{\text{bsf}} \leftarrow G_{\text{new}}, FS_{\text{bsf}} \leftarrow FS_{\text{new}}, S[X] = \text{Score}(X, \mathbf{U}; D)$
            end
        end
    end
end
return G_{bsf}

Our evaluation of FaGGM covers three key aspects, stated as hypotheses:

Fairer Graph Structures. FaGGM learns graph structures with fewer discriminatory paths than those derived using the standard Hill-Climb search algorithm.
Fairer Generated Data. Data generated by FaGGM with fairer graph structures exhibit greater fairness than the original datasets.
Fairer ML Models. ML models built with FaGGM-generated data can achieve higher fairness levels than existing pre-processing bias mitigation techniques.

In our experiments, we first derive the graph structure from the datasets using both standard and fairness-aware Hill-Climbing search algorithms to construct Bayesian Networks. Next, we generate synthetic data from these networks using forward and rejection sampling and divide the sampled data into training and validation sets. We develop ML classifiers with the XGBoost package using the training data, incorporating five-fold cross-validation with reshuffling. Finally, we assess and compare the performance and fairness of these models using the validation set.

Additionally, we conducted a comparative analysis of FaGGM, assessing its fairness and performance with four established pre-processing bias mitigation techniques [8]: Fairness Through Unawareness (FTU), Reweighing (RW), Disparate Impact Remover (DIR), and Learn Fair Representations (LFR).

4.1 Assessing Fairness in Graph Structures

In this experiment, we use a subset of features from the Credit Card Clients dataset [27], which contains credit card default information from Taiwan. This dataset features demographic details along with a binary outcome, default. To mimic real-world interest rates assigned to credit card applicants, we also introduce a synthetic variable, *Interest Rate* [11], drawn from Gaussian distributions so that applicants who defaulted are charged higher interest rates on their credit cards. Women pay higher interest rates than men.

Figure 2 demonstrates with this toy example that the standard Hill-Climbing search identifies the unfair edge between gender and income. In contrast, even when analysing the same dataset, the fairness-aware Hill-Climbing search does not learn this discriminatory pathway in the DAG, indicating its capability to avoid unfair linkages.

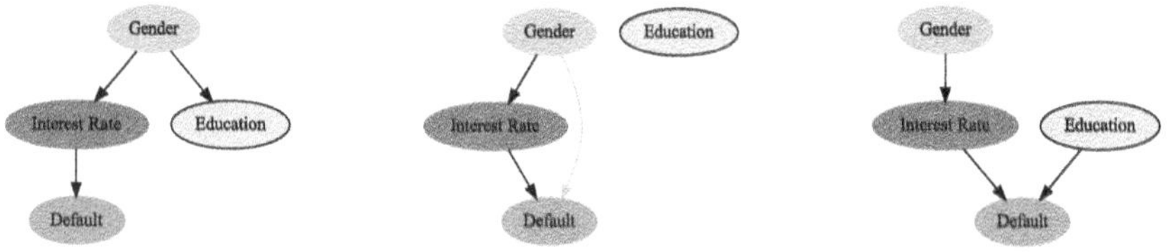

(a) (b) (c)

Fig. 2. DAGs from Credit Card Clients data. *Gender* is the protected attribute, *Default* is the outcome node, and *Interest Rate* is the resolving attribute, which is influenced by the protected attribute in a non-discriminatory way. (a) Initial DAG. (b) DAG from standard Hill-Climbing Search. Edge from Gender to Default is a **direct discriminative** path. (c) The DAG from Fairness-aware Hill-Climbing Search does not contain any discriminative pathways, as pathways that pass through resolving attributes are not considered discriminative. (Color figure online)

4.2 Assessing Fairness of Generated Data

We evaluate the fairness of synthetic data generated by FaGGM using the Mean Difference metric [16], which measures data fairness independently of ML models. MD is defined as $MD = |Pr(y = 1|P = p_1) - Pr(y = 1|P = p_2)|$. This metric compares the probability of an outcome $y = 1$ between two sub-populations p_1 and p_2 within a protected attribute P. We incorporate the absolute value in the formula to prevent negative outcomes, modifying the original definition by [16].

Table 1. Results on **Data Fairness** measured by Mean Difference (MD). Lower values mean greater dataset fairness regarding the protected attribute

Data Type	MD ↓			
	Dutch	Adult	Compas	Law
Original Data	0.299	0.196	0.098	0.057
Synthetic Data without Fairness Regularization	0.290	0.170	0.051	0.027
FaGGM Same Size as Original Data	0.004	0.043	0.020	0.002
FaGGM Double Size of Original Data	0.004	0.026	0.012	0.005
FaGGM Privileged/Unprivileged Equal Size	0.001	0.030	0.011	0.002

Table 1 compares the MD values of the original dataset, a dataset generated by Bayesian Networks without fairness considerations, and three datasets generated by FaGGM: one matching the original dataset in size, one twice the size, and one that balances the sizes of privileged and unprivileged groups within the protected attribute. These latter two datasets highlight FaGGM's capability to generate data as needed, enhance representation for minority groups, and thereby facilitate further analysis or model training.

For all four datasets, the data generated from FaGGM shows reduced bias compared to both the original datasets and those generated by graphical models without fairness adjustments. Importantly, this reduction in bias is maintained as the dataset size increases and distributions change.

4.3 Assessing Fairness of ML Models

In our experiments, we evaluate the changes in data utility and fairness achieved by FaGGM and four existing pre-processing bias-mitigation methods. We measure performance using the F1 Score and AUROC, where higher values signify better predictive capability and data utility.

Fairness is assessed using Statistical Parity Difference [6] $SPD = |Pr(\hat{y} = 1|P = p_1) - Pr(\hat{y} = 1|P = p_2)|$, Disparate Impact [12] $DI = \frac{Pr(\hat{y}=1|P=p_1)}{Pr(\hat{y}=1|P=p_2)}$, and Equal Opportunity [13] $EO = (FPR_{P=p_1} - FPR_{P=p_2}) + (TPR_{P=p_1} - TPR_{P=p_2})$.

The results from Table 2 highlight the **Fairness-Accuracy Trade-off**, a well-documented phenomenon where increasing fairness leads to reduced model accuracy [2,23]. A recent study [8] found that no existing bias mitigation technique can achieve the optimal balance in every scenario.

Our experimental results in Table 2 demonstrate that FaGGM is the most effective method in reducing bias across all four datasets, as measured by the three most recognised fairness metrics. The gain in fairness is accompanied by a small decrease in model accuracy with a 4% to 8% drop in AUROC compared to the best-performing existing bias mitigation method. However, FaGGM achieves over 80% to 97% better fairness metrics (SPD), highlighting its advantages over other existing methodologies.

Furthermore, it is important to highlight that FaGGM uniquely generates new, fairer data points, unlike the four existing pre-processing bias mitigation methods.

Table 2. Bias removal experimentation results

	Data Type	Classification Quality		Fairness		
		F1 Score ↑	AUROC ↑	SPD ↓	DI ↑	EO ↓
Dutch	Original Data	0.829 ± 0.002	0.838 ± 0.002	0.346 ± 0.016	0.472 ± 0.013	0.212 ± 0.016
	FTU	0.801 ± 0.001	0.818 ± 0.001	0.187 ± 0.014	0.663 ± 0.017	0.053 ± 0.008
	RW	**0.811 ± 0.001**	**0.827 ± 0.001**	0.207 ± 0.013	0.668 ± 0.022	0.063 ± 0.009
	DIR	0.803 ± 0.001	0.819 ± 0.001	0.190 ± 0.012	0.648 ± 0.019	0.051 ± 0.007
	LFR	0.343 ± 0.104	0.507 ± 0.047	0.097 ± 0.049	0.000 ± 0.000	0.110 ± 0.045
	FaGGM	**0.724 ± 0.001**	**0.786 ± 0.001**	**0.006 ± 0.000**	**0.979 ± 0.001**	**0.012 ± 0.003**
Adult	Original Data	0.645 ± 0.004	0.756 ± 0.004	0.172 ± 0.014	0.223 ± 0.014	0.098 ± 0.027
	FTU	0.641 ± 0.005	0.753 ± 0.004	0.176 ± 0.010	0.263 ± 0.019	0.126 ± 0.028
	RW	0.623 ± 0.007	0.741 ± 0.004	0.087 ± 0.009	0.752 ± 0.029	0.139 ± 0.023
	DIR	**0.642 ± 0.005**	**0.754 ± 0.004**	0.179 ± 0.008	0.282 ± 0.021	0.128 ± 0.021
	LFR	0.318 ± 0.166	0.567 ± 0.058	0.069 ± 0.068	0.604 ± 0.234	0.086 ± 0.076
	FaGGM	**0.590 ± 0.008**	**0.721 ± 0.005**	**0.031 ± 0.016**	**0.857 ± 0.066**	**0.026 ± 0.017**
Compas	Original Data	0.706 ± 0.006	0.651 ± 0.007	0.184 ± 0.038	0.745 ± 0.041	0.219 ± 0.054
	FTU	**0.712 ± 0.005**	**0.658 ± 0.003**	0.149 ± 0.027	0.757 ± 0.045	0.162 ± 0.051
	RW	0.694 ± 0.009	0.645 ± 0.011	0.053 ± 0.030	0.933 ± 0.042	0.076 ± 0.037
	DIR	0.710 ± 0.009	0.656 ± 0.010	0.153 ± 0.027	0.859 ± 0.049	0.175 ± 0.038
	LFR	0.429 ± 0.258	0.531 ± 0.050	0.055 ± 0.056	0.615 ± 0.342	0.084 ± 0.068
	FaGGM	**0.677 ± 0.017**	**0.630 ± 0.016**	**0.030 ± 0.016**	**0.951 ± 0.026**	**0.058 ± 0.017**
Law	Original Data	0.926 ± 0.012	0.965 ± 0.012	0.059 ± 0.008	0.399 ± 0.056	0.062 ± 0.032
	FTU	0.927 ± 0.010	0.965 ± 0.009	0.055 ± 0.008	0.431 ± 0.062	0.031 ± 0.024
	RW	**0.927 ± 0.008**	0.963 ± 0.009	0.056 ± 0.009	0.443 ± 0.072	0.042 ± 0.031
	DIR	0.932 ± 0.006	**0.971 ± 0.006**	0.056 ± 0.009	0.415 ± 0.071	0.029 ± 0.015
	LFR	0.088 ± 0.081	0.504 ± 0.054	0.137 ± 0.098	0.000 ± 0.000	0.152 ± 0.123
	FaGGM	**0.832 ± 0.009**	**0.897 ± 0.007**	**0.009 ± 0.006**	**0.916 ± 0.052**	**0.023 ± ±0.022**

5 Conclusions and Further Work

This paper introduces FaGGM, a framework designed to generate fair graphical representations and synthetic data. At the heart of FaGGM is an algorithm that discovers graph structures focusing on fairness. It incorporates a fairness-regularization term that promotes fairness to the scoring function, guiding the algorithm to create more equitable graphs. We demonstrate through real datasets that the FaGGM framework effectively reduces bias in the graph, data and downstream ML models. Next, we discuss limitations, applications, and future research directions.

Mixed Data Types. Our current approach is based on Discrete Bayesian Networks, which only handle discrete values, each with a specific probability that

depends on the parent values. Handling continuous or mixed data poses challenges. Future studies could explore integrating fairness regularization terms into Bayesian scoring methods for continuous data, like the Bayesian Gaussian equivalent score (BGe) [17].
Fairness Regularization Parameter. $\lambda_{fairness}$ guides the structure learning algorithm towards achieving a balance between fairness and accuracy, allowing for fine-tuning to reach this balance. Our experiments demonstrate that varying $\lambda_{fairness}$ can adjust fairness levels, but this is not always monotonic and often increases graph complexity. To address this, another parameter could be introduced to the complexity control component of the BIC score in Eq. 3, preventing increased fairness from resulting in more complex graphs.
Incorporating Human Knowledge. The FaGGM framework currently discovers the graph structure based solely on the dataset. Future enhancements could allow for the inclusion of human expertise before the learning process begins. For instance, variables like race and age, which are *immutable*, could be pre-defined so they do not receive any incoming edges. Experts could also identify pathways for the algorithm to avoid. Integrating human expertise with the algorithmic process could produce graphs that better reflect real-world situations and human insights.
Ethics and Social Impact. While FaGGM reduces bias in historical data, it does not eliminate all biases. It targets group fairness, meaning biases on an individual level may persist. Future work could explore integrating individual fairness principles with graphical models. Fairness is a nuanced issue that varies by task and context, requiring thoughtful public discussion. We hope that FaGGM contributes to a greater awareness of machine learning bias and serves as a step forward in achieving equality in AI decision systems.

Competing Interests. There are no competing interests to declare relevant to this article's content.

A Appendix

A.1 Experiment Data

We use five datasets in our experimentation. The Credit Card Clients subset demonstrates that FaGGM returns fairer DAGs. The remaining datasets further show FaGGM's ability to generate fair synthetic datasets while preserving the original predictive power. An overview of the datasets is provided in Table 3.

Credit Card Clients. In our experiments, we use a subset of features from the Credit Card Clients dataset [27], which contains a snapshot of the credit card holders' account information as of September 2005. This dataset features demographic attributes such as gender, education, marital status, credit limit, and payment history information from April to September 2005. The binary

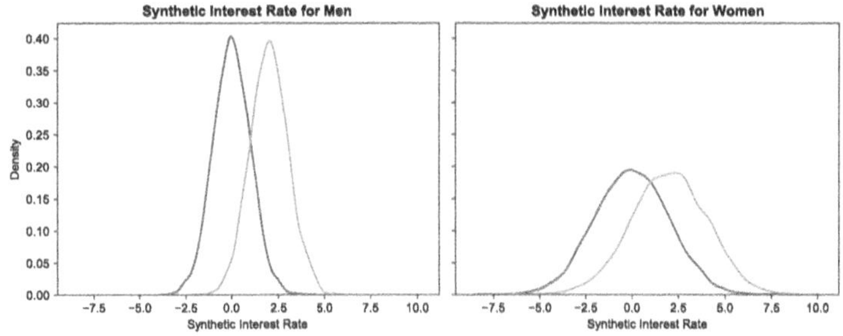

Fig. 3. Distributions of the synthetic Interest Rate variable in the Credit Card Clients dataset.

default flag indicates whether the credit card holder defaulted in October 2005. The protected attribute for our experiments is gender, with males being the privileged class and females being the unprivileged class. The data has 30,000 observations.

To mimic real-world interest rates assigned to credit card applicants, we also introduce a synthetic variable, *Interest Rate* [11], drawn from the following Gaussian distribution so that applicants who defaulted are charged higher interest rates on their credit cards, and women pay higher interest rates than men:

$$\text{For } \textbf{Male} \text{ applicants: } \begin{cases} \textit{Interest Rate} \sim N(0,1) & \text{if } default = 0, \\ \textit{Interest Rate} \sim N(2,1) & \text{if } default = 1. \end{cases}$$

$$\text{For } \textbf{Female} \text{ applicants: } \begin{cases} \textit{Interest Rate} \sim N(0,2) & \text{if } default = 0, \\ \textit{Interest Rate} \sim N(2,2) & \text{if } default = 1. \end{cases}$$

Typically, banks use historical data to assess credit risk and set interest rates, charging higher rates to applicants who resemble past high-risk individuals, as shown in Fig. 3. This approach is not considered discriminatory as it reflects the necessity of managing the additional risk. In this scenario, interest rates serve as a resolving attribute.

The Dutch Census Dataset [10] includes 60,420 observations, with 11 categorical attributes and a binary outcome variable, "occupation", which distinguishes between individuals having occupations with either high (labelled as 1) or low (labelled as 0) income levels. The protected attribute for our experiments is gender, with males being the privileged class and females being the unprivileged class.

The Adult Dataset [1] comprises 32,561 records, with 14 features, a binary "income" variable that differentiates whether an individual earns more than $15,000. The Adult dataset presents a blend of continuous and categorical variables, and we converted the continuous variables into discrete categories for the

Table 3. Experimentation Data Summary

Dataset	Protected Attribute	Privileged Group	Unprivileged Group	Favourable Outcome	Observation Count	Feature Count
Credit Card Clients	Gender	Male	Female	No Default	30,000	3
Dutch Census	Gender	Male	Female	High Income Occupation	60,420	11
Adult	Gender	Male	Female	Income > \$50k	32,561	14
Compas	Race	White	Non-white	No Recidivism	6,173	22
Law School Admission	Race	White	Non-white	Admission	20,527	5

purposes of our experiments. The protected attribute for our experiments is gender, with males being the privileged class and females being the unprivileged class.

The Compas Dataset [22] contains 6,173 observations after excluding records with missing outcome labels, with race as the protected attribute. The outcome labels a binary indicator of whether an individual has committed a re-offence within two years. From a fairness point of view, the desired label is that the individual did not re-offend, labelled 0 in the original data. However, for consistency with other datasets in our study where a value of 1 represents the favourable outcome, we have adjusted the Compas dataset by recoding non-reoffending instances from 0 to 1. The protected attribute for our experiments is race, with the white race being the privileged class and the non-white race being the unprivileged class.

The Law School Admissions Dataset . We construct an admissions scenario for the top 20 US law schools using data from the Law School Admission Council survey [24]. In this scenario, we identify race (categorised as white/non-white) as a protected attribute. The binary outcome variable Y represents admission success, with $Y = 1$ indicating successful admission and $Y = 0$ indicating unsuccessful admission.

The admissions cutoff is derived from medium entry requirements for the top 20 US law schools, calculated as a weighted sum of 60% UGPA (3.86 out of 4.00)[2] and 40% LSAT score (42.6 out of 48)[3], resulting in a combined cutoff score of 19.5. This threshold ensures that 9.8% of the dataset is admitted, consistent with the average admission rate of these institutions.

We include only UGPA, LSAT scores and demographic variables and exclude post-admission variables, such as bar exam results. UGPA and LSAT scores are discretised, and 381 rows with missing values are removed, resulting in a final dataset of 20,597 observations for our experiments.

[2] The average LSAT score of the top 20 US law schools is 172 on a 120–180 scale, and after converting to the 48-point scale, it is 42.6. For more information, visit Best Colleges Research.

[3] The average median UGPA score of top 20 US law schools is 3.86 Law School Admission Statistics.

References

1. Becker, B., Kohavi, R.: Adult Dataset. UCI Machine Learning Repository (1996). https://doi.org/10.24432/C5XW20
2. Berk, R., Heidari, H., Jabbari, S., Kearns, M., Roth, A.: Fairness in criminal justice risk assessments: the state of the art. Sociol. Methods Res. **50**(1), 3–44 (2021)
3. Bouckaert, R.: Bayesian belief networks: from construction to inference. Ph.D. thesis, University of Utrecht (1995)
4. van Breugel, B., Kyono, T., Berrevoets, J., van der Schaar, M.: DECAF: generating fair synthetic data using causally-aware generative networks. In: Advances in Neural Information Processing Systems, vol. 34, pp. 22221–22233 (2021)
5. Calders, T., Kamiran, F., Pechenizkiy, M.: Building classifiers with independency constraints. In: 2009 ICDM Workshop, pp. 13–18. IEEE (2009)
6. Calders, T., Verwer, S.: Three Naive Bayes approaches for discrimination-free classification. Data Min. Knowl. Disc. **21**, 277–292 (2010)
7. Caton, S., Haas, C.: Fairness in machine learning: a survey. ACM Comput. Surv. **56**(7), 1–38 (2024)
8. Chen, Z., Zhang, J.M., Sarro, F., Harman, M.: A comprehensive empirical study of bias mitigation methods for machine learning classifiers. ACM Trans. Softw. Eng. Methodol. **32**(4), 1–30 (2023)
9. Chow, C., Liu, C.: Approximating discrete probability distributions with dependence trees. IEEE Trans. Inf. Theory **14**(3), 462–467 (1968)
10. Dutch Central Bureau for Statistics Volkstelling: Dutch Census Dataset (2001)
11. Fairlearn: Example Notebook: Credit Loan Decisions. https://fairlearn.org/main/auto_examples/plot_credit_loan_decisions.html. Accessed 17 Jan 2024
12. Feldman, M., Friedler, S.A., Moeller, J., Scheidegger, C., Venkatasubramanian, S.: Certifying and removing disparate impact. In: Proceedings of the 21th ACM SIGKDD, pp. 259–268 (2015)
13. Hardt, M., Price, E., Srebro, N.: Equality of opportunity in supervised learning. In: Advances in Neural Information Processing Systems, vol. 29 (2016)
14. Heckerman, D., Geiger, D., Chickering, D.M.: Learning Bayesian networks: the combination of knowledge and statistical data. Mach. Learn. **20**, 197–243 (1995)
15. Hort, M., Chen, Z., Zhang, J.M., Sarro, F., Harman, M.: Bias mitigation for machine learning classifiers: a comprehensive survey. ACM J. Respons. Comput. (2023)
16. Hort, M., Zhang, J.M., Sarro, F., Harman, M.: Fairea: a model behaviour mutation approach to benchmarking bias mitigation methods. In: Proceedings of the 29th ESEC/FSE, pp. 994–1006 (2021)
17. Kitson, N.K., Constantinou, A.C., Guo, Z., Liu, Y., Chobtham, K.: A survey of Bayesian network structure learning. Artif. Intell. Rev. 1–94 (2023)
18. Kusner, M.J., Loftus, J., Russell, C., Silva, R.: Counterfactual fairness. In: Advances in Neural Information Processing Systems, vol. 30 (2017)
19. Liu, T., Qian, Z., Berrevoets, J., van der Schaar, M.: Goggle: generative modelling for tabular data by learning relational structure. In: The Eleventh ICLR (2023)
20. Mehrabi, N., Morstatter, F., Saxena, N., Lerman, K., Galstyan, A.: A survey on bias and fairness in machine learning. ACM Comput. Surv. (CSUR) **54**(6), 1–35 (2021)
21. Newman, M.: Networks, 2nd edn. Oxford University Press, Oxford (2018)
22. propublica.org: Compas Dataset. https://github.com/propublica/compas-analysis

23. Wick, M., Tristan, J.B., et al.: Unlocking fairness: a trade-off revisited. In: Advances in Neural Information Processing Systems, vol. 32 (2019)
24. Wightman, L.F.: LSAC national longitudinal bar passage study. LSAC Research Report Series (1998)
25. Xu, D., Wu, Y., Yuan, S., Zhang, L., Wu, X.: Achieving causal fairness through generative adversarial networks. In: Proceedings of the Twenty-Eighth IJCAI (2019)
26. Xu, D., Yuan, S., Zhang, L., Wu, X.: FairGAN: fairness-aware generative adversarial networks. In: 2018 IEEE Big Data, pp. 570–575. IEEE (2018)
27. Yeh, I.C.: Default of credit card clients. UCI Machine Learning Repository (2016). https://doi.org/10.24432/C55S3H
28. Zemel, R., Wu, Y., Swersky, K., Pitassi, T., Dwork, C.: Learning fair representations. In: International Conference on Machine Learning, pp. 325–333. PMLR (2013)
29. Zhang, L., Wu, Y., Wu, X.: A causal framework for discovering and removing direct and indirect discrimination. arXiv preprint arXiv:1611.07509 (2016)
30. Zhang, L., Wu, Y., Wu, X.: Achieving non-discrimination in data release. In: Proceedings of the 23rd ACM SIGKDD, pp. 1335–1344 (2017)

Shedding Light on Greenwashing: Explainable Machine Learning for Green Ad Detection

Yihan Bao[1(✉)], Abdul Karim Obeid[2], Daniel Angus[2], Julian Bagnara[1], and Christopher Leckie[1]

[1] The University of Melbourne, Parkville, VIC, Australia
yihanb@student.unimelb.edu.au, {julian.bagnara,caleckie}unimelb.edu.au
[2] The University of Queensland, Brisbane, QLD, Australia
{obei,daniel.angus}@qut.edu.au

Abstract. Businesses and organisations often include environmental claims in their product advertisements, but some can be misleading or false. This practice is colloquially known as âĂIJgreenwashingâĂİ, which can erode consumer trust and hinder genuine efforts towards sustainability. To maintain a trustworthy advertising environment, it is crucial to systematically identify and reveal said misleading advertisements. However, with the significant increase in the number of advertisements being served digitally, manual screening of online platforms becomes time-consuming and inconsistent. Therefore, we present an automated system to identify advertisements containing environmentally-friendly claims (i.e., green claims). Unlike previously established models that rely on neural networks, we propose a logistic regression model because it offers improved explainability. Specifically, each word has a coefficient representing how much it influences the modelâĂŹs prediction, and each prediction is associated with its uncertainty to assist further scrutiny. We also compare the features of green advertising between Australia and the US, showing a significant difference in their terminology. The findings demonstrate the potential of machine learning and explainable AI (XAI) in addressing greenwashing and promoting more effective and trustworthy green marketing practices, ultimately fostering a healthier advertising ecosystem.

Keywords: Greenwashing · Logistic Regression · Ad Observatory · Explainable AI · Sustainability

1 Introduction

'Green' products refer to merchandise manufactured by specific techniques that achieve reduced environmental impact identified through particular aspects: for example, recyclability and reusability [14]. As public awareness and concern for environmental sustainability grows, the demand for these products has significantly increased, leading many companies to introduce "green marketing" [7].

M. Gong et al. (Eds.): AI 2024, LNAI 15442, pp. 186–197, 2025.
https://doi.org/10.1007/978-981-96-0348-0_14

Green marketing encompasses a range of strategies and practices to promote products based on their environmental benefits. Consequently, it has become a competitive advantage that improves a company's reputation by satisfying consumers and other stakeholders [13].

However, the rise of green marketing has also led to an increase in "greenwashing". Greenwashing refers to activities where businesses are unclear or lie about their green credentials, or utilise green claims to divert consumers' attention from other activities [6]. This deceptive practice can undermine consumer trust and hinder genuine sustainability efforts. Specifically, some consumers would feel distrust and confusion towards green offerings, potentially affecting their decisions and attitudes in the future [12]. For instance, when consumers realise that they have been misled by false green claims, they may become sceptical of all green marketing, even from genuinely sustainable companies, thereby reducing the overall effectiveness of green marketing campaigns.

To address the issue of greenwashing, it is crucial to develop methods for accurately identifying and verifying green claims in marketing. Traditional approaches to detecting greenwashing often rely on manual verification, which can be time-consuming and inconsistent. Moreover, there has been a significant increase in digital advertisements on social media platforms in the past few years. While this provides opportunities and convenience for the public, it raises challenges regarding transparency in delivery and reporting [15]. For example, misleading green claims can be disseminated quickly and widely, complicating efforts to ensure the accuracy and honesty of environmental claims. As the volume of digital advertising continues to grow, manual verification becomes increasingly impractical. With the advancement of machine learning and natural language processing techniques, there is an opportunity to automate and enhance the detection process, making it more efficient and reliable.

As of 2020, explainability has emerged as a critical component for the wider adoption of AI systems across various sectors. This involves understanding the methods and reasons behind automated decisions to provide users with a sense of safety and trust [5]. In the context of detecting green claims and preventing greenwashing, XAI is crucial because it provides insights into how specific features, such as particular words or phrases, contribute to classifying an advertisement as green or non-green. This transparency is essential for building trust among consumers, businesses, and regulators, as it allows them to verify the legitimacy of the model's predictions and understand the rationale behind them. By ensuring that the AI models used are interpretable, we can better address concerns about the ethical and practical implications of automated decision-making in sensitive areas such as consumer protection and environmental sustainability. Thus, in this paper we address the following research questions:

RQ1: What are the characteristics of green advertisements, and how can we identify them automatically while ensuring explainability?

RQ2: Is there any linguistic difference in green advertisements from different countries?

2 Related Work

2.1 Ad Observatory

Our research is based on two prominent ad observatory projects: the NYU Ad Observatory[1] and the Australian Ad Observatory[2]. The NYU Ad Observatory, established by New York University, aims to provide transparency around the sources and content of online advertisements, helping the public understand the nature and impact of digital advertising. It mainly focuses on political advertisements but also covers various topics of social issues, such as education, healthcare, and environmental protection. The Australian Ad Observatory is an ongoing project of the Australian Research Council (ARC) Centre of Excellence for Automated Decision-Making and Society (ADM+S). Inspired by the NYU Ad Observatory, it aims to investigate the volume and characteristics of online advertising and how particular ads target specific demographics [2]. The project is undertaking various case studies, including scams, gambling advertising, alcohol advertising and green claims.

Both observatory projects involve detecting green claims, but they use different approaches. The NYU observatory embedded machine learning and topic modelling techniques to classify advertisements into categories. However, the explicit algorithms are not disclosed, leading to difficulties in evaluating and explaining the decision-making process. In contrast, the Australian Ad Observatory used a keyword search strategy, which extracts advertisements containing environmental-related terms, such as "environmentally friendly", "sustainable", and "compostable" [1]. While this method is transparent and explainable, several concerns regarding its reliability exist. It inevitably included around 11.8% of advertisements that are not environmentally related, requiring human efforts to clean the dataset [1]. Additionally, we found that some green claims do not contain the predefined keywords, and thus were overlooked. This limitation introduces potential risks and biases. For example, the scalability of the systems is challenged by the rapidly evolving language surrounding sustainability and environmental issues. As public awareness and terminology develop, the static nature of keyword lists may not capture the relevant claims effectively. This can limit the ability of such systems to adapt to new trends or changes in how green claims are communicated, requiring frequent updates to maintain reliability. Furthermore, the keyword search strategy identifies the existence of green claims but does not provide insights into the degree of commitment behind these claims. This method treats all occurrences of green advertisements equally, without differentiating between a superficial mention and a deep, meaningful engagement with sustainability practices.

2.2 Green Claim Detection

In the past, many studies have emphasised the necessity of detecting faulty claims and identified some methods to achieve this. However, very few of them target

[1] https://adobservatory.org/.

[2] https://www.admscentre.org.au/adobservatory/.

the specific context of greenwashing. Chhabra [3] summarised the characteristics of green marketing that can be referred to as a checklist for identifying dubious environmental claims. They also listed companies participating in green marketing, such as McDonalds, Starbucks, and Walmart. Although this provides a framework to monitor potential greenwashing activities, a more procedural approach must be developed to improve the generalisability and efficiency of the detection process.

Several studies have adopted machine learning techniques or social research to filter green advertisements from others. Woloszyn et al. [20] focused on the domain of cosmetics and electronics, selecting 38,742 advertisements from Twitter before manually labelling each as "explicit green claim", "implicit green claim", or "not green claim". They developed an automated algorithm using pre-trained neural networks to perform green detection with the capability of resisting adversarial attacks. Their experiment achieved an F1 score of 0.921 when classifying between "green claim" and "not green claim" and an F1 score of 0.815 when doing the multiclass classification. More recently, Stammbach et al. [11,17] created a dataset in the domain of climate and global warming from sustainability reports, earning calls, and annual reports. They implemented Support Vector Machine (SVM) and neural network models to achieve an F1 score of 0.849 from 2,647 annotated advertisements.

While the previous automated detection strategies performed well on related datasets, the scope of these studies was restricted to specific domains. Therefore, their approach and the developed models may need to be revised to adapt better to other situations. Moreover, neural network models lack interpretability since it is difficult to explain their underlying behaviours [8]. Even though some may be interpretable, they often require specific domain knowledge. Therefore, a framework that automatically detects green advertisements while enabling direct analysis and interpretation of their results is needed.

3 Dataset

Our study utilised two distinct datasets, each comprising advertisements sourced from Facebook. One dataset contains advertisements from the United States (US), while the other includes advertisements from Australia.

The US advertisement dataset was downloaded from the open-source New York University Ad Observatory. It contains both green and non-green advertisements. Green advertisements were directly obtained through a designated link provided by the Ad Observatory project, ensuring they were explicitly labelled green. In contrast, non-green advertisements were randomly selected from the entire pool of ads. Although there is a possibility that a few green ads might be included in the non-green selection, this risk is minimal given that green ads constitute a very small fraction of the total advertisements.

The Australian Ad Observatory extended the functionality of the browser extension developed by 'ProPublica' - an open-sourced tool that collected advertisements on Facebook during the 2016 presidential election in the United States.

Participants were able to install the browser extension to become volunteers in the data donation process, and the extension automatically captured the advertisements that they then observed on Facebook. After that, researchers from the Australian Ad Observatory and the Consumer Policy Research Centre (CPRC) collaborated to perform keyword extraction and data cleaning to identify green advertisements. Despite this effort, there was a concern that some green ads might have been overlooked due to the initial filtering process. To address this, we conducted a preliminary analysis and selected a subset of samples for further modelling. This approach helped refine our dataset, ensuring a more accurate representation of green and non-green advertisements for subsequent analysis.

After discarding duplicate advertisements, our advertisement sample from NYU Ad Observatory totalled 14,444, where 55.3% were labelled as green ads. In the Australian dataset, our advertisement sample totalled 16,931, where 51.3% were green.

4 Methodology

The increasing adoption of green marketing has led to a significant rise in greenwashing, undermining consumer trust and genuine environmental efforts. Traditional manual verification methods are likely inadequate due to the volume and complexity of digital advertisements on social media platforms. Furthermore, existing methods utilising neural networks present additional drawbacks in interpretation. This research aims to develop an explainable machine learning model to automatically detect green claims in social media advertisements from the Australian and US markets. By ensuring transparency in green ad detection, this study seeks to enhance consumer trust and support regulatory efforts in promoting genuine sustainability practices.

4.1 Preprocessing

To ensure the data was clean and suitable for model training, we designed the following preprocessing pipeline. This process aims to standardise the text data, making it more manageable and relevant for subsequent feature extraction and modelling stages. First, we removed named entities to eliminate potential bias introduced by specific brands, individuals, or place names. This step was crucial in focusing on the general linguistic patterns rather than context-specific details. Next, we eliminated all special symbols and punctuation, which could interfere with the text analysis. Following this, we removed stopwords, which are common words like 'and', 'the', and 'is' that carry little information and could dilute the significance of more meaningful words. Finally, we applied stemming to reduce words to their root form, for example, 'running' and 'runs' are treated as 'run'. These steps aim to reduce the dimension of vocabulary in the ad texts, which brings the advantage of effective computation and better generalisation. Although the preprocessing steps may potentially perpetuate their own negative effects on model performance, this is a means to the end of necessarily facilitating explainability analysis [19].

4.2 Logistic Regression Model

Logistic regression is a statistical method focusing on the relative probability (odds ratio) of an event occurring versus not occurring, and is therefore suitable for testing the ad's probability of being green [9]. It predicts the probability of the positive class $P(\hat{y}_i = 1 \mid \boldsymbol{x}_i)$ as:

$$\hat{p}(\boldsymbol{x}_i) = \frac{1}{1 + \exp(-\boldsymbol{x}_i \boldsymbol{w})}$$

where $\boldsymbol{x}_i$ is the input feature (word vectors of the advertisement text) for the ith observation, and $\boldsymbol{w}$ denotes the weight for each feature. If $P(\hat{y}_i = 1 \mid \boldsymbol{x}_i) \geq 0.5$, this advertisement is classified as green.

Table 1. Prediction performance (precision, recall, F1 score, accuracy) of models, on cross-validation and test set

Model	Cross Validation				Test			
	pr	rc	F1	ACC	pr	rc	F1	ACC
Majority Baseline	0.516	1.000	0.681	0.516	0.493	1.000	0.660	0.493
Decision Tree	0.887	0.859	0.873	0.871	0.824	0.795	0.809	0.806
SVM	0.921	0.842	0.880	0.881	0.914	0.833	0.871	0.873
Logistic Regression (LR)	0.923	0.896	0.910	0.908	0.903	0.873	0.888	0.886
LR - Australian	0.980	0.961	0.971	0.971	0.963	0.932	0.948	0.950
LR - the US	0.923	0.928	0.926	0.919	0.874	0.877	0.875	0.864

Table 1 reports the various metrics of each model's performance on the training set using a five-fold cross-validation, as well as their ability to predict unseen samples (test set). Initially, a combined dataset is used to measure the adequacy of the logistic regression model. We compared logistic regression with a majority voting baseline model and other interpretable machine learning models (Decision Tree and SVM), and the results indicate that logistic regression is more predictive.

Next, we designed two logistic regression models sharing identical hyperparameters and tuning, then trained them using advertisements from Australia and the US separately. The results demonstrate a consistently high performance across different runs, indicating the robustness and reliability of our approach. The slight variation in performance could be attributed to differences in the underlying data distribution or specific characteristics of Australian and US green ads. We also observe that the model trained from the Australian dataset achieved the highest performance, the US one had the lowest performance, and the combined model in between.

4.3 Interpretation

The interpretability of logistic regression models is widely recognised, making them a preferred choice for many practical applications where understanding the decision-making process of the model is crucial [4,10,18]. Unlike more complex models, logistic regression provides clear insights into how each feature impacts the outcome through its coefficients. Table 2 shows the top and bottom ten words from the combined model, ranked by their importance in predicting green claims. These coefficients represent how much the presence of each word changes the likelihood of an advertisement being classified as a green claim. For example, the word "environment" has a high coefficient (19.413). This means that if an advertisement includes the word "environment", it significantly increases the chance of the ad being identified as a green claim. Specifically, it increases the likelihood by a factor of $e^{19.413}$, which indicates that "environment" is a very strong indicator of green advertisements. In contrast, words with low or negative coefficients decrease the likelihood of an ad being classified as a green claim.

Table 2. Top and Bottom 10 feature weights for predicting green ads

Feature (word)	Weight	Feature (word)	Weight
environment	19.413	moreabout	−32.570
environ	14.760	moreconfirm	−26.564
nativ	8.569	affili	−23.557
dust	8.262	confirm	−17.297
solar	7.859	feb	−15.301
sustain	7.423	adform	−15.203
protect	7.410	valentin	−10.005
pollut	7.156	ticket	−9.243
coal	7.110	februari	−9.179
discuss	7.091	parliament	−6.199

Recall that the output of the logistic function $\hat{p}(x_i)$ is the probability that a particular instance will be classified as a green ad. Therefore, it directly shows how "confident" the model makes that prediction, i.e., a value close to one shows that the model strongly believes this instance should be a green ad and 0 for a non-green ad. The characteristic of returning a probability from the model predictions enables more intuitive decision-making. Figure 1 shows different prediction outcomes against the model's confidence in predicting ads of the test set from the combined dataset.

We observe that the model produces less error when the predicted probability is near 0 or 1. Therefore, predictions close to these values are more reliable since

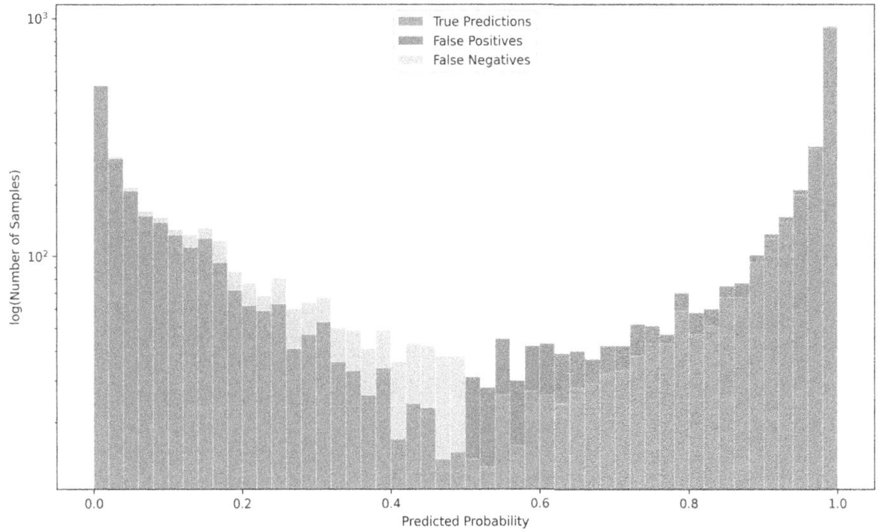

Fig. 1. Prediction errors and correct prediction produced by the combined model as prediction probability is varied

they are unlikely to produce errors. This reliability allows for automated decision-making without requiring additional human review in these cases. In contrast, additional scrutiny is needed for predictions with probabilities around 0.5. These low-confidence predictions can be flagged for human review or further analysis to reduce the error rate in these ambiguous cases. By understanding the confidence level, it is possible to adjust the threshold to manage the trade-off between false negatives (missing actual green ads) and false positives (incorrectly identifying non-green ads as green). The model's capability of providing probability scores builds trust with stakeholders by showing that the model's predictions are not just binary but come with an explanation of certainty. This transparency helps in better understanding and accepting the model's decisions.

4.4 Comparing Features

The next step is to compare the specific models trained on the Australian and the US datasets. This comparison aims to uncover regional differences in the language used to make green claims. By analysing the indicative words identified by each model, we can gain insights into the unique linguistic patterns and marketing strategies employed in different markets. This comparative analysis is crucial for deploying detection methods to specific regions and enhancing the overall understanding of green marketing.

Figure 2 shows the word clouds for the most significant words as green claims. We see that sentences classified as green advertisements are likely to contain some environmental-related keywords. These words can be seen as potential

Fig. 2. Top 50 most significant words of logistic regression models trained from Australian (left) and the US (right) datasets

red flags when they appear prominently in advertisement content. However, we observed contrasting results from two different logistic regression models. The model trained from Australian advertisements highlights environmentally related words as significant predictors, aligning with the previous hypothesis that terms directly referencing environmental benefits or sustainability practices are strong indicators of green claims. In contrast, the model based on US advertisements achieves good predictive performance but relies heavily on features that do not explicitly describe environmental aspects.

We plotted a heatmap (in Fig. 3) to intuitively compare the weight of the two models by highlighting areas of significant divergence through contrasting colours. By comparing the feature coefficients between the Australian and the US models, we observed notable differences in the terms most indicative of green claims in each region. Specifically, the terms ‘worth’, ‘protect’ and ‘environment’ have large weights in the US model but relatively small weights in the Australian model. Conversely, the stemmed terms ‘clean’, ‘green’, ‘climat’, and ‘energi’ have

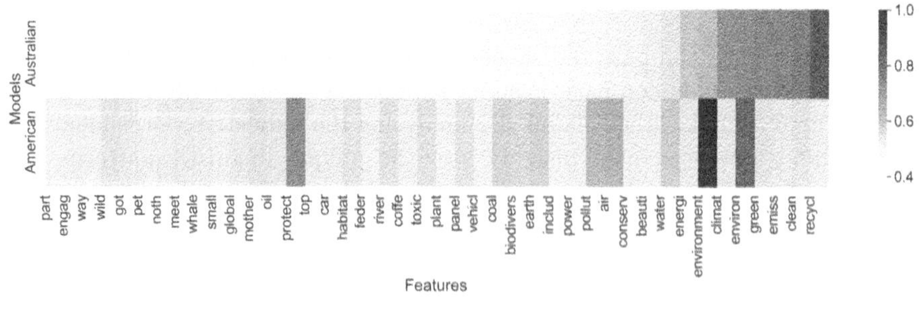

Fig. 3. Feature comparison between Australian and US models

large weights in the Australian model but are less significant in the US model. This could be due to specific regulatory focuses or environmental awareness in each country, leading to different frequencies of words used by organisations.

5 Conclusion

In this study, we addressed the critical issue of greenwashing in digital advertisements by developing machine learning models to identify green claims. Our approach focused on leveraging explainable AI techniques to ensure transparency in the decision-making process of our models. This is crucial for building trust among consumers, businesses, and regulators, enabling them to verify the legitimacy of green claims and better understand the model's predictions.

Using datasets from Australian and US advertisements, we trained three logistic regression models: one on the Australian dataset, one on the US dataset, and one on a combined dataset. All three models achieved high prediction performances, with the combined model demonstrating adequate generalisability across regions. However, region-specific models may still be necessary to capture unique linguistic patterns effectively because different countries may focus on diverse aspects of environmental protection.

Our study contributes to the ongoing efforts to combat greenwashing by promoting transparency and accountability in green marketing. The models and methodologies developed in this research offer practical tools for automatically identifying and verifying green claims, thereby empowering consumers' and regulators' understanding of green advertisements. By continuing to refine these models and approaches, we can contribute to a more transparent and sustainable marketplace, ultimately supporting broader environmental sustainability goals.

5.1 Future Work

Although our model can output the uncertainty of its predictions, we did not utilise this feature. Prediction uncertainty can provide valuable insights into the confidence level of the model's classifications. In future work, this information could be considered for active learning strategies. By focusing on the most ambiguous cases, active learning can help refine the model more efficiently and effectively, potentially leading to better generalisation and robustness in detecting green claims [16].

Finally, while textual content provides important and direct information for determining whether an advertisement is green, it is also essential to consider other information. In this paper, we focused exclusively on the text of advertisements, omitting other potentially valuable features such as images. Visual elements can play a significant role in green marketing, as images can convey sustainability themes and messages that support or enhance textual claims. Future research should integrate multimodal data, combining both text and image analysis, to capture a more comprehensive view of how green claims are presented and perceived. This holistic approach could improve the accuracy and depth of green claim detection, providing a more robust solution for identifying greenwashing practices.

Acknowledgements. We thank Prof Christine Parker for providing background information and directions, and the Consumer Policy Research Centre (CPRC), whose partnership has been instrumental to the success of this project.

References

1. Angus, D., et al.: The Australian ad observatory technical and data report (2024)
2. Burgess, J., Andrejevic, M., Angus, D., Obeid, A.: Australian ad observatory: background paper (2022)
3. Chhabra, M.K.: Green marketing: golden goose or lame duck. Biz Bytes **8**(1), 75–82 (2017)
4. Christoph, M.: Interpretable machine learning: a guide for making black box models explainable. Lulu. com (2019)
5. Confalonieri, R., Coba, L., Wagner, B., Besold, T.R.: A historical perspective of explainable artificial intelligence. Wiley Interdisc. Rev. Data Min. Knowl. Discov. **11**(1), e1391 (2021). https://doi.org/10.1002/widm.1391
6. Consumer Policy Research Centre: The consumer experience of green claims in Australia (2022)
7. Dangelico, R.M., Vocalelli, D.: "green marketing": An analysis of definitions, strategy steps, and tools through a systematic review of the literature. J. Clean. Prod. **165**, 1263–1279 (2017). https://doi.org/10.1016/j.jclepro.2017.07.184
8. Fan, F.L., Xiong, J., Li, M., Wang, G.: On interpretability of artificial neural networks: a survey. IEEE Trans. Radiat. Plasma Med. Sci. **5**(6), 741–760 (2021). https://doi.org/10.1109/TRPMS.2021.3066428
9. Guido, J.J., Winters, P.C., Rains, A.B.: Logistic regression basics. M.Sc. University of Rochester Medical Center, Rochester, NY, vol. 21 (2006)
10. Kost, S., Rheinbach, O., Schaeben, H.: Using logistic regression model selection towards interpretable machine learning in mineral prospectivity modeling. Geochemistry **81**(4), 125826 (2021). https://doi.org/10.1016/j.chemer.2021.125826
11. Leippold, M., Stammbach, D., Webersinke, N., Bingler, J.A., Kraus, M.: Environmental claim detection. In: Proceedings of the 61st Annual Meeting of the Association for Computational Linguistics, pp. 1051–1066. Association for Computational Linguistics (2023). https://doi.org/10.18653/v1/2023.acl-short.91
12. Lim, W.M., Ting, D.H., Bonaventure, V.S., Sendiawan, A.P., Tanusina, P.P.: What happens when consumers realise about green washing? A qualitative investigation. Int. J. Glob. Environ. Issues **13**(1), 14–24 (2013). https://doi.org/10.1504/IJGENVI.2013.057323
13. Miles, M.P., Covin, J.G.: Environmental marketing: a source of reputational, competitive, and financial advantage. J. Bus. Ethics **23**, 299–311 (2000). https://doi.org/10.1023/A:1006214509281
14. Mishra, P., Sharma, P.: Green marketing: challenges and opportunities for business. BVIMR Manag. Edge **7**(1) (2014)
15. Pastor, A., Cuevas, R., Cuevas, Á., Azcorra, A.: Establishing trust in online advertising with signed transactions. IEEE Access **9**, 2401–2414 (2020). https://doi.org/10.1109/ACCESS.2020.3047343

16. Settles, B.: Active learning literature survey. Technical report, University of Wisconsin-Madison Department of Computer Sciences (2009)
17. Stammbach, D., Webersinke, N., Bingler, J.A., Kraus, M., Leippold, M.: A dataset for detecting real-world environmental claims. Center Law Econ. Working Paper Ser. **2022**(07) (2022)
18. Tahirovic, E., Krivic, S.: Interpretability and explainability of logistic regression model for breast cancer detection. In: ICAART (3), pp. 161–168 (2023). https://doi.org/10.5220/0011627600003393
19. Uysal, A.K., Gunal, S.: The impact of preprocessing on text classification. Inf. Process. Manag. **50**(1), 104–112 (2014). https://doi.org/10.1016/j.ipm.2013.08.006
20. Woloszyn, V., Kobti, J., Schmitt, V.: Towards automatic green claim detection. In: Proceedings of the 13th Annual Meeting of the Forum for Information Retrieval Evaluation, pp. 28–34 (2021). https://doi.org/10.1145/3503162.3503163

Beyond Factualism: A Study of LLM Calibration Through the Lens of Conversational Emotion Recognition

Samad Roohi[1](✉), Richard Skarbez[1], and Hien Nguyen[1,2]

[1] La Trobe University, Bundoora, VIC 3086, Australia
{s.roohi,r.skarbez,h.nguyen5}@latrobe.edu.au
[2] Institute of Mathematics for Industry, Kyushu University, Fukuoka, Japan

Abstract. The calibration of large language models (LLMs) is crucial for ensuring their reliability and effectiveness, especially in tasks requiring sophisticated reasoning. While significant research has focused on understanding the calibration of LLMs in factual tasks like question answering, there is a notable gap in assessing their calibration for more complex tasks. This gap is particularly evident in domains such as emotion recognition in conversation (ERC), where the challenge extends beyond basic contextual understanding to interpreting nuanced emotional states integral to human interactions. In this paper, we explore the extent to which state-of-the-art LLMs are well-calibrated for the specific task of ERC. Our findings reveal that these models exhibit poor calibration—specifically, a tendency toward overconfidence—such that their confidence levels do not accurately reflect actual performance. While one can leverage the intrinsic verification capabilities of LLMs to assess the correctness of their predictions to some degree, this does not sufficiently address the significant issue of overconfidence inherent in these models.

Keywords: Emotion recognition · Large language models · Calibration · Textual conversation

1 Introduction

Large language models (LLMs) have exhibited exceptional performance across various natural language processing (NLP) tasks, ranging from text comprehension [22], through commonsense reasoning [23], to code generation [32]. These models have shown significant improvement in generating responses aligned with human instructions and preferences [6]. The enormous parameter architectures of LLMs have empowered them with strong generalization ability, facilitating their application to various downstream tasks. This flexibility has enabled the

Supplementary Information The online version contains supplementary material available at https://doi.org/10.1007/978-981-96-0348-0_15.

M. Gong et al. (Eds.): AI 2024, LNAI 15442, pp. 198–212, 2025.
https://doi.org/10.1007/978-981-96-0348-0_15

transformation of traditionally discriminative tasks, such as emotion recognition in conversation (ERC), into generative tasks, which enhances the scope and application of these methods.

Despite this substantial progress, previous research has shown a persistent issue of overconfidence in model responses to fact-based tasks. In natural language generation, one such issue is hallucination, where the model generates responses that are syntactically correct, fluent, and natural but contain factually incorrect or nonsensical content, or deviate from the provided input [29]. This phenomenon is often attributed to sub-optimal calibration of out-of-the-box LLMs, potentially originating from the model alignment to human instructions and preferences [14,25].

Humans can meaningfully express and communicate the level of uncertainty in language (confidence in information) in various ways [7]. A reliable estimate of predictive uncertainty can help in informed decision-making and serve as an indicator of response accuracy [11]. Within this framework, the calibration of predictive models is essential, as a poorly calibrated model introduces significant challenges in reporting a reliable level of confidence for each prediction.

While significant research has studied the calibration capabilities of LLMs on factual tasks, such as fact-based question answering [14], their performance and calibration capabilities for more advanced tasks-such as emotion recognition in conversation (ERC), which requires complex reasoning on context and human emotion-remain unexplored. Emotion plays a critical role in social communication. Bloom and Lahey defined it as the pragmatic component of language [2].

This study investigates the calibration of state-of-the-art LLMs for the task of emotion recognition in conversation. To do this, we converted the traditionally discriminative task of ERC into a generative problem by crafting appropriate prompts. Subsequently, we evaluated the performance of LLMs in emotion recognition across three publicly accessible datasets. Furthermore, we evaluated the calibration of LLMs using two approaches: verbalized confidence elicitation and logit-based probabilities. In verbalized confidence elicitation, we directly asked each model to provide its confidence in the emotion label they predicted for a query utterance. On the other hand, in logit-based probabilities, we normalized the logit values over the set of emotion labels provided by each dataset for the specific token predicted by the model. Finally, we evaluated the verification capability of LLMs, as introduced in the literature, to examine the performance when the probability of truth—P(True)—is used to verify the models' predictions.

2 Background

Emotion fundamentally underpins a natural communication. It shapes interpersonal dynamics, drives conversation flow, and facilitates understanding of a speaker's intents, concerns, and desires. In critical domains, such as clinical mental health, a reliable emotion recognition can help in early detection of emotional distress [3]. Moreover, in customer service support applications, the ability to accurately interpret emotions facilitates acquisition of realistic feedback from customers, thereby empowering service providers to respond with enhanced effectiveness and empathy [5].

Emotion recognition in conversation (ERC) is the process of automatically identifying the emotional state of an utterance considering its context. As an important development of NLP studies, advanced deep learning techniques, such as recurrent neural networks (RNNs), attention based method, and transformers, have been utilized to improve the accuracy of ERC tasks [10,17]. In the pre-LLM era, the development of ERC systems was constrained by the complex architecture of pipeline designs. This complexity often resulted in models that exhibit tendency a tendency to overfit on a particular dataset and dialogue pattern, thereby limiting their general applicability and robustness.

The advent of LLMs has significantly encouraged generalization of the task of emotion recognition in conversation (ERC) in both open-domain dialogues and task-oriented interactions. Nonetheless, research has shown that although out-of-the-box LLMs show satisfactory performance in zero-shot setting for binary sentiment classification, for more complex affect recognition tasks they are still inferior to traditional supervised learning approaches [33]. To improve the performance of LLMs on ERC, Feng et al. proposed the use of task-specific fine-tuning [9]. [34]. Additionally, Lei et al. proposed InstructERC which employs a retrieval module to concatenate the historical dialog content statement, label statement and emotional domain demonstrations with high semantic similarity [15]. They also improved the reasoning capabilities of the model by adding emotion alignment through speaker identification and emotion prediction tasks. InstructERC outperformed traditional supervised methods. Moreover, to enhance the quality of instructions, Zhang et al. [34] proposed to fine-tune LLMs with context and emotion knowledge extracted from supplementary visual data. Furthermore, Tu et al. [27] showed that integration of a multiple knowledge fusion model, where the knowledge is generated by LLMs (e.g., co-reference, topics, and emotional causes), can improve the effectiveness of emotion recognition.

In factual tasks, such as multiple-choice question answering, LLMs have shown a level of overconfidence in the generated responses. In language generation task, this overconfidence leads to the problem of hallucination, where the model fabricate incorrect responses that are syntactically correct, fluent, and natural, but show factually incorrect or nonsensical content, or deviation from the provided input [19]. Hallucination occurs when the input either lacks definitive answers or necessitates reasoning that extends beyond the model's existing knowledge base.

In prediction systems, a calibrated estimate of predictive uncertainty can help in an informed decision making and serve as an indicator of response validity [11]. Current approaches to the estimation of the confidence level either treats the language model as a black-box system, where the model is prompted to self-assess and report its confidence [1,14,25], or as a probabilistic model [11], where the probability measures computed from logit scores serve to estimate confidence on the output. In deep learning, confidence typically is referred to as the extent to which a model is certain about its prediction. This is often expressed as the probability level that a model assigns to a prediction, where a high confidence score indicates greater certainty. In this context, a well-calibrated confidence score can be considered an indicator of the likelihood of the prediction's accuracy [11]. In a well-calibrated model, confidence levels and uncertainty estimates are related, though they are not strictly identical. High confidence typically suggests low overall uncertainty, but it doesn't specify whether this is due to data clarity (low aleatoric uncertainty) or robust model knowledge (low epistemic uncertainty).

3 Methodos

In this section, we describe the approaches adopted to evaluate the calibration and performance of LLMs in the task of ERC. This involves redefining the discriminative task of ERC as a generative task through prompt-based few-shot learning—specifically, 2-shot in-context learning. We opted for two-shot examples after observing that using only a single emotional example led the model to consistently generate outputs with an emotion label, neglecting the 'neutral' class. By including both an emotional example and a neutral example, we aimed to better prompt the model to consider neutral responses in its outputs. Additionally, this section outlines procedures for evaluating model calibration through two approaches: verbalized confidence elicitation and logit-based model probabilities. Finally, we explain our methodology for employing the probability of truth-P(True)-to verify the model's predictions.

3.1 Emotion Recognition

The goal of ERC is to categorize the emotional dynamics expressed within dialogues into specific emotional categories, such as "happiness" or "sadness." Traditionally, language model-based supervised methods for ERC employ a discriminative framework which typically necessitates extensive training on large task-specific labeled datasets. In contrast, the generative approach reformulates text inputs to align with the specific template requirements of the language model. This restructured input enables the LLM to more effectively predict an emotional label from a specific set of emotion categories.

Despite the remarkable advancements in natural language processing achieved by large language models (LLMs), their performance in zero-shot emotion recognition within conversations exhibits significant limitations. LLMs like GPT-3 and GPT-4 are primarily trained on extensive text corpora without explicit annotations for emotional content, which weakens their ability to accurately detect and interpret emotions in dialogue. They often struggle with capturing subtle emotional nuances, sarcasm, and context-dependent expressions essential for emotion recognition in conversational settings [9]. One significant challenge is the inherent complexity of human emotions, which can be context-dependent and subtle, making it difficult for LLMs to generalize from available data [12]. Furthermore, these models may exhibit biases based on their training data, leading to inconsistent or inaccurate emotion detection, particularly in underrepresented demographics or cultural contexts [31]. The reliance on textual cues alone can also limit performance, as non-verbal signals and tone, which are crucial for emotion interpretation, are absent in text-based conversations. Few-shot learning, on the other hand, enhances performance of ERC by providing the model with a small set of labeled examples that guide it in recognizing patterns specific to the emotional context, improving both accuracy and generalization [30,33]. In this paper, we evaluated the calibration of LLMs in ERC using few-shot learning approaches to obtain a more reliable results.

3.2 Methods for Confidence Estimation

To evaluate the calibration of LLMs, this study employs a dual approach for extracting confidence measures: verbalized confidence elicitation and logit-based confidence estimation. This section outlines the application of these approaches to the specific challenge of emotion recognition within conversational contexts.

Verbalized Confidence Elicitation. Verbalized confidence elicitation refers to the capability of an LLM to express the confidence of their predictions using natural language phrases and sentences [16]. It enables the model to generate an answer along with an articulated level of confidence. The application of verbalized confidence within LLMs provides several advantages. First, it does not require a deep understanding of sophisticated statistical paradigms or the necessity for model fine-tuning. Second, it encourages more natural and intuitive interactions between end-users and LLMs. Third, it accords with the black-box nature of closed-source LLMs, where model logits are not directly accessible.

We first consider the performance of verbalized confidence elicitation in ERC tasks. This involves designing prompts specifically tailored to the unique template of each LLM, instructing the model to concurrently generate both a prediction and its confidence in that prediction (e.g., 'happy with 90% confidence'). The prompts used to elicit model confidence in predictions are depicted in the supplemental materials.

Logit-Based Confidence Estimation. We leveraged token-level logits as a quantitative indicator of a model's confidence in its predictions. Logit probabilities provide a nuanced understanding of model behavior and the certainty of predictions by quantifying the likelihood of the next token in a generating sequence [11,14]. To obtain these data, we designed prompts in accordance with model templates, aimed to restrict model response a singular token chosen from the predefined set of emotion categories as defined by the dataset under investigation. To ensure the model's subsequent token is indeed an emotion label from the specified emotion categories, we explicitly constrained the model's output to the provided set of emotions. For example, for EmoContext, we limited the model outputs to 'others', 'happy', 'sad', 'angry'. This obligation necessitates the model to allocate probability densities to these emotion labels. Subsequently, we extracted the logit values associated with each label and employed the softmax function to normalize these probabilities. The label with the highest probability was then selected as the model's prediction and its probability in the normalized probability vector was used as an indicator of the model's confidence in that prediction (See Algorithm 1.) The probability vector derived from this methodology serves as the basis for evaluating the model's calibration in Sect. 4.

Algorithm 1. Logit-Based Confidence Estimation for Emotion Recognition

Input: LLM $\hat{\pi}$, dataset $\{(x_i, y_i)\}_{i=1}^{N}$ with label space $\mathcal{Y}$
Output: Model predictions $y \in \mathcal{Y}$ with corresponding confidence scores
procedure LogitBasedConfidenceEstimation
    Use the dataset to design prompts according to model templates to elicit single-token responses, constrained to the set of emotion categories $\mathcal{Y}$
    for each prompt in dataset **do**
        let T = model's output token
        if $T \in \mathcal{Y}$ **then**
            Extract logit scores $L = \{l_1, l_2, \ldots, l_n\}$ for each $y_i \in \mathcal{Y}$
            Normalize logits using softmax: $\mathrm{P} = \mathrm{softmax}(L)$
            Find max probability: $p_{\max} = \max(\mathrm{P})$
            let y_{pred} = emotion label corresponding to $p_{\max}$
            return y_{pred} and $p_{\max}$ as model's prediction and confidence
        else
            return Error: Invalid token generated
        end if
    end for
end procedure

3.3 Verification by P(True)

In computational complexity, the process of verification of correctness is generally less complex than that of generation; it is primarily involves assessing truthfulness of given information (True/False), rather than generating contextually appropriate content. A substantial body of research indicates that LLMs demonstrate improved proficiency in verification- and criticism-based tasks compared to generation tasks [14,16,25].

In the context of selective prediction, a response is judged correct if P(True) surpasses a predetermined threshold, and notably, if the model is well-calibrated. Selective prediction based on P(True) using LLMs has been utilized in several tasks including multi-choice and open-world question answering [14]. In this paper, we take a similar approach on the task of ERC to verify the verbalized predictions generated by the model. Specifically, we take the model's predictions produced in the verbalized confidence stage, and ask the model to verify whether a suggested emotional state can accurately convey the emotion state of a query utterance within the given context, constrained to the emotion categories provided by the dataset under investigation. Model output is constrained to 'True' and 'False.' We then extract the logit score generated by the model as P(True), the probability that the model considers a response to a query to be correct. In this context, P(True) acts as a potential indicator of model confidence in the validity of the response. We identify a threshold, and if the P(True) surpasses that, the prediction is accepted. This technique enables the filtering of model output by considering only those predictions that the model asserts with high confidence.

3.4 Evaluation Metrics

In this paper, the performance of in-context learning is evaluated in terms of weighted F1-score. Accuracy measures the overall proportion of correct predictions, while F1-score balances precision and recall, making it more informative for evaluating performance on imbalanced datasets. Furthermore, the overall effectiveness of probabilistic predictions is measured by Brier scores, and the calibration of each model is evaluated using the Expected Calibration Error (ECE). Additional results, including the Area Under the ROC Curve (AUC-ROC), are provided in the supplemental material https://github.com/samadroohi/erc-llm-calibration, along with comprehensive definitions of each metric.

4 Experiments and Results

This section describes the experimental settings and analysis results conducted in this paper. All the supplementary materials and source code for this paper are available in the GitHub repository: https://github.com/samadroohi/erc-llm-calibration.

4.1 Models and Datasets

To assess the performance and calibration efficacy of four open-access large language models (LLMs)-Llama2-7B, Llama2-13B, Mistral-7B, and Zephyr-7B-we conducted a comprehensive experimental analysis. The Llama family models are trained using entirely public datasets without reliance on proprietary and inaccessible data sources [26]. Mistral-7B features an innovative grouped query attention alongside a sliding window attention, allowing for faster inference and longer response sequences [13]. Zephyr is a version of Mistral-7B, fine-tuned on human preferences to enhance its functionality as a conversational assistants [28]. These models are examined on three widely used datasets for emotion recognition in conversation as followings:

- The Multimodal EmotionLines Dataset (MELD) comprises approximately 1,400 dialogues from the TV series *Friends* and private Facebook Messenger conversations, annotated with labels from extended Ekman's basic emotion categories: 'neutral', 'joy', 'sadness', 'anger', 'fear', 'surprise', 'disgust' [20].

- The EmoWOZ dataset features around 11,000 task-oriented human-machine dialogues, annotated with emotion categories of 'neutral', 'fearful', 'dissatisfied', 'apologetic', 'abusive', 'excited', and 'satisfied' [8].
- The EmoContext dataset, published for the emotion detection challenge of SemEval-2019 Task 3, is a dataset with 30,160 three turns dialogues, labeled by one of the emotion classes: 'happy', 'sad', 'angry', or 'others' [4].

4.2 Structure of Dialogues

A dialogue used for prompt generation is a windowed multi-turn conversation between interlocutors, tagged using special tokens. We set the window size to 3, meaning that each input includes a starting utterance and the subsequent two utterances by the speakers as context and the third utterance as query. Let's consider a conversation represented as a sequence $[(s_1, u_1), \cdots, (s_N, u_N)]$, where each pair (s_i, u_i) includes an utterance u_i and its corresponding speaker s_i for $i = 1, \cdots, N$. The goal of ERC is to predict emotion label $y \in \mathcal{Y}$ for the query utterance u_i, considering the context provided by a sequence of d preceding utterances $[u_{i-d}, \cdots, u_{i-1}]$, here $d = 2$.

4.3 Performance of LLMs on ERC

To evaluate the performance of 2-shot ERC tasks using LLMs, we compare their performance against that of traditional discriminative methods, utilizing weighted F1-score metric. This metric is a common measure of model performance when there is an imbalance between the classes. The comparison illustrated in Table 1 shows the performance disparities between LLM-based verbalization techniques and the benchmark state-of-the-art (SOTA) accuracies for discriminative methods. It is evident from the analysis that in-context learning approach shows an average performance but generally lag behind supervised, task-specific models tailored for ERC. In the table, the SOTA performance for discriminative models are marked in green, while superior performance amongst the evaluated LLMs are highlighted in blue outline. Notably, the Mistral-7B model consistently outperforms its LLM counterparts across all datasets included in the study. A detailed analysis of model performance in terms of a confusion matrix is available in supplemental material.

Table 2 depicts the performance of predictions derived from logit probabilities outputted by the three models. The results across datasets reveals that Mistral-7B consistently outperforms the other, followed by Llama-13B, Zephyr, and Llama-7B, which exhibits the lowest performance. One discrepancy is for EmoWoz dataset, where Llama-7B shows superior performance to Llama-13B and Zephyr, though it has smaller parameter size and same prompt is used to produce outputs. Further analysis using confusion matrix revealed that Llama-7B tends to classify the majority of instances as 'neutral' which includes huge part of the dataset distribution and it shows inferior performance for other categories. This strategy inadvertently boosts the overall performance for Llama-7B. Conversely, Llama-13B and Zephyr demonstrate consistent performance

Table 1. Performance Comparison: 2-shot Learning With LLMs Versus Traditional Supervised Discriminative Approach

Model	MELD	EmoWoZ	EmoContext
DialogRNN [18,21]	0.57	0.75	0.76
HRLCE [21]	-	-	0.77
Hidialog [15]	0.67	–	–
COSMIC [8]	0.64	0.77	–
Llama-7B	0.42	0.56	0.53
Llama-13B	0.53	0.43	0.71
Mistral-7B	0.58	0.63	0.72
Zephyr-7B	0.46	0.41	0.50

across different classes when evaluated using the F1-score. Furthermore, discrepancies between the performance of logit-based and verbalized methods can be attributed to several factors, including techniques such as thresholding and sampling employed by LLMs to produce final outputs in the verbalized method. Comparing predictions produced by logit probabilities and verbalized predictions, Mistral demonstrates a higher level of consistency between two methods and still the highest level of performance across all datasets.

Table 2. The Performance of Logit-Based Approach in Terms of F-Score (Weighted)

Model	MELD	EmoWoZ	EmoContext
Llama-7B	0.43	0.61	0.53
Llama-13B	0.46	0.38	0.55
Mistral-7B	0.57	0.65	0.71
Zephyr-7B	0.51	0.56	0.73

4.4 Calibration Efficacy of LLMs

In this section we analyze the calibration efficacy of each LLMs employing verbalized confidence elicitation and logit-based approaches as presented in Table 3. Specifically, we evaluate ECE for verbalized and logit-based approaches alongside the Brier scores (BS) for logit-based. Notably, Llama-7B demonstrates relatively higher ECE scores in both verbalized (0.45) and logit (0.54) forms, indicating a significant calibration discrepancy from other models. Conversely, Mistral-7B shows a notably lower ECE score in logit form (0.28), alongside the minimal BS (0.10) among the models evaluated, suggesting a more reliable value of predictive probability.

Table 3. Brier Scores and ECE Scores for Logit-Based and ECE Scores for Verbalized Approaches

Model	ECE-Verbalized	BS-Logits	ECE-Logits
MELD			
Llama-7B	0.45	0.16	0.54
Llama-13B	0.28	0.15	0.49
Mistral-7B	0.29	0.10	0.28
Zephyr-7B	0.46	0.12	0.38
EmoWoZ			
Llama-7B	0.31	0.10	0.35
Llama-13B	0.46	0.16	0.54
Mistral-7B	0.31	0.08	0.23
Zephyr-7B	0.77	0.10	0.34
EmoContext			
Llama-7B	0.31	0.21	0.39
Llama-13B	0.14	0.21	0.41
Mistral-7B	0.13	0.11	0.18
Zephyr-7B	0.43	0.12	0.20

For the EmoWoZ dataset, a notable divergence is observed with Zephyr-7B recording the highest ECE-verbalized score (0.77) across all datasets and models, which points to a considerable mis-calibration between verbalized confidence and actual outcomes. This discrepancy may rooted in the extra layer of human alignment applied on Zephyr. Nonetheless, its BS (0.10) and ECE-logits (0.34) remain competitive for logit-based approach, suggesting that while the model struggles with verbalized calibration, its logit-based predictions retain a level of reliability.

Within the EmoContext dataset, Mistral-7B stands out with the lowest ECE scores in both verbalized (0.13) and logit-based (0.18) formats, coupled with a low BS (0.11), underscoring its superior calibration efficacy across diverse contexts. This performance indicates a robustness in Mistral-7B's calibration mechanisms, enabling it to maintain reliability across both verbalized and logit-based approaches. Conversely, Llama-13B, while exhibits a relatively low ECE of 0.14 when assessed using verbalized confidence measures, suggesting a degree of calibration in its predictions. However, a significant divergence is observed in its performance under logit-based ECE evaluation, where the score escalates to 0.41. This disparity underscores the inherent potential within Llama-13B in verbalized approach, albeit it noteworthy not to overestimate its superiority to logit-based as ECE for the verbalized approach is computed using the confidence value expressed by the model for single label. Figure 1 illustrates the superior and inferior models in terms of calibration.

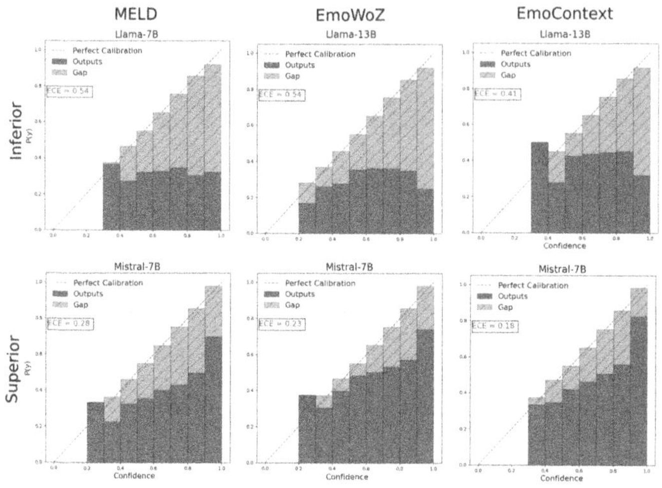

Fig. 1. The calibration diagrams comparing models calibration efficacy: Superior calibration of Mistral-7B versus inferior calibration of Llama family across datasets

4.5 The Performance of P(True)

Using an approach similar to the selective prediction, discussed in Sect. 3, we classified each emotion label, predicted by verbalization technique, as either 'True' (correctly identified) or 'False' (incorrectly identified) using the model itself applying a threshold of 0.5 on the probability of output being 'True'. Our Analysis showed a trade-off between the ability to correctly identify true labels (TPR) and the inclination to validate false labels (FPR) across all datasets. Notably, the Mistral-7B model consistently exhibited high TPRs at the cost of elevated FPRs, suggesting a radical strategy that prioritizes the capture of true instances over the avoidance of false positives. Conversely, models like Llama-7B adopted a more conservative strategy, emphasizing specificity over sensitivity. On the other hand, Zephyr-7B model presents a more balanced profile (e.g., for MELD and EmoWoZ), highlighting its effectiveness in discerning true labels while maintaining a lower rate of false identifications. A detailed analysis of P(True) is available in the supplemental materials.

5 Discussion

Our investigation highlights significant limitation of LLMs in providing calibrated output for ERC task. While LLMs have demonstrated impressive capabilities in generating coherent text and answering factual queries, they are not well-calibrated for more nuanced tasks like ERC. Furthermore, methods such as P(True), while effective for assessing factual information, may fail to fully capture the complex emotional nuances present in conversational contexts. The subtlety of human emotions, often conveyed through sarcasm, irony, or cultural references, cannot be effectively captured by simplistic probabilistic measures.

To address these challenges, it is imperative to explore alternative methods widely used in deep learning that could enhance calibration in ERC tasks. Techniques such as temperature scaling, isotonic regression, and ensemble methods have shown promise in improving model calibration in classification tasks [11]. Temperature scaling, for instance, adjusts the confidence levels of predictions without altering the model's underlying behavior, leading to better-aligned probabilities with actual outcomes. Applying such methods to ERC could help models better estimate the uncertainty associated with their predictions.

Another promising approach is the application of conformal prediction, a framework that provides reliable confidence measures for model predictions by offering valid prediction intervals or sets [24]. Conformal prediction does not make strong distributional assumptions and can adapt to the uncertainty inherent in complex tasks like emotion recognition. By generating prediction sets that contain the true label with a specified probability, conformal prediction can offer a more nuanced understanding of model confidence in ERC tasks.

In summary, our findings suggest that to improve the calibration of LLMs in ERC tasks, we must look beyond traditional methods and embrace advanced techniques prevalent in deep learning. Future research should focus on systematically evaluating these methods within the ERC framework using LLMs, aiming to develop models that are both accurate and well-calibrated in their emotional predictions.

6 Conclusion

In this study, we demonstrated that few-shot emotion recognition in conversations using out-of-the-box large language models (LLMs) can achieve performance close to supervised benchmarks, highlighting their potential to generalize across affective computing tasks without the need for sophisticated architectures or fine-tuning.

Our findings reveal that while neither verbalized confidence elicitation nor logit-based confidence estimation achieve perfect calibration, LLMs tend to be overconfident in their predictions. The Mistral-7B model showed modest calibration, with the Llama family models performing better with verbalized outputs and Mistral and Zephyr models better with logit-based outputs. This divergence likely stems from alignment with human instructions, introducing biases towards human preferences. Additionally, using P(True) for prediction verification yielded moderate success.

We also found that tailoring prompts to each model's specific requirements significantly improves ERC performance. However, LLMs performed suboptimally on datasets like EmoWoz without explicit category definitions, due to its unique categorization strategy.

While our study offers new insights into LLM performance and calibration in ERC tasks, several limitations remain. Future research should explore more advanced architectures, alternative calibration methods, and a broader range of LLMs. Addressing these challenges could improve the feasibility of using general-purpose LLMs as reliable substitutes for traditional ERC systems.

References

1. Amayuelas, A., Pan, L., Chen, W., Wang, W.: Knowledge of knowledge: exploring known-unknowns uncertainty with large language models. arXiv:2305.13712 [cs] (2023)
2. Bloom, L., Lahey, M.: Language development and language disorders (1978). https://eric.ed.gov/?id=ED150613, publisher: ERIC
3. Casas, J., et al.: Enhancing conversational agents with empathic abilities. In: Proceedings of the 21st ACM International Conference on Intelligent Virtual Agents. IVA '21, New York, NY, USA, pp. 41–47. Association for Computing Machinery (2021). https://doi.org/10.1145/3472306.3478344
4. Chatterjee, A., Narahari, K.N., Joshi, M., Agrawal, P.: SemEval-2019 Task 3: EmoContext contextual emotion detection in text. In: May, J., Shutova, E., Herbelot, A., Zhu, X., Apidianaki, M., Mohammad, S.M. (eds.) Proceedings of the 13th International Workshop on Semantic Evaluation, Minneapolis, Minnesota, USA, pp. 39–48. Association for Computational Linguistics (2019). https://doi.org/10.18653/v1/S19-2005, https://aclanthology.org/S19-2005
5. Chen, D., Zhengwei, H., Yiting, T., Jintao, M., Khanal, R.: Emotion and sentiment analysis for intelligent customer service conversation using a multi-task ensemble framework. Clust. Comput. (2023). https://doi.org/10.1007/s10586-023-04073-z
6. Chung, H.W., et al.: Scaling instruction-finetuned language models. arXiv:2210.11416 [cs] (2022). https://doi.org/10.48550/arXiv.2210.11416
7. Cosmides, L., Tooby, J.: Are humans good intuitive statisticians after all? Rethinking some conclusions from the literature on judgment under uncertainty. Cognition **58**(1), 1–73 (1996). https://doi.org/10.1016/0010-0277(95)00664-8, https://www.sciencedirect.com/science/article/pii/0010027795006648
8. Feng, S., et al.: EmoWOZ: a large-scale corpus and labelling scheme for emotion recognition in task-oriented dialogue systems. arXiv:2109.04919 [cs] (2022). https://doi.org/10.48550/arXiv.2109.04919
9. Feng, S., Sun, G., Lubis, N., Zhang, C., Gašić, M.: Affect recognition in conversations using large language models. arXiv:2309.12881 [cs] (2023). https://doi.org/10.48550/arXiv.2309.12881
10. Ghosal, D., Majumder, N., Gelbukh, A., Mihalcea, R., Poria, S.: COSMIC: COmmonSense knowledge for eMotion identification in conversations. In: Findings of the Association for Computational Linguistics: EMNLP 2020, pp. 2470–2481. Association for Computational Linguistics (2020). https://doi.org/10.18653/v1/2020.findings-emnlp.224, https://aclanthology.org/2020.findings-emnlp.224
11. Guo, C., Pleiss, G., Sun, Y., Weinberger, K.Q.: On calibration of modern neural networks. In: International Conference on Machine Learning, pp. 1321–1330. PMLR (2017). http://proceedings.mlr.press/v70/guo17a.html
12. Huang, X., et al.: Emotion Detection for conversations based on reinforcement learning framework. IEEE Multimedia **28**(2), 76–85 (2021). https://doi.org/10.1109/MMUL.2021.3065678, conference Name: IEEE MultiMedia
13. Jiang, A.Q., et al.: Mistral 7B (2023). https://arxiv.org/abs/2310.06825v1

14. Kadavath, S., et al.: Language models (mostly) know what they know. arXiv:2207.05221 [cs] (2022). https://doi.org/10.48550/arXiv.2207.05221
15. Lei, S., Dong, G., Wang, X., Wang, K., Wang, S.: InstructERC: reforming emotion recognition in conversation with a retrieval multi-task LLMS framework. arXiv:2309.11911 [cs] (2023). https://doi.org/10.48550/arXiv.2309.11911
16. Lin, Z., Trivedi, S., Sun, J.: Generating with confidence: uncertainty quantification for black-box large language models. arXiv:2305.19187 [cs, stat] (2023). https://doi.org/10.48550/arXiv.2305.19187
17. Majumder, N., et al.: MIME: MIMicking emotions for empathetic response generation. arXiv:2010.01454 [cs] (2020). https://doi.org/10.48550/arXiv.2010.01454
18. Majumder, N., Poria, S., Hazarika, D., Mihalcea, R., Gelbukh, A., Cambria, E.: DialogueRNN: an attentive RNN for emotion detection in conversations. In: Proceedings of the AAAI Conference on Artificial Intelligence, vol. 33, pp. 6818–6825 (2019). issue: 01
19. McKenna, N., Li, T., Cheng, L., Hosseini, M., Johnson, M., Steedman, M.: Sources of hallucination by large language models on inference tasks. In: Bouamor, H., Pino, J., Bali, K. (eds.) Findings of the Association for Computational Linguistics: EMNLP 2023, pp. 2758–2774. Association for Computational Linguistics, Singapore (2023). https://doi.org/10.18653/v1/2023.findings-emnlp.182, https://aclanthology.org/2023.findings-emnlp.182
20. Poria, S., Hazarika, D., Majumder, N., Naik, G., Cambria, E., Mihalcea, R.: MELD: a multimodal multi-party dataset for emotion recognition in conversations. arXiv:1810.02508 [cs] (2019). https://doi.org/10.48550/arXiv.1810.02508
21. Poria, S., Majumder, N., Mihalcea, R., Hovy, E.: Emotion recognition in conversation: research challenges, datasets, and recent advances. IEEE Access **7**, 100943–100953 (2019). https://doi.org/10.1109/ACCESS.2019.2929050, https://ieeexplore.ieee.org/document/8764449?signout=success, conference Name: IEEE Access
22. Radford, A., Wu, J., Child, R., Luan, D., Amodei, D., Sutskever, I.: Language models are unsupervised multitask learners. OpenAI blog **1**(8), 9 (2019). https://insightcivic.s3.us-east-1.amazonaws.com/language-models.pdf
23. Rajani, N.F., McCann, B., Xiong, C., Socher, R.: Explain Yourself! leveraging language models for commonsense reasoning. arXiv:1906.02361 [cs] (2019)
24. Shafer, G., Vovk, V.: A tutorial on conformal prediction. J. Mach. Learn. Res. **9**(3) (2008)
25. Tian, K., et al.: Just ask for calibration: strategies for eliciting calibrated confidence scores from language models fine-tuned with human feedback. In: Bouamor, H., Pino, J., Bali, K. (eds.) Proceedings of the 2023 Conference on Empirical Methods in Natural Language Processing, pp. 5433–5442. Association for Computational Linguistics, Singapore (2023). https://doi.org/10.18653/v1/2023.emnlp-main.330, https://aclanthology.org/2023.emnlp-main.330
26. Touvron, H., et al.: LLaMA: open and efficient foundation language models (2023). https://arxiv.org/abs/2302.13971v1
27. Tu, G., Liang, B., Qin, B., Wong, K.F., Xu, R.: An empirical study on multiple knowledge from ChatGPT for emotion recognition in conversations. In: Bouamor, H., Pino, J., Bali, K. (eds.) Findings of the Association for Computational Linguistics: EMNLP 2023. pp. 12160–12173. Association for Computational Linguistics, Singapore (2023). https://doi.org/10.18653/v1/2023.findings-emnlp.813, https://aclanthology.org/2023.findings-emnlp.813
28. Tunstall, L., et al.: Zephyr: direct distillation of LM alignment (2023). arXiv:abs/2310.16944v1

29. Xiao, Y., Wang, W.Y.: On hallucination and predictive uncertainty in conditional language generation. arXiv:2103.15025 [cs] (2021). https://doi.org/10.48550/arXiv.2103.15025
30. Xu, Y., Zeng, Z., Shen, Z.: Efficient cross-task prompt tuning for few-shot conversational emotion recognition. arXiv:2310.14614 [cs] (2023)
31. Yohanes, D., Putra, J.S., Filbert, K., Suryaningrum, K.M., Saputri, H.A.: Emotion detection in textual data using deep learning. Procedia Comput. Sci. **227**, 464–473 (2023). https://doi.org/10.1016/j.procs.2023.10.547, https://www.sciencedirect.com/science/article/pii/S1877050923017143
32. Zhang, S., Chen, Z., Shen, Y., Ding, M., Tenenbaum, J.B., Gan, C.: Planning with large language models for code generation. arXiv:2303.05510 [cs] (2023)
33. Zhang, W., Deng, Y., Liu, B., Pan, S.J., Bing, L.: Sentiment analysis in the era of large language models: a reality check. arXiv:2305.15005 [cs] (2023). https://doi.org/10.48550/arXiv.2305.15005
34. Zhang, Y., Wang, M., Tiwari, P., Li, Q., Wang, B., Qin, J.: DialogueLLM: context and emotion knowledge-tuned LLaMA models for emotion recognition in conversations (2023). https://doi.org/10.48550/arXiv.2310.11374, http://arxiv.org/abs/2310.11374

Ensuring Fairness in Stochastic Multi-armed Bandit Problems for Effective Group Recommendations

Rei Ozaki and Atsuyoshi Nakamura(✉)

Graduate School of Information Sience and Technology, Hokkaido University, Sapporo, Japan
{rei.ozaki,atsu}@ist.hokudai.ac.jp

Abstract. We study online batch recommendation in which one item is recommended to each user and the reward is returned from the user at each round. Users are assumed to be partitioned into groups and the task of the recommender system is to decide the probability distribution of the items for each group of users to send the next round of recommendations. The objective of the recommender system is to maximize the cumulative reward summed up for all users. In this paper, we formalize the FAIR-BATCH-MAB problem (Multi-Armed Bandit) as the above online batch recommendation with fairness constraint: the number of recommendations of each item must be at least [the specified minimum recommendation ratio of the item] $\times$ #[user] $\times t - \alpha$ in any round t, where α is an unfairness tolerance constant. We show that this problem is represented as linear programming using an unknown set of click rate $\{\mu_{i,j}\}$ if the reward for item-i recommendation to a user of the group j is 1(clicked) with probability $\mu_{i,j}$, and 0(not clicked) with probability $1 - \mu_{i,j}$. We define regret by the expected total cumulative reward difference from the cumulative reward of the optimal solution of the linear programming. We propose the FBO (Fair Batch Optimizer) algorithm using the bandit algorithm as click rate estimators for the FAIR-BATCH-MAB problem. In simulation experiments based on a real-world dataset, we demonstrate that the performance of the FBO algorithm combined with Thompson Sampling is close to the performance of the optimal solution of the linear programming and outperforms the FBO algorithm combined with UCB or sample mean.

Keywords: Multi-armed Bandit · Batch Optimization · Fair Recommendation

1 Introduction

The classical stochastic Multi-Armed Bandit (MAB) problem considers a game of one player who repeatedly pulls one of k arms and receives a reward generated according to an unknown and arm-dependent distribution over some real number

M. Gong et al. (Eds.): AI 2024, LNAI 15442, pp. 213–227, 2025.
https://doi.org/10.1007/978-981-96-0348-0_16

range. In the game, the player aims to maximize cumulative reward within a time horizon T. The stochastic MAB problem is widely used in practical optimizations such as recommender systems [10], retail, and crowd-sourcing [13]. To maximize the cumulative reward, the player must pull sub-optimal arms as few times as possible, which causes unfair service for arm providers in applications like recommender systems and crowd-sourcing.

In this paper, we study a kind of recommendation problem called a FAIR-BATCH-MAB problem in which a system repeats the following round; the system recommends one item to each user and receives a reward (response) from the user. Users are assumed to be partitioned into groups of similar users from whom the same reward is expected for any item recommendation; rewards of the users in group j for item i are independently generated according to Bernoulli distribution with parameter $\mu_{i,j}$. (Reward 1 is obtained if a user in group j clicks the recommendation link of item i, and reward 0 is obtained otherwise, and a group-j user clicks item i with probability $\mu_{i,j}$.) The objective of the problem is the maximization of expected cumulative rewards summed over all the users under α-batch-*fairness* constraint that the ratio of pulling each arm by any round must be at least (a predetermined ratio) $-\ \alpha$. A specific application of FAIR-BATCH-MAB involves periodic recommendation systems such as direct mail delivery, online advertising delivery, and matching applications. In such applications, minimum exposure ratios can be set based on the payment amount or registration fee for each item.

The FAIR-BATCH-MAB problem is a problem of optimizing $\{N_{i,j}(t)\}$ to maximize the total cumulative reward $\sum_t \sum_i \sum_j \mu_{i,j} N_{i,j}(t)$, where $N_{i,j}(t)$ is the number of recommendations of item i to the users in group j at round t. If we know the true click rate set $\{\mu_{i,j}\}$, this optimization problem can be written as linear programming, which we call a FAIR-BATCH-MAB optimization problem. Since $\{\mu_{i,j}\}$ is unknown, such an optimal solution cannot be used in practice. We define regret of an algorithm for the FAIR-BATCH-MAB problem by the total cumulative reward difference from that for the solution of the FAIR-BATCH-MAB optimization problem.

For the FAIR-BATCH-MAB problem, we propose an algorithm called FBO (Fair-Batch Optimizer), which maximizes the instantaneous predicted total reward for the estimated click rates at each round t while satisfying the α-batch-fairness constraint. FBO with sample mean estimator (SM), which estimates click rate as (the number of clicks)/(the number of recommendations), the items that were clicked few by group-j users at the initial rounds, are never recommended to the group-j users at the later rounds, thus click rate estimator (CRE) that promotes some exploration can achieve better performance. As such estimators, we propose popular bandit policies UCB [1], KL-UCB [3], and Thompson Sampling (TS) [14].

According to our experimental results by the simulations based on the real-world dataset called the Books dataset, we confirmed (1) the optimal expected cumulative reward is significantly improved by demographic-feature-group-based recommendation compared to the recommendation that treats all the users uniformly, (2) the total cumulative reward optimization under α-batch-fairness-constraint is done by the FBO algorithm depending on the parameter α; larger α causes greater total cumulative reward and more recommendation concentration to popular items, and (3) the perfromance of FBO(TS) is close to that of the optimal solution of the FAIR-BATCH-MAB optimization problem, and it outperforms FBO(UCB) and FBO(SM) in terms of total cumulative reward and regret.

The structure of this paper is as follows. First, we discuss related work later in this section. Then, the detailed problem settings are described in Sect. 2. In Sect. 3, we propose an FBO (Fair-Batch-Optimizer) algorithm combined with several click rate estimators. We demonstrate the effectiveness of our proposed algorithm in Sect. 4, and conclude in Sect. 5.

Related Works

The first main result on the finite-time stochastic MAB problem is presented in the paper by Lai and Robbins [8] which established a lower bound on the cumulative reward for the problem. They proposed a policy that uses the upper confidence bound on expected rewards as the arm selection criterion. Around 2000, Auer et al. [2] demonstrated that the UCB policy, which is a more straightforward and intuitive arm selection strategy, can achieve logarithmic regret over time. Since then, MAB has been applied in various contexts, and numerous models have been proposed.

Among those, fairness in the MAB domain has become a significant concern. Gillen et al. [4] introduced the concept of fairness in the adversarial linear bandit setting, ensuring that arms with similar contexts must be selected with similar probability. Joseph et al. [6] considered the concept of meritocratic fairness, ensuring that good arms with higher expected rewards are selected more. Wang et al. [15] proposed the merit-based fairness of exposure constraints under which the optimal selection probability of an arm must be proportional to the merit function value of its expected reward. Reward and fairness regrets are defined as cumulative expected reward and selection distribution differences from the optimal selection distribution, respectively, and regret upper bounds for UCB-based and Thompson-Sampling-based algorithms are analyzed. All of these studies aim to ensure fairness among arms. We also deal with the fairness among arms based on the FAIR-MAB problem [12], which imposes a minimum exposure ratio r_i for each arm i, where $\{r_i\}_{i=1}^{K}$ are given beforehand.

We study fair group-based batch recommendation using bandit indices as click rate estimators. Our algorithm FBO can use any bandit index such as UCB [1], KL-UCB [3] and Thompson Sampling [14]. Our problem setting is similar to the transportation problem setting of online advertising delivery [9], and also similar to direct mail problem [11] in the viewpoint of batch recommendation using user's similarity. The difference from those previous studies is that our algorithms satisfy a fairness constraint, but the algorithms in the earlier studies do not.

Algorithm 1. BATCH-MAB Procedure

Parameters: N: number of users,
k: number of items,
u: number of groups,
$\{m_1, m_2, \dots, m_u\}$: set of the numbers of users m_i in the ith group,
T: number of rounds

for $t = 1, 2, \dots, T$ **do**
  for $j \in [u]$ **do**
    Set $\{N_{i,j}(t)\}_{i=1}^{k}$ under the constraint that $\sum_{i=1}^{k} N_{i,j}(t) = m_j$
  end for
  Recommend item i to $N_{i,j}(t)$ users in group j for $(i, j) \in [k] \times [u]$
  Receive reward $R_{i,j}(t) \sim B(N_{i,j}(t), \mu_{i,j})$ for $(i, j) \in [k] \times [u]$
end for

2 Problem Settings

Let $[n]$ denote $\{1, 2, \dots, n\}$ for any positive integer n. There are N users and k items (arms). Each user belongs to one of u groups, and the number of users in the jth group is m_j. Thus, $N = \sum_{j=1}^{u} m_j$ holds. At each round $t \in [T]$, a recommender system decides the number of users $N_{i,j}(t) (\leq m_j)$ for each item $i \in [k]$ and group $j \in [u]$, and recommends the item i to the $N_{i,j}(t)$ users in the group j. Note that $\sum_{i=1}^{k} N_{i,j}(t) = m_j$ must hold. Then, the recommender system receives a reward $R_{i,j}(t)$ from each recommendation of item i to $N_{i,j}(t)$ users in group j, where $R_{i,j}(t)$ is a random variable generated according to binomial distribution $B(N_{i,j}(t), \mu_{i,j})$. (Each user in group j clicks on the recommended item i with probability $\mu_{i,j}$, and the reward 1 is received if the user clicks and reward 0 is received otherwise.) Note that the recommender system does not know the click probabilities $\mu_{i,j}$ ($(i, j) \in [k] \times [u]$). We also assume that the random variables $R_{i,j}(t)$ at different rounds t are independent. We call this process the BATCH-MAB procedure (Algorithm 1), and an algorithm following this procedure is called the BATCH-MAB algorithm.

Let $N_i(t)$ denote the total number of item-i-recommended users at round t, that is, $N_i(t) = \sum_{j=1}^{u} N_{i,j}(t)$. We design an algorithm that satisfies the following fairness constraint.

Definition 1 (α-batch-fairness). Given fairness constraint vector $r = (r_1, r_2, \dots, r_k) \in [0, 1/k)^k$, and an unfairness tolerance constant $\alpha \geq 0$, a BATCH-MAB algorithm $\mathcal{A}$ is said to be *α-batch-fair* if

$$\lfloor r_i N t \rfloor - \sum_{s=1}^{t} N_i(s) \leq \alpha$$

for all $(i, t) \in [k] \times [T]$.

Note that α-batch-fairness is an extension of α-fairness defined by Patil et al. [12] to our repeated batch setting.

The problem we study in this paper is described as follows.

Problem 1 (FAIR-BATCH-MAB problem). Given a fairness constraint vector $r = (r_1, r_2, \ldots, r_k) \in [0, 1/k)^k$ and an unfairness tolerance constant $\alpha \geq 0$, design a BATCH-MAB algorithm that maximizes the expected cumulative reward $\sum_{t=1}^{T} \sum_{i=1}^{k} \sum_{j=1}^{u} \mathbb{E}[R_{i,j}(t)]$.

2.1 Concepts of Regret

The optimal algorithm for the FAIR-BATCH-MAB problem is the algorithm with $\{N_{i,j}(t)\}_{(i,j,t)\in[k]\times[u]\times[T]}$ that is the solution of the following problem.

Problem 2 (FAIR-BATCH-MAB optimization problem). In the setting of the FAIR-BATCH-MAB problem, solve the optimization problem for the set of variables $\{N_{i,j}(t)\}_{(i,j,t)\in[k]\times[u]\times[T]}$ to

$$\begin{aligned}
\text{maximize} \quad & \sum_{t=1}^{T} \sum_{i=1}^{k} \sum_{j=1}^{u} \mu_{i,j} N_{i,j}(t) & \\
\text{s.t.} \quad & \sum_{i=1}^{k} N_{i,j}(t) = m_j & \text{for } (j,t) \in [u] \times [T] \\
& \sum_{s=1}^{t} N_i(s) \geq \lfloor r_i N t \rfloor - \alpha & \text{for } (i,t) \in [k] \times [T].
\end{aligned}$$

Let $\{N_{i,j}^*(t)\}_{(i,j,t)\in[k]\times[u]\times[T]}$ denote the solution of the FAIR-BATCH-MAB optimization problem. Note that this solution cannot be used for a recommendation system because the click rates $\mu_{i,j}$ for $(i,j) \in [k] \times [u]$ are unknown for the system. Using the optimal solution $\{N_{i,j}^*(t)\}_{t=1}^{T}$, the regret for the FAIR-BATCH-MAB problem is defined as follows.

Definition 2 (Regret for the FAIR-BATCH-MAB problem). Given fairness constraint vector $r = (r_1, r_2, \ldots, r_k)$, and an unfairness tolerance constant $\alpha \geq 0$, the regret $\mathrm{Reg}(T)$ of a FAIR-BATCH-MAB algorithm with $\{N_{i,j}(t)\}_{t=1}^{T}$ is defined as:

$$\begin{aligned}
\mathrm{Reg}(T) &= \sum_{t=1}^{T} \sum_{i=1}^{k} \sum_{j=1}^{u} \mu_{i,j} N_{i,j}^*(t) - \sum_{t=1}^{T} \sum_{i=1}^{k} \sum_{j=1}^{u} \mathbb{E}[R_{i,j}(t)] \\
&= \sum_{t=1}^{T} \sum_{i=1}^{k} \sum_{j=1}^{u} \mu_{i,j} \left(N_{i,j}^*(t) - \mathbb{E}[N_{i,j}(t)]\right)
\end{aligned}$$

The above Definition 2 takes fairness into account by considering the difference from the cumulative reward obtained using the optimal solution $\{N_{i,j}^*(t)\}_{t-1}^{T}$ that satisfies α-batch-fairness.

Algorithm 2. FBO(CRE): Fair-Batch-Optimizer

Input: CRE: Estimator of $\{\mu_{i,j}\}_{(i,j)\in[k]\times[u]}$ (Click Rate Estimator)

$\tilde{N}_{i,j}(0), \tilde{R}_{i,j}(0) \leftarrow 0$ for $(i,j) \in [k] \times [u]$
for t=1,2,...,T **do**
  if $t = 1$ **then**
    $N_{i,j}(t) \leftarrow \dfrac{N}{k}$ for $(i,j) \in [k] \times [u]$
  else
    $\{N_{i,j}(t)\}_{(i,j)\in[k]\times[u]} \leftarrow \text{OptInstSol}\left(\{\hat{\mu}_{i,j}\}_{(i,j)\in[k]\times[u]}, \{\tilde{N}_{i,j}(t-1)\}_{(i,j)\in[k]\times[u]}\right)$
  end if
  Recommend item i to $N_{i,j}(t)$ users in group j for $(i,j) \in [k] \times [u]$
  Receive reward $R_{i,j}(t)$ for $(i,j) \in [k] \times [u]$
  $\tilde{N}_{i,j}(t) \leftarrow \tilde{N}_{i,j}(t-1) + N_{i,j}(t)$, $\tilde{R}_{i,j}(t) \leftarrow \tilde{R}_{i,j}(t-1) + R_{i,j}(t)$
  $\hat{\mu}_{i,j} \leftarrow \text{CRE}(\{\tilde{N}_{i,j}(t)\}_{(i,j)\in[k]\times[u]}, \{\tilde{R}_{i,j}(t)\}_{(i,j)\in[k]\times[u]})$
end for

3 Algorithm

Let $\tilde{N}_{i,j}(t)$ denote the cumulative number of recommendations of item i to users in group j by round t, and let $\tilde{R}_{i,j}(t)$ denote the cumulative reward for the item-i recommendations to the group-j users by round t, that is, $\tilde{N}_{i,j}(t) = \sum_{s=1}^{t} N_{i,j}(s)$ and $\tilde{R}_{i,j}(t) = \sum_{s=1}^{t} R_{i,j}(s)$.

The pseudocode of the proposed algorithm FBO (Fair Batch Optimizer) is shown in Algorithm 2. The algorithm FBO needs CRE click rate estimators, and at each round t, it calculates the set of estimated click rates $\{\hat{\mu}_{i,j}\}_{(i,j)\in[k]\times[u]}$ by CRE. Then, FBO solves the following instantaneous FAIR-BATCH-MAB optimization problem using the estimated click rate set and the cumulative recommendation number set $\{\tilde{N}_{i,j}(t-1)\}_{(i,j)\in[k]\times[u]}$.

Problem 3 (Instantaneous FAIR-BATCH-MAB optimization problem). In the setting of the FAIR-BATCH-MAB problem, at each time $t \in [T]$, given the set of estimated click rates $\{\hat{\mu}_{i,j}\}_{(i,j)\in[k]\times[u]}$ and the set of cumulative numbers of item-i recommendations to the group-j users $\{\tilde{N}_{i,j}(t-1)\}_{(i,j)\in[k]\times[u]}$ by round $t-1$, solve the optimization problem for the set of variables $\{N_{i,j}(t)\}_{(i,j)\in[k]\times[u]}$ to

$$\begin{aligned}
\text{maximize} \quad & \sum_{i=1}^{k}\sum_{j=1}^{u} \hat{\mu}_{i,j} N_{i,j}(t) & \\
\text{s.t.} \quad & \sum_{i=1}^{k} N_{i,j}(t) = m_j & \text{for } j \in [u] \\
& \sum_{j=1}^{u} \left(N_{i,j}(t) + \tilde{N}_{i,j}(t-1)\right) \geq \lfloor r_i N t \rfloor - \alpha & \text{for } i \in [k].
\end{aligned}$$

We define the function $\text{OptInstSol}\left(\{\hat{\mu}_{i,j}\}_{(i,j)\in[k]\times[u]}, \{\tilde{N}_{i,j}(t-1)\}_{(i,j)\in[k]\times[u]}\right)$ as the function returning the optimal values of $\{N_{i,j}(t)\}_{(i,j)\in[k]\times[u]}$ for this problem. At round $t \geq 2$, FBO sets $\{N_{i,j}(t)\}_{(i,j)\in[k]\times[u]}$ to the optimal solution returned by OptInstSol $\left(\{\hat{\mu}_{i,j}\}_{(i,j)\in[k]\times[u]}, \{\tilde{N}_{i,j}(t-1)\}_{(i,j)\in[k]\times[u]}\right)$. At round $t = 1$, FBO sets it to $\{m_j/k\}_{(i,j)\in[k]\times[u]}$, which satisfies $\sum_{i=1}^{k} N_{i,j}(1) = m_j$ and $N_i(1) = N/k$.

Let $\bar{\mu}_{i,j}(t)$ denote the sample mean of rewards($\in \{0,1\}$) for item i and an individual user in group j obtained by round t, that is,

$$\bar{\mu}_{i,j} = \frac{\tilde{R}_{i,j}(t)}{\tilde{N}_{i,j}(t)}.$$

We consider the following five CREs that calculate $\hat{\mu}_{i,j}$, which is used to calculate $N_{i,j}(t+1)$, from $\tilde{N}_{i,j}(t)$ and $\tilde{R}_{i,j}(t)$ for each $(i,j) \in [k] \times [u]$.

U: Uniform

$$\hat{\mu}_{i,j} = 1$$

SM: Sample Mean

$$\hat{\mu}_{i,j} = \bar{\mu}_{i,j}$$

UCB: Upper Confidence Bound (UCB1 [1])

$$\hat{\mu}_{i,j} = \bar{\mu}_{i,j} + \sqrt{\frac{\log(Nt)}{2\tilde{N}_{i,j}(t)}}$$

KL-UCB: Kullback-Leibler (KL) divergence UCB [3]

$$\hat{\mu}_{i,j} = \max\left\{\mu_{i,j} : 2\tilde{N}_{i,j}(t)d\left(\bar{\mu}_{i,j}, \mu_{i,j}\right) \geq \log(Nt)\right\},$$

where $d(p,q)$ is KL divergence of Bernoulli distribution with parameter p from Bernoulli distribution with parameter q, that is,

$$d(p,q) = p \ln\frac{p}{q} + (1-p)\ln\frac{1-p}{1-q}.$$

TS: Thompson Sampling [14]

$$\hat{\mu}_{i,j} \sim \mathrm{Beta}(\tilde{R}_{i,j}(t)+1, \tilde{N}_{i,j}(t) - \tilde{R}_{i,j}(t)+1),$$

where $\mathrm{Beta}(\alpha,\beta)$ is a beta distribution with parameters α and β. The above distribution is the posterior distribution for the prior distribution $\mathrm{Beta}(1,1)$ after $\tilde{R}_{i,j}(t)$ clicks for $\tilde{N}_{i,j}(t)$ recommendations.

4 Experiments

In this section, we analyze our proposed algorithm's expected reward and regret. First, we begin by introducing the baseline and explaining the dataset setup.

4.1 Dataset

In our experiments, two datasets, *Books dataset* [7] and *MovieLens 1M Dataset* [5], are used. *Books dataset* includes book information (book ID, title, author, publication year, etc.), user information (user ID, location, age), and rating information (user ID, book ID, rating (0–10)), while *MovieLens 1M Dataset* includes movie information (movie ID, title, genres), user information (user ID, gender, age, occupation, zip-code), and rating information (user ID, movie ID, rating (1–5)). Gender is denoted by "M" for male and "F" for female, and is treated as 0 for male and 1 for female. Age is categorized into the following ranges: 1 ("Under 18"), 18 ("18 – 24"), 25 ("25 – 34"), 35 ("35 – 44"), 45 ("45 – 49"), 50 ("50 – 55"), and 56 ("56+").

The set of ground truth click rates $\{\mu_{i,j}\}_{(i,j)\in[k]\times[u]}$ used in our experiments is constructed as follows.

Books dataset: First, we sorted books by the number of ratings in descending order and selected the top k books. We then extracted all the users who rated these k books. Set N to the number of the extracted users. Next, we partitioned N users into u groups by u-means clustering of their age feature data. Finally for each $(i, j) \in [k] \times [u]$, we set $\mu_{i,j}$ to the proportion of ratings at least 9 to all the book-i ratings by group-j users.

MovieLens 1M Dataset: First, we did multi-hot encoding of all 3706 movies based on their genres and then divided them into k groups by k-means clustering. We then extracted all the users who rated the movies in these k groups. Set N to the number of the extracted users. Next, we partitioned these N users into u groups using u-means clustering based on their age and gender feature data. Finally, for each $(i, j) \in [k] \times [u]$, we set $\mu_{i,j}$ to the proportion of ratings of at least 5 to all movie group-i ratings by group-j users.

The difference between the above click-rate-model constructions from the two datasets is whether items are grouped or not. MovieLens 1M dataset has a good feature (genres) for grouping items for recommendation. Item grouping is desirable for our proposed method because the same item might be repeatedly recommended by the optimal recommendation in our problem setting, and we can select items that have not been recommended by the recommended item group in the case of the item group recommendation.

We constructed four click rate sets bCRS-k-u from the Books dataset and one click rate set mCRS-k-u from MovieLens 1M dataset by the above procedure using different parameters k and u. The number of users N and those in the groups $m_1, m_2, \ldots, m_u$ in the five sets are shown in Table 1.

All the constructed click rate $\{\mu_{i,j}\}_{(i,j)\in[k]\times[u]}$ except bCRS-100-1 are shown by heat maps in Fig. 1 (bCRS-100-5, bCRS-100-10) or by tables in Table 2 (bCRS-10-5, mCRS-10-10). Figure 1 shows a heatmap of click rates $\{\mu_{i,j}\}_{(i,j)\in[k]\times[u]}$ for the settings $(k, N, u) = (100, 10951, 5)$ and $(k, N, u) = (100, 10951, 10)$ with the click rate set bCRS-100-5 and bCRS-100-10. The row represents items i, and the column represents user groups j. The color scale indicates that higher click rates are shown in darker red, and lower click rates are shown in darker blue. The visualization of click rates through the heatmap helps to highlight preference patterns for items across various age groups. For instance, arm $(i, j) = (5, 1)$ is shown in dark red, indicating that users in those groups consistently rate specific items highly, showing a strong preference compared to other user groups. This analysis demonstrates that grouping users by age is effective in capturing their preferences.

4.2 Settings

Using the four sets of click rates in Table 1, we conduct the following three experiments. Books dataset only is used in the first two experiments, and both datasets are used in the last performance comparison experiment. In all the experiments, the ith component of the fairness constraint vector r_i is set to $1/2k$ for all $i \in [k]$. In the experiment on grouping effect (Sect. 4.3), we solved the FAIR-BATCH-MAB optimization problem for each $(u, \alpha) \in \{1, 5, 10\} \times \{1000\ell | \ell \in \{0\} \cup [8]\}$ using Python-MIP library. In the experiments on fairness constraint satisfaction (Sect. 4.4) and algorithms' performance comparison (Sect. 4.5), we run FBO(CRE) for CRE=SM and CRE=U, SM, UCB, KL-UCB, TS, respectively,

Table 1. The number of users N and those in the groups $m_1, m_2, \ldots, m_u$ in the four click rate sets used in our experiments

data sets	name	k	N	u	$m_1, m_2, \ldots, m_u$
Books dataset	bCRS-10-5	10	4788	5	1134, 1034, 884, 559, 1177
	bCRS-100-1	100	10951	1	10951
	bCRS-100-5			5	2093, 2805, 2881, 1274, 1898
	bCRS-100-10			10	698, 1252, 709, 1886, 1232, 946, 1778, 313, 698, 1439
MovieLens 1M Dataset	mCRS-10-10	10	6040	10	1538, 946, 338, 805, 222, 550, 380, 855, 558, 298

Table 2. The click rates in the click rate sets bCRS-10-5 and mCRS-10-10

(a) bCRS-10-5										
Age Range	item 1	item 2	item 3	item 4	item 5	item 6	item 7	item 8	item 9	item 10
1–25 years old	0.182800	0.234380	0.256760	0.289770	0.247790	0.159090	0.183100	0.111110	0.305260	0.023610
26–32 years old	0.184400	0.166670	0.280000	0.234740	0.323350	0.089890	0.130840	0.125000	0.300810	0.005740
33–41 years old	0.128380	0.310610	0.201390	0.234380	0.304090	0.073680	0.217690	0.147060	0.169640	0.008280
42–53 years old	0.135800	0.317310	0.262500	0.242270	0.363010	0.150000	0.215690	0.144930	0.306670	0.009090
54–82 years old	0.222220	0.456520	0.291670	0.220780	0.301080	0.136360	0.346150	0.266670	0.268290	0.004900
(b) mCRS-10-10										
(Gender, Age Range)	item G1	item G2	item G3	item G4	item G5	item G6	item G7	item G8	item G9	item G10
(F and M, 50-55)	0.265260	0.179790	0.196500	0.182010	0.219030	0.228560	0.135150	0.242140	0.190140	0.209350
(M, 18-24)	0.302330	0.240940	0.228320	0.218290	0.276620	0.257870	0.155280	0.273920	0.233810	0.231510
(F,18-24)	0.270010	0.243600	0.218370	0.216660	0.249010	0.212870	0.174650	0.262100	0.262280	0.200900
(F and M, 56-)	0.282500	0.190190	0.207250	0.192560	0.232830	0.265010	0.151470	0.274880	0.194760	0.212120
(F and M, -18)	0.324040	0.239850	0.241440	0.217310	0.269320	0.277930	0.197720	0.285500	0.197130	0.243870
(F and M, 45-49)	0.266690	0.233810	0.196790	0.204640	0.230530	0.229050	0.150840	0.244870	0.244200	0.194490
(F, 25-34)	0.329840	0.272100	0.239190	0.234500	0.282210	0.266610	0.173760	0.277870	0.252080	0.208950
(F, 35-44)	0.269820	0.202640	0.190630	0.182090	0.225540	0.237410	0.141010	0.242240	0.204430	0.220650
(M, 25-34)	0.268090	0.224200	0.194710	0.191000	0.227560	0.234950	0.140880	0.259680	0.255950	0.210500
(M, 35-44)	0.238850	0.215850	0.188710	0.172880	0.192250	0.228190	0.123380	0.267740	0.229100	0.185350

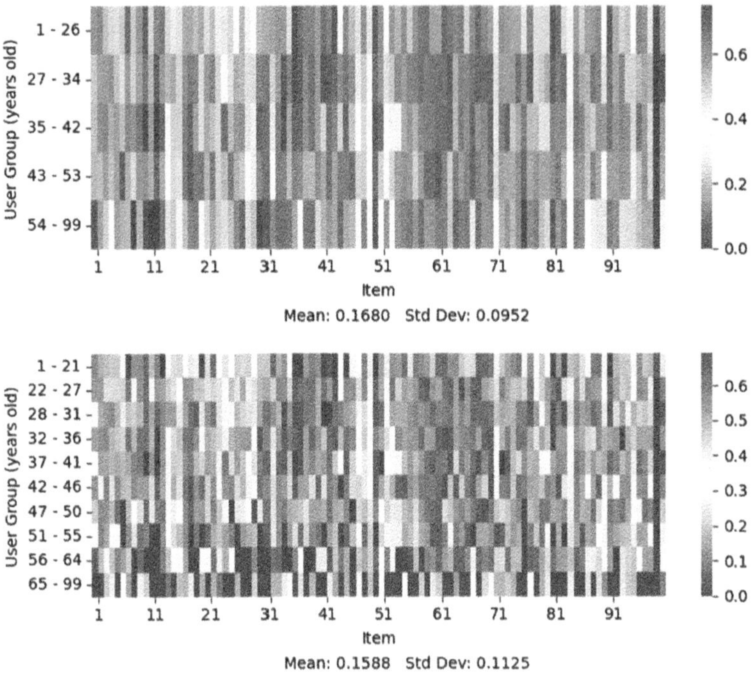

Fig. 1. The heatmap of the click rates in the settings $(k, N, u) = (100, 10951, 5)$ and $(k, N, u) = (100, 10951, 10)$, with the click rate sets bCRS-100-5 ($u = 5$) (above) and bCRS-100-5 ($u = 5$) (below)

generating $\{0, 1\}$-rewards according to the Bernoulli distribution with the click rate in the corresponding click rate sets. Note that the implementation of the OptInsSol function in the FBO algorithm is also done using Python-MIP library.

4.3 Result on Grouping Effect

First, to demonstrate the effectiveness of grouping based on biographical information, we compare the cumulative expected rewards $\sum_{t=1}^{T} \sum_{i=1}^{k} \sum_{j=1}^{u} \mu_{i,j} N_{i,j}^{*}(t)$ of different $u = 1, 5$ and 10 for the solution $\{N_{i,j}^{*}(t)\}_{t=1}^{T}$ of the FAIR-BATCH-MAB optimization problem in the case with $k = 100$, $N = 10951$ and $T = 100$ using click rate sets bCRS-100-1, bCRS-100-5 and bCRS-100-10, respectively.

Line graphs of cumulative expected reward over the unfairness tolerance constant α are shown in Fig. 2. We can see the grouping effect for the Books dataset; at every α value, the cumulative expected reward for $u = 10$ is greatest, that for $u = 5$ is second greatest and that for $u = 1$ is smallest. In all three line graphs, the cumulative expected rewards increase over the α-range from 0 to 6000 because increasing α weakens the fairness constraints. The α greater than 6000 does not affect the optimal solution of the FAIR-BATCH-MAB optimization problem for the settings of $k = 100$ of the Books dataset.

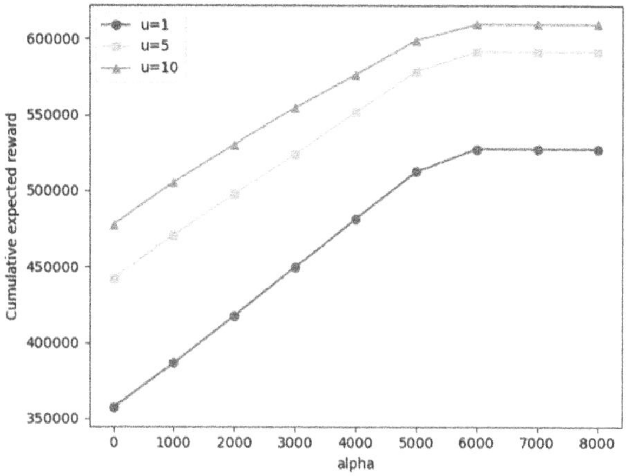

Fig. 2. Line graphs of cumulative expected reward over the unfairness tolerance constant α for the three click rate sets of different numbers of groups bCRS-100-1 ($u = 1$), bCRS-100-5 ($u = 5$) and bCRS-100-10 ($u = 10$)

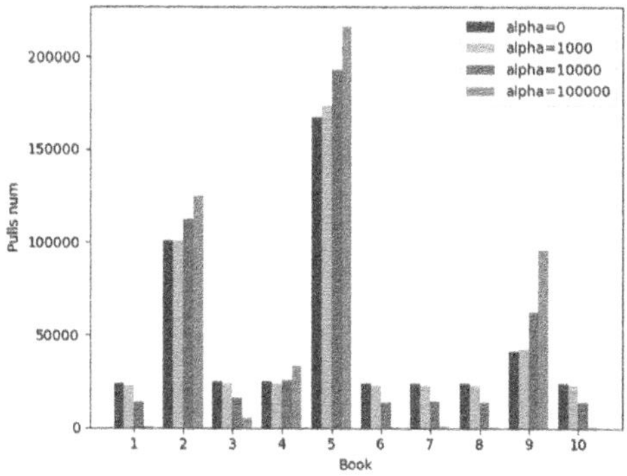

Fig. 3. Number of item-i recommendations averaged over 10000 runs for $i = 1, 2, \ldots, 10$ in the cases with $\alpha = 0, 1000, 10000, 100000$ in the setting of $(k, N, u, T) = (10, 4788, 5, 100)$ and the click rate set bCRS-10-5

4.4 Result on Fairness Constraint Satisfaction

Next, we examine how fairness is ensured for each item (book). For $\alpha = 0, 1000, 10000, 100000$, we run FBO(SM) 1000 times for the setting $(k, N, u, T) = (10, 4788, 5, 100)$ by simulating received rewards using click rate sets bCRS-10-5, and count the number of recommendations of each item averaged over 10000 runs. The click rate in this setting is shown in the Table 2.

The result is shown in Fig. 3. We can confirm that α-batch-fairness is satisfied in all the cases with $\alpha = 0, 1000, 10000, 100000$ at round T by checking

$$\sum_{s=1}^{T} N_i(s) \geq \lfloor r_i N T \rfloor - \alpha = \lfloor 0.05 \times 4788 \times 100 \rfloor - \alpha = 23940 - \alpha$$

for all $i = 1, 2, \ldots, 10$, that is, by checking that the number of recommendations of each item is at least $23940, 22940, 13940, -76060$ for $\alpha = 0, 1000, 10000, 100000$, respectively. As α increases, the recommendations concentrate on high-click-rate items more, but not on one item only because the highest-click-rate items of the five groups are not the same item.

4.5 Result on Algorithms' Performance Comparison

Finally, we compare the performances of FBO(CRE) algorithms with CRE = U, SM, UCB, KL-UCB and TS in terms of cumulative reward and regret defined in Sect. 2. For each CRE = U, SM, UCB, KL-UCB, TS, we run FBO(CRE) 100 times for the setting $(k, N, u, T) = (100, 10951, 5, 1000)$ by simulating received rewards using click rate sets bCRS-100-5. Additionally, we use the setting $(k, N, u, T) = (10, 6040, 10, 1000)$ by simulating received rewards using the click rate set mCRS-10-10. The click rate in this setting is shown in the Table 2.

The obtained cumulative rewards for each CRE are averaged over 100 runs.

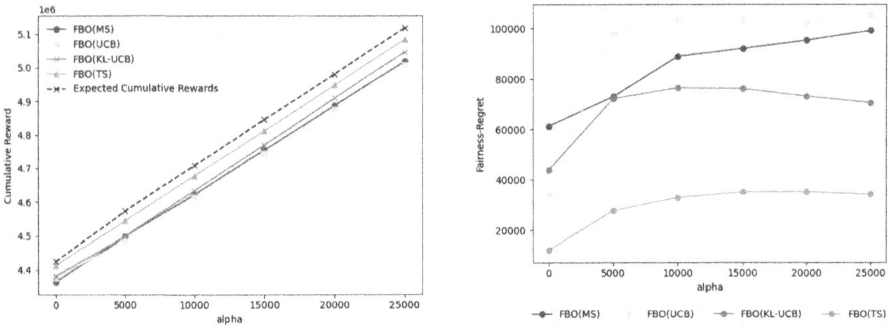

Fig. 4. Line graphs of cumulative rewards (left) and regrets (right) of FBO(CRE) over the unfairness tolerance constant α in the setting $(k, N, u, T) = (100, 10951, 5, 1000)$ and the click rate set bCRS-100-5 for CRE = SM, UCB, KL-UCB, TS

The results for the bCRS-100-5 click rate set are shown in Fig. 4 except for FBO(U). For the setting of this experiment, the performance of FBO(TS) is best, that of FBO(KL-UCB) is second best, and FBO(SM) performs better than FBO(UCB). Since FBO(U) uses constant $\hat{\mu}_{i,j}$, its cumulative reward is steady around 1.8×10^6 regardless of the value of α. Therefore, it is not shown on the left (right) figure in Fig. 4 because its cumulative reward (regret) is too

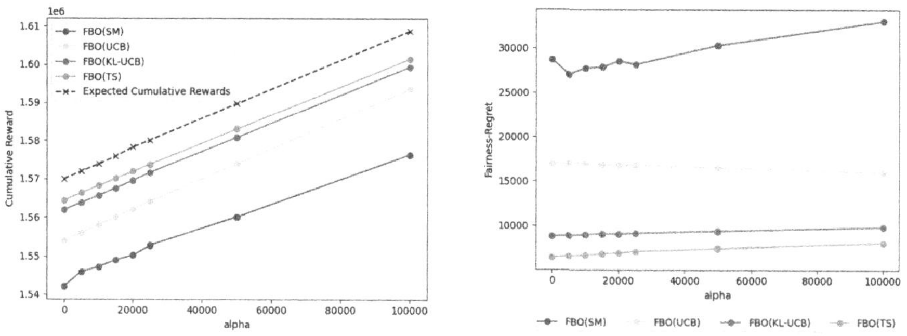

Fig. 5. Line graphs of cumulative rewards (left) and regrets (right) of FBO(CRE) over the unfairness tolerance constant α in the setting $(k, N, u, T) = (10, 6040, 10, 1000)$ and the click rate set mCRS-10-10 for CRE $=$ SM, UCB, KL-UCB, TS

low (high) to show with the cumulative reward (regret) of the other algorithms. The reason why FBO(UCB) performs worse than FBO(SM) is guessed that the term for exploration $\sqrt{\frac{\log(Nt)}{2\tilde{N}_{i,j}(t)}}$ is too large for $T = 1000$ in this setting. The line graph of the optimal expected cumulative reward is also shown in the left figure, and we can see that the performance of FBO(TS) is very close to the optimal performance. As α increases, the fairness constraint becomes weaker, which increases the optimal cumulative reward. Thus, regrets increase less than linearly in α even though cumulative rewards increase linearly in α.

The result for the mCRS-10-10 click rate set is shown in Fig. 5 except for FBO(U). For the setting of this experiment, the performance of FBO(TS) is best, and that of FBO(KL-UCB) is second best, but unlike the experiment using the bCRS-100-5 click rate set, FBO(UCB) performs better than FBO(SM). Since FBO(U) uses constant $\hat{\mu}_{i,j}$, its cumulative reward is steady around 1.3×10^6 regardless of the value of α. Therefore, it is not shown on the left (right) in Fig. 5 figure because its cumulative reward (regret) is too low (high) to show with the cumulative reward (regret) of the other algorithms.

The Table 3 shows the cumulative reward for $\alpha = 0$ in two settings. Comparing the results, it is evident that FBO(SM) has a larger standard deviation than the other three algorithms, indicating instability in its performance. Since the recommendation by FBO(CRE) assigns item i to user group j for the pair (i, j) of high click rate $\hat{\mu}_{i,j}$ estimated by CRE, item i is never recommended to user group j for a pair (i, j) of low-estimated click rate $\hat{\mu}_{i,j}$. Crick rates $\hat{\mu}_{i,j}$ estimated by SM do not change unless item i is recommended to user group j, thus FBO(SM) performs poorly when some crick rates are estimated very low in the early rounds.

In contrast, UCB and TS are estimators used in bandit problems and explore to correct falsely low-estimated crick rates, resulting in a smaller standard deviation compared to FBO(SM).

Table 3. Comparison of cumulative reward analysis results at $\alpha = 0$ for each algorithm.

click rate set	algorithm	mean	std dev	min	max
bCRS-100-5	FBO(U)	1830609	14246	1799678	1871757
	FBO(SM)	4364130	49837	4149193	4411927
	FBO(UCB)	4391598	1914	4387174	4396008
	FBO(KL-UCB)	4381060	2827	4376669	4385267
	FBO(TS)	**4413243**	**1786**	**4408126**	**4417459**
mCRS-10-10	FBO(U)	1326995	2743	1316876	1334540
	FBO(SM)	1452077	15180	1485200	1571389
	FBO(UCB)	1553846	**1206**	1549746	1557658
	FBO(KL-UCB)	1561599	1216	**1560068**	1563908
	FBO(TS)	**1564375**	1371	1558079	**1568617**

5 Conclusion

We introduced the concept of a FAIR-BATCH-MAB algorithm that performs batch-based group recommendations under fairness constraints. The algorithm can achieve fairness by ensuring minimum exposure constraints on each item when solving instantaneous batch optimization problems at each round. In practical applications, such an algorithm is particularly beneficial for periodic mass deliveries such as direct mail campaigns and online advertising. Moving forward, critical challenges involve ensuring equity among groups and restricting the same item recommendation to the same user. Addressing these challenges could lead us to propose more practical algorithms.

Acknowledgment. This work was supported by JSPS KAKENHI Grant Number JP24H00685.

References

1. Auer, P.: Using confidence bounds for exploitation-exploration trade-offs. J. Mach. Learn. Res. **3**(null), 397–422 (2003)
2. Auer, P., Cesa-Bianchi, N., Fischer, P.: Finite-time analysis of the multiarmed bandit problem. Mach. Learn. **47**(2–3), 235–256 (2002). https://doi.org/10.1023/A:1013689704352
3. Garivier, A., Cappé, O.: The kl-ucb algorithm for bounded stochastic bandits and beyond. In: Kakade, S.M., von Luxburg, U. (eds.) Proceedings of the 24th Annual Conference on Learning Theory. Proceedings of Machine Learning Research, Budapest, Hungary, vol. 19, pp. 359–376. PMLR (2011). https://proceedings.mlr.press/v19/garivier11a.html
4. Gillen, S., Jung, C., Kearns, M., Roth, A.: Online learning with an unknown fairness metric. In: Bengio, S., Wallach, H., Larochelle, H., Grauman, K., Cesa-Bianchi, N., Garnett, R. (eds.) Advances in Neural Information Processing Systems. vol. 31.

Curran Associates, Inc. (2018). https://proceedings.neurips.cc/paper_files/paper/2018/file/50905d7b2216bfeccb5b41016357176b-Paper.pdf
5. Harper, F.M., Konstan, J.A.: The movielens datasets: history and context. ACM Trans. Interact. Intell. Syst. (TiiS) **5**(4), 19 (2015)
6. Joseph, M., Kearns, M., Morgenstern, J.H., Roth, A.: Fairness in learning: Classic and contextual bandits. In: Lee, D., Sugiyama, M., Luxburg, U., Guyon, I., Garnett, R. (eds.) Advances in Neural Information Processing Systems. vol. 29. Curran Associates, Inc. (2016). https://proceedings.neurips.cc/paper_files/paper/2016/file/eb163727917cbba1eea208541a643e74-Paper.pdf
7. Books dataset. https://www.kaggle.com/datasets/elvinrustam/books-dataset. Accessed 01 May 2024
8. Lai, T., Robbins, H.: Asymptotically efficient adaptive allocation rules. Adv. Appl. Math. **6**(1), 4–22 (1985). https://doi.org/10.1016/0196-8858(85)90002-8
9. Langheinrich, M., Nakamura, A., Abe, N., Kamba, T., Koseki, Y.: Unintrusive customization techniques for web advertising. Comput. Networks **31**(11–16), 1259–1272 (1999). https://doi.org/10.1016/S1389-1286(99)00033-X
10. Li, L., Chu, W., Langford, J., Schapire, R.E.: A contextual-bandit approach to personalized news article recommendation. In: Proceedings of the 19th International Conference on World Wide Web. WWW '10, New York, NY, USA, pp. 661–670. Association for Computing Machinery (2010).https://doi.org/10.1145/1772690.1772758
11. Nakamura, A.: A ucb-like strategy of collaborative filtering. In: Phung, D.Q., Li, H. (eds.) Proceedings of the Sixth Asian Conference on Machine Learning, ACML 2014, Nha Trang City, Vietnam, November 26–28, 2014. JMLR Workshop and Conference Proceedings, vol. 39. JMLR.org (2014). http://proceedings.mlr.press/v39/nakamura14.html
12. Patil, V., Ghalme, G., Nair, V., Narahari, Y.: Achieving fairness in the stochastic multi-armed bandit problem. J. Mach. Learn. Res. **22**(1) (2021)
13. Rangi, A., Franceschetti, M.: Multi-armed bandit algorithms for crowdsourcing systems with online estimation of workers' ability. In: Proceedings of the 17th International Conference on Autonomous Agents and MultiAgent Systems. AAMAS '18, pp. 1345–1352. International Foundation for Autonomous Agents and Multiagent Systems, Richland, SC (2018)
14. THOMPSON, W.R.: On the likelihood that one unknown probability exceeds another in view of the evidence of two samples. Biometrika **25**(3-4), 285–294 (1933). https://doi.org/10.1093/biomet/25.3-4.285
15. Wang, L., Bai, Y., Sun, W., Joachims, T.: Fairness of exposure in stochastic bandits. In: Meila, M., Zhang, T. (eds.) Proceedings of the 38th International Conference on Machine Learning. Proceedings of Machine Learning Research, vol. 139, pp. 10686–10696. PMLR (2021). https://proceedings.mlr.press/v139/wang21b.html

Human Decision-Making Concepts with Goal-Oriented Reasoning for Explainable Deep Reinforcement Learning

Chris Lee[1(✉)], Eduardo Benitez Sandoval[1], and Francisco Cruz[1,2]

[1] University of New South Wales, Sydney, Australia
christopher.lee1956@student.unsw.edu.au
[2] Universidad Central de Chile, Santiago, Chile

Abstract. Recently, the development and integration of Artificial Intelligence (AI) has accelerated and been popularized widely throughout modern society. AI is becoming a powerful tool ranging from leisurely use to critical applications. However, due to the black-box nature of some AI approaches such as Deep Reinforcement Learning (DRL), complex AI algorithms now face growing concerns of trust in ethical and responsible decision-making. EXplainable Artificial Intelligence (XAI) is a subfield of AI focused on deriving interpretable information from incomprehensible statistics to generate explanations for an AI's decisions. This paper proposes an architecture that combines 2 XAI techniques, Testable Concept Activation Vectors (TCAV) and Reward Decomposition, to create goal-oriented explanations. The XAI approach is tested in a simulated movement prediction environment where a DRL agent is trained to represent different human concepts and goal prioritizations; we can confidently distinguish those concepts between agents in a human-centric framework. Results obtained demonstrate our method allows users to insert their own high-level thinking into XAI and use it to generate explanations.

Keywords: Artificial Intelligence · Explainable Artificial Intelligence · Neural Networks · Reinforcement Learning · Reward Decomposition

1 Introduction

The XAI ecosystem has yet to reach a philosophical approach towards explanation that will allow users of all knowledge levels to understand and interact with the explaining agent conversationally. The process of 'explanation and understanding' has been widely studied to conceptualize how the brain works during these social interactions [4,20], but has yet to be properly adopted in the current ecosystem. Currently, popular XAI techniques such as Lime [13] or GradCAM [17] generate meaningful explanations yet often assume that users can interpret their visual or statistical information. Another research gap is post-hoc capability that maintains faithfulness towards the AI algorithm it explains and is

M. Gong et al. (Eds.): AI 2024, LNAI 15442, pp. 228–240, 2025.
https://doi.org/10.1007/978-981-96-0348-0_17

more evident in eXplainable Reinforcement Learning (XRL) methods, including Reward Decomposition [8]. Das and Rad explain that post-hoc capability means "the XAI algorithm doesn't know the internal operations and model architectures," able to be applied to already trained and accurate AI algorithms [3]. If the XAI has no knowledge of the internal architecture, then concerns of faithful explanations arise, questioning whether XAI reasoning actually matches the AI's reasoning.

In this implementation, we developed an initial system inspired by Testable Concept Activation Vectors (TCAV), aiming to mimic human-human interaction where the mutual agreement of concept representations is reached and used for explanation [9]. In the context of our research, concepts are defined as high-level reasoning that is composed of a set of basic reasonings like 'moving closer towards a coordinate'. Furthermore, our study addresses the limitations of post-hoc application by adapting Hybrid Reward Architecture (HRA), to represent different goal-types and use it to estimate the reward function of the DRL method [16]. Then using HRA to approximate Reward Decomposition [8], faithful goal-oriented explanations are created that do not require any knowledge of the hidden layers inside the AI algorithm.

We aim to create an interactive system which allows the end-user to determine whether or not their own conceptual framework is reflected by the AI decision-making. Additionally, our focus was to develop an XAI that could reflect the many layers and goal-prioritizations in decision-making and how they weighed against each other towards the final decision, then communicate them in a coherent and written explanation. Ultimately allowing people of all knowledge levels to understand the decisions made by AI benefitting its integration into modern society.

2 Related Work

2.1 Testable Concept Activation Vectors (TCAV)

TCAV uses directional derivatives to quantify the degree to which a user-defined concept is important to a classification result, being completely post-hoc [9]. Kim et al. [9] emphasizes the significance of human-centricity through this approach, training a high-level human-intuitive concept vector with user input. Currently, this is the only technique that successfully utilizes user-defined datasets to construct the basis of the explanation. Additionally, if the XAI does not represent the concept to the user's satisfaction, they have the ability to retrain the TCAV, mimicking back-and-forth communication.

Its advantage is in enabling users with no AI knowledge, the ability to plug in their own reasoning framework, solely through data that they define representative, and use that for explanation. For example, if a child believes that a zebra is most recognizable due to its striped patterning, they can create a dataset of all striped animals or other striped patterns that they know of, insert it into the XAI and see whether both the TCAV and AI model agrees.

To train our TCAV, we provide a user-defined dataset where they can create the data themselves in the Pygame engine, saved into a CSV file with each column referencing:

'Shooter's x coordinate, shooter's y coordinate, projectile's x coordinate, projectile's y coordinate, player's current x coordinate, player's current y coordinate, player's initial x coordinate, player's initial y coordinate, angle of shot taken, shot success'

The TCAV will then be represented in the engine allowing the user to see the concept in live runtime and has the opportunity to retrain the same concept until they are satisfied. However, accurate concept representation by the TCAV is heavily limited by the user's own interpretation and whether their provided dataset is actually representative of the concept. Then for goal-type alignment, different coefficients of the generated concept vector are aggregated to determine the total significance of each goal-type towards the concept. Similar to Shapley Q-Values [18], the significances will act as a global critic towards the Reward Decomposition component. Due to the human-centric focus and interactivity of TCAV with the user, we prioritized it as a major component of our method, as it has the ability to represent high-level concepts that is both examined and defined by the user.

2.2 Reward Decomposition

Juozapaitis et al. [8] proposes the theory of Reward Decomposition, which takes the Q-Values [19] and decomposes them into semantically meaningful reward types where actions are explained by the trade-offs between them. To meaningfully decompose those rewards, the DRL method will have to function through an HRA where the individual reward functions will each then represent a specific goal-type.

Our Reward Decomposition implementation will run post-hoc when the Deep Q-Learning Neural Network (DQN) [11] agent is in its testing state, meaning no further training of the target network and its weights, experience replay will still be required to train a neural network at each step. In the field of XAI and even many DRL techniques, Reward Decomposition serves as the basis of many goal-oriented implementations in XRL, breaking down a complex environment and reward function into much simpler and easily managed rewards. By itself, it is not inherently post-hoc, but our aim was to determine how effective it would be if implemented that way.

In real-life decision-making, we typically have overarching goals that encompass a diverse variety of factors and this is computed through Q-Values in DRL. However, one broad and singular value/reason can easily be interpreted as a poor explanation that needs to be broken down into smaller components and weighed against each other. That is the goal of Reward Decomposition, one reason should rarely ever be enough.

The difference in our implementation is that HRA training will occur post-hoc/after DQN training and will estimate the Q-values during testing runtime using a simplified state representation with column headers; *'Player's x coordinate, player's y coordinate, aimer's x coordinate, aimer's y coordinate, aimer's aiming angle'* and the Q-Values at each state with actions; *'Move_ right, Move_ left, Aim_ right, Aim_ left, Shoot, Nothing'.* There will be 2 stages of estimation, the first would estimate the individual Q-Value of each action, then to predict the rest of the Q-value table using that estimation. The difference in predicted against actual Q-Values will act as our measure of significance with lower accuracy meaning less significance. Additionally, we evaluate how well HRA agents represented the reward-type using r-squared, derived from each regression model, to act as our confidence of the significance of each goal-type.

3 Methods

We present the architecture for our XAI technique, combining TCAV with Reward Decomposition. The overall focus of our XAI is an interactive explanation system that allows users to reach mutual understanding with the XAI and allow it to be applied post-hoc on any already trained DRL method. Assessed and achieved through mutual agreement of the representation of different concepts between user and XAI. Our aim is to be able to distinguish concepts and goal-prioritizations of different DRL agents and explain them to people of all knowledge levels.

3.1 Scenario

The scenario in which we trained our DRL agent and ultimately conducted testing of our XAI implementation was through motion prediction where the DRL agent's goal was to successfully aim a projectile at a moving target. We used DQN as our DRL method, a state-of-the-art implementation that has inspired much of DRL research presently and previous research supports the utilization of Reward Decomposition to explain DQN models [7]. Our aiming agent will try to shoot a projectile at a target, moving in a predefined route that correlates with each hierarchy of decision-making [4], with aims of dodging the projectile. The rewarding system can be broken down and grouped into goal-types, enabling the scale of each goal-type to be tuned to have different agents with different prioritzations e.g.: *'Positioning'* will focus on decreasing x and y coordinate differences between aimer and target whilst *'Aiming'* will encompass decreasing angle differences between aimed and actual. The movement prediction scenario is shown in the red-dotted frame in Fig. 1 with the aimer at the top, target at the bottom and aimer's aim displayed through the red line attached.

3.2 Combining TCAV with Reward Decomposition

The TCAV XAI results will define how effectively it represents the concept and how significant each goal-type is towards that concept. On the other-hand, the Reward Decomposition XAI results demonstrate how confident it is in stating the significance of each goal-type towards the decision. By combining the results of both XAI's as shown in Fig. 1, our TCAV global critic will then analyze the alignment of the goal-types to determine, for the decision to be explained, the confidence that the decision was made using this conceptual reasoning. For example, 'we have 40% confidence that this decision was made using the neurological layer of decision-making'.

The advantage of combining TCAV with Reward Decomposition, instead of other popular XAI techniques, is that both XAI techniques followed the same framework; high-level thinking through either a concept or overarching reward, and break it down into smaller individual components. Meanwhile, they can also go from simple and one-dimensional reasoning to complex and multi-dimensional reasoning, as long as the array of goal types is calibrated between the two. As seen in Fig. 1, four goal-types are used for alignment, but this number is mutable, as long as it matches the two. Still, TCAV can be combined with other XAI techniques and is not just limited to Reward Decomposition, such as saliency mapping, however more processing of the explanations between each are necessary.

4 Results and Discussion

4.1 Deep Q-Learning Neural Network (DQN)

In our experimental setup, we trained 12 individual DQN agents that displayed different behaviours at different magnitudes due to the fine tuning of our reward function and training in various scenarios. Our training scenarios were based on the concepts from the decision-making hierarchy with *Neurological*, *Social* and *Cultural* [4], along with the goal-types of *Base*, *Positioning*, *Aiming* and *Stalling*. In our scenario and through the decision-making hierarchy framework; *Neurological* represents a biological reaction to the environment, therefore, aiming directly at target, *Social* represents basic prediction of the target's forward movement without knowledge of the aiming agent's goals and *Cultural* represents prediction of the target's movement if it had knowledge of the aiming agent's goals and movement patterns. Additionally, to enact the different goal-types and prioritizations of *Positioning*, *Aiming* and *Stalling*, we inflated the reward of reducing coordinate difference, angle difference and shot rewarding by 2 respectively.

By analyzing the differences in learning patterns through inspection of episodic rewarding and other predefined goal measurements such as average positional difference, we could predict how our XAI would perform and what the results would contain. It is important to consider than innately from the environment, some goal-types would inherently be more significant than others and it is more noticeable and vital if the weaker ones are most significant. Each agent was trained until the previous 50 episode running reward either exceeded 300 or if it ran out of memory.

Through human observation of our DQN performances, we can derive explanations of goal-oriented behaviour even before application of our XAI, to see if those Reward Decomposition explanations align with human-intuition. Our

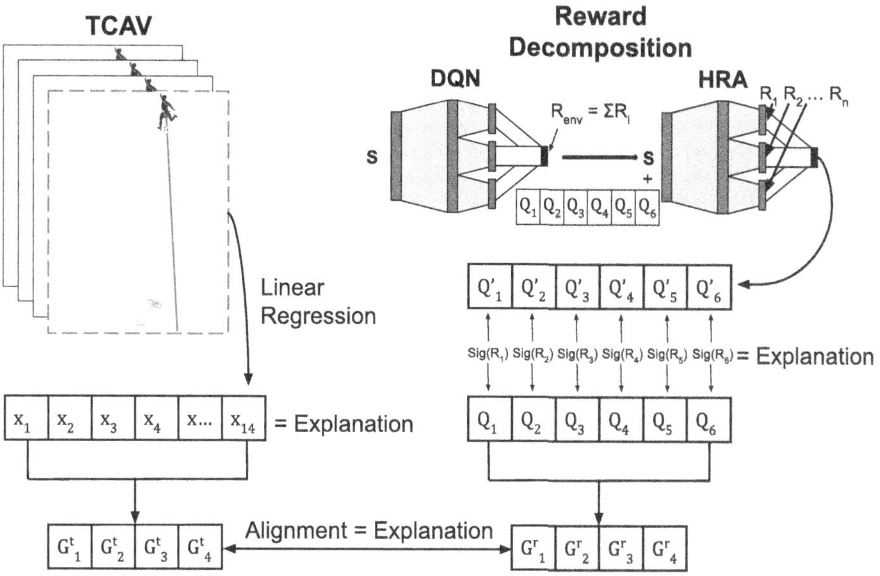

Fig. 1. Combined TCAV and Reward Decomposition architecture. The TCAV component, on the left, will take a representation of the environment and train a linear regression model with selected column headers as the variables. The coefficients of each variable act as the significance and can be used for explanation. Reward decomposition, on the right, trains an HRA to approximate the Q-Values of the DQN agent. It then combines the difference between approximated and actual, with the coefficient of determination of each reward regression model to use as the significance for our explanation. Finally, the explanations between both components will be aligned, with TCAV as the global critic based on the high-level concept. The coefficients of the TCAV and estimate difference + confidence of the HRAs will be grouped into identical goal-types which can then be aligned directly for final explanation.

results show that modifying the reward function to favour different goal-types can significantly affect the performance of our DQN agents. For our social agents, although our 'aiming' agent reached completion faster, Fig. 2 demonstrates that the 'positioning' agent performed much better as it reaches completion with a much steeper gradient of learning. Additionally, we can infer that overall our social agents naturally prefer the goal-type of 'positioning' as it had the best performance, meaning that even though it might be another social agent such as 'Social Stalling'. We predict that the positioning goal-type to hold the second-most significance towards overall behaviour. On the contrary, 'Social Stalling' would still have 'stalling' as the most significant goal-type, compared to other social agents that rank it last on the hierarchy.

Additionally, Fig. 3 showcases the different performances of each decision-making layer with the base reward function. Only the social agent had consistent and positive learning during training even if it had the lowest starting running reward. This suggests that in our scenario, a balance between complexity and

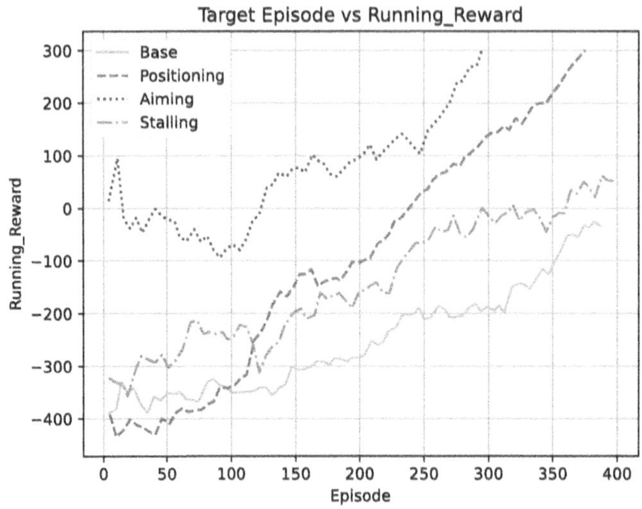

Fig. 2. 50 Episode running reward of social layered DQN agents during training, showcasing the difference in performance of each reward function.

performance was necessary as although our cultural and neurological agents started with higher running rewards, learning was not consistent. Our findings conclude that decision-making layers will behave differently in RL agents, the more complex layers should perform better, but eventually overcomplication could deteriorate performance.

4.2 TCAV Results

TCAV acts like a global critic towards our Reward Decomposition results, determining based on those results how much alignment there is towards the high-level TCAV concept. However, in order to ensure that there is satisfactory representation of that concept, we must still evaluate both statistically and through human observation the TCAV's understanding of that concept in its own linguistics. Critiquing an explanation through a high-level concept is unwise if the understanding of that concept is not mutual.

Initially, to evaluate whether or not our TCAV variables will sufficiently represent the environment and the concepts, we evaluate relevance through whether or not the shots taken were a hit based on the CSV variables. Our base dataset achieved an accuracy of 70%, with concept sets of neurological at 93%, social at 65% and cultural at 84%. Overall, we can safely assume that the defined variables will sufficiently represent the environment to determine shot success and can differentiate between concepts. Another notable observation is that our TCAVs typically define data as non-concept, having much less true positives and lower positive recall compared to true negatives and vice-versa. Poor performance can be attributed to an imbalance within the testing dataset itself as our

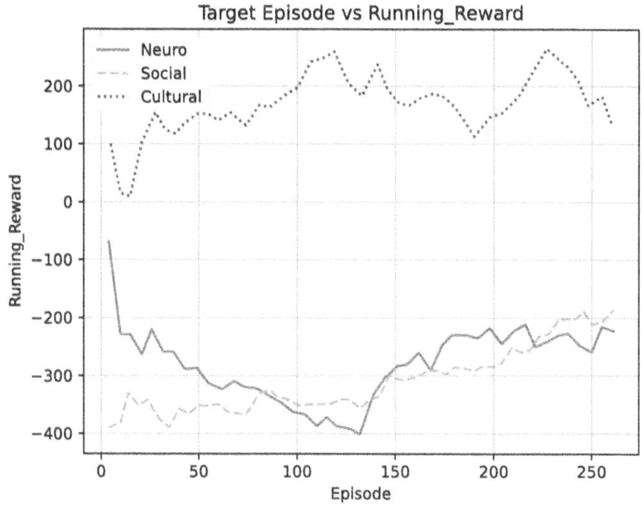

Fig. 3. 50 Episode running reward of base reward function DQN agents during training, showcasing the difference in performance of each decision-making layer.

weighted averages for both recalls was much more even. This could also demonstrate a large similarity between the datasets, meaning our variables could not differentiate or that inherently, those similarities exist in human-intuition and as a result, were captured.

Table 1 demonstrates the significance of each goal-type for each decision-making hierarchy, highlighting that each layer can have different goal-prioritizations. We observe that the Neurological layer prioritizes 'Positioning' the most, whilst the other 2 layers prioritize 'Aiming'. Now with the different goal-type prioritizations for each concept we can use this as a global critic to the Reward Decomposition component.

Table 1. Goal significances for each neurological, social and cultural TCAV.

TCAV List			
TCAV Concept	Positioning (%)	Aiming (%)	Stalling (%)
Neurological	42	40	18
Social	31	44	26
Cultural	27	48	24

4.3 Reward Decomposition Results

The majority of explanation will be conducted through the Reward Decomposition component of our XAI, allowing us to derive goal-oriented reasoning behind why a specific action was taken at a certain state. Based on the experimental setup, overarching goal bias and influence should differ at different magnitudes between agents, however goal-oriented reasoning for a specific decision can be maintained between any agent.

Analyzing the confidence of each HRA when predicting the Q-Values of the action it represents, we observe a rankable hierarchy between goal-types. For 'Social Base Confidence' all confidences sit around 39%. However, it is noticeable that the goal-type hierarchy is 'aiming', 'stalling' then 'positioning'. Comparing the different decision-making layers together, we derive that neurological and cultural agents yield the same goal-type hierarchy with social being the exception. Our XAI is also able to reflect reward inflation and its effect on the goal-type hierarchy as for example our 'Social Aiming' agents held 'aiming' at the top of its hierarchy compared to the others holding their respectives.

Difference significance is drawn after the predicted Q-Value of action is used to predict the rest of the Q-Value table. The difference between significance and confidence is that confidence aims to explain the overall preferred behaviour of the agent, whilst significance is of a particular action at state. Our findings show that it is possible to have a difference between the said two. Previously, our 'Social Aiming' agent had the highest confidence in 'Aiming' but actually held the least significance, meaning it was the least impactful towards the final decision. By using the confidence and significance results drawn by our Reward Decomposition XAI, we can align that with our TCAV results so that without knowing which DQN agent we are explaining, we can classify from those results combined which agent it is.

4.4 Combined Final Explanation

The objective of combining our TCAV results with our Reward Decomposition results is to align decomposed Q-Value's goal-prioritization with a user's reasoning framework. Our TCAV results represent the user's reasoning framework and act as a global critic towards our goal-oriented Reward Decomposition results.

$$Align(goal\_type) = RD\_conf(goal\_type) \times TCAV\_sig(goal\_type) \quad (1)$$
$$\times (1/RD\_sig(goal\_type)) \quad (2)$$

The above formula will be calculated for each agent to determine specific goal-type prioritization and finally concept alignment. We will take the alignment confidence of the most prioritized goal-type for that concept to use as our critic. Looking through Table 2, our TCAV defines that 'aiming' is the most influential goal-type for social decision-making. That is also true for the Reward Decomposition result of our 'Social Aiming' agent's overall behaviour. Therefore, for this specific decision, we take the alignment confidence of 'Aiming' in the Social TCAV column in the 'Social Aiming' row.

Hence our final explanation follows:

"For our social aiming agent, we have 29% and 30% confidence that 'positioning' and 'stalling' respectively were the goals the agent prioritized towards choosing inaction when it had off aim and was not optimally positioned. Overall the agent would prefer to optimize its aim but with 31% confidence we believe the agent found it suboptimal. Hence we can say with 39% confidence that this decision belonged to social decision-making"

However, if we wanted to use our Neurological TCAV or Cultural TCAV as the global critic instead, then the explanation for action at state would differ stating that *'we can say with 35% confidence that this decision belonged to neurological decision-making'* or *'we can say with 43% confidence that this decision belonged to cultural decision-making'*. These results are drawn by using the 'aiming' column in the other TCAVs as they are acting as the critic. Hence, the actual action itself might align more with another concept despite overall behaviour representing a chosen concept.

Table 2. Goal alignment confidences (%) between TCAV (columns) and Reward Decomposition (rows) results of select DQN agents (P = Positioning, A = Aiming, S = Stalling). In bold are the goal alignments one would use for selecting TCAV as that was the most significant goal for the concept. Some RL agents would have better alignment with a concept that it was not trained to reflect because alignment factors more than just overall behaviour.

	Neuro			Social			Cultural		
	P	A	S	P	**A**	S	P	**A**	S
Neuro Base	**30**	29	12	22	**31**	17	19	**24**	16
Social Base	**83**	51	14	61	**56**	20	53	**61**	19
Social Positioning	**61**	38	25	45	**41**	36	39	**45**	34
Social Aiming	**38**	35	17	28	**39**	24	24	**43**	22
Social Stalling 1	**39**	39	17	29	**43**	24	25	**47**	23
Social Stalling 2	**39**	48	17	29	**52**	25	25	**57**	23
Cultural Base	**55**	48	19	40	**53**	28	35	**57**	26

Ultimately, our combined TCAV and Reward Decomposition XAI results allowed us to generate a literate and comprehensive explanation of the decision we wished to explain, manifesting the different goal biases of the DQN agent overall and the most influential goals at that specific state. Our XAI could then correlate that explanation to a user-defined concept to see how influential that concept was. If we want to further improve our results, that would be by improving the individual components of TCAV and Reward Decomposition and addressing their limitations.

5 Conclusion

This paper proposed a post-hoc technique that created goal-oriented explanations for Deep Reinforcement Learning (DRL) methods whilst also integrating user-interactivity, allowing the ability to apply user-defined high-level concepts for explanation. We found that it was possible to insert interactivity into already existing eXplainable Artificial Intelligence (XAI) techniques like Reward Decomposition, with Testable Concept Activation Vectors (TCAV) acting as a global critic upon the explanation. Furthermore, we converted the traditionally non-post-hoc Reward Decomposition into post-hoc, through Hybrid Reward Architecture (HRA) estimation of the reward function. Our method sacrifices some confidence and trustworthiness of the explanation; however, it enables the method for more diverse real-world applicability as it does not require the knowledge of the inner-workings of the DRL method it explains.

By combining a very human-intuitive and interactive XAI method in Testable Concept Activation Vectors (TCAV) with Reward Decomposition, our XAI generated literate and comprehensive explanations for users of all knowledge levels to be able to understand. Breaking down complex Q-Values into more human-intuitive goals whilst maintaining faithfulness towards the Deep Q-Learning Neural Network (DQN) method it aimed to explain. Our overall research suggests that it is possible to create interactive XAI systems that can be applied to any Artificial Intelligence (AI) algorithm, mimicking human-human conversation with back-and-forth communication for domains such as human-robot interaction. We are creating a gateway for further XAI research, benefiting the rapid integration of black-box AI and diminishing the fears surrounding it. Foreseen potential uses of our research include applications such as self-driving vehicles, which rely heavily on the movement of external agents within a complex environment, made critical by the ethical implications if an accident were to occur due to unforeseen behaviour.

6 Future Work

In the future, we plan to broaden the application of the proposed amalgamated TCAV-Reward Decomposition XAI to reflect a larger range of concepts or even pair TCAV with other XAI techniques. Either in the same scenario or to cover more DRL tasks to generalize to more diverse real-world applications, such as multi-agent scenarios, to further support our findings. TCAV performance is limited to the variables we define as able to represent different concepts. If we can implement a dynamic algorithm that generates variables that best fit each individual concept, a larger range of concepts can be captured.

Furthermore, additional research can be conducted to optimize the estimation of different reward functions in Hybrid Reward Architecture (HRA), to achieve higher accuracy in prediction of Q-Values and increase our confidence in explanation. However, there are limitations concerning computational time and memory within a dynamic environment of DRL. Our HRA currently consists of multiple linear regression models that train off a simplified state-representation during testing runtime. We foresee the possibility to improve this through using neural networks instead of linear models and storing image data rather than a simple CSV file for the environment state. However, with several thousands of states being stored in just a few seconds, extended periods of runtime will demand a significant amount of storage. If we can develop an efficient way to manage this, there is potential for substantial improvement.

Disclosure of Interests. Authors have no conflict of interest to declare.

References

1. Beyret, B., Shafti, A., Faisal, A.A.: Dot-to-dot: explainable hierarchical reinforcement learning for robotic manipulation. https://arxiv.org/pdf/1904.06703.pdf, arXiv preprint (2019)
2. Craig, C.: Understanding perception and action in sport: How can virtual reality technology help? Sports Technology, Research Gate (2014)
3. Das, A., Rad, P.: Opportunities and Challenges in Explainable Artificial Intelligence (XAI): A Survey. https://arxiv.org/abs/2006.11371v2, arXiv preprint (2020)
4. Dazeley, R., Vamplew, P., Foale, C., Young, C., Aryal, S., Cruz, F.: Levels of explainable artificial intelligence for human-aligned conversational explanations. https://arxiv.org/abs/2107.03178v1, arXiv preprint (2021)
5. Dellaert, F.: Annual Review of Control, Robotics, and Autonomous Systems. Robotics, and Autonomous Systems, Annual Reviews, Annual Review of Control (2021)
6. Grimm, C., Singh, S.: Learning independently-obtainable reward functions. https://arxiv.org/pdf/1901.08649v3, arXiv preprint (2019)
7. Heuillet, A., Couthouis, F., Díaz-Rodríguez, N.: Explainability in deep reinforcement learning. https://arxiv.org/pdf/2008.06693v4, arXiv preprint (2020)
8. Juozapaitis, Z., Koul, A., Fern, A., Erwig, M., Doshi-Velez, F.: Explainable reinforcement learning via reward decomposition. In: International Joint Conference on Artificial Intelligence. A Workshop on Explainable Artificial Intelligence (2019). https://par.nsf.gov/servlets/purl/10159391
9. Kim, B., Wattenberg, M., Gilmer, J., Wexler, C.C.J., Viegas, F., Sayres, R.: Interpretability beyond feature attribution: quantitative testing with concept activation vectors (TCAV). https://arxiv.org/pdf/1711.11279v5, arXiv preprint (2018)
10. Lin, Z., Zhao, L., Yang, D., Qin, T., Yang, G., Liu, T.Y.: Distributional reward decomposition for reinforcement learning. https://arxiv.org/pdf/1911.02166v1, arXiv preprint (2019)
11. Mnih, V., et al.: Human-level control through deep reinforcement learning. Nature **518**, 529–533 (2015) (2015). https://doi.org/10.1038/nature14236
12. Petsiuk, V., Abir Das, K.S.: RISE: Randomized input sampling for explanation of black-box models. https://arxiv.org/pdf/1806.07421v4, arXiv preprint (2018)
13. Ribeiro, M.T., Singh, S., Guestrin, C.: "Why Should I Trust You?" Explaining the predictions of any classifier. https://arxiv.org/pdf/1602.04938v3, arXiv preprint (2016)
14. Schroeter, N., Cruz, F., Wermter, S.: Introspection-based explainable reinforcement learning in episodic and non-episodic scenarios. https://arxiv.org/pdf/2211.12930v1, arXiv preprint (2022)

15. Schwarting, W., Alonso-Mora, J., Rus, D.: Planning and decision-making for autonomous vehicles. Robotics, and Autonomous Systems, Annual Reviews, Annual Review of Control (2018)
16. Seijen, H.V., Fatemi, M., Romoff, J., Laroche, R., Barnes, T., Tsang, J.: Hybrid reward architecture for reinforcement learning. In: Advances in Neural Information Processing Systems Vol.30 (NIPS). Curran Associates, Inc. (2017)
17. Selvaraju, R.R., Cogswell, M., Das, A., Vedantam, R., Parikh, D., Batra, D.: Grad-CAM: visual explanations from deep networks via gradient-based localization. https://arxiv.org/pdf/1610.02391v4, arXiv preprint (2019)
18. Wang, J., Zhang, Y., Kim, T.K., Gu, Y.: Shapley Q-value: a local reward approach to solve global reward games. https://arxiv.org/pdf/1907.05707.pdf, arXiv preprint (2022)
19. Watkins, C., Dayan, P.: Technical note: Q-learning. Mach. Learn. **8**, 279–292 (1992). https://doi.org/10.1007/BF00992698
20. Wright, G.H.V.: Explanation and Understanding of Action. La Philosophie Scandinave **35**(135), 127–142 (1981). Revue Internationale de Philosophe (1981)

Towards Explainable Deep Learning for Non-melanoma Skin Cancer Diagnosis

Anh Le Van[1(✉)], Karin Verspoor[2], Thomas Brett Kirk[3], and Andy Song[2]

[1] RMIT Vietnam, 702 Nguyen Van Linh, District 7, Ho Chi Minh City, Vietnam
anh.levan@rmit.edu.vn

[2] RMIT University, 124 La Trobe Street, Melbourne, VIC 3000, Australia

[3] Curtin University, Kent Street, Bentley, WA 6102, Australia

Abstract. Skin cancer is a global health concern, but early diagnosis can be costly and challenging. An automated system is essential to prevent fatalities. This study focuses on non-melanoma skin cancer (NMSC), the most prevalent and one of the most serious health problems in the world, whereas the majority of computer-aided skin cancer diagnosis systems primarily target melanoma. Basal cell carcinoma (BCC), a kind of NMSC, is expected to overtake all other cancers in the near future. This study presents a highly accurate and automated diagnosis of NMSC. In addition to developing a classification model to differentiate NMSC from other skin diseases, we investigate the explainability of the model, since models should be accompanied by intuitive, persuasive, and consistent explanations for clinicians to accept the diagnosis. The model was trained and tested on the International Skin Imaging Collaboration (ISIC) archive and tested using the ISIC and derm7pt data sets, achieving high accuracy, sensitivity, and specificity. To deconstruct the prediction output, the model is studied using explainability methods that include gradient-based and perturbation-based approaches.

Keywords: XAI · ResNet50 · Skin cancer diagnosis · Non-melanoma skin cancer classification · Occlusion · Saliency · Integrated Gradients · GradientSHAP

1 Introduction

Melanoma and non-melanoma skin cancers (NMSC) are the two main types of skin cancer, and both continue to be major global health concerns [11]. Non-melanoma skin cancer (NMSC) is one of the five most common types of cancer

This work is supported by a well-known hospital. All datasets we worked with have been approved for public release and usage.

M. Gong et al. (Eds.): AI 2024, LNAI 15442, pp. 241–254, 2025.
https://doi.org/10.1007/978-981-96-0348-0_18

worldwide. In 2018, more than one million cases were reported, and this figure is projected to increase to about 2 million by 2040 [18]. NMSC, which refers to all types of skin cancer that are not melanoma, in which basal cell carcinoma (BCC) and squamous cell carcinoma (SCC) represented 95% of their types [18].

Although NMSC is often straightforward to treat and has good outcomes, if not discovered and treated early, it can progress to ulceration and metastases throughout the body [13]. This can have a severe impact on the patient's well-being and quality of life. The skin lesion is diagnosed primarily visually, beginning with a clinical evaluation based on symptoms and signs, followed by a dermoscopic study and, if necessary, a biopsy and histological examination [12]. Dermoscopy is a non-invasive method for analyzing pigmented melanocytic lesions, with high accuracy, but requires expertise in skin cancer, which requires dermatologists for verification and confirmation [3].

Deep learning models have been gradually adopted in healthcare in recent years to assist healthcare professionals, accelerating the efficiency and consistency of disease diagnosis [4,6,13]. The use of automated lesion classification can be beneficial in both assisting clinicians in their daily clinical routine and offering access to inexpensive and lifesaving diagnostics. A large amount of research has been done on dermoscopic images to identify skin lesions [13]. Some recent models have reached the same level or even surpassed medical specialists in terms of accuracy [8].

Although a digital camera can photograph skin lesions, little research is available on how skin lesions can be classified using these clinical images. Furthermore, the application of these advanced techniques in healthcare settings is still limited. One of the key reasons is the lack of confidence in machine-generated diagnosis, as dermatologists are unable to comprehend the rationale of deep learning models, hence not openly accepting the model output.

In critical fields, particularly medicine, it is crucial that explanations support model outputs. Diagnostic processes usually require far more information than just a binary prediction. Firstly, medical experts may have concerns that black-box models may be biased in some way. The ability to understand the rationale of the model is important for clinicians to make evidence-based decisions and to justify these conclusions to their patients and colleagues. Secondly, the European Union's General Data Protection Regulation (GDPR, Article 15) states that patients have the right to request an explanation of how a model makes a decision [23]. Although this is not a ban, the implementation of this legislation can prevent black-box models from being widely adopted. Unless the mechanism of the model's decision can be revealed, we may not be able to leverage the full potential of deep learning to benefit the medical domains.

To address the above, this study investigates explainable learning for the diagnosis of NMSC using an optimized ResNet50 model in the International Skin Imaging Collaboration (ISIC) archive and the Atlas of Dermoscopy dataset (derm7pt) [27]. To explain the predictions, Integrated Gradient (IG) [42], Occlusion [45], Saliency [40], Input*Gradient [39] and GradientSHAP [31] are investigated. This work takes a considerable step further in its clinical use by incorporating explainability and working in collaboration with dermatologists.

2 Related Work

2.1 Deep Learning in Skin Cancer Diagnosis

Deep learning, in particular, convolutional neural networks (CNNs), are neural networks with convolution layers that have proven to be powerful for tasks like image classification, object detection, and target identification. CNNs can learn the relationship between input and class labels and perform end-to-end classification. The learning process can be established from scratch or through transfer learning. In skin cancer image diagnosis, transfer learning is recommended due to data availability constraints. Research in skin lesions classification has employed well-known CNN models, such as AlexNet [28], VGG [40], GoogLeNet [43], DenseNet [24], ResNet [21], EfficientNet [44] for early detection of highly fatal melanoma [37], or binary classification of malignant and benign lesions [2,12]. Given the diverse nature of skin diseases, it is common to encounter multiple-class categorization tasks [15,17,19]. In addition, some researchers designed new CNN-based architectures for the skin lesion classification task, e.g. InSiNet [36], SCDNet [35]. It is hard to compare the performance of various classifiers because each study used distinct data sets for training and testing, as well as different metrics. Many studies have demonstrated the ability of deep learning to perform at least on par with dermatologists in the classification of melanomas [9,22,33]. There has been less study on the classification of NMSC using machine learning or deep learning. A recent study fine-tuned an EfficientNet-B6 to recognize skin lesions from melanoma and non-melanoma in the International Skin Imaging Collaboration (ISIC) 2019 and 2020 challenge datasets and achieved a ROC AUC of approximately 0.97 [41]. A quantitative study did not find significant differences in sensitivity or specificity between expert and deep learning methods [38].

2.2 Explainable Deep Learning Neural Networks

Explainability in deep learning image analysis can be achieved through post hoc heatmap visualization approaches. These methods generate a heatmap from an already trained neural network without altering how or what the model learned. They provide insight into model decisions by estimating the attribution scores for each input feature using gradient-based or perturbation-based methods. Gradient-based approaches assign a score to every input pixel based on the gradient of the output function with respect to the input. A large gradient indicates that a small change in the input has a great effect on the output.

In our study four gradient-based methods are involved, namely *Saliency* [40], *Input*Gradient* [39], *Integrated Gradients* [42], and *GradientSHAP* [31]. The simplest method for computing input attribution is through **Saliency**, also known as *Vanilla Gradients*. This technique assumes a linear relationship between the output and the input, where gradients act as the coefficients of each feature, reflecting their relative significance. **Input*Gradient** is an enhanced saliency method that refines saliency maps by multiplying the gradients by their corresponding input features. **Integrated Gradients** (IG) offers a more comprehensive explanation by averaging gradients. It begins with a reference point, typically a black or white image or an average of these, which contributes no information to the model's final decision. The method then calculates the integral of the gradients of the output along the shortest path between the reference and the input, revealing the contribution of each input to the output.

GradientSHAP uses gradient techniques to compute Shapley (SHAP) [32] values in cooperative game theory. The SHAP values represent the gradient expectations along the path between a reference and an input. Unlike the Integrated Gradients approach, it adds Gaussian noise to each input multiple times and randomly selects a reference from the reference distribution. GradientSHAP can thus be considered as an estimate of IG computed by estimating the gradient expectations for different baselines.

Furthermore, we also examine one of the early perturbation-based model explanations, *Occlusion analysis*, to compute feature attribution [45]. A gray square, in particular, covers a small area of the input. As the square moves across the input, the probability scores for a given class are measured. A heat map depicts the variations in the likelihood of the class score. The resulting heat map can be used to determine whether the model accurately detects the location of a picture that is crucial to the model's output.

2.3 Explainable Deep Learning in Skin Cancer Image Diagnosis

Several studies have been conducted on deep learning and Explainable AI (XAI) for dermoscopic skin cancer diagnosis, both for melanoma and non-melanoma skin cancer. Melanoma detection, on the other hand, is dominant because of the high rate of metastasis of the disease. Regarding the perturbation-based approaches, Li *et al.* utilized occlusion analysis for explainability, and the patterns highlighted by the method are consistent with dermatologists' opinions [30]. In terms of gradient-based approaches, the Class Activation Map (CAM), Gradient CAM (Grad-CAM) and their variations are commonly used [14–16,20,34]. These researchers reported that these methods can capture the most significant input features for future inferences. There was a substantial correlation between the clinical characteristics used by dermatologists and the visual explanation techniques.

However, recent work also reveals a significant issue with Grad-CAM, highlighting that it sometimes highlights regions of an image that the model did not use for prediction [1]. Despite extensive research on black-box models, their inner mechanisms and reasoning remain unclear. There is no consensus on good explanation characteristics or evaluation methods, leaving XAI with significant challenges in understanding and evaluating explanations. Therefore, in this study, we investigated the gradient-based algorithms mentioned above and evaluated their efficacy on our model.

3 Methodology

3.1 Datasets

In this study, two publicly available datasets are used, the ISIC Archive[1] and the Atlas of Dermoscopy 7-point criteria skin database (derm7pt)[2] [27]. Both data sets contain clinical and dermoscopic images of skin lesions, together with other metadata. The ISIC Archive was used to train and evaluate the performance of the chosen model, while the derm7pt was employed for assessing the model's generalizability.

In terms of image acquisition, clinical images are captured by digital cameras such as mobile phones, tablets, point-and-shoot cameras, etc., while dermoscopic images are taken by microscopes, which usually have approximately 10–15 magnification[3]. The images primarily originate from the United States, Europe, and Australia. The training input images have one of the five original labels from the ISIC Archive: basal cell carcinoma (BCC), squamous cell carcinoma (SCC), melanoma (MEL), seborrheic keratosis (SK) and nevus (NEV). The latter two are the most prevalent benign skin lesions, while the first three are cancerous. The images are categorized into two groups: BCC and SCC, as non-melanoma skin cancer (*NMSC*); the remaining three types into the (*Other*) group. The number of images in each category is presented in Table 1.

The derm7pt data set is not used in training, as it is used to assess the model's generalizability. To ensure consistency with the training data set, the derm7pt was also filterred and only images with original labels of BCC, SK and different types of nevus were selected. Table 1 shows the number of images in each category of each data set. The derm7pt dataset includes an equal number of clinical and dermoscopic images, which were presented together in a single column.

Table 1. The number of images in each category of the ISIC Archive and derm7pt

Category	Diagnosis	ISIC Archive		derm7pt
		Clinical	Dermoscopic	Cli/Derm
NMSC	BCC	40	3,377	42
	SCC	23	665	
Other	MEL	51	5,900	268
	NEV	20	20,754	575
	SK		1,561	45
Total		134	32,391	930

[1] Available from https://pypi.org/project/isic-cli/.
[2] Available from https://derm.cs.sfu.ca.
[3] https://dermnetnz.org/topics/image-acquisition-in-dermatology, [Online; accessed 10-Aug-23].

Data Pre-processing and Augmentation

The chosen ISIC images were first divided into three subsets, i.e., 22,040 images for training, 6280 images for validation, and 4071 images for testing. Approximately 15% of the lesion in each set was NMSC and 85% Other. We employ several image-enhancement techniques to enhance the generalizability of our models. Before training, all images were randomly cropped, scaled to 224 × 224 pixels and horizontally flipped with a probability of 50% to increase diversity, as there is no guarantee in the clinical setting that the image is taken under exactly the same conditions. Figure 1 shows some representative instances in the training set. The data were grouped into batches of 64, converted to tensor form, and normalized using the mean and standard deviation of the pretrained ResNet50 on Imagenet. During the loading phase, the Imbalance Dataset Sampler samples all classes equally (Fig. 2). The technique oversamples the low frequent classes, while undersamples the high frequent ones.

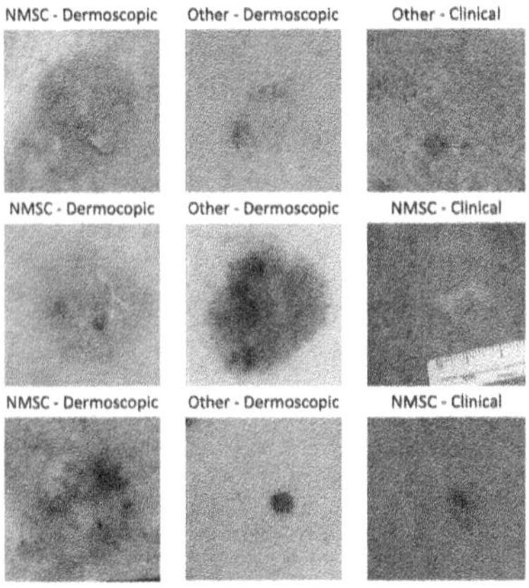

Fig. 1. Examples of NMSC images and images of the "Other" category in the training data set, containing both dermoscopic and clinical images

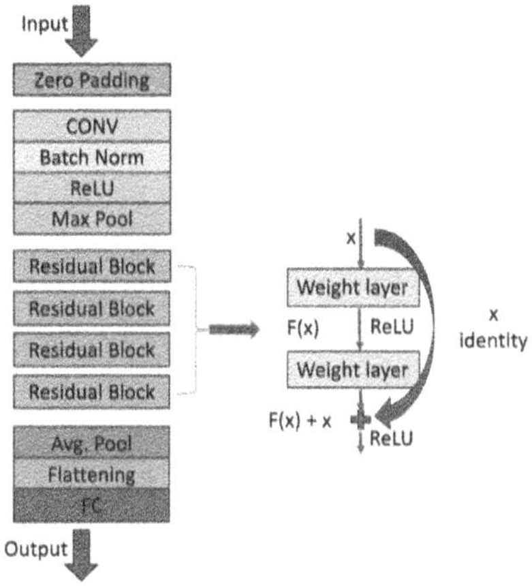

Fig. 2. ResNet50 model architecture and residual learning block (RLB) architecture. Illustration of RLB from *He et al. (2016), Fig. 2, page 2* [21]

3.2 Experiment Setup

For the classification task, we decided to fine-tune a ResNet50 model that was pre-trained on the ImageNet dataset. This choice was based on earlier research, which showed that ResNet50 achieved the highest average F1 score and precision in skin lesion detection tasks among other residual network designs [10,26]. We froze the lower-level layers and trained only the high-level portion of the convolution layers. The model is built in PyTorch and trained for 100 epochs with the Adam optimizer. The initial learning rate is set to 3×10^{-3} and decays every 10 epochs. Cross-entropy loss is used as the loss function and ReLU is the activation function. For evaluation of the performance of the classification of the imbalance, we group the prediction results into a confusion matrix and measure with four common metrics, namely accuracy (Acc), precision (Pre), recall (Rec), and F1-score, defined below. The F1 score is the harmonic mean of the precision and recall.

TP: True Positive, TN: True Negative, FP: False Positive, FN: False Negative

$$Acc = \frac{\text{TP} + \text{TN}}{\text{Total Predictions}} \qquad Pre = \frac{\text{TP}}{\text{TP} + \text{FP}} \qquad Rec = \frac{\text{TP}}{\text{TP} + \text{FN}}$$

To explore the explainability of the model, we utilize an open library, Captum[4], to implement all 5 methods introduced previously. For each input feature, its attribution score is calculated, which indicates how much each feature contributes to the final output of the model. We visualize the attribution scores by displaying them as a saliency map (or heatmap) and overlaying the map on the original image. Lastly, we had a meeting to perform a preliminary assessment of the explainability heat maps with two experienced dermatologists.

4 Results and Discussion

We optimized the pre-trained ResNet50 to detect NMSC from the ISIC Archive dataset. The model has a total of 23,512,130 parameters, with 100 epochs of training on a NVIDIA T4 GPU, which takes 9.7 h. In the ISIC validation set, the best accuracy was 88.6%. The model's effectiveness was then assessed using the ISIC test set and the derm7pt. With the second dataset, we further evaluated the model's generalization capabilities on different image modalities. The confusion matrix for different combinations of predicted and actual diagnoses for the ISIC test set is shown in Fig. 3 and Table 2 shows the accuracy, precision, recall and F1 score, calculated based on the confusion matrix.

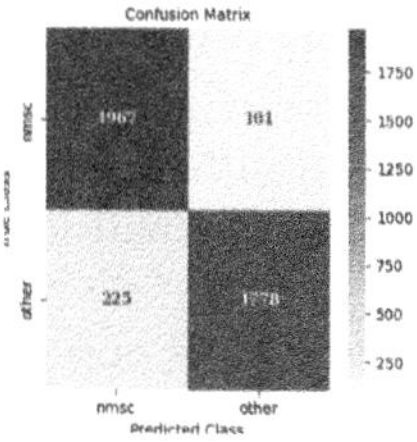

Fig. 3. Confusion Matrix

Table 2. Model performance evaluation on the ISIC testing set

Category	Precision	Recall	F-score
NMSC	0.90	0.95	0.92
Other	0.95	0.89	0.92
Accuracy			0.92
Macro Average	0.92	0.92	0.92
Weighted Average	0.92	0.92	0.92

[4] https://captum.ai/.

4.1 Model Performance

Overall, the performance of the trained classifier is promising, indicating that early NMSC detection is possible with the presented model. The training model performs very well on a common laptop when it comes to real-time inference. On average, the model takes about 0.1 seconds per image in the test set, demonstrating the effectiveness and efficiency of diagnosis in a real clinical environment. Furthermore, the general accuracies of the test in both dermoscopic and clinical images were 92% in the ISIC test set and 88. 5% in the derm7pt dataset, indicates that the model generalizes well to unseen data. The precision and recall of the NMSC classification are 90% and 95%, respectively, while both the macro average and the weighted average of these metrics are 92%. When it comes to medical diagnosis, precision and recall are crucial factors. High precision ensures the reliability of a positive diagnosis. High recall indicates that the model can effectively and efficiently identify all cases of a disease, reducing the chances of missing important diagnoses. The F1 score is 89% reflecting a more precise evaluation of the overall performance of the model. Although minor in number, all four clinical images in the test are correctly classified. In a broader context, our ResNet50 classifier outperforms Sharma *et al.*'s reported mean performance of ML algorithms in the diagnosis of NMSC [38] in terms of specificity and falls short in terms of sensitivity, although Sharma *et al.* did not provide a complete list of the results of their study. Compared to the most recent report on NMSC detection accuracy using artificial intelligence and dermoscopic images, our classifier is comparable to the average accuracy of 86.80% ± 12.05% achieved by 44 other studies [13].

As shown in Table 3, the results of the derm7pt dataset are separated for clinical and dermoscopic images, to examine the performance of the classifier based on a different image type for the same lesion. Despite the fact that the training set contains only 78 clinical photos, the model can learn features that may be present in both types of images. So, it can achieve an overall accuracy of 82% in unseen clinical data. Compared to clinical images, the performance of the model in dermoscopic images is better on all three metrics. Additionally, for the same clinical cases, dermoscopic images generally lead to a higher accuracy compared to clinical images. The disparity in accuracy between the two modalities may be related to the lack of sufficient clinical images in the training set. Since there are relatively few clinical images in the training phase, the model cannot learn solid features and is unable to generalize to new images as a result. Hence, it is expected that significantly increasing the amount of clinical data could lead to a better classification of these types of images.

As indicated by the dermatologists involved in this study, they commonly use dermoscopy to diagnose and visualize non-melanoma skin cancer, with clinical images augmenting dermoscopic images. Little research is available on the detection of cancers using clinical images alone. In addition to the lack of study, researchers use different datasets, prepare their own training/test set, and determine the number of classes to use. There is little common ground on how to evaluate the performance of these systems together. In comparison, Ballerini *et*

Table 3. ResNet50 Performance, plus additional reference results

Dataset	Accuracy	Sensitivity	Specificity
ISIC Archive	92%	86%	91%
Mean performance arma *et al.* [38]		89.2% (95% CI: 87.0–91.3)	81.1% (95% CI: 74.5–87.8)
Foltz *et al.* [13]	86.80% ± 12.05%		
derm7pt/Clinical	82%	40%	84%
derm7pt/Dermoscopic	95 %	62%	96%

al. used a kNN model to classify five common classes of skin lesions, including two non-melanomas, achieving an overall accuracy of 74% [5]. A study by Brinker *et al.* compared the performance of ResNet50 trained on dermoscopic images for the classification of clinical skin lesions with that of 145 dermatologists [7]. In the task of classifying 100 clinical images of melanoma and benign nevi, humans achieved an overall sensitivity of 89.5% and a specificity of 64. 4%, while machines exhibited the same sensitivity and higher specificity (69.2%). Some research constructs multimodal models in which dermoscopic and clinical data are combined with a loss function [27] or models are trained individually and then combined via an ensemble for prediction [15]. Kawahara *et al.* reported that the sensitivity and specificity of their combined models are 60. 4% and 91% in derm7pt, while the ensemble study by Ge *et al.* differentiated cancerous from non-cancerous moles with precision 82% on a private MoleMap dataset.

4.2 Model Explanability

In terms of explainability, five XAI techniques are studied for producing saliency maps that can show how important the regions of the input image are to the model output. The model properly classified images were from the ISIC Archive testing subset and the derm7pt with a confidence level varying from 0.9–0.97 for BCC, 0.92–0.99 for SCC, 0.87–0.97 for MEL, 0.6–0.96 for SK and 0.64–0.94 for NEV. Figure 4 illustrates some images of skin lesions and the five saliency maps.

From Fig. 4, we can observe that the heat map generated by occlusion produces more relevant regions while other approaches highlight pixel features. Amongst gradient-based methods, *Saliency* produces the most fine pixel-wise explanations, resulting in the most noisy heatmaps. As expected, the noise is reduced on images generated by *Input*Gradient, Integrated Gradients, and GradientSHAP*, respectively. In the research of "sanity checks for saliency maps", Adebayo *et al.* checked whether some XAI methods provide insight into a model or not by randomizing the model's weights and the training data and observing the effect on the saliency maps. The result showed that Saliency passed the tests, while Input*Gradient and IG failed, and GradientSHAP was not on their tested list [1]. The validation of these approaches should be evaluated before they are applied in clinical settings.

Our expert human assessors selected occlusion as the most "interpretable" method in terms of human interpretability because the information presented to users is not as complex as the gradient-based method. They pointed to the model's identification of specific relevant characteristics of NMSC, such as tiny blood vessels around the edge, are highlighted, which certainly adds trust in the model by our dermatologists. The granularity of the information in each pixel can limit the readability of saliency maps created by gradient-based techniques. Regarding applicability, gradient-based approaches require access to the neural network gradients, whereas occlusion can apply to any "black-box" machine learning model structure. A gradient-based method can provide the importance of every input pixel in a single forward and backwardpass, while a perturbation-based method requires numerous forward passes. Each perturbed input requires a model reevaluation, and the output solely represents the importance of the relevant perturbed feature [25]. Generalization of all possible perturbations is impractical because of the prohibitive cost.

Evaluating XAI methods is crucial to ensure their reliability and effectiveness in providing interpretable results. Our evaluation of the saliency maps shows the first step toward clinical application. Explainable models should be evaluated from both the AI technique and the perspective of a medical professional [29]. Engaging domain experts to validate the methods in clinical settings is essential. Their insights are valuable in understanding the clinical relevance of explanations and identifying potential limitations.

5 Limitations and Conclusions

Non-melanoma skin cancer, NMSC, the most common skin cancer, is a major public health concern. Automated diagnosis of NMSC, especially early-stage detection, is important to mitigate its health risk and economic cost. Currently, there are no baseline models or methods that support NMSC diagnosis. In this paper, we present the initial steps in helping dermatologists automate the diagnosis of NMSC with accurate, consistent, and interpretable results. Our exploration provides evidence that deep learning coupled with XAI techniques can be effective. The performance of the model is clinically practical to differentiate two types of carcinoma from other skin conditions, including melanoma. The key characteristics of carcinoma aligned with the factors identified as relevant by experts are highlighted through the occlusion-based explanability method.

This study lays the foundation for substantial future work, both from the modeling and explainability perspective. However, the model exclusively trained on images of individuals with predominantly fair skin, primarily because the original data set consisted mainly of images of fair-skinned individuals with infections. This poses several challenges and pitfalls regarding the model's performance, generalizability, and diversity. The data set's focus on fairly toned individuals resulted in a lack of diversity, leading to biased and inaccurate predictions. The model's capacity to generate precise forecasts for individuals with diverse skin tones may be limited, leading to lower performance and mistakes in real-world situations. The model perpetuated biases, resulting in inequitable treatment for those with other skin tones in medical evaluations.

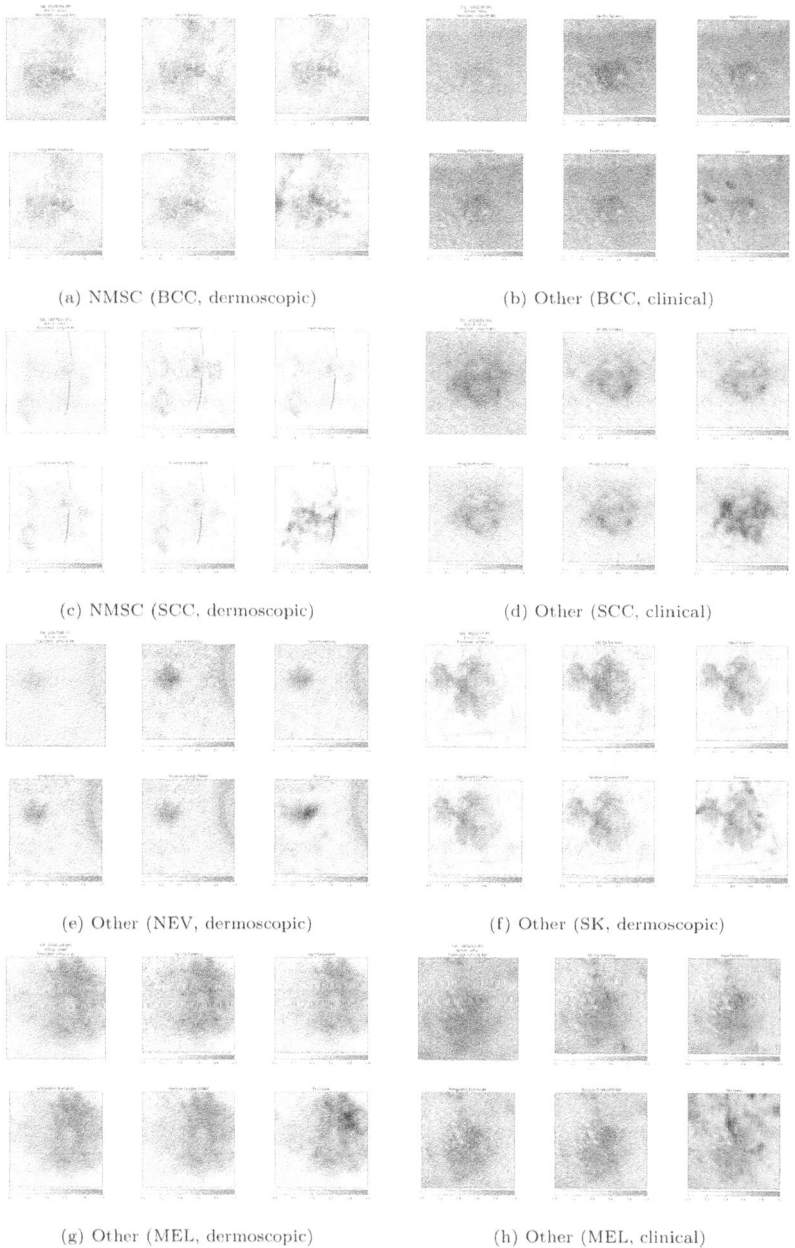

(a) NMSC (BCC, dermoscopic)

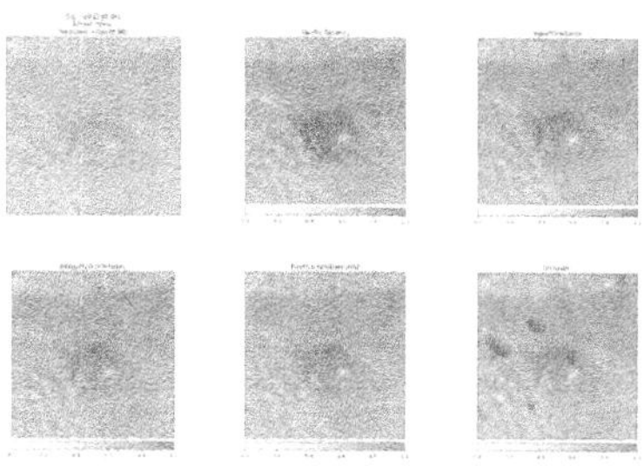

(b) Other (BCC, clinical)

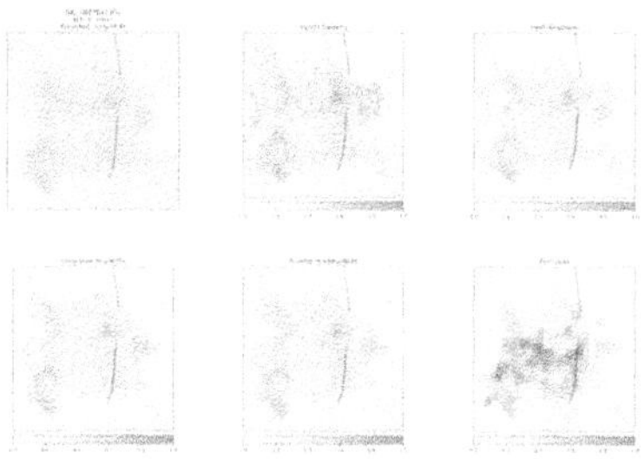

(c) NMSC (SCC, dermoscopic)

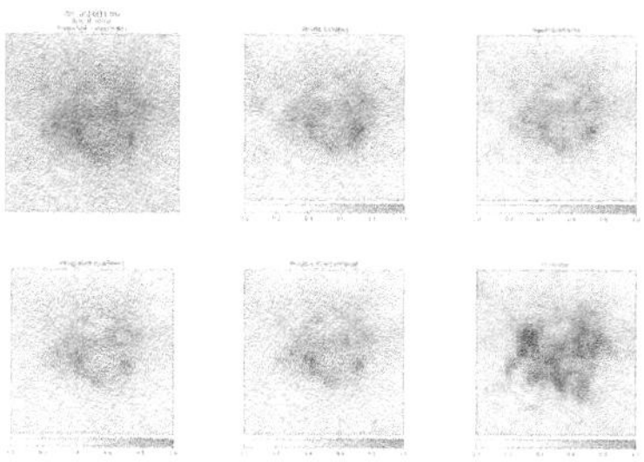

(d) Other (SCC, clinical)

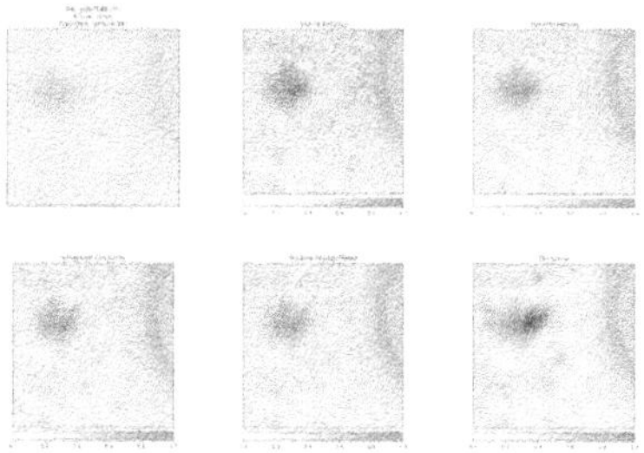

(e) Other (NEV, dermoscopic)

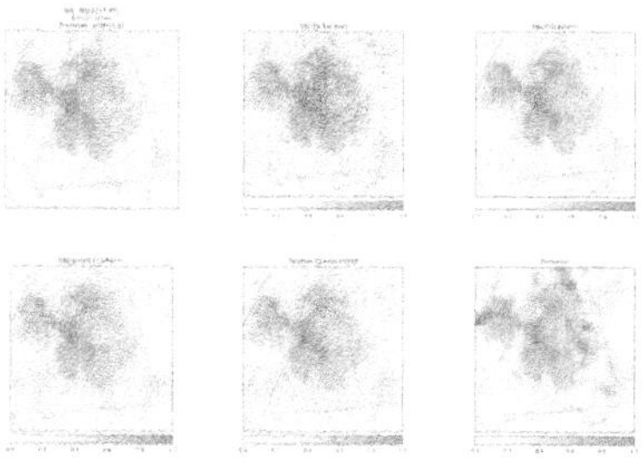

(f) Other (SK, dermoscopic)

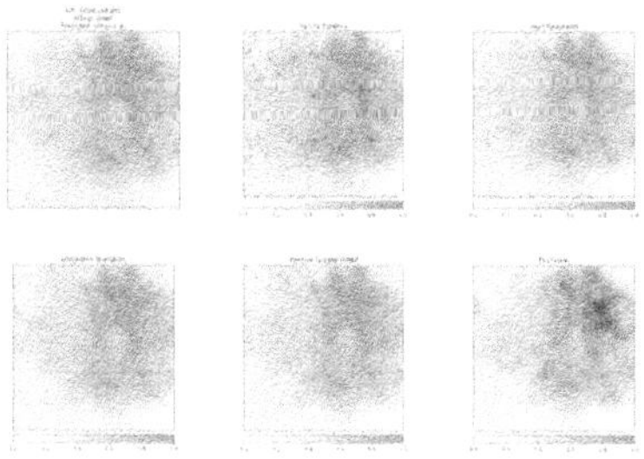

(g) Other (MEL, dermoscopic)

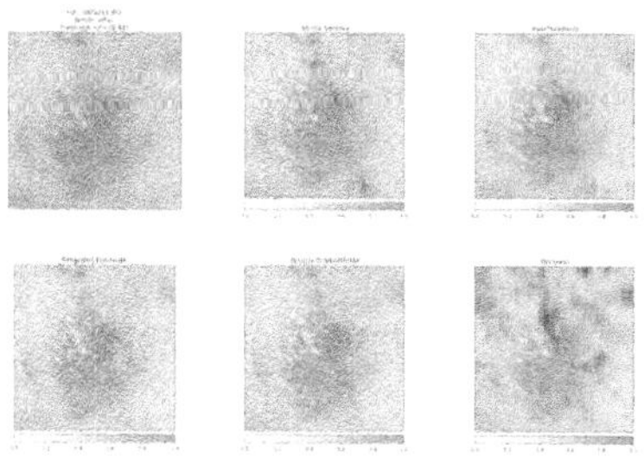

(h) Other (MEL, clinical)

Fig. 4. Illustrative XAI visualizations are provided for six images from the ISIC testing set. Each subfigure, labeled from (a) to (h), consists of two rows: the first displays the original image along with saliency maps from Vanilla Saliency and Input*Gradients, while the second row presents maps generated by Integrated Gradients, GradientSHAP, and Occlusion. These saliency maps highlight the most important pixels contributing to the prediction. The greater the presence of green and darker shades in the saliency maps, the more crucial those pixels or regions are in determining the model's prediction (Color figure online)

Furthermore, the collection lacks a substantial quantity of clinical images, as there are merely 134 of them, representing less than 1% of the entire data set. This shortcoming raises issues regarding the model's capability to accurately classify clinical images that it has not previously encountered. To improve the usefulness of the research, it is essential to address the lack of clinical photos in the training set. Additional categories of skin cancer data and medical photos need to be incorporated to enhance the versatility of the model.

In addition, pre-trained models derived from skin lesion datasets such as DermNet will be explored. Regarding XAI methods, we will investigate specific carcinoma-related patterns so that they can be highlighted better with the methods. Furthermore, to fully and systematically evaluate XAI approaches, evaluation metrics such as Co-12 and human-based approaches will be explored. Overall, further work will aim to assist dermatological practice in both diagnostic accuracy and acceptance by dermatologists.

References

1. Adebayo, J., Gilmer, J., Muelly, M., et al.: Sanity checks for saliency maps. In: Advanced Neural Information Processing Systems, vol. 31 (2018)
2. Ameri, A.: A deep learning approach to skin cancer detection in dermoscopy images. J Biomed. Phys. Eng. **10**(6), 801 (2020)
3. Argenziano, G., Soyer, H.P.: Dermoscopy of pigmented skin lesions-a valuable tool for early. Lancet Oncol. **2**(7), 443–449 (2001)
4. Aung, Y.Y., Wong, D.C., Ting, D.S.: The promise of AI: a review of the opportunities and challenges of AI in healthcare. Br. Med. Bull. **139**(1), 4–15 (2021)
5. Ballerini, L., Fisher, R.B., Aldridge, B., Rees, J.: A color and texture based hierarchical k-NN approach to the classification of non-melanoma skin lesions. Color Med. Image Anal. 63–86 (2013)
6. Banegas-Luna, A.J., Peña-García, J., Iftene, A., et al.: Towards the interpretability of machine learning predictions for medical applications targeting personalized therapies: A cancer case survey. Int. J. Mol. Sci. **22**(9), 4394 (2021)
7. Brinker, T.J., Hekler, A., Enk, A.H., et al.: A convolutional neural network trained with dermoscopic images performed on par with 145 dermatologists in a clinical melanoma image classification task. Eur. J. Can. **111**, 148–154 (2019)
8. Brinker, T.J., Hekler, A., Enk, A.H., et al.: Deep learning outperformed 136 of 157 dermatologists in a head-to-head dermoscopic melanoma image classification task. Eur. J. Can. **113**, 47–54 (2019)
9. Brinker, T.J., Hekler, A., Enk, A.H., et al.: Deep neural networks are superior to dermatologists in melanoma image classification. Eur. J. Can. **119**, 11–17 (2019)
10. Budhiman, A., Suyanto, S., Arifianto, A.: Melanoma cancer classification using resnet with data augmentation. In: 2019 International Seminar on Research of IT and Intelligent Systems (ISRITI), pp. 17–20. IEEE (2019)
11. Ciążyńska, M., Kamińska-Winciorek, G., Lange, D., et al.: The incidence and clinical analysis of non-melanoma skin cancer. Sci. Rep. **11**(1), 4337 (2021)
12. Esteva, A., Kuprel, B., Novoa, R.A., et al.: Dermatologist-level classification of skin cancer with deep neural networks. Nature **542**(7639), 115–118 (2017)
13. Foltz, E.A., Witkowski, A., Becker, A.L., et al.: AI applied to non-invasive imaging modalities in identification of nonmelanoma skin cancer: a systematic review. Cancers **16**(3), 629 (2024)

14. Gamage, L., Isuranga, U., Meedeniya, D., et al.: Melanoma skin cancer identification with explainability utilizing mask guided technique. Electronics **13**(4), 680 (2024)
15. Ge, Z., Demyanov, S., Chakravorty, R., Bowling, A., Garnavi, R.: Skin disease recognition using deep saliency features and multimodal learning of dermoscopy and clinical images. In: Descoteaux, M., Maier-Hein, L., Franz, A., Jannin, P., Collins, D.L., Duchesne, S. (eds.) MICCAI 2017, Part III. LNCS, vol. 10435, pp. 250–258. Springer, Cham (2017). https://doi.org/10.1007/978-3-319-66179-7_29
16. Giavina-Bianchi, M., Vitor, W.G., Fornasiero de Paiva, V., et al.: Explainability agreement between dermatologists and five visual explanations techniques in DNNs for melanoma AI classification. Front. Med. **10**, 1241484 (2023)
17. Girdhar, N., Sinha, A., Gupta, S.: Densenet-II: an improved DNN for melanoma cancer detection. Soft. Comput. **27**(18), 13285–13304 (2023)
18. Girvalaki, C., Cardone, A., Weinert, P., John, S.: Non-melanoma skin cancer as an occupational disease. What is the impact on the society and the welfare system? J. Health Inequal. **6**(2), 153–159 (2020)
19. Han, S.S., Kim, M.S., Lim, W., et al.: Classification of the clinical images for benign and malignant cutaneous tumors using a deep learning algorithm. J Investigat. Dermatol. **138**(7), 1529–1538 (2018)
20. Hauser, K., Kurz, A., Haggenmueller, S., et al.: Explainable AI in skin cancer recognition: a systematic review. Eur. J Cancer **167**, 54–69 (2022)
21. He, K., Zhang, X., Ren, S., Sun, J.: Deep residual learning for image recognition. In: CVPR, pp. 770–778 (2016)
22. Hekler, A., Utikal, J.S., Enk, A.H., et al.: Deep learning outperformed 11 pathologists in the classification of histopathological melanoma images. Eur. J. Can. **118**, 91–96 (2019)
23. Holzinger, A., Biemann, C., Pattichis, C.S., Kell, D.B.: What do we need to build explainable AI systems for the medical domain? preprint arXiv:1712.09923 (2017)
24. Iandola, F., Moskewicz, M., Karayev, S., Girshick, R., Darrell, T., Keutzer, K.: DenseNet: implementing efficient convnet descriptor pyramids. arXiv preprint arXiv:1404.1869 (2014)
25. Ivanovs, M., Kadikis, R., Ozols, K.: Perturbation-based methods for explaining deep neural networks: a survey. Pattern Recogn. Lett. **150**, 228–234 (2021)
26. Kassani, S.H., Kassani, P.H.: A comparative study of deep learning architectures on melanoma detection. Tissue Cell **58**, 76–83 (2019)
27. Kawahara, J., Daneshvar, S., Argenziano, G., Hamarneh, G.: Seven-point checklist and skin lesion classification using multitask multimodal neural nets. IEEE J. Biomed. Health Inform. **23**(2), 538–546 (2019)
28. Krizhevsky, A., Sutskever, I., Hinton, G.E.: Imagenet classification with deep convolutional neural networks. Commun. ACM **60**(6), 84–90 (2017)
29. Le, P.Q., Nauta, M., Van, B.N., et al.: Benchmarking eXplainable AI: a survey on available toolkits and open challenges. In: International Joint Conference on AI (2023)
30. Li, X., Wu, J., Chen, E.Z., Jiang, H.: From deep learning towards finding skin lesion biomarkers. In: 2019 41st International Conference of the IEEE Engineering in Medicine and Biology Society, pp. 2797–2800. IEEE (2019)
31. Lundberg, S.M., Lee, S.I.: Consistent feature attribution for tree ensembles. arXiv preprint arXiv:1706.06060 (2017)
32. Lundberg, S.M., Lee, S.I.: A unified approach to interpreting model predictions. In: Advanced Neural Information Processing Systems, vol. 30 (2017)

33. Maron, R.C., Weichenthal, M., Utikal, J.S., et al.: Systematic outperformance of 112 dermatologists in multiclass skin cancer image classification by convolutional neural networks. Eur. J. Can. **119**, 57–65 (2019)
34. Mridha, K., Uddin, M.M., Shin, J., et al.: An interpretable skin cancer classification using optimized CNN for a smart healthcare system. IEEE Access (2023)
35. Naeem, A., Anees, T., Fiza, M., et al.: SCDNet: a deep learning-based framework for the multiclassification of skin cancer using dermoscopy images. Sensors **22**(15), 5652 (2022)
36. Reis, H.C., Turk, V., Khoshelham, K., Kaya, S.: Insinet: a deep convolutional approach to skin cancer detection and segmentation. Med. Biol. Eng. Comput. 1–20 (2022)
37. Sauter, D., Lodde, G., Nensa, F., et al.: Deep learning in computational dermatopathology of melanoma: a technical systematic literature review. Comput. Biol. Med. **163**, 107083 (2023)
38. Sharma, A.N., Shwe, S., Mesinkovska, N.A.: Current state of machine learning for non-melanoma skin cancer. Arch. Dermatological Res. **314**(4), 325–327 (2022)
39. Shrikumar, A., Greenside, P., Shcherbina, A., Kundaje, A.: Not just a black box: learning important features through propagating activation differences. preprint arXiv:1605.01713 (2016)
40. Simonyan, K., Zisserman, A.: Very deep convolutional networks for large-scale image recognition. Preprint arXiv:1409.1556 (2014)
41. SM, J., P, M., Aravindan, C., Appavu, R.: Classification of skin cancer from dermoscopic images using DNN architectures. Multimedia Tools Appl. **82**(10), 15763–15778 (2023)
42. Sundararajan, M., Taly, A., Yan, Q.: Axiomatic attribution for deep networks. In: International Conference on ML, pp. 3319–3328. PMLR (2017)
43. Szegedy, C., Liu, W., Jia, Y., et al.: Going deeper with convolutions. In: CVPR, pp. 1–9 (2015)
44. Tan, M., Le, Q.: EfficientNet: rethinking model scaling for convolutional neural networks. In: International Conference on ML, pp. 6105–6114. PMLR (2019)
45. Zeiler, M.D., Fergus, R.: Visualizing and understanding convolutional networks. In: Fleet, D., Pajdla, T., Schiele, B., Tuytelaars, T. (eds.) ECCV 2014. LNCS, vol. 8689, pp. 818–833. Springer, Cham (2014). https://doi.org/10.1007/978-3-319-10590-1_53

Machine Learning and Data Mining

Localization System Enhanced with CDLPE: A Low-Cost, Resilient Map-Matching Algorithm

Yanyan Wang(✉), Hailu Jia, Yu Pan, and Hongxia Bai

Baidu.com Times Technology (Beijing) Co., Ltd, Beijing, China
{wangyanyan04,jiahailu}@baidu.com

Abstract. Accurate and robust localization is a critical component in intelligent vehicles, playing a significant role in route planning and efficient navigation. There is a rising trend towards affordable positioning solutions that use common vehicular sensors like GPS, IMU, and cameras to improve navigation accuracy. This paper presents a comprehensive, low-cost localization framework with a lightweight map. The framework introduces two key novelties. Firstly, we propose a method known as the Cross-Dimensional Lane and Pose Estimator (CDLPE), designed to effectively resist scenarios with poor satellite signals. Additionally, our system delivers a reliable localization service by effectively integrating matching results and capitalizing on the benefits of the sensors used, coupled with the understanding of the environment. We have verified the robustness of our method under different driving scenarios. Compared to the classical Iterative Closest Point (ICP) algorithm, the lane identification accuracy has improved by 4.42% and 9.23% during normal and weak satellite signal conditions, respectively. Videos in: https://youtu.be/DsYXSeWQhWc.

Keywords: Intelligent Vehicle · Map-matching · Navigation · Localization System · Low-cost

1 Introduction

Localization is a critical technology that helps intelligent vehicles determine their precise position and orientation within the environment. This process is essential for path planning, obstacle avoidance, and ensuring safe and efficient navigation (Fig. 1).

Vehicle localization typically involves the integration of various sensors such as Global Navigation Satellite System (GNSS), LiDAR, radar, and cameras, along with sophisticated algorithms to process sensor data, and may also utilize map data and other information sources to accurately determine the vehicle's position and orientation. In recent years, the approach of using high-precision maps and Lidar for auxiliary positioning has been quite common due to its high accuracy, with [1] serving as a prime illustration. However, due to the

M. Gong et al. (Eds.): AI 2024, LNAI 15442, pp. 257–272, 2025.
https://doi.org/10.1007/978-981-96-0348-0_19

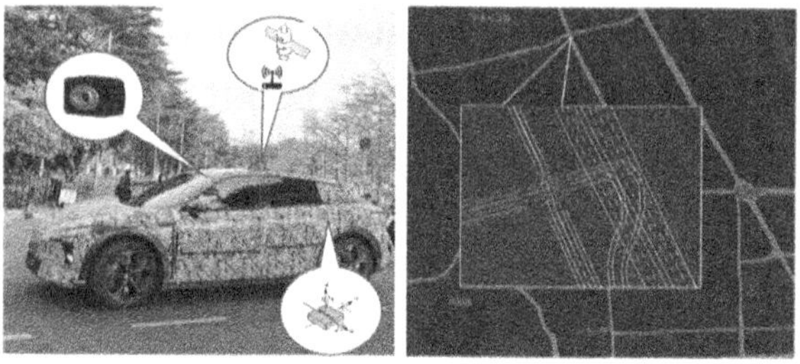

Fig. 1. Test Vehicle Configuration and Lightweight Map

high maintenance cost of large-scale 3D high-precision maps, and with the rapid improvement of image-based algorithm accuracy due to the development of deep learning, people are increasingly inclined to use a low-cost positioning solution that combines lightweight maps such as [4] and images with other commonly used vehicular sensors such as GPS, IMU for precise positioning. Each sensor possesses distinct attributes: GNSS is effective in open spaces, IMU consistently computes position, orientation, and speed but is prone to integration drift, and cameras necessitate road features and a good field of view. Maps not only assist in positioning but also provide prior driving information, significantly aiding in the integration of data from all sensors.

This paper introduces a robust and precise localization system for intelligent vehicle navigation in both urban and highway settings. Our work adroitly amalgamates data from an array of sensors with environmental knowledge and employs an innovative map matching algorithm on a lane-level light map. The system is designed to consistently provide accurate location services even in complex scenarios such as tunnels, intersections, and multi-layered roads. In summary, our main contributions are as follows:

1. We present an accurate and low-cost complete localization framework for intelligent vehicle, which effectively integrates map-matching results and leverages the advantages of multiple sensors with the understanding of environment.
2. We have developed a map matching algorithm that effectively resists poor GNSS signal scenarios and significantly enhances the accuracy of lane association. This method, called the Cross-Dimensional Lane and Pose Estimator (CDLPE), represents a substantial improvement in the field.
3. Our localization system has been rigorously tested on vehicles navigating crowded urban streets on a daily basis. The results demonstrate that our system consistently provides promising localization outcomes across a variety of driving scenarios.

2 Related Works

In this section, we contextualize our contributions within relevant sub-fields.

Low-Cost Vehicular Localization Systems. With the advancement of computer vision, utilizing low-cost cameras [33] is the direction of development for intelligent vehicles. In earlier works, it's common to utilize 3D point cloud based high-definition map [2,3,34] in localization. These works [21,22,32] usually transfer the image data to the same format by specific algorithm firstly and than exploit rigid registration algorithms such as Iterative Closest Point(ICP) [5] to estimate pose. Due to the high maintenance costs of high-definition maps, people [6–8] are now more inclined to use semantics representation which are more robust against illumination variation and seasonal changes [4] from data in vector-format maps for positioning, such as lane marking, curb, and so on. To leverage additional sensor inputs, some learning-based methods [29,30] try to utilize a coarse GPS location and gravity direction. However, multi-sensor fusion by filter methods has been long leveraged to build accurate and reliable localization systems [10,15,31]. Our work falls into the category of vision-based localization, integrating vision systems in a loosely coupled manner with the inertial navigation system (INS).

Map Matching Algorithms. An critical issue in map-matching methods is that the detected feature can falsely correspond with a one stored on a offline map in the lateral direction. To tackle this problem, RANSAC algorithm [23,24], probabilistic description [25,26] and tracking methods [27] are applied, and several researches believe ego-lane identification is important, [13,28] explored lane-change detection system. [9,10] divide the location task into several parts, including road level, lane level and ego-lane level localization. [12] proposed two distinct map matching algorithms fused by EKF. [11,16] present approaches for lane-level localization in a coarse-to-fine fashion. Some works [17,19,35] explored deep learning algorithms. Nonetheless, the effectiveness of most skills is highly dependent on the quality of perception and GNSS inputs in contributing to the pose estimation process. Thus, this work introduces the Cross-Dimensional Lane and Pose Estimator (CDLPE), a method that optimally utilizes previous maps and visual data by cross-dimensionally estimating the vehicle's pose with different algorithms and fusing the result adaptively, thereby bolstering the system's resilience in scenarios of poor visibility or weak GNSS signals.

3 Approach

3.1 System Overview

Figure 2 shows the block diagram of the proposed vehicle localization system. It utilizes a GPS, IMU, built-in wheel speed sensor, a single front camera, a digital map and the planned route. The system combines GPS, dead reckoning using INS feedback and wheel odometry, and multi-layer map matching algorithms through an error state Kalman filter (ESKF). The sensor fusion module uses a unicycle kinematic model, and ESKF helps avoid parameter singularity and ensures their linearization. Like many public researches, the system also uses a neural network for lane lines detection and tracking, fitting the detected lines to a cubic curve, then calibrated using multi-frame detection results. These curves are cross-validated with the map to filter out unstable results, playing a crucial role in map matching.

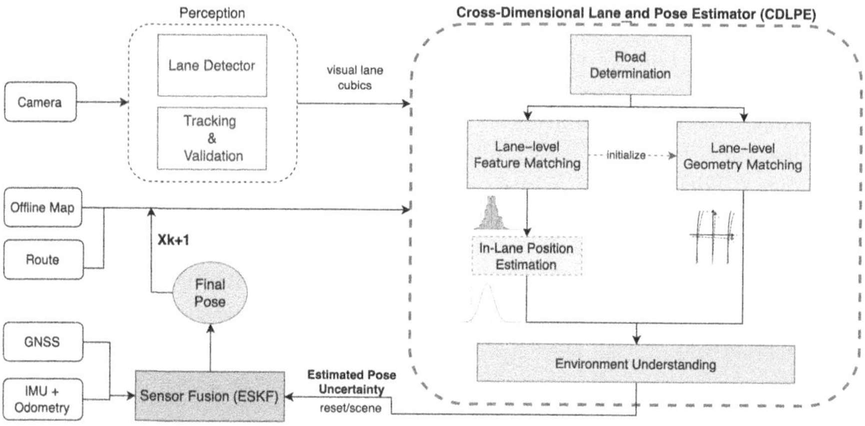

Fig. 2. Overview of the Proposed Localization System Enhanced with CDLPE

3.2 Cross-Dimensional Lane and Pose Estimator (CDLPE)

As shown in the Fig. 2, CDLPE cross-dimensionally estimates its ego-lane and pose in multiple modules. The algorithm starts from matching at road level, which employs a standard forward algorithm in Hidden Markov Modeling (HMM), in which the assumptions and innovative parts of HMM algorithm modeling are detailed in Appendix 6.1. Then, in order to make better use of features from different dimensions, the lane-level matching of features and geometry are run independently with different algorithms. Finally, the environment understanding module determine the output pose as well as estimation confidence.

Lane-Level Feature Matching. This module aims to rely on rough initial positioning information, located road segment, lane-level map data as a priori, and combined with perception lane lines features to determine the approximate location of the vehicle.

As Fig. 3a shows, in order to determine the approximate location, our method divide the target road into several columns. The grey dashed lines draw the divided lateral grids in the road, with each grid measuring 20 cm in width, and the red line shows the one-dimensional coordinate system within the road. We use $D = \{d_1, d_2, ..., d_k\}$ to represent the distance to left side of road boundary for each grey line, i.e., d_i is the x-coordinate of the left grey line of the grid in the one-dimensional coordinate system on the diagram.

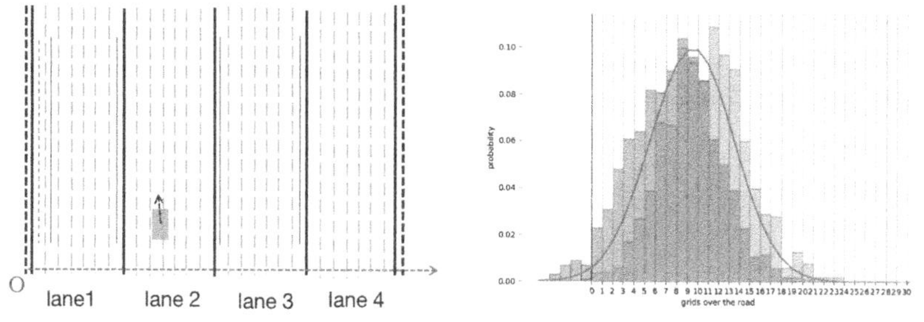

(a) (b)

Fig. 3. (a) The grey dashed lines draw the lateral grids in the road, with each grid measuring 20 cm in width, and the red line shows the one-dimensional coordinate system within the lane; (b) The blue and orange bars represent probability distribution of prediction $\bar{p}_{i,t}$ and measurement $p(x_i|z_t)$ respectively. The red curve denotes final fitted Gaussian function (Color figure online)

This module applies a histogram filter, which approximates the posteriors by decomposing the state space into finitely many spaces, and representing the cumulative posterior for each region by a single probability value [14]. Its formula is as follows

$$\bar{p}_{i,t} = \sum_j p(X_t = x_i | u_t, X_{t-1} = x_j) p_{j,t-1} \tag{1}$$

$$p_{i,t} = \eta p(z_t | X_t = x_i) \bar{p}_{i,t} \tag{2}$$

where $p_{i,t}$ represents the belief of each state x_i at time t, u_t is the control input, and z_t is the measurement vector. Equations (1) and (2) correspond to the prediction and update step. As the Fig. 3a shows, the $p_{i,t}$ means the probability of that vehicle is located in the lateral grid i.

In our work, the prediction step is performed based on the knowledge of odometry velocity, meanwhile according to Bayes theorem, the update step can be rewritten as

$$p_{i,t} = \eta p(z_t | X_t = x_i) \bar{p}_{i,t} = \eta \frac{p(X_t = x_i | z_t) p(z_t)}{p(X_t = x_i)} \bar{p}_{i,t} \tag{3}$$

in which $p(X_t = x_i) = \frac{d_i^{right} - d_i^{left}}{width_{road}}$, and $p(X_t = x_i | z_t)$ is calculated by fusing all the measurement probabilities. Meanwhile, $p(z_t)$ can be merged into the normalization factor η, thus we can derive

$$p_{i,t} = \bar{\eta} \frac{p(X_t = x_i | z_t)}{p(X_t = x_i)} \bar{p}_{i,t} \tag{4}$$

The problem is abstracted as a one-dimensional positioning problem with discrete and limited status, where state x_i means the vehicle is located in the lateral grid i. The lane line position data, given as points (latitude and longitude) on the map, is fitted with a cubic polynomial to match the perception. As Fig. 3a shows, for ease of subsequent computations, the leftmost lane line is used as the origin, adjusting the intercepts accordingly, i.e., d_i always represent the distance to left side of road boundary.

Next, we introduce the process of prediction and update corresponding to Equation (1) and (4).

The Prediction Step. The prediction step is to predict on the new belief of the state with the historical distribution of the filter and the motion model. Since the state only considers horizontal localization, we calculate the new lateral position of the car on the road at the current time t based on the posterior distribution fitted at the time $t-1$ and the odometry velocity on lateral direction $velocity_x$.

$$\mu_i = \mu_{i-1} + velocity_x * \Delta t \tag{5}$$

$$\bar{p}_{i,t} \sim N(\mu_i, \sigma_{i-1}^2) \tag{6}$$

which can be interpreted as a slight movement based on the probability distribution within the road of the previous moment.

The Update Step. There are only two states of a vehicle in a certain grid, i.e., it is either within the grid or not. Therefore, in this solution, static binary Bayes filtering is used to calculate the probability $p(x_i|z_{1:k,t})$, which indicates the confidence of vehicle being in the ith grid under k observations at time t (For simplicity, we ignore the symbol t representing the time).

$$l_k^i = l_{k-1}^i + \log\frac{p(x_i|z_k)}{1-p(x_i|z_k)} - \log\frac{p(x_i)}{1-p(x_i)} \tag{7}$$

in which l is the logistic probability, i.e. $l_k^i = \log\frac{p(x_i|z_{1:k})}{1-p(x_i|z_{1:k})}$. After obtain l_m^i by the above Equation (7), it can be derived that the probability of vehicle is on the grid i under the measurements $z_1 : z_k$ at time t, as shown in the following equation:

$$p(x_i|z_{1:k}) = 1 - \frac{1}{1+e^{l_k^i}} \tag{8}$$

Our k observations include lane line types, lane line distances, the number of lane lines, etc. Here, taking the number of lanes detected on the left and right side of the vehicle as an example of observation as z_{num}, we illustrate the calculation of observation probability $p(x_i|z_{num})$ in Equation (7). When the vehicle observes n_l lane lines on the left and n_r lane lines on the right through the camera, and w, δ, N denotes the road width, the variance of this observation and the total lane number of this road respectively, the probability can be expressed by Equation (9):

$$p(x_i|z_{num}) = \frac{1}{4w}\int_{d_{i-1}}^{d_i} [1 + erf(\frac{y - d_{max(n_l-1,0)}}{\sqrt{2}\delta})]\,[1 - erf(\frac{y - d_{min(N-n_r,n-1)}}{\sqrt{2}\delta})]\,dy \tag{9}$$

This method utilizes the concept of filtering and offers impressive scalability. It allows for the seamless integration of new observational data into the existing model.

In-Lane Pose Estimation. After deriving the grid probability distribution $p(x_i)$, the vehicle's position must be determined. In ambiguous situations, this is achieved using an optimization solution. We employ the Ceres Solver, a non-linear optimization library, to fit a standard Gaussian function to our probability distribution, as Fig. 3b shows. The cost function can be articulated through the subsequent equations.

$$J = \sum^{i}(p(x_i) - f(x)) \tag{10}$$

$$\text{s.t.} \quad f(x) = \exp(-\frac{(x-\mu)^2}{2var^2}) \tag{11}$$

This estimated pose serves as both the output for in-lane positioning and the prior position for the next moment's prediction process.

Lane-Level Geometry Matching. This section focuses on achieving centimeter-level positioning and maintaining stability based on geometric relationships, unaffected by accidental incorrect ego-lane determination in feature matching. The plan is divided into two parts: the first part addresses the data association problem between perceived and prior map lane lines data. The second part obtains accurate pose optimization based on the established association relationship.

Data Association. This part abstracts the data association problem between perception and map lane lines into a classic maximum matching solving problem of a weighted bipartite graph. It calculates weights from several lane lines data features that affect positioning accuracy and assigns solutions as a whole to ensure the optimal matching set is directly obtained during the data association stage. The key point is to identify and eliminate erroneous data effectively. Consider a weighted bipartite graph G = (S, T; E) with perception lane lines as vertices S and map lane lines as vertices T, and the edges E each have a cost weight $weight(i,j)$. The weight can be calculated by Equation (12)

$$weight(i,j) = c_i * (1 - \frac{error_{i,j}}{\sum_i error_{i,j}}) \tag{12}$$

where $error_{i,j}$ represents the value of error between perception line i and map line j in terms of direction and distance, and c_i is the correction coefficient, which is related to the properties of the perception-acquired lane line itself.

After obtaining the association weight matrix, the Hungarian matching algorithm is directly used to solve it to obtain the maximum matching. This solution can maintain good correlation accuracy when there are partial misdetections in the perception data such as ground cracks or potential problems in the map such as temporary road repairs.

Pose Optimization. For vehicle pose correction using perception and map data matching results, it's necessary to convert the map lane lines data to the body system, which involves projection transformation. Equidistant and equiangular projections can both introduce errors. As the vehicle moves further from the initial reference point, these projection errors increase. To mitigate this, our approach uses the vehicle's current position as the reference point for converting map lane lines data to the body system, as shown in Fig. 4, the map and perceived red lines are drawn under the current vehicle pose assumption.

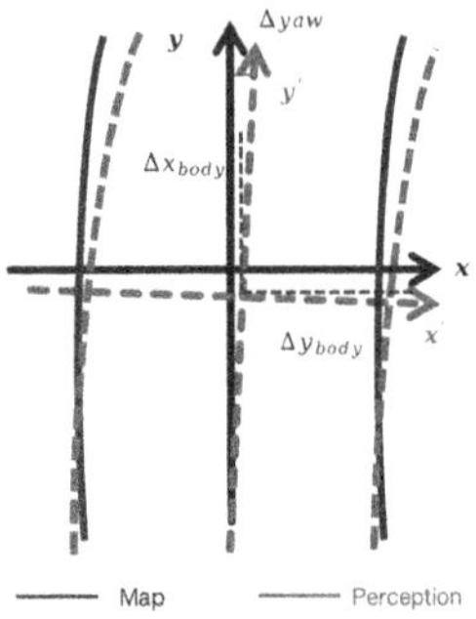

Fig. 4. The map and perceived lines

Then, we solve the pose error based on the data association results of the map data and the perceived lane lines equation in the body system. The state to optimize is the error of the vehicle pose, defined as

$$X = [\Delta x_{body}, \Delta y_{body}, \Delta yaw]^T \tag{13}$$

, in which Δx_{body}, Δy_{body} and Δyaw represent lateral, longitudinal and orientation error respectively. The residual term comes from the fitness of the curve equation of the lane lines perceived from the points from the prior map as well as the Mahalanobis distance between the map and the perceived object.

Environment Understanding. This module consists of uncertainty estimation and scene recognition. Uncertainty estimation calculates the value $UncV$ for the fusion algorithm to determine the variance while fusing matching result. This uncertainty stems from matching uncertainty and map inaccuracy, as (14) shows.

$$UncV = \lambda * (UncV_{geometry} + UncV_{feature}) + UncV_{map} \tag{14}$$

$UncV_{feature} = \frac{1}{\sum_i p(i)} \sum_i p(i)(d_{optimal} - d_i)^2$ indicates the uncertainty in the estimated probability distribution across road grids in feature matching. On the other hand, $UncV_{geometry}$ represents the variance value of the optimized longitude, latitude and yaw value in geometry matching module, given by the Ceres solver in the tangent space. Note that the position with lower uncertainty will be output as matching result.

Currency issues like changes in lanes, width, and heading are common in maps. To detect potential anomalies, we measure this uncertainty value by the difference between the failed-matched confident visual detection results and the map data.

$$UncV_{map} = \frac{1}{N_{failed}} \sum_{i}^{N_{failed}} [\frac{abs(yaw_i - yaw_{map})}{\lambda * velocity} + \frac{abs(offset_i - offset_{map})}{width}] \tag{15}$$

As a supplement to the use of $UncV$, the scene recognition identifies whether it is a scene that has a significant impact on map matching or need a special strategy in fusion.

In essence, the road determination module outputs the vehicle's current located road and recommended position, useful for navigation planning and deviation assessment in extreme scenarios. The lane-level feature matching calculates the vehicle's probability distribution within the road grids, using Bayesian methods based on determined road and perception data. It estimates the vehicle's approximate location, mitigating lane matching errors during significant GPS deviation via lane lines features. The lane-level geometry matching focuses on producing precise and stable within-lane pose, which remains unaffected by errors in feature-based matching after successful initialization. The environment understanding module not only guarantees the algorithm's applicability in complex road networks, but also provides uncertainty information for fusion algorithm to select the estimated pose from matching.

4 Evaluation Results

4.1 Experimental Setup and Preliminary Performance

Our algorithm is road-tested daily within the city by our intelligent vehicles. Each test vehicle is equipped with a front camera, GNSS receiver, IMU and vehicle's wheel odometry. To construct the ground truth value, we have additionally assembled NovAtel devices. The test set consists of approximately 300 km of data in Beijing, shown in Fig. 5. We have provided a preliminary performance estimate based on manual statistics, as shown in Table 1 below. Qualitative and quantitative experiment will further assess the accuracy and robustness of the method.

Table 1. The performance of various scenarios in Beijing testset

Scenario	Distance	Lateral RMS < 1.0 m
Highway/Expressway	145 km	0.999
Urban with servere shelter	42 km	0.988
Tree-lined road	23 km	0.992
Tunnel	53 km	1.0
Under stacked road	25 km	0.975
Ordinary/open urban road	98 km	0.994

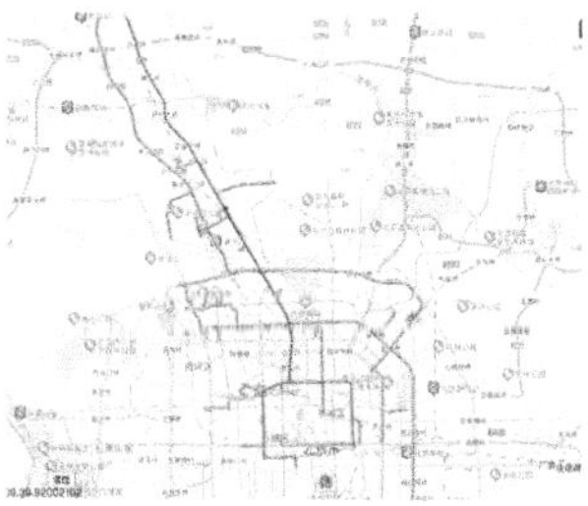

Fig. 5. Beijing testset

4.2 Quantitative Analysis

We evaluate the performance of our localization approach against the traditional ICP method [20] using our Beijing test set. Utilizing the perception detection of lane line points and map lane line sample points, we replaced the CDLPE module with ICP for matching. For fair comparison, the lane line type is utilized as an expanded dimension in the ICP method. To explicitly present the contributions from various sources, we conducted experiments separately under ordinary GPS and RTK technologies and represent the test results in four unique modes. 1. Our system with RTK and ICP; 2. Our system with RTK and CDLPE; 3. Our system with GPS and ICP; 4. Our system with only GPS and CDLPE. In Table 2, we display the quantitative results on both regular and weak GNSS roads.

Table 2. The comparison of different localization methods based on pose and lane accuracy

Scenes	GNSS	Method	RMS. Lat(m)		RMS(Longi.m)		Yaw(deg)	Ego Lane
			max	mean	max	mean	mean	Accuracy
Regular	RTK	ICP	4.942	0.544	4.863	0.871	1.740	0.9530
		CDLPE	**0.746**	**0.175**	**1.794**	**0.414**	**0.692**	**0.9965**
	GPS	ICP	7.540	0.746	5.008	1.424	2.658	0.9312
		CDLPE	**1.598**	**0.213**	**2.379**	**0.631**	**0.978**	**0.9877**
Weak GNSS	RTK	ICP	11.214	1.565	13.621	1.424	3.138	0.8426
		CDLPE	**1.174**	**0.289**	**4.658**	**1.296**	**1.025**	**0.9641**
	GPS	ICP	12.057	1.947	15.362	1.876	4.357	0.8206
		CDLPE	**1.975**	**0.485**	**5.680**	**1.684**	**1.687**	**0.9394**

Simultaneously, to depict the evaluation accuracy more clearly, we've chosen a continuous 1200-s data segment randomly from the regular scenes. Figure 6 presents the individual results of the lateral and longitudinal errors from the aforementioned four methods when the sampling rate is decreased from 20 Hz to 1 Hz. Note that longitudinal errors can be reduced effectively as well due to the effective update of the state in the positioning direction.

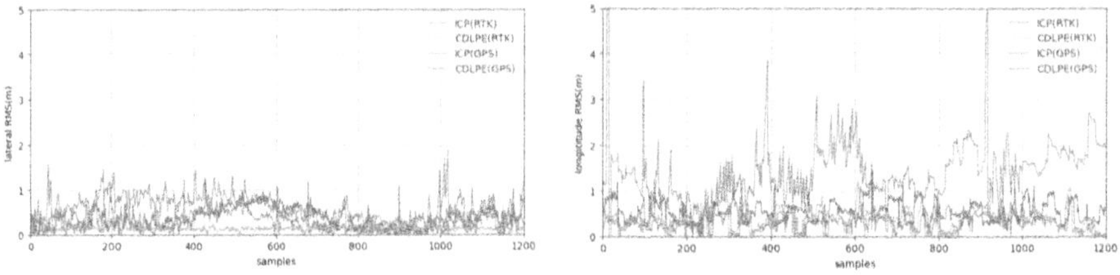

(a) lateral (b) longititude

Fig. 6. Comparison of errors set by different experimental methods. (Color figure online)

4.3 Ablation Study

An accurate localization ensures that the projection of map elements on image is completely consistent with the semantic perception. Projection results in different scenes in test set are shown in Fig. 8. Sub-figures d, e, f depict a typical turning process, where longitudinal errors become lateral ones and GNSS has a margin of error. Our algorithm ensures a smooth and accurate correction at intersections, preventing wrong lane placements, by utilizing matching result in multiple dimensions. Sub-figures g and h represent weak GNSS signal scenarios like tunnels and underpasses. Despite GNSS drift (light blue block), our

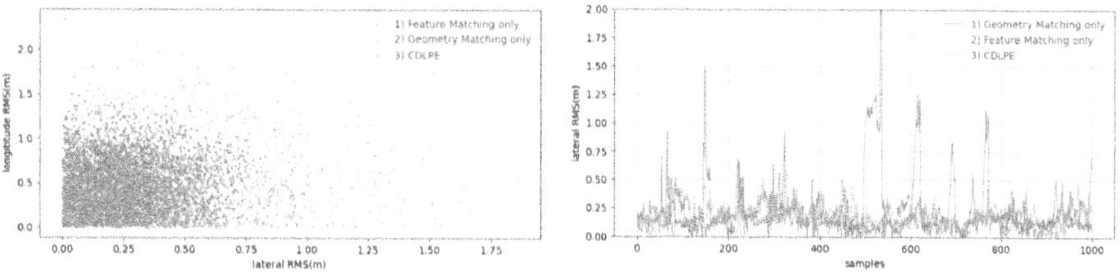

(a) 2-dimensional distribution (b) lateral

Fig. 7. Errors of different modules in the ablation experiment.

stable fused result (green block) is achieved through scene understanding and map-matching verification (Fig. 7).

4.4 Qualitative Result

An accurate localization ensures that the projection of map elements on image is completely consistent with the semantic perception. Projection results in dif-

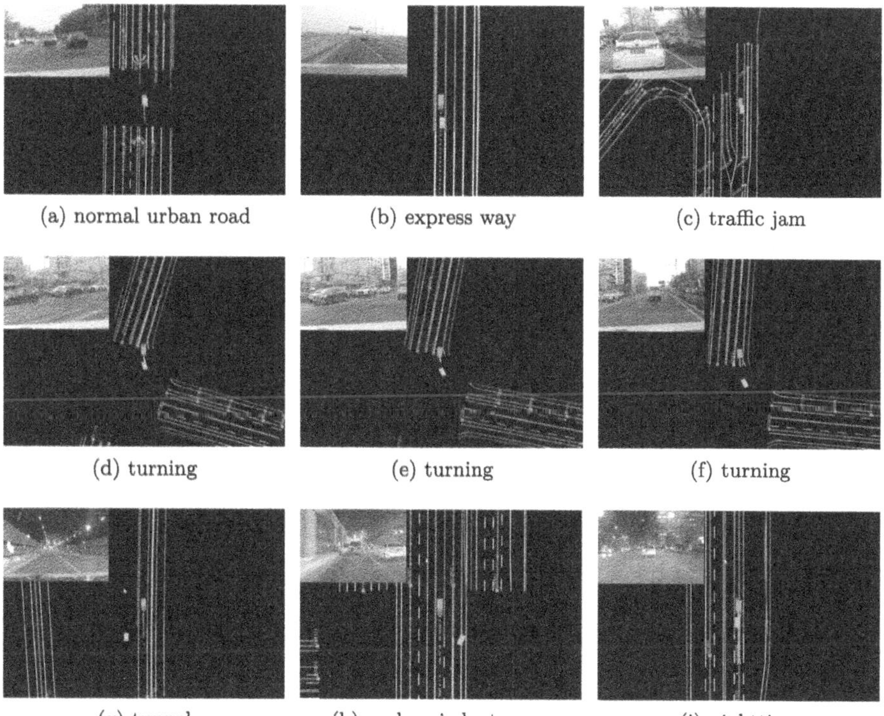

(a) normal urban road (b) express way (c) traffic jam

(d) turning (e) turning (f) turning

(g) tunnel (h) under viaduct areas (i) nighttime

Fig. 8. The top left corner represents the projected lane lines on the image. The figures illustrate the simulation interface of the algorithm within the global coordinate system. In this system, the lane lines are signified in red, the final positioning result is marked by the green block, the RTK result is denoted by the sky-blue block, and the map-matching result is displayed by the red block. (Color figure online)

ferent scenes in test set are shown in Fig. 8. Sub-figures d, e, f depict a typical turning process, where longitudinal errors become lateral ones and GNSS has a margin of error. Our algorithm ensures a smooth and accurate correction at intersections, preventing wrong lane placements, by utilizing matching result in multiple dimensions. Sub-figures g and h represent weak GNSS signal scenarios like tunnels and underpasses. Despite GNSS drift (light blue block), our stable fused result (green block) is achieved through scene understanding and map-matching verification.

5 Conclusion and Future Works

In this paper, we present a comprehensive, end-to-end, vision-based vehicle localization framework. Our framework is further enhanced by a novel map-matching algorithm that we've developed, named Cross-Dimensional Lane and Pose Estimator (CDLPE). We have demonstrated the reliability of our algorithm in a variety of challenging driving conditions. Moreover, our high level of accuracy and resilience to weak signal interference has been confirmed through experimental comparisons of pose and lane accuracy under varying GNSS signal strengths. Lightweight maps often have limitations in terms of absolute accuracy. Therefore, in future work, it would be necessary for the algorithm to estimate these mapping errors in real-time for precise positioning.

A Appendices

A.1 Road-Level Localization

Problem Definition. Given a trajectory X, the goal of road level matching is to find the correspondence between each trajectory point in X to a road segment in G, also known as *Link*.
Definition 1: A road network G consists of a set of road segment r, in which the road segment $r_i = [p_1, p_2, p_3, ..., p_m]$ point polyline representing a road segment curve.
Definition 2: A trajectory X, is a sequence of the history pose, denoted by $X = [x_1, x_2, x_3, ..., x_n]$:

$$x_i = [x, y, \theta, d, b]^T, i = 1, 2, ..., n \tag{16}$$

Each x_i is specified by its longitudinal and lateral coordinates of the position x, y, angle θ between the position point and the road segment, the distance d from the position point to the road segment. Note that if the position point is on the road, then d is 0. b represents the confidence of the position point, which is estimated by sensor fusion.

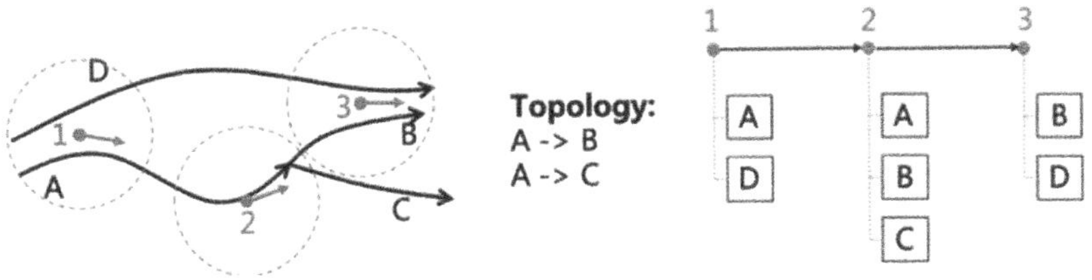

Fig. 9. A, B, C, D represent road segments, and there are different road network observations at point 1, 2, and 3.

Modeling. This work employs a standard forward algorithm in Hidden Markov Modeling (HMM) due to its ability to handle uncertainty in GPS data, consider the sequence of observations and maintain multiple hypotheses. As shown in Fig. 9, the algorithm calculates the posterior probability of each observed link at each point 1, 2, and 3. To ensure the real-time performance of the algorithm, this algorithm only calculate the forward probability of the chains in the sliding window, and maintain a sliding window with a certain length (Fig. 10).

It should be noted that in order to utilize the historical information in a better way, this program defines three states:

1. INIT-SELECT: In initial stage, the algorithm ignore historical information and keep all the topology to maintain multiple hypotheses.
2. STABLE-SELECT: In stable derivation stage, the cumulative probability of the link at the time of discard will be used as the prior probability in subsequent calculations. Meanwhile, the past topology branches with a low probability will be reduced.
3. INTERRUPT: In intermediate unstable stage, the historical road probability will be ignored.

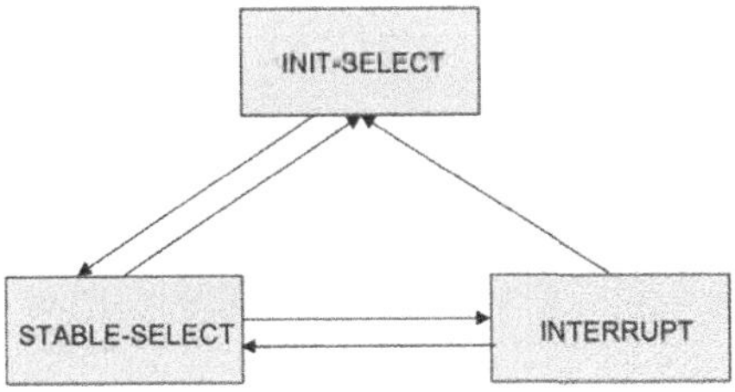

Fig. 10. Three states in road-level matching algorithm

Emission Probability is determined by three observations:

$$P_e = \lambda_1 P_d(X, r_j) + \lambda_2 P_\theta(X, r_j) + \lambda_3 P_{vision}(r_j) \tag{17}$$

In which assumed that there are m road segments, noted as $r_j, j = 1, 2, ..., m$, where the value of $\lambda_{i=1,2,3}$ was pre-trained offline.

$P_d(X, r_j)$ represents the measurement probability between the current trajectory sequence and the road segment in distance $d_i(x_i, r_j)$, where b_i denotes the position confidence.

$$P_d(X, r_j) = \frac{1}{n} \sum_{i=1,2,\ldots,n} e^{-kd_i(x_i, r_j)/b_i} \tag{18}$$

$P_\theta(X, r_j)$ is the probability between the current trajectory sequence and the road segment in angle $\theta_i(r_j)$.

$$P_\theta(X, r_j) = \frac{1}{n} \sum_{i=1,2,\ldots,n} cos(\theta_i(r_j)) \tag{19}$$

$P_{vision}(r_j)$ is the measurement probability of the matching degree of perception result and map knowledge.

$$P_{vision}(r_j) = \frac{1}{n} \sum_{i=1,2,\ldots,n} p_i \tag{20}$$

Transition Probability between road segments are governed by road topology, thus the setting of the transition probability depends on the connectivity of road networks, i.e. if there is a direct topological connection between two road segments, the value of transition probability is 1, otherwise it is 0. In our algorithm, the topological relationship between all the candidate segments in the sliding window will be merged into an adjacency matrix T_f for chain probability calculation, as shown in Fig. 11.

T=

	A	B	C	D
A	1	1	1	0
B	0	1	0	0
C	0	0	1	0
D	0	0	0	1

Fig. 11. T_{ij} represents the transition probability from $link_i$ to $link_j$

In this road-level part, the longitude and latitude of the center point of the recommended lane $pos_{recommend}$on the planned route of the located road is used as the output of the vehicle position, which serves as a reference position for fusion in extremely harsh scenarios where GNSS signals are severely drifting and perception fails. Meanwhile, the subsequent lane-level positioning will be based on the link id output from this step.

References

1. Nayak, A., Cattaneo, D., Valada, A. RaLF: flow-based global and metric radar localization in LiDAR maps. arXiv:abs/2309.09875 (2023). https://api.semanticscholar.org/CorpusID:262045190
2. Zhang, J., Singh, S. LOAM : lidar odometry and mapping in real-time. In: Robotics: Science And Systems Conference (RSS), pp. 109–111 (2014)
3. Mur-Artal, R., Tardos, J.: ORB-SLAM2: an open-source slam system for monocular, stereo, and RGB-D cameras. IEEE Trans. Robot. **33**, 1255–1262 (2017). https://doi.org/10.1109/TRO.2017.2705103
4. Poggenhans, F., et al.: Lanelet2: a high-definition map framework for the future of automated driving. In: 2018 21st International Conference On Intelligent Transportation Systems (ITSC), pp. 1672-1679 (2018)
5. Besl, P., McKay, N.: A method for registration of 3-D shapes. IEEE Trans. Pattern Anal. Mach. Intell. **14**, 239–256 (1992)
6. Schreiber, M., Knöppel, C., Franke, U.: LaneLoc: lane marking based localization using highly accurate maps. In: 2013 IEEE Intelligent Vehicles Symposium (IV), pp. 449–454 (2013)
7. Xiao, Z., Jiang, K., Xie, S., Wen, T., Yu, C., Yang, D.: Monocular vehicle self-localization method based on compact semantic map. In: 2018 21st International Conference On Intelligent Transportation Systems (ITSC), pp. 3083–3090 (2018)
8. Lu, Y., Huang, J., Chen, Y., Heisele, B.: Monocular localization in urban environments using road markings. In: 2017 IEEE Intelligent Vehicles Symposium (IV), pp. 468–474 (2017)
9. Choi, K., Suhr, J., Jung, H.: In-lane localization and ego-lane identification method based on highway lane endpoints. J. Adv. Transp. **2020**, 1–16 (2020)
10. Kasmi, A., Laconte, J., Aufrere, R., Denis, D., Chapuis, R.: End-to-end probabilistic ego-vehicle localization framework. IEEE Trans. Intell. Veh. **6**, 146–158 (2021)
11. Guo, C., Lin, M., Guo, H., Liang, P., Cheng, E.: Coarse-to-fine semantic localization with HD map for autonomous driving in structural scenes (2021)
12. Asghar, R., Garzón, M., Lussereau, J., Laugier, C.: Vehicle localization based on visual lane marking and topological map matching. In: 2020 IEEE International Conference on Robotics and Automation (ICRA), pp. 258–264 (2020)
13. Wu, Z., Li, J., Yu, J., Zhu, Y., Xue, G., Li, M.: L3: sensing driving conditions for vehicle lane-level localization on highways. In: IEEE INFOCOM 2016 - The 35th Annual IEEE International Conference On Computer Communications, pp. 1–9 (2016)
14. Thrun, S., Burgard, W., Fox, D.: Probabilistic Robotics. MIT Press (2005)
15. Heidenreich, T., Spehr, J., Stiller, C.: LaneSLAM - simultaneous pose and lane estimation using maps with lane-level accuracy (2015)
16. Deng, L., Yang, M., Hu, B., Li, T., Li, H., Wang, C.: Semantic segmentation-based lane-level localization using around view monitoring system. IEEE Sens. J. **19**, 10077–10086 (2019)
17. Liu, D., Cui, Y., Guo, X., Ding, W., Yang, B., Chen, Y.: Visual localization for autonomous driving: mapping the accurate location in the city maze. In: 2020 25th International Conference On Pattern Recognition (ICPR), pp. 3170–3177 (2021)
18. Engel, N., Hoermann, S., Horn, M., Belagiannis, V., Dietmayer, K.: DeepLocalization: landmark-based self-localization with deep neural networks. In: 2019 IEEE Intelligent Transportation Systems Conference (ITSC), pp. 926–933 (2019)

19. Yan, L., Cui, Y., Chen, Y., Liu, D. :Hierarchical attention fusion for geo-localization (2021)
20. Rusinkiewicz, S., Levoy, M.: Efficient variants of the ICP algorithm. In: Proceedings Third International Conference On 3-D Digital Imaging And Modeling, pp. 145–152 (2001)
21. Qin, T., Zheng, Y., Chen, T., Chen, Y., Su, Q.: A light-weight semantic map for visual localization towards autonomous driving. In: 2021 IEEE International Conference On Robotics and Automation (ICRA), pp. 11248–11254 (2021)
22. Zuo, X., Geneva, P., Yang, Y., Ye, W., Liu, Y., Huang, G.: Visual-inertial localization with prior LiDAR map constraints. IEEE Robot. Autom. Lett. **4**, 3394–3401 (2019)
23. Xiao, Z., Yang, D., Wen, T., Jiang, K., Yan, R.: Monocular localization with vector HD map (MLVHM): a low-cost method for commercial IVs. Sensors. **20** (2020). https://www.mdpi.com/1424-8220/20/7/1870
24. Suhr, J., Jang, J., Min, D., Jung, H.: Sensor fusion-based low-cost vehicle localization system for complex urban environments. IEEE Trans. Intell. Transp. Syst. **18**, 1–9 (2016)
25. Kümmerle, J., Sons, M., Poggenhans, F., Kühner, T., Lauer, M., Stiller, C.; Accurate and efficient self-localization on roads using basic geometric primitives. In: 2019 International Conference On Robotics And Automation (ICRA), pp. 5965–5971 (2019)
26. Wang, H., Xue, C., Zhou, Y., Wen, F., Zhang, H.: Visual semantic localization based on HD map for autonomous vehicles in urban scenarios. In: 2021 IEEE International Conference On Robotics And Automation (ICRA), pp. 11255–11261 (2021)
27. Wilbers, D., Merfels, C., Stachniss, C.: Localization with sliding window factor graphs on third-party maps for automated driving. In: 2019 International Conference On Robotics And Automation (ICRA), pp. 5951–5957 (2019)
28. Yan, W., et al.: Ego lane estimation using visual information and high definition map. In: 2023 IEEE/ION Position, Location And Navigation Symposium (PLANS), pp. 603–608 (2023)
29. Svärm, L., Enqvist, O., Kahl, F., Oskarsson, M.: City-scale localization for cameras with known vertical direction. IEEE Trans. Pattern Anal. Mach. Intell. **39**, 1455–1461 (2017)
30. Zeisl, B., Sattler, T., Pollefeys, M.: Camera pose voting for large-scale image-based localization. In: 2015 IEEE International Conference On Computer Vision (ICCV), pp. 2704–2712 (2015)
31. Hsu, C., Lin, N.: A Visual SLAM Based-Method for Vehicle Localization. SAE Technical Paper (2024)
32. Zhang, H., Xie, C., Toriya, H., Shishido, H., Kitahara, I.: Vehicle localization in a completed city-scale 3D scene using aerial images and an on-board stereo camera. Remote Sensing. **15** (2023). https://www.mdpi.com/2072-4292/15/15/3871
33. Dong, X., Cappuccio, M.: Applications of Computer Vision in Autonomous Vehicles: Methods, Challenges and Future Directions (2023)
34. Xia, X., Bhatt, N., Khajepour, A., Hashemi, E.: Integrated inertial-LiDAR-based map matching localization for varying environments. IEEE Trans. Intell. Veh. **8**, 4307–4318 (2023)
35. Sarlin, P., et al.: Orienternet: visual localization in 2D public maps with neural matching. In: Proceedings of the IEEE/CVF Conference on Computer Vision and Pattern Recognition, pp. 21632–21642 (2023)

FocDepthFormer: Transformer with Latent LSTM for Depth Estimation from Focal Stack

Xueyang Kang[1,2(✉)], Fengze Han[3], Abdur R. Fayjie[1], Patrick Vandewalle[1], Kourosh Khoshelham[2], and Dong Gong[4]

[1] Department of ESAT, KU Leuven, Leuven, Belgium
[2] Faculty of Engineering IT, The University of Melbourne, Melbourne, Australia
alexander.kang@tum.de
[3] EI, Faculty, Technical University of Munich, Munich, Germany
[4] EI Faculty, The University of New South Wales, Kensington, Australia

Abstract. Most existing methods for depth estimation from a focal stack of images employ convolutional neural networks (CNNs) using 2D or 3D convolutions over a fixed set of images. However, their effectiveness is constrained by the local properties of CNN kernels, which restricts them to process only focal stacks of fixed number of images during both training and inference. This limitation hampers their ability to generalize to stacks of arbitrary lengths. To overcome these limitations, we present a novel Transformer-based network, FocDepthFormer, which integrates a Transformer with an LSTM module and a CNN decoder. The Transformer's self-attention mechanism allows for the learning of more informative spatial features by implicitly performing non-local cross-referencing. The LSTM module is designed to integrate representations across image stacks of varying lengths. Additionally, we employ multi-scale convolutional kernels in an early-stage encoder to capture low-level features at different degrees of focus/defocus. By incorporating the LSTM, FocDepthFormer can be pre-trained on large-scale monocular RGB depth estimation datasets, improving visual pattern learning and reducing reliance on difficult-to-obtain focal stack data. Extensive experiments on diverse focal stack benchmark datasets demonstrate that our model outperforms state-of-the-art approaches across multiple evaluation metrics.

Keywords: Transformer · Attention · Recurrent Network · Focal Stack · LSTM · Early CNN Kernel · Depth Estimation

1 Introduction

With the advancement of deep neural networks (DNNs) and the availability of high-volume data, the challenging task of depth estimation from monocular images [34] has seen significant success on benchmark datasets [35]. However, the

M. Gong et al. (Eds.): AI 2024, LNAI 15442, pp. 273–290, 2025.
https://doi.org/10.1007/978-981-96-0348-0_20

focal stack depth estimation problem, which is distinct from monocular depth estimation [13,21,28], stereo depth or disparity estimation [1], and multi-frame depth estimation [3], yet has not received as much attention in the research community.

DDepth estimation using focus and defocus techniques involves predicting the depth map from a captured *focal stack* of the scene, which consists of images taken at different focal planes [16]. This problem, also known as depth of field control [2], typically uses images obtained with a light field camera [14]. Traditional methods [15] rely on handcrafted sharpness features, but they frequently struggle in textureless scenes. To enhance feature extraction, Convolutional Neural Networks (CNNs) have been used to predict depth maps from focal stacks [5]. Models like DDFFNet [5], AiFDepthNet [8], and DFVNet [6] leverage in-focus cues, while DefocusNet [17] learns permutation invariant defocus cues, or Circle-of-Confusion (CoC). While these methods use 2D or 3D convolutions to represent visual and focal features, they are limited to processing focal stacks with a fixed number of images, which restricts their generalization to stacks of arbitrary length.

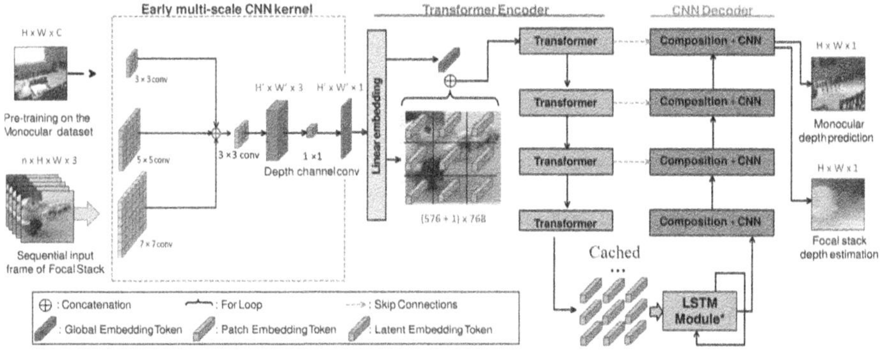

Fig. 1. The overview of our proposed network, FocDepthFormer, is presented with its core components: the Transformer encoder, the recurrent LSTM module, and the CNN decoder. Preceding the Transformer encoder, early-stage multi-scale convolutional kernels are depicted within the dashed line. The resulting multi-scale feature maps are concatenated and subjected to spatial and depth-wise convolution. Subsequently, the fused feature map of a image stack is divided into patches, which are then individually projected by a linear embedding layer into tokens. A red token represents a global embedding token mapped from the entire image and is summed with each individual patch embedding token. (Color figure online)

In this paper, we introduce FocDepthFormer, a novel network for depth estimation from focal stacks that combines LSTM and Transformer architectures. The core component is a module consisting of a Transformer encoder [27], an LSTM-based recurrent module [26] applied to latent tokens, and a CNN decoder.

The Transformer and LSTM models process spatial and stack information separately. Unlike CNNs, which are restricted to local representation, the Transformer encoder captures visual features with a larger receptive field. Given that focal stacks may have arbitrary and unknown numbers of images, we use the LSTM in the latent feature space to fuse focusing information across the stack for depth prediction. This approach differs from existing focal stack depth estimation methods [6,8,17] and monocular depth estimation methods based on CNNs or Transformers [36], which typically handle inputs with a fixed number of images. Specifically, we compactly fuse activated token features via the recurrent LSTM module after the Transformer encoder, enabling the model to handle focal stacks of arbitrary lengths during both training and testing, thereby providing greater flexibility. Before inputting data into the Transformer, we employ an early-stage convolutional encoder with multi-scale kernels [22] to capture low-level focus/defocus features across different scales. Considering the limited availability of focal stack data, our model enhances its representation of scene features through pre-training on monocular depth estimation datasets. Meanwhile, the recurrent LSTM module facilitates the model's ability to accommodate varying numbers of input images in focal stack.

The main contributions of this work are as follows:

- We introduce a novel Transformer-based network model for depth estimation from focal stack images. The model uses a Vision Transformer encoder with self-attention to capture non-local spatial visual features, effectively representing sharpness and blur patterns. To accommodate an arbitrary number of input images, we incorporate a recurrent LSTM module. This structural flexibility allows for pre-training with monocular depth estimation datasets, thereby reducing the reliance on focal stack data.
- To fuse the stack features, we utilize the LSTM and implement a grouping operation to manage recurrent complexity across tokens, avoiding an increase in complexity as the token count grows with larger stack sizes. This is accomplished by applying the LSTM exclusively to a subset of activated embedding tokens while maintaining the information on other non-activated tokens through averaging aggregation.
- We propose the use of multi-scale kernels in an early-stage convolutional encoder to enhance the capture of low-level focus/defocus cues at various scales.

2 Related Work

Depth from Focus/Defocus. Depth estimation from focal stacks relies on discerning relative sharpness within the stack of images for predicting depth. Traditional machine learning methods [16] treat this problem as an image filtering and stitching process. Johannsen *et al.* [31] provide a comprehensive overview of methods addressing the challenges posed by light field cameras, laying a foundation for research in this direction. More recently, CNN-based approaches have emerged in the context of focal stacks. DDFFNet [5] introduces the first end-to-end learning model trained on the DDFF 12-Scene dataset. DFVNet [6] utilizes the first-order derivative of volume features within the stack. AiFNet [8] aims to bridge the gap between supervised and unsupervised methods, accommodating both ground truth depth and its absence. Barratt *et al.* [12] formulate the problem as an inverse optimization task, utilizing gradient descent search to simultaneously recover an all-in-focus image and depth map. DefocusNet [17] exploits the Circle-of-Confusion, a defocus cue determined by focal plane depth, for generating intermediate defocus maps in the final depth estimation. Anwar *et al.* [32] leverage defocus cues to recover all-in-focus images by eliminating blur in a single image. Recently, the DEReD model [37] learns to estimate both depth and all-in-focus (AIF) images from focal stack images in a self-supervised way by incorporating the optical model to reconstruct defocus effects. Gur and Wolf [29] present depth estimation from a single image by leveraging defocus cues to infer disparity from varying viewpoints.

Attention-Based Models. The success of attention-based models [7] in sequential tasks has led to the rise of the Vision Transformer for computer vision tasks. The Vision Transformer represents input images as a series of patches (16×16). While this model performs well in image recognition compared to CNN-based models, a recent study [22] demonstrates that injecting a small convolutional inductive bias in early kernels significantly enhances the performance and stability of the Transformer encoder. In the context of depth estimation, Ranftl *et al.* [18] utilize a Transformer-based model as the backbone to generate tokens from images, assembling these tokens into an image-like representation at multiple scales. DepthFormer [36] merges tokens at different layer levels to improve depth estimation performance. The latest advancement in this domain, the Swin Transformer [30], achieves a larger receptive field by shifting the attention window, revealing the promising potential of the Transformer model.

Recurrent Networks. Recurrent networks, specifically LSTM [26], have found success in modeling temporal distributions for video tasks such as tracking [25] and segmentation [24]. The use of LSTM introduces minimal computational overhead, as demonstrated in SliceNet [10], where multi-scale features are fused for depth estimation from panoramic images. Some recent works [23] combine LSTM with Transformer for language understanding via long-range temporal attention.

3 Method

Given a focal stack $\mathbf{S}$, containing N images ordered from near to far by focus distance, denoted as $\mathbf{S} = (\mathbf{x}_i)_{i=1}^{N}$, where $\mathbf{x}_i \in \mathbf{R}^{H \times W \times 3}$ represents each single image, our objective is to generate a single depth map $\mathbf{D} \in \mathbf{R}^{H \times W \times 1}$ for a stack of images. In contrast to the vanilla Transformer [27], we initially encode each image $\mathbf{x}$ using an *early-stage multi-scale kernel-based convolution* $\mathcal{F}(\cdot)$. This convolution ensures the multi-scale feature representation $\mathbf{x}'$ for the focal stack images. Subsequently, the *transformer encoder* $g(\cdot)$ processes the feature maps by transforming them into a series of ordered tokens that share information through self-attention. The self-attention weights between the in-focus features and blur features, encode spatial feature information from each input image. The *recurrent LSTM module* sequentially processes cached latent tokens from different frames of a focal stack and fuses them along the stack dimension. The LSTM module learns the stack feature fusion process within the latent space. Our attention design combined with LSTM enhances the model's capability to handle an arbitrary number of input images. The final disparity map is decoded (denoted as $d(\cdot)$) from the fusion features, utilizing the aggregated tokens from all images in a focal stack.

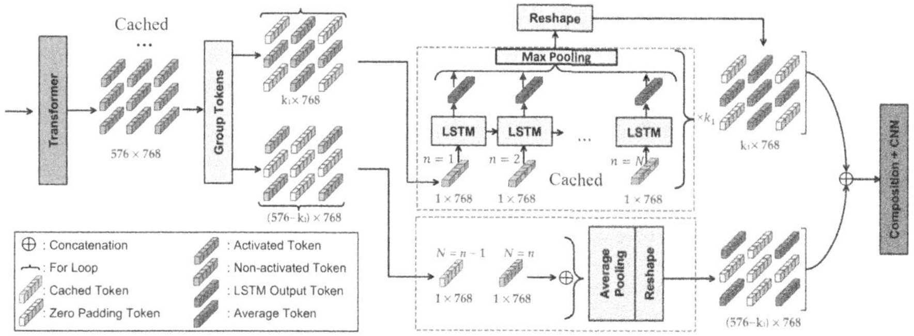

Fig. 2. To illustrate the LSTM module in our network, the initial step involves grouping the all cached output tokens from the Transformer encoder into activated and non-activated tokens. These two groups are then individually processed, with activated tokens undergoing LSTMs followed by max pooling and non-activated tokens undergoing average pooling. Following this, the output tokens undergo reshaping and concatenation before being fed into the CNN decoder for predicting the depth map.

3.1 Early-Stage Encoding with Multi-scale Kernels

To capture low-level focus and defocus features at different scales, we employ an early-stage convolutional encoder with multi-scale kernels, which is different from methods using fixed size kernel convolution stem before the Transformer [22]. As illustrated in Fig. 1, the early-stage encoder utilizes three convolutional kernels to generate multi-scale feature maps $f_m(\mathbf{x}), \{m = 1, 2, 3\}$. All feature maps are concatenated and merged into the feature map $\mathbf{x}' \in \mathbf{R}^{H' \times W' \times 1}$ through spatial convolution, followed by 3×3 and 1×1 convolution on the feature map depth channel:

$$\mathbf{x}' = \mathcal{F}(\mathbf{x}) = \text{Conv}(\text{Concat}(f_m(\mathbf{x}))), \quad (1)$$

where m ranges from 1 to 3. Feature concatenation after convolutions with multiple kernel sizes preserves fine-grained details of features across varying depth scales. The first module from the left in Fig. 1, comprising parallel multi-scale kernel convolutions followed by depth-wise convolution, ensures the model has a large receptive field even beyond the 7×7 kernel size. This facilitates capturing more defocus features while preserving intricate details.

3.2 Transformer with LSTM

Transformer encoder. The Transformer depicted in Fig. 1, denoted as $g(\cdot)$, operates on the feature maps $\mathbf{x}'$ derived from preceding early-stage multi-scale convolutions to produce a sequence of tokens $(\mathbf{t'}_p)_{p=1}^{k}$:

$$\mathbf{t'}_1, \mathbf{t'}_2, ..., \mathbf{t'}_k = g(\mathbf{x'}). \tag{2}$$

Specifically, the early-stage kernel CNNs and Transformer encoder sequentially process the focal stack images, caching and concatenating the feature maps of a specified stack of images into $\mathbf{x}'$. Initially, a linear embedding layer divides the feature maps $\mathbf{x}'$ into k patches of size 16×16. Thus, $\mathbf{x'}_p \in \mathbf{x'}, p = 1, 2, ..., k$, is projected by a linear embedding layer (MLP) into corresponding embedding tokens $(\mathbf{l}_p)_{p=1}^{k}$, each token having a dimension of 768 (576 in total). All tokens of a complete stack N are cached into $(\mathbf{l}_p)_{p=1}^{k} \times N$ before LSTM for simultaneous fusion. The Transformer's *Position Embedding* encodes the positional information of image patches in a sequential order from the top-left of the image. An MLP layer generates the Global Embedding Token (Fig. 1) by mapping the entire image into a global token and subsequently adding each individual patch embedding token. Each linear embedding token is projected into three vectors - Query, $\mathbf{l}_Q$; Key, $\mathbf{l}_K$; and Value, $\mathbf{l}_V$ via a weight matrix $\mathbf{W}^{(\cdot)}$ with dimensions d_Q, d_K, and d_V respectively. Queries, Keys, and Values are processed in parallel through Multi-Head Attention (MHA) units.

$$\mathrm{MHA}(\mathbf{l}_Q, \mathbf{l}_K, \mathbf{l}_V) = (\mathrm{head}_1 \oplus ... \oplus \mathrm{head}_N)\mathbf{W}^O, \tag{3}$$

$$\mathrm{head}_i = \mathrm{softmax}\left(\frac{\mathbf{l}_Q \mathbf{W}^{\mathbf{l}_Q} \mathbf{l}_K \mathbf{W}^{\mathbf{l}_K}}{\sqrt{d_k}}\right) \mathbf{l}_V \mathbf{W}^{\mathbf{l}_V}, \tag{4}$$

where $\mathbf{W}^{\mathbf{l}_Q} \in \mathbf{R}^{d_m \times d_V}$, $\mathbf{W}^{\mathbf{l}_K} \in \mathbf{R}^{d_m \times d_K}$, $\mathbf{W}^{\mathbf{l}_V} \in \mathbf{R}^{d_m \times d_V}$, and $\mathbf{W}^O \in \mathbf{R}^{d_m \times d_V}$. Following the Multi-Head-Attention modules within the encoder $g(\mathbf{x}')$, the resulting tokens $(\mathbf{t'}_p)_{p=1}^{k}$ capture features that distinguish focus and defocus cues among different stack image patches at the same spatial location within the image. This capability is illustrated in Fig. 3. Consequently, the embedding space emphasizes sharper, more in-focus features of the image patches.

LSTM module. To ensure our model's flexibility in handling stacks of arbitrary lengths, in contrast to the fixed lengths used in existing methods [6], we employ an LSTM to progressively integrate sharp features across the stack. The LSTM treats patch embedding tokens at the same image position as sequential features along the stack dimension. Corresponding feature tokens $(\mathbf{t'}_p)_{p=1}^{k}$ from stack images at stack number N are sequentially ordered spatially and fed into LSTM

(a) Focus at front. (b) Attention of the (a). (c) Focus at back. (d) Attention of (c).

Fig. 3. Comparison of Transformer attention between the two left column images. Cropped image patches within green and orange boxes in (a) and (c) serve as query inputs to compute the self-attention map over the entire input image, respectively. In (b) and (d), the attention maps on the left and right sides of the green line illustrate the attention outputs of the green and orange boxes, respectively. This demonstrates the model's capability to selectively attend to both foreground and background areas, distinguishing between focus and defocus cues. (Color figure online)

modules arranged in the original spatial image order. At each position, each LSTM module incrementally integrates the latent token $\mathbf{t'}_p$ from a stack. This approach differs from existing models constrained to a 3D volume stack with a predefined and fixed size [6]. Importantly, sequential processing in the latent space after a shared encoder for stack images ensures manageable complexity in practice.

Preceding the LSTM modules, tokens associated with image patches from a single frame are classified into activated and non-activated tokens, indicating the level of informativeness of the features. The L_2 norm, denoted as $\|\cdot\|$, of each embedding token is compared against a threshold of 0.4, as depicted in Fig. 2. Specifically, within a single frame, only tokens surpassing this threshold are considered activated and forwarded to the LSTM, totaling k_1 tokens. This approach significantly reduces the computational load of the LSTM by processing only a subset of the latent tokens:

$$(\mathbf{t}_p^n, \mathbf{h}_p^n) = LSTM(\mathbf{t'}_p^n, \mathbf{h}_p^{n-1}, c^n), \tag{5}$$

this is a single LSTM layer expression above, where $p = 1, 2, ..., k_1$, and n is the frame index number of a stack. Here we set the number of hidden layers of the LSTM equal to the stack size N. The memory cell c undergoes continuous updates at each step, influenced by the input $\mathbf{t'}_p^n$ and the hidden state $\mathbf{h}$. Subsequently, all LSTM layer outputs are combined via max pooling $\max \mathbf{t'}_p^1, ..., \mathbf{t'}_p^n$ to obtain $\mathbf{t'}_p$. For non-activated tokens $(\mathbf{t'}_p), p = k+1, ..., k$, an averaging operation is performed with corresponding cached tokens from the previous step at the same embedding position. Finally, the two groups of output tokens are arranged in the original input embedding order, yielding the final fused tokens $(\mathbf{t}_p)_{p=1}^k$.

CNN Decoder. Our decoder $d(\cdot)$ adopts the methodology introduced by Ranftl *et al.* [18], utilizing Transpose-convolutions to integrate feature maps following LSTMs. Additionally, the decoder incorporates feature maps from the i-th layer

$g_i(\mathbf{x}')$ of the encoder through skip connections, as illustrated in Fig. 1. Ultimately, decoder $d(\cdot)$ outputs the depth prediction.

$$\hat{D} = d((\mathbf{t}_p)_{p=1}^{k}, g_i(\mathbf{x}')), \quad i = \{1, 2, 3\}. \tag{6}$$

3.3 Training Loss

Our training loss comprises a sum of the Mean Squared Error (MSE) loss, denoted as $\mathcal{L}_{MSE}$, and a sharpness regularizer $\mathcal{L}_{log}$ weighted by α:

$$\mathcal{L}_{total} = \mathcal{L}_{MSE}(\hat{D}, D) + \alpha\mathcal{L}_{\log}(\delta_{\boldsymbol{\Delta}\hat{D}}, \delta_{\boldsymbol{\Delta}D}), \tag{7}$$

where D represents the ground truth depth, and $\hat{D}$ indicates the predicted depth. $\boldsymbol{\Delta}$ denotes the Laplacian operator applied to the predicted and ground truth depth images, respectively. δ represents the variance of the depth image. The regularization term is formulated as $\log(\delta_{\boldsymbol{\Delta}\hat{D}}/\delta_{\boldsymbol{\Delta}D})$. Pixel blurriness due to out-of-focus effects can be described by the Circle-of-Confusion (CoC),

$$CoC = \frac{1}{2r}\frac{f^2}{N(z-d_f)}\left|1 - \frac{d_f}{z}\right|, \tag{8}$$

where N denotes the f-number, defined as the ratio of the focal length to the effective aperture diameter, and r is the CMOS pixel size. d_f denotes the focus distance of the lens, and z represents the distance from the lens to the target object. Generally, the range of z is $[0, \infty]$, though in practice, it is always bounded by lower and upper limits. The model aims to learn a depth map from focus and defocus features of stack images.

3.4 Pre-training with Monocular Depth Prior

Focal stack datasets are often limited in size due to the high cost and challenges involved in data collection. To address data scarcity and fully exploit the Transformer's potential, we optionally pre-train the Transformer encoder on widely available monocular depth dataset like NYUv2 [19], to enhance spatial representation learning further, yet without pre-training, the model performance is still superior over baseline models, as exhibited in following experiment section.

4 Experiments

Datasets. We extensively evaluated our model using four benchmark focal stack datasets: DDFF 12-Scene [5], Mobile Depth [11], LightField4D [9], and FOD500 [6] (Synthetic dataset). As Mobile Depth has no depth ground truth, so only visual comparison results are provided. Additionally, our model supports pre-training on the monocular RGB-D dataset NYUv2 [19]. Specifically, we conducted separate training on DDFF 12-Scene and FOD500 for subsequent experiments, while Mobile Depth and LightField4D were used to assess the model's

Table 1. Summary of evaluation datasets.

Dataset	Image source	GT type	Cause of defocus
DDFF 12-Scene [5]	Real	Depth	Light-field settings
Mobile Depth [11]	Real	—	Real
LightField4D [9]	Real	Disparity	Light-field settings
FOD500 [6]	Synthetic	Depth	Synthesis blendering

generalizability directly after pre-training on DDFF 12-Scene. Qualitative and quantitative evaluation on FOD500, and quantitative metric evaluation of Light-Field4D are provided in the supplementary part. Additionally, the more qualitative results of each dataset are available in the supplementary part, please refer to it. Finally, a comprehensive summary of the evaluation datasets, including their individual properties (real or synthetic), with or without the depth ground truth is presented in Table 1, along with defocus cause.
Implementation Details. For the evaluation on DDFF 12-Scene, we conducted training experiments using our model with and without pre-training on NYUv2 [19], presenting results for both scenarios. We employed a patch size of 16×16 and an image size of 384×384 for the Transformer. Our network utilizes the Adam optimizer with a learning rate of 1×10^{-4} and a momentum of 0.9. The regularization scalar α in Eq. (7) is set to 0.2. In terms of hardware configuration, all training and tests below were conducted on a single Nvidia RTX 2070 GPU with 8 GB of vRAM.

Metric Evaluation. In this work, we perform quantitative evaluation using the following metrics: Root Mean Squared Error (RSME), logarithmic Root Mean Squared Error (logRSME), relative absolute error (absRel), relative squared error (sqrRel), Bumpiness (Bump), and accuracy threshold at three levels (1.25, 1.25^2, and 1.25^3).

Runtime. We evaluated the runtime of the proposed method and baseline approaches by executing them on focal stacks from DDFF 12-Scene. Our FocDepthFormer processes a stack of 10 images sequentially in 15 ms, averaging 2 ms per image. In comparison, DDFFNet [5] requires 200 ms per stack under the same conditions, and DFVNet [6] performs in the range of 20–30 ms.

4.1 Comparisons to the State-of-the-Art Methods

DDFFNet [5] and DefocusNet [17] lacked pre-trained weights; therefore, we utilized their open-source codebases to train the networks from scratch. Notably, DefocusNet [17] offers two architectures, and we chose the "PoolAE" architecture due to its consistently good performance for comparison. Conversely, for AiFNet [8] and DFVNet [6], we employed the pre-trained weights provided by the authors to conduct the evaluations. We also provide the visual results of the

Table 2. Evaluation results on DDFF 12-Scene. The best results are denoted in Red while Blue indicates the second-best. $\delta = 1.25$.

Model	RMSE↓	logRMSE↓	absRel↓	sqrRel↓	Bump↓	$\delta\uparrow$	$\delta^2\uparrow$	$\delta^3\uparrow$
DDFFNet [5]	2.91e−2	0.320	0.293	1.2e−2	0.59	61.95	85.14	92.98
DefocusNet [17]	2.55e−2	0.230	0.180	6.0e−3	0.46	72.56	94.15	97.92
DFVNet [6]	2.13e−2	0.210	0.171	6.2e−3	0.32	76.74	94.23	98.14
AiFNet [8]	2.32e−2	0.290	0.251	8.3e−3	0.63	68.33	87.40	93.96
Ours (w/o Pre-training)	2.01e−2	0.206	0.173	5.7e−3	0.26	78.01	95.04	98.32
Ours (w/ Pre-training)	1.96e−2	0.197	0.161	5.4e−3	0.23	79.06	96.08	98.57

Table 3. Metric evaluation results on "additional" set of LightField4D dataset. The best results are denoted in Red, while Blue indicates the second-best.

Model	RMSE↓	logRMSE↓	absRel↓	Bump↓	$\delta(1.25)\uparrow$
DDFFNet	0.431	0.790	0.761	2.93	44.39
DefocusNet	0.273	0.471	0.435	2.84	48.73
DFVNet	0.352	0.647	0.594	2.97	43.54
AiFNet	0.231	0.407	0.374	2.53	55.04
Ours	0.237	0.416	0.364	1.54	58.90

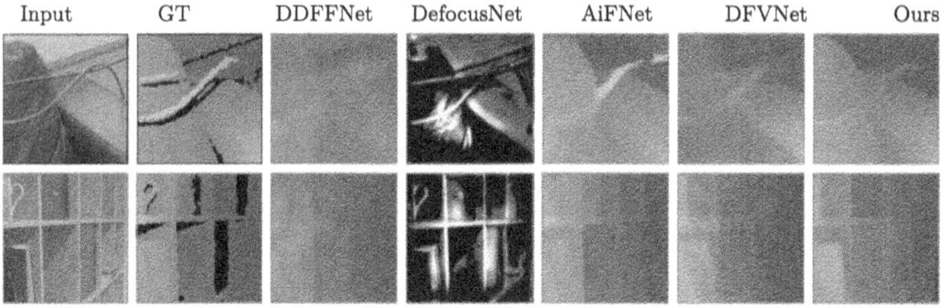

Fig. 4. Qualitative evaluation of our model on DDFF 12-Scene dataset

latest DEReD [37] model in supplementary part, as the publicly available code is not complete for implementation.

Results on DDFF 12-Scene. Table 2 presents the quantitative evaluation results of our model on the DDFF 12-Scene dataset. As the ground truth for the "test set" is not publicly available, we adhere to the standard evaluation protocol used in other comparative works, assessing the models on the "validation set" as per the split provided by DDFFNet [5]. Moreover, we demonstrate that our model, trained on DDFF-12, performs robustly on other completely unseen datasets, highlighting its generalization ability and mitigating concerns of overfitting. Furthermore, the qualitative results are provided in Fig. 4.

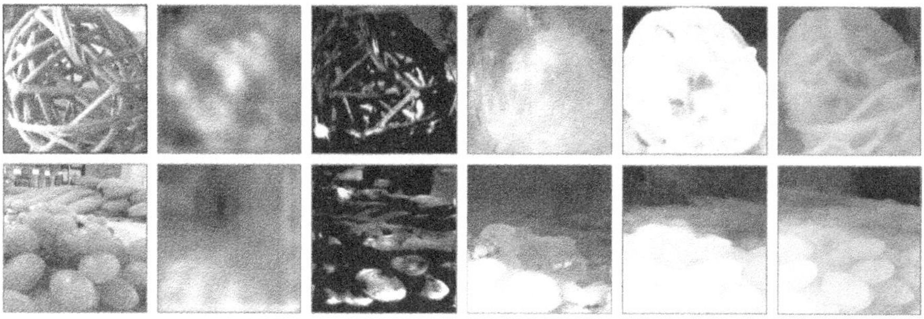

Fig. 5. Qualitative evaluation of our model on Mobile Depth dataset

Table 4. Metric evaluation results on FOD500 test dataset. Here the first 400 FOD500 focal stacks are used for training, following the standard setting from DFVNet [6]. The best results are denoted in Red, while Blue indicates the second-best. $\delta = 1.25$.

Model	RMSE↓	logRMSE↓	absRel↓	sqrRel↓	Bump↓	δ ↑	δ^2 ↑	δ^3 ↑
DDFFNet [5]	0.167	0.271	0.172	3.56e−2	1.74	72.82	89.96	96.26
DefocusNet [17]	0.134	0.243	0.150	3.59e−2	1.57	81.14	93.31	96.62
DFVNet [6]	0.129	0.210	0.131	2.39e−2	1.44	81.90	94.68	98.05
AiFNet [8]	0.265	0.451	0.400	4.32e−1	2.13	85.12	91.11	93.12
Ours (w/o Pre-training)	0.121	0.203	0.129	2.36e−2	1.38	85.47	94.75	98.13

Results on Mobile Depth, LightField4D, and FOD500. We provide the qualitative results on Mobile Depth in Fig. 5, and the depth is not available for this dataset for metric evaluation. Regarding LightField4D, quantitative and qualitative results are provided in Table 3 and Fig. 6 respectively. Table 4 presents the quantitative evaluation of our model on the synthetic FOD500 dtaset and LightField4D. For FOD500, we use the last 100 image stacks from the dataset for test, while the initial 400 image stacks are reserved for training. DDFFNet and DefocusNet are re-trained on FOD500 from scratch. The results highlight the consistent superiority of our model across all metrics when compared to the baseline methods. Notably, in our experiments, we observed that pre-training on NYUv2 did not provide significant benefits, likely due to the gap between synthetic and real data.

We present the quantitative results for cross-dataset evaluation of our model on the LightField4D dataset in Table 3. Our model achieves a comparable performance in terms of accuracy (58.90%) on this completely unseen dataset.

4.2 Cross Dataset Evaluation

To evaluate the generalizability of our model, it is initially trained on DDFF 12-Scene and subsequently evaluated on the Mobile Depth and LightField4D

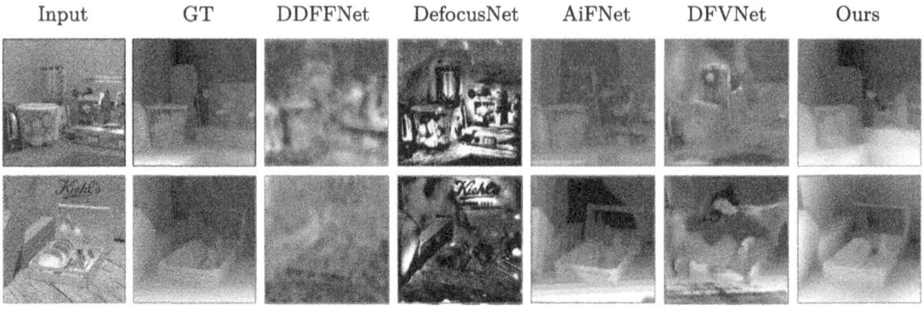

Fig. 6. Qualitative evaluation of our model on LightField4D dataset

datasets. The Mobile Depth dataset poses a challenge as it comprises 11 aligned focal stacks captured by a mobile phone camera, each with varying numbers of focal planes and lacking ground truth. Results on the Mobile Depth dataset are illustrated in Fig. 5, showcasing the model's ability to preserve sharp information for depth prediction in complex scenes.

4.3 Ablation Study

We conduct a comprehensive ablation study to evaluate the key components of our proposed model architecture. Additional ablation experiments are detailed in the supplementary materials.

Multi-scale Early-Stage Kernels: Table 5 presents a comparison of different early CNN kernel design configurations in conjunction with our Transformer encoder (ViT). The effectiveness of the proposed *early-stage multi-scale kernels* encoder is evident, demonstrating robust performance. Conversely, omitting multi-scale kernels or forgoing subsequent convolutions after in-parallel convolutions leads to a degradation in model performance.

Table 5. Results of different designs of the early-stage convolution.

	RMSE↓	absRel↓	Bump↓
early kernel size at 3×3	2.18e−2	0.216	0.31
early kernel size at5×5	2.20e−2	0.214	0.32
early kernel size at7×7	2.13e−2	0.192	0.35
multi-scale combination kernels	**2.01e−2**	**0.173**	**0.26**

LSTM for Handling Arbitrary Stack Length: Our proposed LSTM-based method exhibits flexibility in processing focal stacks of arbitrary lengths, a feature distinguishing it from designs that limit inputs to fixed lengths. To illustrate the advantages of our LSTM-based model, we conducted experiments

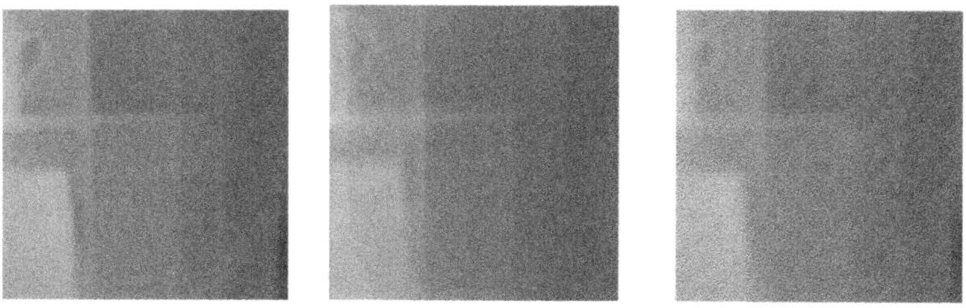

(a) Ours (b) LSTM+CNN (c) Transformer

Fig. 7. Different model structure design comparisons

Table 6. RMSE for evaluation of LSTM compared to DFVNet.

Model	2 Frames	4 Frames	6 Frames	8 Frames	10 Frames
Ours-10F	3.2e−2	**2.61e−2**	**2.18e−2**	**2.16e−2**	**2.04e−2**
DFVNet-10F	—	—	—	—	2.43e−2
DFVNet-#F	**2.97e−2**	2.70e−2	2.52e−2	2.47e−2	2.43e−2

using DDFF 12-Scene. The results, including the RMSE comparison between our model and DFVNet [6], are presented in Table 6. Initially, we trained our model and DFVNet using 10-frame (10F) stacks (*i.e.*, Ours-10F and DFVNet-10F). During testing, DFVNet-10F is constrained and cannot process stacks with fewer than 10 frames. Due to fixed stack size requirements during training and testing, DFVNet must be retrained for different stack sizes (DFVNet-#F). In contrast, our model is trained once on 10-frame stacks and can be tested with varying numbers of stack images. The focal stack images are ordered based on focus distances. Despite our model's initial performance being inferior to DFVNet, the learning curve of LSTM indicates rapid convergence (Fig. 7).

The LSTM module enables our network to incrementally fuse each image from the focal stack, enhancing the model ability to accommodate varying focal stack lengths. Figure 8 depicts the fusion process of the ordered input images of one stack. As the images from the stack are given sequentially, starting from the in-focus plane close to the camera, the model can fuse the sharp in-focus features from a sweep of various frames to attain a final all-in-focus prediction depth map at the bottom right, form near to far focus distances.

Our network comprises the early-stage multi-scale CNN, the Transformer, the LSTM module, and the decoder. We present a summary of each module parameter size, and the inference time, From Table 7, we can see the main time consumption is allocated on the Transformer module and decoder, which shows the potential to reduce the model size further can be achieved, *e.g.*, by using MobileViT as an encoder. Our proposed LSTM over the latent feature representation, has a size of only around one-third of Transformer encoder, furthermore, the shallow early multi-scale kernel CNNs0 is in quite a small size with only 0.487 million parameters and fast processing time around 1ms. The time model

(a) 1st image. (b) 5th image. (c) 10th image.

(d) Depth prediction of 1st input frame. (e) Fusion prediction of 2 input frames. (f) Fusion prediction of 3 input frames.

Fig. 8. The top row is the input, and the bottom is the output disparity map. The final disparity map is at the bottom right. The red rectangle highlights the incremental fusion results of depth in the background (Color figure online)

size and time complexity summary table further justify our model's compact and efficient design, where the recurrent LSTM module has benefits both in memory size and computational complexity in the design. The main inference time for a single image processing is from the Transformer Encoder, which can be attributed mainly to the attention computation of multiple self-attention heads, and CNN decoder.

Loss Function: Table 8 presents the results obtained using three distinct loss functions: Mean Squared Error (MSE), Mean Absolute Error (MAE), and Regularized MSE (MSE with gradient regularization). The findings consistently show that MSE outperforms MAE in terms of overall performance. Notably, incorporating the gradient regularizer contributes to achieving the highest accuracy, as evidenced by the bumpiness metric (0.26).

Pre-training: Table 9 illustrates the impact of pre-training on the performance of our proposed method. The results demonstrate that pre-training enhances the

Table 7. Summary of module size, and inference time for processing one image, for the whole stack, the total time is calculated from a whole stack of images' processing.

Module	Params size	Inference Time
Early multi-scale kernel CNNs	0.487M	0.001 s
Transformer	42.065M	0.006 s
LSTM	15.247M	0.003 s
CNN Decoder	16.856M	0.005 s
Total	74.655M	0.015

Table 8. Evaluation for our model with different losses.

	RMSE↓	absRel↓	Bump↓
MSE loss	2.94e−2	0.280	0.50
MAE loss	3.76e−2	0.372	0.62
MSE + Gradient loss	**2.01e−2**	**0.173**	**0.26**

Table 9. Pre-training contribution comparisons.

	Pre-training	RMSE↓	logRMSE↓	absRel↓	Bump↓
Ours	✗	2.01e−2	0.206	0.173	0.26
	✓	**1.96e−2**	**0.197**	**0.161**	**0.19**
DFVNet	✗	2.13e−2	0.210	0.171	0.32
	✓	2.57e−2	0.233	0.184	0.49

model's capabilities, leveraging the compact design with Transformer and LSTM modules. Even without pre-training, our model achieves competitive results. Notably, attempts to apply pre-training to DFVNet using a stack created from repeated monocular images of NYUv2 did not yield improvements and, in some cases, led to performance degradation due to the data modality gap.

5 Conclusion

We introduce the FocDepthFormer model tailored for depth estimation from focal stack. At the core of our network is a Transformer encoder coupled with a recurrent LSTM module in the latent space, allowing the model to effectively capture spatial and stack information independently. Our approach demonstrates flexibility in accommodating varying focal stack lengths. A significant drawback lies in the heightened model complexity associated with the vanilla Transformer architecture. More efficient attention design techniques like Mamba can be explored as the future work.

References

1. Godard, C., Mac Aodha, O., Brostow, G.J.: Unsupervised monocular depth estimation with left-right consistency. In: Proceedings of the IEEE Conference on Computer Vision and Pattern Recognition, pp. 270–279 (2017)
2. Pentland, A.P.: A new sense for depth of field. IEEE Trans. Pattern Anal. Mach. Intell. 523–531 (1987)
3. Schonberger, J.L., Frahm, J.M.: Structure-from-motion revisited. In: Proceedings of the IEEE Conference on Computer Vision and Pattern Recognition, pp. 4104–4113 (2016)
4. Ramamonjisoa, M., Firman, M., Watson, J., Lepetit, V., Turmukhambetov, D.: Single image depth prediction with wavelet decomposition. In: Proceedings of the IEEE/CVF Conference on Computer Vision and Pattern Recognition, pp. 11089–11098 (2021)
5. Hazirbas, C., Soyer, S.G., Staab, M.C., Leal-Taixé, L., Cremers, D.: Deep depth from focus. In: Jawahar, C.V., Li, H., Mori, G., Schindler, K. (eds.) ACCV 2018. LNCS, vol. 11363, pp. 525–541. Springer, Cham (2019). https://doi.org/10.1007/978-3-030-20893-6_33
6. Yang, F., Huang, X., Zhou, Z.: Deep depth from focus with differential focus volume. In: Proceedings of the IEEE/CVF Conference on Computer Vision and Pattern Recognition, pp. 12642–12651 (2022)
7. Vaswani, A., et al.: Attention is all you need. In: Advances in Neural Information Processing Systems, vol. 30 (2017)
8. Wang, N.H., et al.: Bridging unsupervised and supervised depth from focus via all-in-focus supervision. In: Proceedings of the IEEE/CVF International Conference on Computer Vision, pp. 12621–12631 (2021)
9. Honauer, K., Johannsen, O., Kondermann, D., Goldluecke, B.: A dataset and evaluation methodology for depth estimation on 4D light fields. In: Lai, S.-H., Lepetit, V., Nishino, K., Sato, Y. (eds.) ACCV 2016. LNCS, vol. 10113, pp. 19–34. Springer, Cham (2017). https://doi.org/10.1007/978-3-319-54187-7_2
10. Pintore, G., Agus, M., Almansa, E., Schneider, J., Gobbetti, E.: SliceNet: deep dense depth estimation from a single indoor panorama using a slice-based representation. In: Proceedings of the IEEE/CVF Conference on Computer Vision and Pattern Recognition, pp. 11536–11545 (2021)
11. Benavides, F.T., Ignatov, A., Timofte, R.: PhoneDepth: a dataset for monocular depth estimation on mobile devices. In: Proceedings of the IEEE/CVF Conference on Computer Vision and Pattern Recognition, pp. 3049–3056 (2022)
12. Barratt, S., Hannel, B.: Extracting the depth and all-in-focus image from a focal stack. In: Proceedings of the IEEE International Conference on Computer Vision, pp. 3451–3459 (2015)
13. Hornauer, J., Belagiannis, V.: Gradient-based uncertainty for monocular depth estimation. In: Avidan, S., Brostow, G., Cissé, M., Farinella, G.M., Hassner, T. (eds.) ECCV 2022. LNCS, vol. 13680, pp. 613–630. Springer, Cham (2022). https://doi.org/10.1007/978-3-031-20044-1_35
14. Liu, C., Qiu, J., Jiang, M.: Light field reconstruction from focal stack based on Landweber iterative scheme. In: Mathematics in Imaging, pp. MM2C–3. Optica Publishing Group (2017)
15. Suwajanakorn, S., Hernandez, C., Seitz, S.M.: Depth from focus with your mobile phone. In: Proceedings of the IEEE Conference on Computer Vision and Pattern Recognition, pp. 3497–3506 (2015)

16. Xiong, Y., Shafer, S.A.: Depth from focusing and defocusing. In: Proceedings of IEEE Conference on Computer Vision and Pattern Recognition, pp. 68–73. IEEE (1993)
17. Maximov, M., Galim, K., Leal-Taixé, L.: Focus on defocus: bridging the synthetic to real domain gap for depth estimation. In: Proceedings of the IEEE/CVF Conference on Computer Vision and Pattern Recognition, pp. 1071–1080 (2020)
18. Ranftl, R., Bochkovskiy, A., Koltun, V.: Vision transformers for dense prediction. In: Proceedings of the IEEE/CVF International Conference on Computer Vision, pp. 12179–12188 (2021)
19. Silberman, N., Hoiem, D., Kohli, P., Fergus, R.: Indoor segmentation and support inference from RGBD images. In: Fitzgibbon, A., Lazebnik, S., Perona, P., Sato, Y., Schmid, C. (eds.) ECCV 2012. LNCS, vol. 7576, pp. 746–760. Springer, Heidelberg (2012). https://doi.org/10.1007/978-3-642-33715-4_54
20. Cho, J., Min, D., Kim, Y., Sohn, K.: DIML/CVL RGB-D dataset: 2M RGB-D images of natural indoor and outdoor scenes. arXiv preprint arXiv:2110.11590 (2021)
21. Godard, C., Mac Aodha, O., Firman, M., Brostow, G.J.: Digging into self-supervised monocular depth prediction. In: The International Conference on Computer Vision (ICCV) (2019)
22. Xiao, T., Singh, M., Mintun, E., Darrell, T., Dollár, P., Girshick, R.: Early convolutions help transformers see better. Adv. Neural. Inf. Process. Syst. **34**, 30392–30400 (2021)
23. Hutchins, D., Schlag, I., Wu, Y., Dyer, E., Neyshabur, B.: Block-recurrent transformers. arXiv preprint arXiv:2203.07852 (2022)
24. Xu, N., et al.: YouTube-VOS: sequence-to-sequence video object segmentation. In: Ferrari, V., Hebert, M., Sminchisescu, C., Weiss, Y. (eds.) ECCV 2018. LNCS, vol. 11209, pp. 603–619. Springer, Cham (2018). https://doi.org/10.1007/978-3-030-01228-1_36
25. Nwoye, C.I., Mutter, D., Marescaux, J., Padoy, N.: Weakly supervised convolutional LSTM approach for tool tracking in laparoscopic videos. Int. J. Comput. Assist. Radiol. Surg. **14**(6), 1059–1067 (2019)
26. Hochreiter, S., Schmidhuber, J.: Long short-term memory. Neural Comput. **9**(8), 1735–1780 (1997)
27. Dosovitskiy, A., et al.: An image is worth 16x16 words: transformers for Image recognition at scale. In: ICLR (2021)
28. Meng, X., Fan, C., Ming, Y., Yu, H.: CORNet: context-based ordinal regression network for monocular depth estimation. IEEE Trans. Circuits Syst. Video Technol. (2021)
29. Gur, S., Wolf, L.: Single image depth estimation trained via depth from defocus cues. In: Proceedings of the IEEE/CVF Conference on Computer Vision and Pattern Recognition, pp. 7683–7692 (2019)
30. Liu, Z., et al.: Swin transformer: hierarchical vision transformer using shifted windows. In: Proceedings of the IEEE/CVF International Conference on Computer Vision, pp. 10012–10022 (2021)
31. Johannsen, O., et al.: A taxonomy and evaluation of dense light field depth estimation algorithms. In: Proceedings of the IEEE Conference on Computer Vision and Pattern Recognition Workshops, pp. 82–99 (2017)
32. Anwar, S., Hayder, Z., Porikli, F.: Deblur and deep depth from single defocus image. Mach. Vis. Appl. **32**(1), 1–13 (2021)
33. Kang, X., Yuan, S.: Integrated visual-inertial odometry and image stabilization for image processing. In: Google Patents, US Patent App. 18/035,479 (2023)

34. Guo, X., Li, H., Yi, S., Ren, J., Wang, X.: Learning monocular depth by distilling cross-domain stereo networks. In: Ferrari, V., Hebert, M., Sminchisescu, C., Weiss, Y. (eds.) ECCV 2018. LNCS, vol. 11215, pp. 506–523. Springer, Cham (2018). https://doi.org/10.1007/978-3-030-01252-6_30
35. Geiger, A., Lenz, P., Urtasun, R.: Are we ready for autonomous driving? The kitti vision benchmark suite. In: 2012 IEEE Conference on Computer Vision and Pattern Recognition, pp. 3354–3361. IEEE (2012)
36. Agarwal, A., Arora, C.: DepthFormer: multiscale vision transformer for monocular depth estimation with global local information fusion. In: 2022 IEEE International Conference on Image Processing (ICIP), pp. 3873–3877. IEEE (2022)
37. Si, H., et al.: Fully self-supervised depth estimation from defocus clue. In: Proceedings of the IEEE/CVF Conference on Computer Vision and Pattern Recognition, pp. 9140–9149 (2023)

TSI: A Multi-view Representation Learning Approach for Time Series Forecasting

Wentao Gao[1(✉)], Ziqi Xu[2], Jiuyong Li[1], Lin Liu[1], Jixue Liu[1], Thuc Duy Le[1], Debo Cheng[1], Yanchang Zhao[3], and Yun Chen[3]

[1] University of South Australia, Adelaide, SA 5095, Australia
gaowy014@mymail.unisa.edu.au
[2] CSIRO, Melbourne, VIC, Australia
ziqi.xu@data61.csiro.au
[3] CSIRO, Canberra, ACT, Australia
{yanchang.zhao,yun.chen}@data61.csiro.au

Abstract. As the growing demand for long sequence time-series forecasting in real-world applications, such as electricity consumption planning, the significance of time series forecasting becomes increasingly crucial across various domains. This is highlighted by recent advancements in representation learning within the field. This study introduces a novel multi-view approach for time series forecasting that innovatively integrates trend and seasonal representations with an Independent Component Analysis (ICA)-based representation. Recognizing the limitations of existing methods in representing complex and high-dimensional time series data, this research addresses the challenge by combining TS (trend and seasonality) and ICA (independent components) perspectives. This approach offers a holistic understanding of time series data, going beyond traditional models that often miss nuanced, nonlinear relationships. The efficacy of TSI model is demonstrated through comprehensive testing on various benchmark datasets, where it shows superior performance over current state-of-the-art models, particularly in multivariate forecasting. This method not only enhances the accuracy of forecasting but also contributes significantly to the field by providing a more in-depth understanding of time series data. The research which uses ICA for a view lays the groundwork for further exploration and methodological advancements in time series forecasting, opening new avenues for research and practical applications.

Keywords: Time series forecasting · Representation learning · ICA

1 Introduction

Time series forecasting holds a pivotal role in machine learning and statistical analysis, particularly salient in domains such as financial market analytics [1],

M. Gong et al. (Eds.): AI 2024, LNAI 15442, pp. 291–302, 2025.
https://doi.org/10.1007/978-981-96-0348-0_21

meteorological prognostication [2], and the forecasting of energy demands [3]. For instance, within the ambit of meteorological forecasting, the precision of time series predictions is paramount in mitigating the impacts of natural catastrophes, including torrential rains, droughts, and tempests [4]. These phenomena, representing significant global challenges, profoundly affect agriculture, water resource management, and ecological systems [5]. Concomitant with the exacerbation of global climate changes, the development of efficacious methodologies for the prediction of these varied natural calamities through time series analysis has emerged as an exigent imperative [8].

Traditional time series prediction methods like ARIMA [9], ETS model [10] and Wavelet Transform model [11], often struggle to handle the nonlinear features and dynamic changes of time series data. Emerging machine learning based prediction models provide new approaches to tackle these challenges, including transformer based models like Informer [12], FEDformer [13]. Yet, transformer models, despite their proficiency in many areas like weather [7], often fall short in explicitly delineating the underlying dynamics they capture, an area where representation learning methods show greater potential.

An alternative approach to improve prediction performance in time series forecasting involves learning representations. Commonly, a representation learning based method decomposes time series data into trend and seasonal components [14,15], effectively capturing patterns and predictive information. However, these methods might miss crucial information in complex, high-dimensional data due to their inherent representational limitations. To counter this, Independent Component Analysis (ICA) is used to extract independent source representations, revealing subtle and nonlinear relationships in the data, thus capturing key hidden features [16,17]. This ICA method addresses the gaps in traditional time series analysis, offering deeper insights for forecasting.

Table 1. Trend/Seasonal Decomposition vs ICA

	Trend/Seasonal	ICA
Focus	Periodicity	Hidden structures
Strength	Intuitive; Ideal for clear cycles	Unveils complex, non-linear patterns
Limitation	Overlooks non-cyclic features	Computationally intensive

The above described trend and seasonal decomposition approach maps data into latent spaces based on set patterns, which may have the risk of mislabelling key information as noise due to their reliance on repetitive cycles. Conversely, ICA effectively identifies distinct data components but may miss specific patterns, such as trends and seasonality. As Table 1 suggests, these are complementary. Despite this, their integration has not been explored in literature.

This study introduces an innovative hybrid of TS and ICA. By fusing trend, seasonal, and independent elements recognized through ICA, we have developed a multi-view time series forecasting model. This model captures both overarching trends and complex nonlinear dynamics. In comparison with contemporary advanced methods on several benchmark datasets, our approach demonstrates enhanced performance in time series representation and forecasting.

The main contributions of this study include:

1. Introducing a novel multi-view approach for time series forecasting which integrates trend and seasonal representation with an ICA derived representation. This approach provides a holistic understanding of time series data, encapsulating a comprehensive perspective on temporal data characteristics and complex non-linear patterns for enhanced predictive accuracy.
2. The efficacy and broad applicability of the TSI model are evaluated through experimentation across a diverse array of benchmark datasets. The proposed method exhibits superior performance, surpassing current state-of-the-art models in multivariate time series forecasting.

2 TSI Representation Learning Approach

2.1 Problem Formulation

In this study, we examine a time series of dimension m, denoted as $(\mathbf{x}_1, \ldots, \mathbf{x}_T) \in \mathbb{R}^{T \times m}$. Our objective is to use historical data spanning h steps to predict the future k steps of the time series, denoted as $Y^* = g(X)$, where $X \in \mathbb{R}^{h \times m}$ represents the input historical data and $Y^* \in \mathbb{R}^{k \times m}$ represents the predicted future values. In this research, we focus on enhancing the performance of the predictive function $g(\cdot)$ by extracting deep feature representations $H = f(X)$ from the historical data X, where $H \in \mathbb{R}^{h \times d}$ maps the m-dimensional raw signals into a d-dimensional latent space at each timestamp.

To achieve this goal, we develop a nonlinear embedding function $f(\cdot)$ that not only captures the complex patterns within the historical data but also enhances the predictive model $g(\cdot)$ by integrating advanced feature representations H, thereby improving predictions for future time steps. Specifically, we utilize the final step feature representation H_h from the historical data as an enriched input to the predictive model, allowing predictor to make more accurate predictions based on a rich contextual understanding. This approach not only optimizes the feature extraction process but also improves the overall predictive framework's performance through refined feature representations.

2.2 TSI Feature Representation

In the realm of time series analysis, unraveling the intricate patterns and underlying factors influencing trends is crucial for accurate forecasting and interpretation. The TSI feature Decomposition approach integrats two powerful analytical perspectives: Trend Seasonality (TS) analysis and Independent Component Analysis (ICA). This multi-view method is shown in Fig. 1 aims to provide a more nuanced and detailed understanding of time series data, especially in

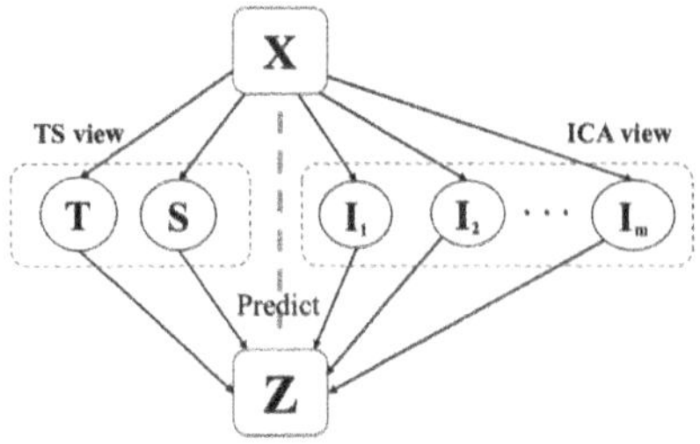

Fig. 1. Schematic of TSI, the proposed multi-view approach for time series forecasting

contexts where complex environmental and climatic factors play a significant role. By integrating the broad, overarching insights offered by TS analysis with the granular, independent factors revealed through ICA, this approach not only captur the 'essence of the data,' which includes fundamental components such as prominent trends and seasonality, and delves deeper into hidden patterns through ICA's ability to separate high-order statistical dependencies, such as subtle yet impactful cyclical fluctuations and random perturbations.

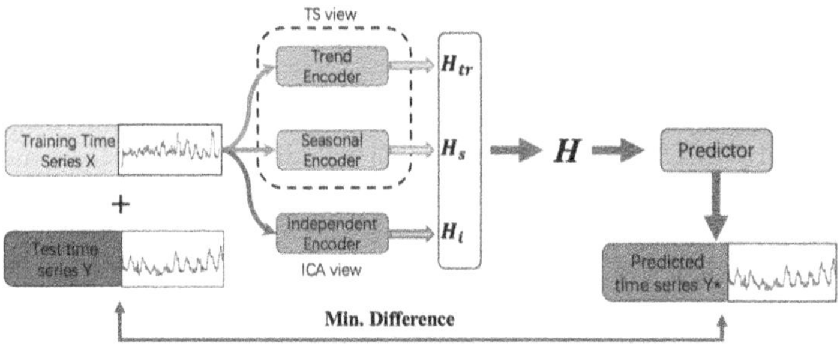

Fig. 2. The proposed multi-view time series forecasting model, incorporating Trend, Seasonal, and ICA encoders. The model's objective is to obtain a forecasted time series Y^* that has the smallest difference from the test time series Y.

ICA View. Independent components (denoted as $I_1, \ldots, I_m$) directly recover from the observed time series data (X). These independent factors represent fine-grained, high-frequency variations that are typically not captured by broader trend or seasonal patterns [18]. ICA decomposes the time series into statistically independent components [19], uncovering potential latent variables that generate the observed data. For example, these variables are often associated with specific environmental or climatic factors, such as natural phenomena or human activities.

TS View. By isolating the trend (T) and seasonality (S) components, significant long-term changes in the time series can be interpreted as responses to latent factors. For instance, a consistent upward trend might reflect the prolonged impact of global warming, while regular seasonal shifts could be linked to cyclical climate changes. These insights at various views provide a backdrop for identifying relationships and attributing certain changes or patterns to climate shifts or natural cycles.

Objective. By integrating TS and ICA analyses, we establish a multi-view approach. This framework combines the TS view on overarching trends and seasonal patterns with the detailed exploration of independent components provided by the ICA view, enabling a comprehensive understanding and prediction of complex dynamics within the data. Together, these will enhance our understanding of the mechanisms behind these predictions.

2.3 The Proposed Multi-view Approach

In this section, we introduce the TSI, which is focused on enhancing the diversity and robustness of feature representation in time series analysis.

The multivariate time series data X, is first decomposed into its constituent components: the Trend H_{tr}, Seasonality H_s, and Independent Components H_i.

Following the schematic of TSI shown in Fig. 1, the proposed overall process for feature representation learning with TSI is presented in Fig. 2. H is conceptualized as the aggregate of trend, seasonality, and independent components. The comprehensive feature representation can be formulated as follows:

$$H = [H_{tr}; H_s; H_i] \tag{1}$$

where [;] denotes the concatenation operation.

After we have obtained a well-trained feature representation H, which captures the nonlinear relationships of the original data, we can directly apply this representation to linear regression for predictive purposes. To predict future k steps values Y^* within the last time step t's representation H_t, we employ the following model:

$$\mathbf{Y^*}_{t+1:t+k} = f(H_t) \tag{2}$$

where $f(\cdot)$ denotes a linear regression model that incorporates an L2 regularization term, utilizing the feature representation H_t as its input to forecast the forthcoming values $\mathbf{Y}^*$. Specifically, this approach leverages the strength of linear regression models in handling continuous data predictions, while the L2 regularization term helps mitigate the risk of overfitting by penalizing large coefficients, ensuring a more generalized model performance (Fig. 3).

2.4 Trend and Seasonal Representation

we are obtain trend and seasonality as most time series representation work does.

$$H_{tr} = \frac{1}{M+1} \sum_{j=0}^{M} \Omega(G_j, 2^j) \tag{3}$$

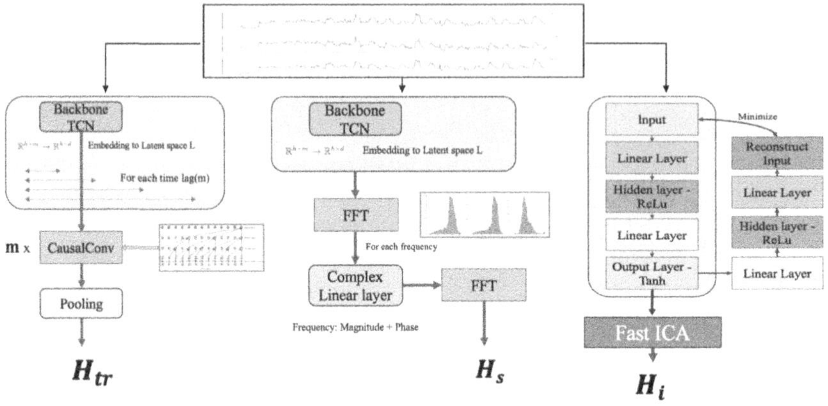

Fig. 3. Architectural Overview of the Time Series Decomposition Model. The left block extracts the trend component H_{tr} using a Temporal Convolutional Network (TCN) and pooling. The middle block captures the seasonal component H_s via FFT and a complex linear layer to encode frequency and phase. The right block extracts the independent component H_i using a fully connected network with activations designed to minimize reconstruction error. This model decomposes time series into distinct features for robust representation.

Here, Ω denotes the specialized causal convolution function [21]. The final output H_{tr} emerges as a synthesis of contributions from individual layers, averaged to formulate a composite yet distinctive representation of the time series trend.

To further refine our model and bolster its ability to distinguish different patterns, we employ a time-domain contrastive loss, inspired by the Momentum Contrast (MoCo) framework [20], which enhances our feature representations by effectively differentiating positive and negative sample pairs.

For robust seasonal feature extraction in time series analysis, we adopt a Fourier Transform (FT)-based decomposition, as suggested by Oppenheim et al. (2009) and recently by [13,15]. This technique facilitates the dissection of time series into constituent seasonal components by projecting the data into the frequency domain.

Building upon [15], a learnable Fourier transformation layer is introduced to encourage nuanced interactions between different frequency components. This is achieved by assigning a distinctive set of complex-valued weights to each frequency, allowing for a tailored enhancement of the seasonal patterns present in the time series data. Our framework integrates this technique into our model's architecture.

The frequency domain interactions and subsequent transformations can be expressed through the following formulation:

$$(H_s)_{i,l} = \mathrm{IFT}\left\{\sum_{n=1}^{N} P_{i,n,l} \odot \mathrm{FT}\{Q_{i,n}\} + B_{i,l}\right\} \tag{4}$$

Here, $(H_s)_{i,l}$ represents the matrix element of the extracted seasonal feature at the i-th observation and l-th frequency component. The operators IFT and FT denote the Inverse Fourier Transform and Fourier Transform respectively. $P_{i,n,l}$ symbolizes the transformation coefficients tailored to each frequency, while $B_{i,l}$ corresponds to the bias term incorporated within the transformation. The original time series data before transformation is $Q_{i,n}$, with N being the number of elements considered.

When implementing trend and seasonal encoder, we apply time-domain contrastive loss to each time series sample. In the context of our study, we adopted a triad of data augmentation strategies, namely scaling, shifting, and jittering [14]. Each technique is probabilistically activated with a chance of 50%. More detailed information are described in source code https://github.com/Wentao-Gao/TSI-forcasting.

2.5 Independent Representation

From the ICA perspective, we believe that independent representations offer a more detailed and nuanced understanding of data dynamics compared to the broader insights provided by the TS view.

This study explores the application of nonlinear Independent Component Analysis (nICA) for extracting latent, independent sources from high-dimensional datasets. Traditional linear ICA approaches often struggle with datasets characterized by complex, nonlinear interdependencies, underscoring the need for nonlinear mappings for effective source separation.

In our framework, we assume the time series data adhere to the general nonlinear mixing model as defined by [19]. Here, X_m represents the m-th observed variable among n total variables, expressed as:

$$X_m = f_m\left(\sum_{j=1}^{n} a_{mj} I_j\right), \quad m = 1, \ldots, n \tag{5}$$

Here, m indexes the observed variables, with I_j denoting the source signals and a_{mj} the mixing coefficients. The source signals I_j, for $j = 1, 2, \ldots, n$, undergo linear mixing followed by a transformation through the nonlinear function f_m to produce the observed variables X_m.

To approximate this nonlinear mapping, we utilize deep neural networks within the framework of Variational Autoencoders (VAEs). VAEs are known for their effectiveness in learning latent representations of data, featuring an encoder and a decoder that map input data to a latent space and then reconstruct the input data from this space, respectively. Specifically, the encoder $f_{\text{encoder}_{\theta_e}}$ transforms the observed data $\mathbf{X}$ into a latent representation, which the decoder $f_{\text{decoder}_{\theta_d}}$ then attempts to reconstruct back to the original data. This reconstruction process is not a mere replication but is achieved through learning the intrinsic structure of the data, closely aligning with the goals of nonlinear ICA.

Our model employs a VAE-like structure to approximate the complex nonlinear mapping f and its inverse f^{-1}, with the encoder mapping the observed mixed signals $\mathbf{X}$ to a latent space that reflects the linear mixing components. The decoder then attempts to reconstruct the observed signals from this latent representation, training the entire model to minimize the reconstruction error-similar to a traditional VAE but with an emphasis on learning a latent representation that mirrors the linear mixing signals.

To enhance the model's capacity to unveil independent components, we incorporate regularization techniques such as L1 regularization, promoting sparsity within the latent representation. This sparsity is essential for fostering independence among latent variables, a fundamental aspect of ICA.

Upon obtaining the linear mixing signals **AI**, we proceed to the next step using FastICA [19]:

$$H_i = FastICA(f_{\text{decoder}}(\mathbf{X})) \tag{6}$$

This step is critical, leveraging FastICA's powerful capability to recover independent source signals from the linearly mixed signals **AI**. In this manner, we not only utilize the deep learning model's ability to handle nonlinear relationships but also augment the separation of independent components via FastICA, further accentuating the principles of independence and non-Gaussianity, which are core to ICA.

In conclusion, by amalgamating the principles of VAEs with FastICA, we have achieved good performance for recovering independent source signals from mixed observations. This approach not only boosts the capacity to process nonlinear mixing models but also, by introducing sparsity-inducing regularization, ensures a congruence between the latent space and the linear mixing signals. Sparsity-inducing regularization encourages the majority of elements in the latent representation to approach zero, retaining only those components critical for reconstructing the linear mixing signals. This compact representation facilitates more effective recovery of independent source signals from linear mixing signals in subsequent steps, such as applying FastICA. We anticipate that this method will unlock new insights and application potentials in fields like time series data analysis and complex signal processing.

3 Experiment

The overarching aim of our research is to learn a representation for time series data that is not only comprehensive and meaningful but also exhibits robustness, thereby facilitating enhanced forecasting tasks. Our experimental design is focused on validating the representational strength of our model across a suite of benchmark datasets in forecasting. To ensure an equitable evaluation, we strictly adhere to the experimental framework as delineated in CoST [15], and TS2Vec [14].

The described process involves first deploying a trained model to convert time series data into a TSI representation, which captures the essential characteristics and patterns of the data. This TSI representation is then used as the basis for training a ridge regression model. The objective of this two-stage approach is to forecast future time steps, denoted as L. By focusing initially on the extraction of

Table 2. Multivariate forecasting result. The best results are highlighted in bold.

Methods		Unsupervised Representation Learning								End-to-end Forecasting					
		TSI		TS2Vec		TNC		CoST		Informer		LogTrans		TCN	
	L	MSE	MAE	MSE	MAE	MSE	MAE	MSE	MAE	MSE	MAE	MSE	MAE	MSE	MAE
ETTh1	24	**0.371**	**0.418**	0.590	0.531	0.708	0.592	0.386	0.429	0.577	0.549	0.686	0.604	0.583	0.547
	48	**0.422**	**0.454**	0.624	0.555	0.749	0.619	0.437	0.464	0.685	0.625	0.766	0.757	0.670	0.606
	168	**0.618**	**0.567**	0.762	0.639	0.884	0.699	0.643	0.582	0.931	0.752	1.002	0.846	0.811	0.680
	336	**0.777**	**0.664**	0.931	0.728	1.020	0.768	0.812	0.679	1.128	0.873	1.362	0.952	1.132	0.815
	720	**0.919**	**0.753**	1.063	0.799	1.157	0.830	0.970	0.771	1.215	0.896	1.397	1.291	1.165	0.813
ETTh2	24	**0.350**	**0.432**	0.423	0.489	0.612	0.595	0.447	0.502	0.720	0.665	0.828	0.750	0.935	0.754
	48	**0.566**	**0.571**	0.619	0.605	0.840	0.716	0.699	0.637	1.457	1.001	1.806	1.034	1.300	0.911
	168	**1.541**	**0.952**	1.845	1.074	2.359	1.213	1.549	0.982	3.489	1.515	4.070	1.681	4.017	1.579
	336	1.773	**1.032**	2.194	1.197	2.782	1.349	**1.749**	1.042	2.723	1.340	3.875	1.763	3.460	1.456
	720	2.062	**1.085**	2.636	1.370	2.753	1.394	**1.971**	1.092	3.467	1.473	3.913	1.552	3.106	1.381
ETTm1	24	**0.242**	**0.322**	0.453	0.444	0.522	0.472	0.246	0.329	0.323	0.369	0.419	0.412	0.363	0.397
	48	**0.320**	**0.376**	0.592	0.521	0.695	0.567	0.331	0.386	0.494	0.503	0.507	0.583	0.542	0.508
	96	**0.370**	**0.414**	0.635	0.554	0.731	0.595	0.378	0.419	0.678	0.614	0.768	0.792	0.666	0.578
	288	**0.452**	**0.473**	0.693	0.597	0.818	0.649	0.472	0.486	1.056	0.786	1.462	1.320	0.991	0.735
	672	**0.601**	**0.563**	0.782	0.653	0.932	0.712	0.620	0.574	1.192	0.926	1.669	1.461	1.032	0.756
ETTm2	24	**0.113**	**0.233**	0.180	0.293	0.185	0.297	0.122	0.244	0.173	0.301	0.389	0.537	0.180	0.324
	48	**0.168**	**0.293**	0.244	0.350	0.264	0.360	0.183	0.305	0.303	0.409	0.538	0.642	0.204	0.327
	96	**0.266**	**0.377**	0.360	0.427	0.389	0.458	0.294	0.394	0.365	0.453	0.912	0.757	3.041	1.330
	288	**0.700**	**0.638**	0.723	0.639	0.920	0.788	0.723	0.652	1.047	0.804	1.334	0.872	3.162	1.337
	672	**1.607**	**0.987**	1.753	1.007	2.164	1.135	1.899	1.073	3.126	1.302	3.048	1.328	3.624	1.484
Exchange	24	**0.105**	**0.260**	0.108	0.252	0.105	0.236	0.136	0.291	0.611	0.626	0.734	0.756	2.483	1.327
	48	0.165	0.330	0.200	0.341	**0.162**	**0.270**	0.250	0.387	0.680	0.644	0.837	0.812	2.328	1.256
	168	0.442	0.542	0.412	0.492	**0.397**	**0.480**	0.924	0.762	1.097	0.825	1.012	0.837	2.372	1.279
	336	**0.808**	**0.743**	1.339	0.901	1.008	0.866	1.744	1.063	1.672	1.036	1.659	1.081	3.113	1.459
	720	**1.121**	**0.880**	2.114	1.125	1.989	1.063	2.160	1.209	2.478	1.310	1.941	1.127	3.150	1.458
Weather	24	**0.293**	**0.354**	0.307	0.363	0.320	0.373	0.298	0.360	0.335	0.381	0.435	0.477	0.321	0.367
	48	**0.357**	**0.407**	0.374	0.418	0.380	0.421	0.359	0.411	0.395	0.459	0.426	0.495	0.386	0.423
	168	**0.464**	**0.490**	0.491	0.506	0.479	0.495	0.464	0.491	0.608	0.567	0.727	0.671	0.491	0.501
	336	**0.497**	**0.517**	0.525	0.530	0.505	0.518	0.497	0.517	0.702	0.620	0.754	0.670	0.502	0.507
	720	0.533	0.542	0.556	0.552	**0.519**	**0.525**	0.533	0.542	0.831	0.731	0.885	0.773	0.598	0.508
Ave.		**0.634**	**0.556**	0.818	0.632	0.912	0.668	0.743	0.603	1.152	0.779	1.339	0.921	1.558	0.880

the TSI and then rigorously training the ridge regression model, the methodology aims to develop a robust forecasting model capable of leveraging the intricate features within the time series' latent space.

3.1 Experimental Setup

Datasets. Our research utilizes six distinct, publicly available real-world datasets for comprehensive experimentation. The **ETT (Electricity Transformer Temperature)** dataset [12] includes two subsets with hourly data (ETTh) and one with 15-minute intervals (ETTm), comprising six power load indicators. The **Weather** dataset[1] encompasses hourly data from nearly 1,600

[1] https://www.ncei.noaa.gov/data/local-climatological-data/.

U.S. locations, featuring 11 climatic variables. **ExchangeRate**[2] contains the daily exchange rates of eight foreign countries from 1990 to 2016, including Australia, Britain, Canada, Switzerland, China, Japan, New Zealand, and Singapore. We consider all countries' value for multivariate forecasting.

Results. Our investigation presents a comprehensive comparison across several benchmarks for time series forecasting, employing a diverse array of methods including Ours, TS2Vec [14], TNC [22], CoST [15], Informer [12], LogTrans [23], and TCN [24]. The performance evaluation is conducted over predicted horizons L of 24, 48, 168, 336, and 720.(for ETTm1, ETTm2: 24, 48, 96, 288, 672). The results are shown in Table 2.

Across all datasets, our approach demonstrates superior performance, particularly at longer predicted horizons, indicating a robust capacity for capturing long-term dependencies within the time series data. For instance, in the ETTh1 dataset at a horizon of 720, our method achieves an MSE reduction of approximately 12% and an MAE improvement of nearly 9% compared to the next best method, CoST. On the Exchange dataset at $L = 168$, our model shows an MSE improvement of over 55% and an MAE reduction of about 42% when contrasted with the baseline TS2Vec method, underscoring the efficacy of our approach in more volatile financial time series.

Our model's average performance shows an MSE of 0.634 and an MAE of 0.556, which reflects an overall improvement of 22.5% in MSE and 20% in MAE against the averaged results of all other models. This enhancement is consistent across various datasets, demonstrating the method's generalizability and robustness.

As for End-to-end Forecasting, the proposed method consistently outperforms other advanced models like Informer and LogTrans, offering a compelling alternative for both short-term and long-term forecasting scenarios. Specifically, in the ETTm2 dataset at a horizon of 672, our approach achieves a substantial decrease in MSE and MAE by 33% and 28%, respectively, compared to the Informer model.

These findings indicate that our model is not only proficient in capturing and forecasting complex temporal dynamics but also demonstrates significant advancements in unsupervised representation learning for time series data. The results point towards our method's potential in providing more accurate, reliable, and computationally efficient forecasts, establishing a new benchmark in the field.

4 Conclusion

In conclusion, this study introduces the innovative TSI model, applying a comprehensive view to capture trends, seasonality, and independent components for time series forecasting. Our model excels in accuracy and offers insights into

[2] https://github.com/laiguokun/multivariate-time-series-data.

the complexity of time series data. Empirical evidence from various datasets confirms the TSI model's superiority, particularly in long-term forecasting, outperforming existing methods in MSE and MAE. The ablation study highlights the effectiveness of integrating TSI components, surpassing individual representations in multivariate tasks. The proposal of integrating TS and ICA and the demonstrated superior performance of TSI is a significant step forward, blending TS and ICA analytical perspectives to enrich the understanding of time series data and establishing new standards for future research and practical applications.

In future research, we will continue to explore the potential of ICA in time series forecasting, particularly in the context of causal analysis [6] [25] [26]. We aim to leverage causal inference to gain a deeper understanding of the structure within time series data, enhancing predictive performance and uncovering hidden causal relationships. This direction promises to bring new breakthroughs to time series forecasting and advance both the theory and practical applications in the field.

References

1. Doe, J., Smith, J.: Advanced time series forecasting in financial markets using deep learning models. J. Finan. Anal. **4**(2), 123–145 (2022)
2. Johnson, A., Lee, S.-H.: State-of-the-art meteorological forecasting through deep learning techniques. J. Meteorol. Res. **35**(5), 500–518 (2021)
3. Kumar, A., Fernandez, M.: Innovations in energy demand forecasting with machine learning. Energy Policy **49**, 284–292 (2022)
4. Chen, M., Roberts, A.: Predictive models for natural disaster mitigation using time series data. J. Disaster Manag. **12**(3), 78–89 (2021)
5. Garcia, R., O'Neill, B.: Impact of climate change on agriculture and ecological systems: a time series analysis approach. Int. Clim. Change Strateg. Manag. **12**(4), 430–442 (2020)
6. Cheng, D., et al.: Instrumental variable estimation for causal inference in longitudinal data with time-dependent latent confounders. In: AAAI 2024, vol. 38, no. 10, pp. 11480–11488 (2023)
7. Gao, W., et al.: A deconfounding approach to climate model bias correction. arXiv preprint arXiv:2408.12063 (2024)
8. Rahman, A., White, T.: Time series forecasting of climate phenomena and the imperative of predictive methodologies. Clim. Dyn. **60**(1), 15–35 (2023)
9. Box, G.E.P., Jenkins, G.M.: ARIMA model. J. Time Ser. Anal. (1960). Introduced the ARIMA model for time series forecasting. The model is represented as ARIMA (p, d, q). https://www.ncbi.nlm.nih.gov
10. Brown, R.G.: Exponential smoothing for predicting demand. Arthur D. Little Inc., Cambridge, Massachusetts (1956). First suggestion of exponential smoothing in statistical literature. https://legacy.library.ucsf.edu
11. Daubechies, I.: Ten Lectures on Wavelets. Society for Industrial and Applied Mathematics, Philadelphia (1992). Comprehensive guide on wavelets, covering various aspects including multiresolution analysis and orthonormal bases. https://www.semanticscholar.org

12. Zhou, H., et al.: Informer: beyond efficient transformer for long sequence time-series forecasting. In: The Thirty-Fifth AAAI Conference on Artificial Intelligence, AAAI 2021, Virtual Conference, vol. 35, no. 12, pp. 11106–11115. AAAI Press (2021)
13. Zhou, T., Ma, Z., Wen, Q., Wang, X., Sun, L., Jin, R.: FEDformer: frequency enhanced decomposed transformer for long-term series forecasting. arXiv preprint arXiv:2201.12740 (2022)
14. Yue, Z., et al.: TS2Vec: towards universal representation of time series. arXiv preprint arXiv:2106.10466 (2022)
15. Woo, G., Liu, C., Sahoo, D., Kumar, A., Hoi, S.: CoST: contrastive learning of disentangled seasonal-trend representations for time series forecasting. arXiv preprint arXiv:2202.01575 (2022)
16. Oja, E., Kiviluoto, K., Malaroiu, S.: Independent component analysis for financial time series. In: Proceedings of the IEEE 2000 Adaptive Systems for Signal Processing, Communications, and Control Symposium (Cat. No.00EX373), pp. 111–116 (2000). https://doi.org/10.1109/ASSPCC.2000.882456
17. Lu, C.-J., Lee, T.-S., Chiu, C.-C.: Financial time series forecasting using independent component analysis and support vector regression. Decis. Support Syst. **47**(2), 115–125 (2009). Elsevier. https://doi.org/10.1016/j.dss.2009.02.001
18. Forootan, E., Kusche, J., Talpe, M., et al.: Developing a complex independent component analysis (CICA) technique to extract non-stationary patterns from geophysical time series. Surv. Geophys. **39**, 435–465 (2018). https://doi.org/10.1007/s10712-017-9451-1
19. Hyvärinen, A., Oja, E.: Independent component analysis: algorithms and applications. Neural Networks **13**(4-5), 411–430 (2000)
20. He, K., Fan, H., Wu, Y., Xie, S., Girshick, R.: Momentum contrast for unsupervised visual representation learning. arXiv preprint arXiv:1911.05722 (2020)
21. Oord, A.V.D., et al.: WaveNet: a generative model for raw audio. arXiv preprint arXiv:1609.03499 (2016)
22. Tonekaboni, S., Eytan, D., Goldengerg, A.: Unsupervised representation learning for time series with temporal neighborhood coding. arXiv preprint arXiv:2106.00750 (2021)
23. Li, S., et al.: Enhancing the locality and breaking the memory bottleneck of transformer on time series forecasting. arXiv preprint arXiv:1907.00235 (2020)
24. Lea, C., Vidal, R., Reiter, A., Hager, G.D.: Temporal convolutional networks: a unified approach to action segmentation. arXiv preprint arXiv:1608.08242 (2016)
25. Du, X., Li, J., Cheng, D., Liu, L., Gao, W., Chen, X.: Estimating Peer Direct and Indirect Effects in Observational Network Data. arXiv preprint arXiv:2408.11492 (2024)
26. Bica, I., Alaa, A. M., van der Schaar, M.: Time series deconfounder: estimating treatment effects over time in the presence of hidden confounders. arXiv preprint arXiv:1902.00450 (2020)

Climate Downscaling Monthly Coastal Sea Surface Temperature Using Convolutional Neural Network and Composite Loss

Chen Wang[1](✉), Erik Behrens[1], Hui Ma[2], Gang Chen[2], and Victoria Huang[1]

[1] The National Institute of Water and Atmospheric Research, Wellington, New Zealand
{chen.wang,erik.behrens,victoria.huang}@niwa.co.nz
[2] Centre for Data Science and Artificial Intelligence, Victoria University of Wellington, Wellington, New Zealand
{hui.ma,aaron.chen}@ecs.vuw.ac.nz

Abstract. Climate downscaling bridges the gap between coarse-resolution General Circulation Model (GCM) outputs and the fine-resolution data needed for regional assessments. Traditional dynamic and statistical downscaling methods face limitations in computational efficiency and accuracy. Leveraging recent advancements in deep learning, particularly convolutional neural networks (CNNs), we propose an improved method for downscaling coastal sea surface temperature (SST). Our approach introduces a novel composite loss function combining Mean Squared Error with perceptual loss, effectively capturing the high seasonal variations typically in the coastal zones. Additionally, we redesign the YNet CNN to incorporate historical monthly mean SST, enhancing its performance. Comprehensive experiments using real-world SST GCM data and observational monthly SST data show our method significantly outperforms existing techniques in coastal zones up to 50 km offshore. Case studies in New Zealand further demonstrate the reduced errors achieved by our method.

Keywords: Seasonal climate forecast · Deep learning

1 Introduction

Climate downscaling is a critical process in climate science that aims to bridge the gap between coarse-resolution outputs from General Circulation Models (GCMs) and the fine-resolution data required for regional and local climate impact assessments [1,7]. GCMs, which are used for climate predictions and projections, typically operate at spatial resolutions on the order of 100 km or more [5], and are regularly generated globally to understand climate change over

M. Gong et al. (Eds.): AI 2024, LNAI 15442, pp. 303–315, 2025.
https://doi.org/10.1007/978-981-96-0348-0_22

short/long lead time [12]. However, many applications, such as farmed marine fishery production [15], and extreme weather/climate events preparedness [9], require climate information at much finer resolutions. Therefore, climate downscaling is employed to enhance the spatial resolution and accuracy of GCM outputs (i.e., low-resolution climate predictions over lead time), making them more relevant and useful for these applications. One particularly important application of climate downscaling is in the prediction of monthly coastal sea surface temperature (SST) [15]. Accurate high-resolution coastal SST predictions are essential for forecasting aquaculture production yield [15], predicting coastal storms and sea breezes [9], and protecting the coastal marine environment [6].

The problem of climate downscaling poses significant challenges, including the computational intensity of dynamic downscaling and the limitations of traditional statistical methods. Dynamic downscaling, which uses physical models to simulate climate processes at finer scales, is computationally intensive but can capture complex interactions and non-linearities [21]. Traditional statistical downscaling methods, such as Quantile Mapping (QM) [10,13], rely on empirical relationships between large-scale variables and local observations. These methods assume stationary relationships and require substantial high-quality historical data, limiting their ability to fully utilize GCM outputs and capture non-linear dependencies. Both approaches have inherent limitations that impact the performance of downscaling.

Recent advancements in deep learning have opened new avenues for improving climate downscaling [3,4,7,14,17,19]. Deep learning models, such as convolutional neural networks (CNNs), show great promise in capturing complex, non-linear relationships, automatically extracting relevant features, and providing high-resolution daily precipitation from coarse inputs [3,4,7,14,17,19]. These models can leverage vast amounts of historical data and auxiliary variables (e.g., topography) to improve the accuracy and reliability of the downscaling results [4,11]. In these studies, the *Mean Squared Error* (MSE) is commonly used as a loss function. However, MSE tends to smooth out the values of climate variables, which can reduce the overall error but makes it less suitable for handling coastal SST, where high seasonal variations are common due to proximity to land. Additionally, MSE is not effective for downscaling to specific zones, such as coastal areas near residential regions, which are of particular interest to environmental scientists.

The goal of this paper is to address the limitations of MSE for downscaling coastal Sea Surface Temperature (SST). We aim to improve the accuracy of downscaled coastal SST predictions by proposing a new composite loss function and an effective, modified CNN based on YNet [11].

1. We introduce a novel composite loss function, for the first time in literature, that combines the MSE loss with the perceptual loss for the climate downscaling problem. This perceptual loss function, suitable for handling coastal SST with high seasonal variations, improves downscaling accuracy in specific coastal SST zones.

2. We redesigned the YNet [11], particularly the fusion layers, to include monthly mean SST to capture SST periodic properties, which significantly improves its performance, compared to other state-of-the-art methods.
3. We conducted experiments comparing our method to existing approaches using real-world SST GCM data with a 6-month lead time and observational monthly SST data. Our results show superior performance in coastal zones up to 50 km offshore. A case study over two sites in New Zealand highlights the significantly smaller errors achieved by our method.

2 Related Work

The first CNN-based climate downscaling work was by Vandal et al. [19], who introduced a stacked Super-Resolution Convolutional Neural Network (SRCNN) for downscaling daily precipitation. They used the output of one SRCNN as the input for the next, creating a stacked enhancement process. Building on this, Sharma et al. [17] proposed a modified SRCNN that uses interpolated daily precipitation via bicubic interpolation, improving performance over the original SRCNN. Passarella et al. [14] introduced a fast SRCNN (FSRCNN) using low-resolution inputs with smaller convolution layers and additional feature extraction and nonlinear mapping layers after deconvolutions, enhancing SRCNN performance. Cheng et al. [3] presented a Laplacian Pyramid Super-Resolution Network (LapSRN) with a residual dense block (RDB) for upsampling. LapSRN leverages the Laplacian pyramid structure and RDB to fully extract hierarchical features, effectively capturing and reconstructing fine details in high-resolution daily precipitation.

All the above machine learning driven climate downscaling methods do not utilize GCM-simulated data as predictors. Instead, they first upscale observational data and then use these as predictors. In other words, the input (low resolution) and the output (high resolution) come from the same data sources, where patterns or structures can be easily captured by traditional interpolations such as bilinear interpolation. As highlighted in [4,11], this simplifies the problem, making it less challenging and interesting to ocean modellers. To address these challenges, Liu et al. [11] introduced YNet, featuring an encoder-decoder architecture enhanced with residual learning through skip connections and fusion layers. Instead of using deconvolutional layers to increase the resolution, YNet employs a more efficient yet equally effective sub-pixel convolutional neural network, as introduced by Shi et al. [18]. Motivated by YNet, a YNet variation [4] is introduced that utilizes several residual attention blocks (e.g., channel attention blocks and spatial attention blocks). Same as Ynet, they also concatenate topography data for the climate downscaling problem on daily precipitation data. These two methods have been proposed to address the daily rainfall downscaling problem, similar to all previously discussed downscaling methods in this section. Since their rainfall data is daily, they have a large amount of data for training the deep CNNs.

Our problem focuses on SST data, which is only available on a monthly basis. Therefore, YNet, which requires less training data and does not include attention blocks, is chosen as a suitable architecture for this study, compared to the YNet attention model. Furthermore, the MSE loss used in YNet and its attention variations is not tailored to our specific interests in coastal SST zones. As a result, a composite loss function is needed to better address our interests.

3 Preliminaries

Following [7], we define climate downscaling problem as a specialized Single Image Super Resolution (SISR) problem. Unlike typical SISR, the downscaling problem involves multiple unique features:

1. Low-resolution data and high-resolution data are obtained from different sources (e.g., GCMs and historical observational datasets).
2. GCMs usually use multiple ensembles to address uncertainties, resulting in multiple channels of unbounded variables.
3. Climate data often exhibit periodic properties (e.g., stationary monthly mean values, called monthly climatology).

Let $L^{i,l,e} \equiv \{\ell^{i,l,e}_{j,k}\}$ be a low-resolution SST from GCMs at initialization date i, lead time l, and ensemble number e at grid point (j,k) for longitude j and longitude k. Lead time l refers to the number of months after the initialization date i. In our context, lead time l can range from 0 to 5 months. Let $H^{i,l,e} \equiv \{h^{i,l,e}_{j,k}\}$ be a high-resolution observational SST at initialization date i, lead time l, and ensemble number e at grid point (j,k).

Let $C^m \equiv \{c^m_{j,k}\}$ be a historical high-resolution auxiliary climate variable at grid point (j,k) (e.g., monthly historical observational mean SST for month m, where $m = i + l$). The goal of the problem is to transform these inputs into a predicted high-resolution SST, $\hat{H}^{i,l,e} (\equiv \{\hat{h}^{i,l,e}\})$, as the output. The super-resolution mapping function G, parameterized by θ, represents the CNN model used for this transformation. The relationship between these variables is defined as:

$$\hat{H}^{i,l,e} = G(L^{i,l,e}, C^m, \theta)$$

The objective of the climate downscaling is to determine the mapping function G that minimizes a loss, such as the MSE, between the predicted high-resolution SST $\hat{H}^{i,l,e}$ and the observational high-resolution SST $H^{i,l,e}$ [4,11]. Therefore, this machine learning problem is to find an optimal $\theta^\star$ such that the MSE defined in Eq. 1 is minimised:

$$MSE = \frac{1}{N} \sum_{i,l,e,j,k} \| h^m_{j,k} - \hat{h}^{i,l,e}_{j,k} \|^2_2 \tag{1}$$

where $\hat{h}^{i,l,e}_{j,k} \in \hat{H}^{i,l,e}_{j,k}$ represents a predicted high-resolution SST value at grid point (j,k) for initialisation month i, lead time l, and ensemble number e, while $h^m_{j,k} \in H^m_{j,k}$ denotes an observational high-resolution SST at month $m = i + l$ at grid point (j,k). Note that, there are multiple $\hat{h}^{i,l,e}_{j,k}$ with distinct initialisation months and ensemble numbers at grid point (j, k) are compared against to one observational SST $h_{j,k}^{i,l,e}$.

By minimising the MSE in Eq. 1, the downscaling problem aims to produce high-resolution SST forecasts that closely match the observational SST values, thereby improving the resolution of GCM outputs for climate impact studies and decision-making processes.

4 A Composite Loss Function for Climate Downscaling Coastal SST

To design a loss function that captures high seasonal variations and improves the accuracy of coastal SST zones, we introduce a new composite loss that combines the MSE (defined in Eq. 1) with perceptual loss. The perceptual loss is utilized to capture anomalies against the historical monthly mean, effectively encoding high seasonal variations into the loss function. Additionally, the perceptual loss is employed to capture the complex structures of coastal zones, such as coastlines, thereby reducing errors around these areas. This perceptual loss is computed using a pre-trained deep CNN, such as VGG-19 originally trained on the ImageNet dataset, and is defined as:

$$L_{\text{perceptual}}(\theta) = \frac{1}{N} \sum_{i,l,e} \sum_{h} \|\phi_v(h_{j,k}^{m} - c_{j,k}^{m}) - \phi_v(\hat{h}_{j,k}^{i,l,e} - c_{j,k}^{m})\|_2^2$$

Here, ϕ_v represents the feature map obtained from the v-th layer of the VGG network. The perceptual loss measures the difference between the feature representations of the predicted high-resolution SST anomalies with respect to the corresponding monthly mean (i.e., $\hat{h}_{j,k}^{i,l,e} - c_{j,k}^{m}$) as well as the observational SST anomalies with respect to the corresponding monthly mean (i.e., $h_{j,k}^{m} - c_{j,k}^{m}$).

To leverage the advantages of both MSE defined in Eq. 1 and perceptual loss, we combine them into a single objective function. The composite loss function is formulated as:

$$L_{\text{composite}}(\theta) = \alpha L_{\text{pixel}}(\theta) + \beta L_{\text{perceptual}}(\theta)$$

where $L_{\text{pixel}}(\theta)$ is the pixel-based loss and $L_{\text{perceptual}}(\theta)$ is the perceptual loss. The weights α and β are hyperparameters that control the relative importance of each loss term. The downscaling coastal SST problem then becomes:

$$\theta^* = \arg\min_{\theta} L_{\text{composite}}(\theta)$$

5 Modified YNet Architecture for Coastal SST

Section 4 introduces a new loss function designed to handle high seasonal variations in coastal zones. To learn the periodic monthly SST properties, we employ a Ynet-based architecture tailored for effective downscaling coastal SST, as shown

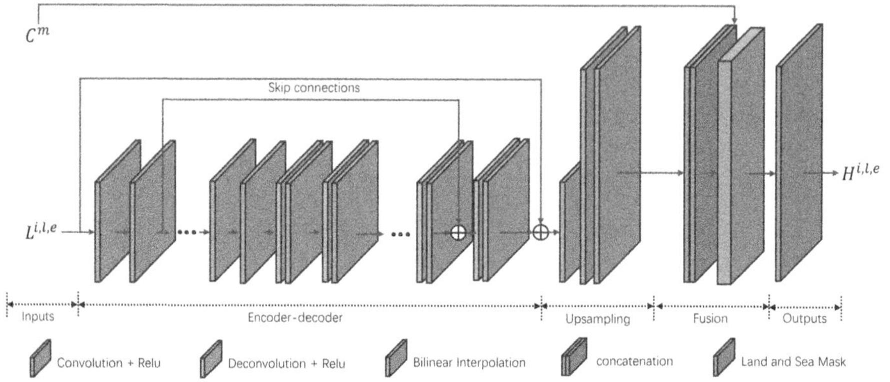

Fig. 1. Modified YNet Architecture: Integrates historical monthly mean SST data into the fusion layer. Excludes bilinear interpolation on raw input into the fusion layer. Reduces encoder-decoder layers from 30 to 10.

in Fig. 1. This architecture comprises three primary components: an encoder-decoder, an upsampling, and a fusion components.

The encoder-decoder part comprises convolutional and deconvolutional networks, functioning as a feature extractor. It processes "noisy" low-resolution SST inputs to generate "cleaner" low-resolution inputs. This component is crucial in refining the input data and preparing it for subsequent stages. The upsampling part mainly utilizes bilinear interpolation followed by a convolutional layer, enhancing the interpolation with learned parameters for improved performance. The final component, the fusion section, integrates high-resolution monthly mean SST as an additional channel to the upsampling output. This integration is crucial for learning the periodic properties of monthly SST in the final output. An ablation study, detailed in Sect. 6, investigates the importance of including the monthly mean SST.

The most significant modifications to the original YNet architecture for downscaling coastal monthly SST involve integrating historical monthly mean SST data into the fusion layer to capture periodic SST properties, unlike the original YNet, which includes high-resolution topography for daily precipitation. We also simplify YNet by excluding the fusion of bilinear interpolation on the raw input $L^{i,l,e}$, since it's already included on the "clean" SST from the encoder-decoder. Additionally, we reduce the number of layers in the encoder-decoder from 30 to 10, enhancing computational efficiency and addressing data volume issues.

6 Experiment

To investigate the significance of incorporating monthly mean SST, we compare the YNet-F and YNet models in Sect. 6.3. YNet-F includes monthly mean SST in its fusion part, unlike YNet. We also compared YNet-F with YNet-F CompLoss to evaluate the impact of the composite loss, where YNet-F CompLoss is trained with a composite loss function, while YNet and YNet-F use MSE. Several baselines from recent works, namely Bilinear Interpolation (BI) [15] and Quantile Mapping (QM) [10] are included into the comparison. BI neither bias-corrects nor calibrates the SST, treating forecasts as "pseudo-observations" that mimic the statistics and teleconnections of actual observations. In contrast, QM and our approaches aim to bias-correct the forecasts using actual observations.

Additionally, we examined the effect of including and excluding raw input interpolation in the fusion layer. Including raw input interpolation makes the model untrainable. Details are provided in Sect. 6.5

We followed the original hyperparameter settings of YNet [11], except for the number of encoder-decoder layers. Testing different layer configurations (30, 20, 10, and 6) for the encoder-decoder part, we found that 10 layers provided the best performance, while the original 30 layers made the network untrainable.

6.1 Datasets

The problem investigated in this paper considers downscaling problem from low resolution (1°, i.e., roughly 100 km) monthly SST to high resolution (1/20°, roughly 5 km) monthly SST with lead time up to 5 months.

We use the high resolution NZ20 SST data [2], developed by NIWA[1] based on a high-resolution (1/20°) two-way nested model. This model significantly reduces temperature biases and effectively captures the oceanic circulation around New Zealand. The high-resolution SST region of NZ20 extends from 142.8°E to 152°W and from 59°S to 22°S, comprising 1304 by 1004 grid cells. NZ20 renalysis data on SST is available from Jan 1993 to Dec 2019 on a monthly basis. An example of NZ20 data on SST in February 2019 is shown in Fig. 2 (a).

We source the low-resolution SST data from SEAS5 [8], ECMWF's[2] fifth generation seasonal forecast system, provided through the Copernicus Climate Change Service [5]. This data covers the same time period and geographical domains required for training. The ECMWF SST data is made of monthly hindcast and forecast data from 1993 to 2022, respectively. The hindcast and the forecast[3] data are made of 25 and 51 ensemble members, each of which has varied initialisation conditions in the forecast system with 5 month lead time. An example of SEAS5 data on SST with lead time = 0 (i.e., initial time) in February 2019 is shown in Fig. 2 (b).

6.2 Pre-processing Datasets

Our study area, ranging from 142.8°E to 152°W and from 59°S to 22°S, captures all regions of interest to New Zealand/Australia Ocean Modellers. Both the NZ20

[1] NIWA, the National Institute of Water and Atmospheric Research, New Zealand.
[2] ECMWF, the European Centre for Medium-Range Weather Forecasts.
[3] We call these 'forecast' hereafter in the paper for simplicity.

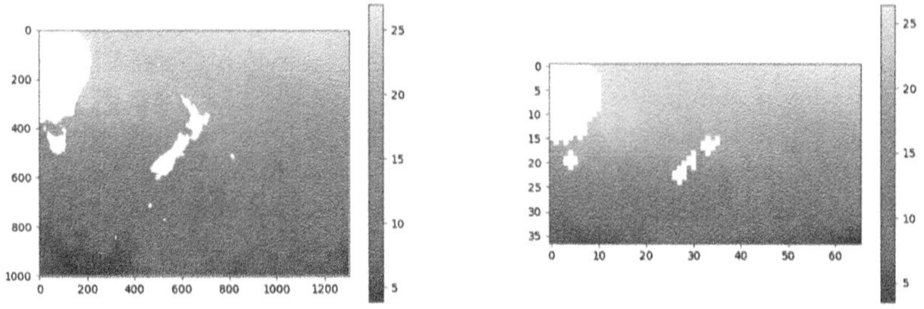

(a) NZ20 data

(b) SEAS5 data

Fig. 2. Examples of SST Datasets in February 2019

data and the SEAS5 data on SST are trimmed to this specified area for a common period from January 1993 to December 2019. To accommodate the presence of NaN values in the landmasses of New Zealand and Australia, we have applied a nearest-neighbor interpolation technique [20] to enable gradient calculations when training our YNet. Furthermore, SST values of SEAS5 data and NZ20 originally in Fahrenheit or Celsius are both converted to degrees Celsius and then normalized to a range of [0, 1] using a min-max normalization technique.

For testing purposes, we isolated a one-year segment from the NZ20 and SEAS5 data, covering August 2018 to July 2019. This segment is further divided into six subsets, each with identical lead times, to evaluate performance across increasing lead times. The selection of July 2019 as the end date of testing ensures that its forecast data with a 5-month lead time aligns with the available data up to December 2019. The period, from January 1993 to July 2018, is used for training. Additionally, this training dataset serves to develop SST climatology (i.e., monthly mean SST) with the help of the Climate Data Operators (CDO) tool [16]. Please refer to (Fig. 3) for examples of monthly climatology data and pre-processed SEAS5 data.

To create training and testing pairs, for every ensemble in SEAS5, we pair the forecast with initial date i with forecast lead time of l month with the NZ20 data and climatology data on date $i+l$. Let E be the total number of ensembles and M be the total number of months, we end up with $E \times L \times M$ pairs.

6.3 Performance Comparison

In this study, we evaluate the performance of several methods, including BI [15], QM [10], YNet [11], YNet-F, and YNet-F CompLoss, over lead times ranging from 0 to 5 months. Mean Absolute Error (MAE) is used to measure their performance, as it is the performance indicator commonly used in many climate downscaling works [4,7,11].

Figure 4 compares the MAEs over six lead times for all methods across three zones: coastal sea areas up to 25KM offshore, 50KM offshore, and the entire

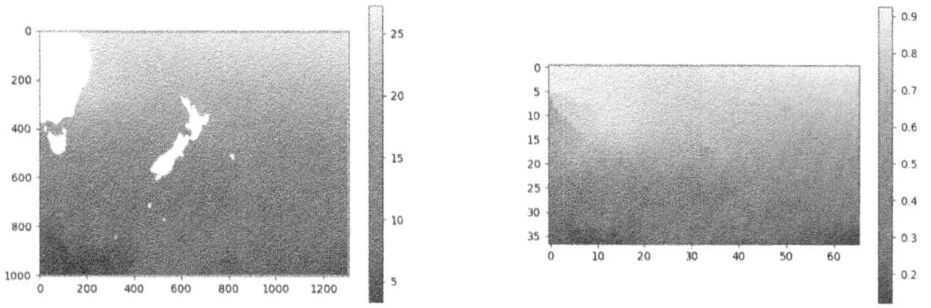

(a) Summer (February) Monthly climatology of SST (in °C)

(b) Pre-processed SEAS5 data (interpolated and normalised) on February 2019

Fig. 3. Examples of Pre-processed datasets

oceanic zone. The BI method consistently exhibits the highest MAE values across all lead times with significant variability, indicating its lowest downscaling accuracy among all competing methods.

YNet, without fusing historical monthly mean SST, does not outperform the QM method. However, incorporating the monthly mean SST into the fusion layer significantly improves the performance of YNet. As shown in Fig. 4, YNet-F outperforms the QM method, demonstrating the effectiveness of this modification.

YNet-F CompLoss further improves on YNet-F, achieving the highest accuracy characterized by the lowest MAE values and smallest error bars, especially in the coastal ocean zones up to 25KM and 50KM offshore. This improvement is clearly illustrated in the right-hand panels of Fig. 4, which provide a zoomed-in view of the corresponding zones shown in the left-hand panels. These results indicate that our composite loss effectively enhances downscaling performance in coastal oceanic zones. For all methods (except BI), the standard deviation shows a gradual increase in MAE upon increasing the lead time, which is attributed to the increasing forecast errors in the GCM data.

6.4 Comparison Methods for Selected Coastal Sites Case Study

To visualise how downscaled SST errors (with respect to the observational SST) vary with small differences in MAE (as discussed in Sect. 6.3) using YNet-F and YNet-F CompLoss, we selected two coastal SST locations in New Zealand-Pelorus Sound, and Little Pigeon Bay (see Fig. 5) as a case study to further evaluate their performance. These sites are crucial due to their significant contributions to New Zealand's economy, particularly in the marine and aquaculture sectors. For instance, Pelorus Sound is renowned for its mussel aquaculture, producing approximately 80,000 metric tons annually and generating over $204 million NZD [22].

Figure 6 presents the testing results of downscaled SST errors (in Celsius Degrees) in February 2019 at a lead time of 0 (February is chosen due to the

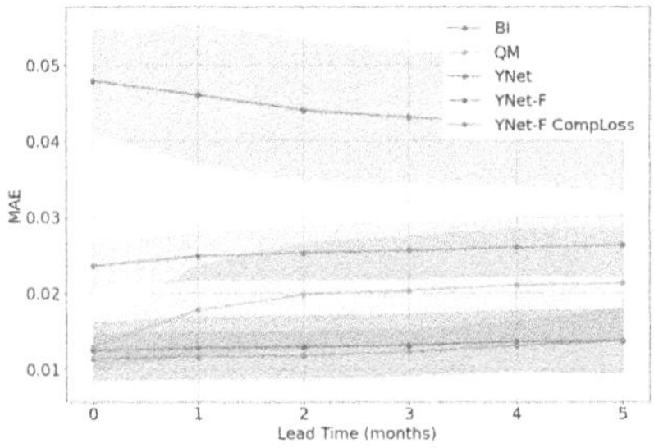

(a) 25KM Offshore

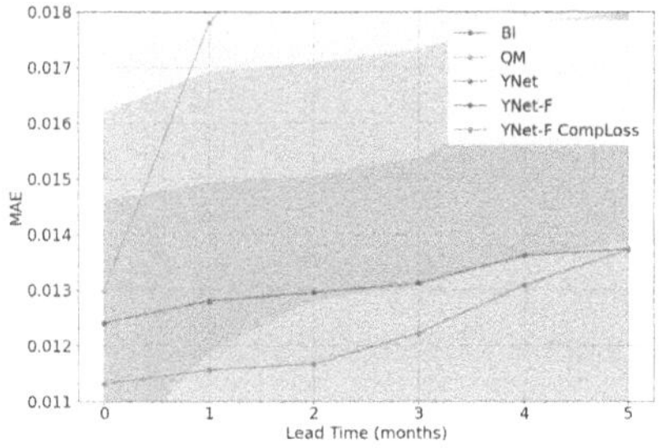

(b) 25KM Offshore (Zoomed-In View)

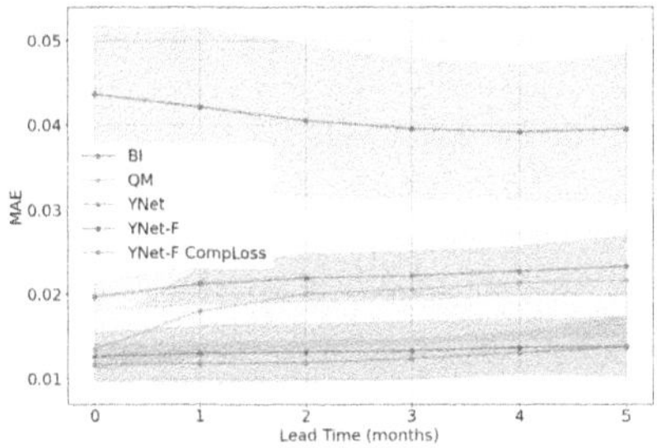

(c) 50KM Offshore

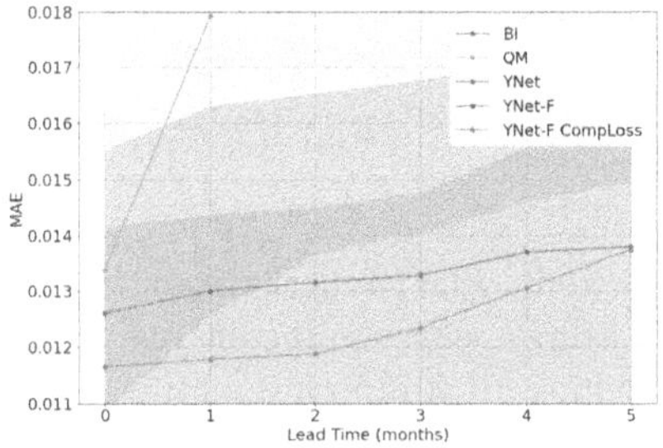

(d) 50KM Offshore (Zoomed-In View)

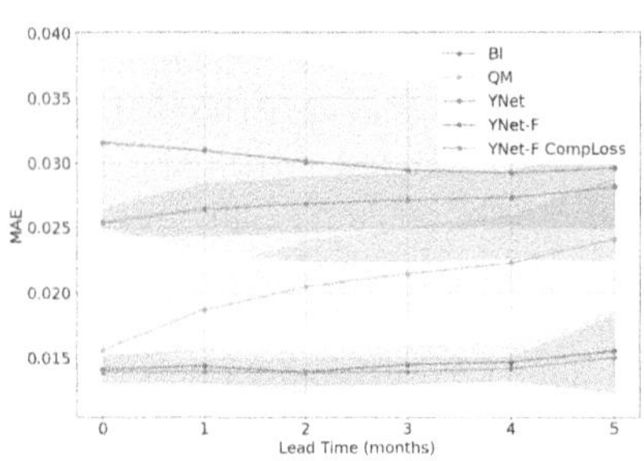

(e) Whole Oceanic Zone

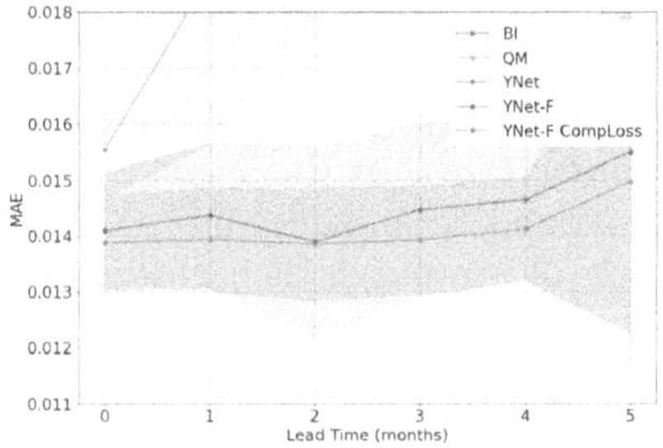

(f) Whole Oceanic Zone (Zoomed-In View)

Fig. 4. MAE by Methods over 5 Lead Times with Standard Deviation

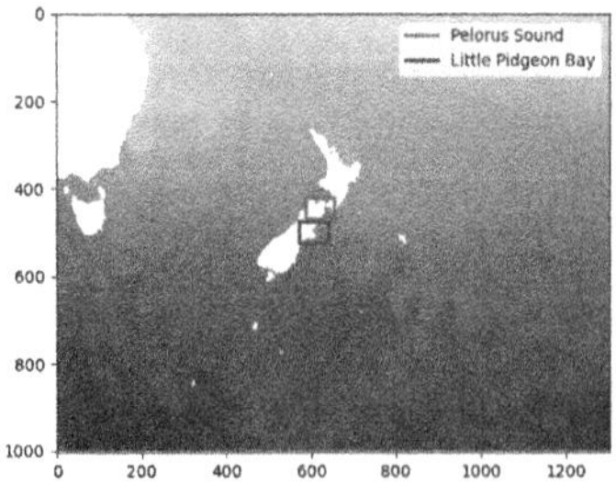

Fig. 5. Two Sites in New Zealand Selected for Our Case Study

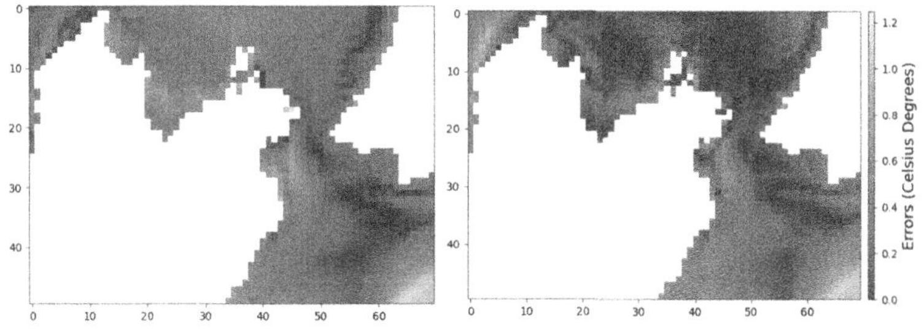

(a) Pelorus Sound (Left: YNet-F, Right: YNet-F ComLoss)

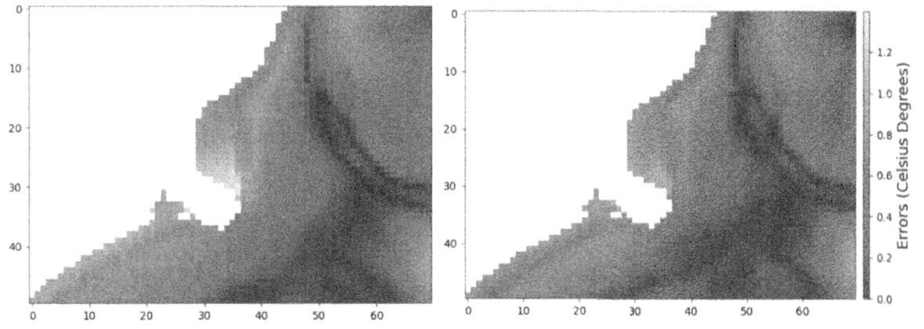

(b) Little Pidgeon Bay (Left: YNet-F, Right: YNet-F ComLoss)

Fig. 6. Downscaled SST Errors (in Celsius Degrees) at Two Coastal Oceanic Zones in New Zealand. Darker Colors Indicate Smaller Errors.

high SST variation during summer). The left figures show the results of YNet-F, while the right figures show the results of YNet-F CompLoss. For the Pelorus Sound zone, YNet-F CompLoss introduces smaller errors compared to YNet-F, as evidenced by the darker areas along the coastlines between the South Island and North Island of New Zealand. This pattern is consistent in Little Pigeon Bay, where YNet-F CompLoss results in lower errors compared to YNet-F. This further indicates that the CompLoss method is more reliable in estimating coastal SST, providing a more accurate representation of the coastal SST conditions. This improved accuracy made our YNet-F CompLoss method more useful in practice, such as the prediction of mussel aquaculture meat yield [15].

6.5 Exclusion of Raw Input Interpolation in Fusion

Figure 7 (a) and (b) illustrate the MAE recorded over 50 epochs for two variants of YNet-F: one including and one excluding raw input interpolation in the fusion layer. It is evident that including raw input interpolation renders the model untrainable. Consequently, the original YNet from the literature should be modified to exclude raw input interpolation in the fusion layer to better suit our problem.

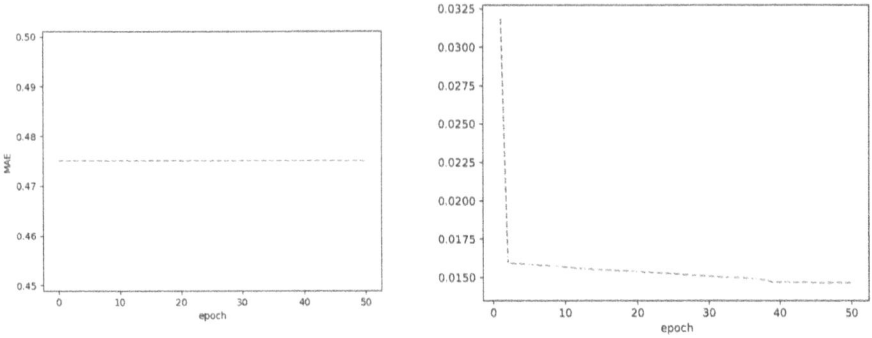

(a) Including the fusion of raw inputs interpolation makes the neural network not trainable

(b) Excluding the fusion of raw inputs interpolation makes the neural network trainable

Fig. 7. Examples of two training curves over epochs

7 Conclusions

In this paper, we addressed the limitations of traditional downscaling methods for predicting coastal sea surface temperature (SST). We proposed a novel composite loss function that combines Mean Squared Error with perceptual loss, specifically designed to capture the high seasonal variability inherent in coastal zones. Additionally, we redesigned the YNet CNN to incorporate monthly mean SST, which significantly enhances its performance. Our comprehensive experiments using real-world SST GCM data and observational monthly SST data demonstrated that our method significantly outperforms existing techniques in coastal zones up to 50 km offshore. The case study in New Zealand further validated the effectiveness of our approach, showing notably reduced errors.

References

1. Bailie, T., Koh, Y.S., Rampal, N., Gibson, P.B.: Quantile-regression-ensemble: a deep learning algorithm for downscaling extreme precipitation. In: Proceedings of the AAAI Conference on Artificial Intelligence, vol. 38, pp. 21914–21922 (2024)
2. Behrens, E., Hogg, A.M., England, M.H., Bostock, H.: Seasonal and interannual variability of the subtropical front in the New Zealand region. J. Geophys. Res. Oceans **126**(2), e2020JC016412 (2021)
3. Cheng, J., Kuang, Q., Shen, C., et al.: Reslap: generating high-resolution climate prediction through image super-resolution. IEEE Access **8**, 39623–39634 (2020)
4. Chiang, C.H., Huang, Z.H., Liu, L., Liang, H.C., et al.: Climate downscaling: a deep-learning based super-resolution model of precipitation data with attention block and skip connections. arXiv preprint arXiv:2403.17847 (2024)
5. Copernicus Climate Change Service: Seasonal forecast monthly statistics on pressure levels. https://doi.org/10.24381/cds.0b79e7c5 (2018). Accessed 14 May 2024

6. Garcia-Soto, C., Cheng, L., Caesar, L., Schmidtko, S., et al.: An overview of ocean climate change indicators: sea surface temperature, ocean heat content, ocean ph, dissolved oxygen concentration, arctic sea ice extent, thickness and volume, sea level and strength of the amoc. Front. Mar. Sci. **8**, 642372 (2021)
7. Jin, H., Jiang, W., Chen, M., Li, M., Bakar, K.S., Shao, Q.: Downscaling long lead time daily rainfall ensemble forecasts through deep learning. Stoch. Env. Res. Risk Assess. **37**(8), 3185–3203 (2023)
8. Johnson, S.J., Stockdale, T.N., Ferranti, L., et al.: SEAS5: the new ECMWF seasonal forecast system. Geosci. Model Dev. **12**(3), 1087–1117 (2019)
9. Køltzow, M.A., Iversen, T., Haugen, J.E.: The importance of lateral boundaries, surface forcing and choice of domain size for dynamical downscaling of global climate simulations. Atmosphere **2**(2), 67–95 (2011)
10. Li, M., Jin, H.: Development of a postprocessing system of daily rainfall forecasts for seasonal crop prediction in Australia. Theoret. Appl. Climatol. **141**(3), 1331–1349 (2020)
11. Liu, Y., Ganguly, A.R., Dy, J.: Climate downscaling using YNet: a deep convolutional network with skip connections and fusion. In: Proceedings of the 26th ACM SIGKDD, pp. 3145–3153 (2020)
12. Merryfield, W.J., Baehr, J., Batté, L., Becker, E.J., et al.: Current and emerging developments in subseasonal to decadal prediction. Bull. Am. Meteor. Soc. **101**(6), E869–E896 (2020)
13. Michelangeli, P.A., Vrac, M., Loukos, H.: Probabilistic downscaling approaches: application to wind cumulative distribution functions. Geophys. Res. Lett. **36**(11) (2009)
14. Passarella, L.S., Mahajan, S., Pal, A., Norman, M.R.: Reconstructing high resolution ESM data through a novel fast super resolution convolutional neural network (FSRCNN). Geophys. Res. Lett. **49**(4), e2021GL097571 (2022)
15. Rampal, N., et al.: Seasonal forecasting of mussel aquaculture meat yield in the Pelorus sound. Front. Mar. Sci. **10**, 1195921 (2023)
16. Schulzweida, U., Kronblueh, L., Budich, R.G.: CDO: climate data operators (2019)
17. Sharma, S.C.M., Mitra, A.: RESDEEPD: a residual super-resolution network for deep downscaling of daily precipitation over India. Environ. Data Sci. **1**, e19 (2022)
18. Shi, W., Caballero, J., Huszár, F., Totz, J., et al.: Real-time single image and video super-resolution using an efficient sub-pixel convolutional neural network. In: Proceedings of the IEEE CVPR, pp. 1874–1883 (2016)
19. Vandal, T., Kodra, E., Ganguly, S., et al.: DeepSD: generating high resolution climate change projections through single image super-resolution. In: Proceedings of the 23rd ACM SIGKDD, pp. 1663–1672 (2017)
20. Xing, Y., Song, Q., Cheng, G.: Benefit of interpolation in nearest neighbor algorithms. SIAM J. Math. Data Sci. **4**(2), 935–956 (2022)
21. Yoshikane, T., Yoshimura, K.: A downscaling and bias correction method for climate model ensemble simulations of local-scale hourly precipitation. Sci. Rep. **13**(1), 9412 (2023)
22. Zealand, A.N.: A sector overview with key facts and statistics for 2022 (2022). https://drive.google.com/file/d/1-Emyq4uIIt1qYcpyw10ZGu2a-xeFdJGr/view. new Zealand sustainable aquaculture

DBSSM: Deep BERT-Based Semantic Skill Matching from Resumes to a Public Skill Taxonomy

Haohui Chen[1(✉)], Claire Mason[1], Qinyong Wang[2], and Yanchang Zhao[1]

[1] Data61, Commonwealth Scientific and Industrial Research Organisation, Alexandria , Australia
Haohui.chen@csiro.au
[2] Baidu, Inc, Beijing, China

Abstract. The ability to extract skill information from unstructured text documents, such as resumes and job advertisements, can deliver efficiencies for recruitment and training and insights for policy and decision makers. The task requires the ability to differentiate between the thousands of broad and narrow skills required across the contemporary labour market. We describe the development of a Deep BERT-based Semantic Skill Matching (DBSSM) classifier that responds to this challenge by incorporating a combination of complementary classification algorithms, domain-expert fine tuning of weights and thresholds and labour market information contained in the European Skills, Competencies, Qualifications and Occupations (ESCO) taxonomy. The performance of the DBSSM classifier was tested on a dataset of anonymized resumes and found to achieve a Mean Average Precision at 1 (MAP@1) of 0.413 and an Average F1 (AF1) score of 0.206, exceeding the performance of alternative approaches. As well as supporting efficient candidate shortlisting (for recruitment) the classifier has applications for job seekers, education and training providers, employers, policy makers and researchers.

Keywords: Skill Classification · Recruitment Technology · Semantics Matching · Deep Learning · Labour Market Intelligence

1 Introduction

1.1 Background

The integration of advanced technologies like AI, biotechnology, and robotics into the global labor market demands continuous workforce upskilling and reskilling [1]. Despite abundant skills-related information on platforms such as job advertisements and resumes, its unstructured nature complicates analysis. This study introduces the Deep BERT-based Semantic Skills Matching (DBSSM) classifier, aligning skills-related text with the European Skills, Competencies,

M. Gong et al. (Eds.): AI 2024, LNAI 15442, pp. 316–328, 2025.
https://doi.org/10.1007/978-981-96-0348-0_23

Qualifications and Occupations (ESCO) taxonomy [3,13], a comprehensive catalog of over 13,000 skills. The DBSSM classifier automates skill data extraction, improving recruitment, training, and employment support by providing insights into labor market demand and workforce skills [13,22].

1.2 Related Work

Skill data extraction from resumes is gaining attention in academia and industry. Existing tools like Oracle Taleo [15], resumeworded.com [18], and livecareer.com [11] often lack transparency and detailed methodologies. Lightcast's job ad database uses exact matching against a skills dictionary to track skill demand trends [1,5], but struggles with granular skill data extraction from unstructured texts.

Machine learning methods for skills classification, such as those by Wowczko [24], Shi [20], and Zhao [26], often miss the semantic context. Pre-trained embeddings have been used to understand skills' semantic context in resumes and job ads [7,19], focusing on matching and clustering. NLP advancements like BERT [6] and Sentence-BERT [17] offer improvements. BERT analyzes words in their bidirectional context using the Transformer model [23], while Sentence-BERT modifies it with a Bi-Encoder architecture for efficient sentence-level comparisons [10]. However, mapping resume skills onto established taxonomies using these models remains unexplored. Our study addresses this by applying Extreme Multi-label Classification (XMC) to align resume skills with the ESCO taxonomy, creating a detailed training dataset for future research.

Several skill taxonomies exist, including O*NET [16], ESCO [22], and LinkedIn's proprietary taxonomy [2]. Our study focuses on the ESCO taxonomy due to its comprehensive skill representation and validation in multiple countries.

1.3 Challenges and Solutions

Matching resume skills to the 13,485 entities in the ESCO taxonomy [22] presents four challenges:

- **Taxonomy Size and Complexity:** Requires an extreme multi-label classifier.
- **Lack of Training Dataset:** Necessitates creating a context-rich dataset.
- **Semantic Understanding:** Context-aware classification is needed for ambiguous skill terms.
- **Threshold Setting:** Balancing precision and recall for skill mapping.

The DBSSM classifier addresses these challenges through:

- **Staged Candidate Skill Generation:** Combines keyword matching and semantic classification algorithms.
- **Threshold Fine-Tuning:** Uses domain expertise for accurate skill mapping.
- **Occupational Skill Profile Leverage:** Adjusts thresholds using occupational information from resumes.

1.4 Contributions

This study develops and validates the DBSSM classifier, enhancing skill extraction from large datasets using deep learning for semantic context. It outperforms traditional keyword matching through expert-driven threshold adjustments and occupation-skill mapping, providing labor market intelligence applicable in recruitment, employment, research, and policy.

2 Methodology

2.1 Overview of the DBSSM Classifier

The DBSSM classifier development process is depicted in Fig. 1. The left side (A) shows high-level steps, while the right side (B) illustrates the process for determining skills thresholds in more detail. The sequential steps are:

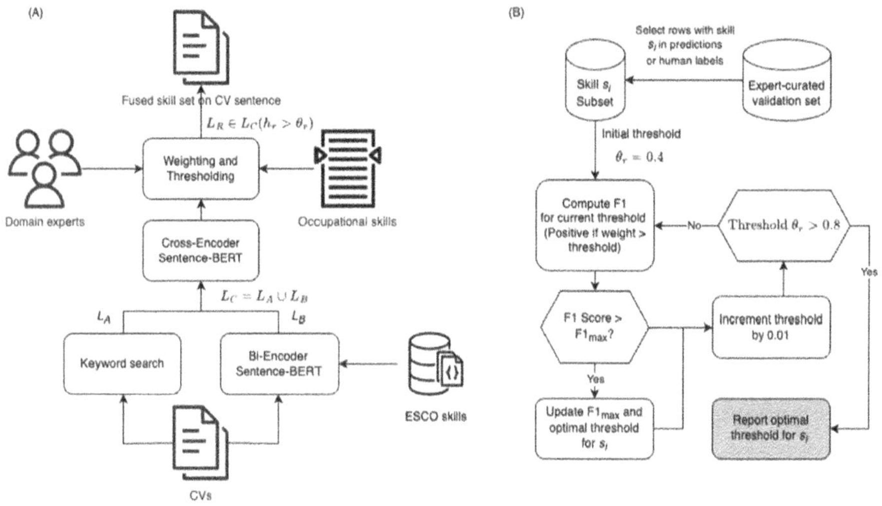

Fig. 1. Overview of DBSSM development. Panel A provides a high-level structure of the entire algorithm. Panel B details the flowchart for computing the optimal threshold for each skill, using an expert-curated validation set for fine-tuning. For further details on the fine-tuning process, see Sect. 3.2.

Text preprocessing. Common NLP practices, including sentence tokenization, stopwords removal, stemming, and lemmatization.

Keyword search. Exact match of ESCO skill labels (and their alternative labels) is performed on the CV sentences. All matched skills form a candidate skill list L_A.

Bi-Encoder. Measures semantic similarity between sentences and skill definitions using Sentence-BERT [17], taking the skills whose similarity scores are above a predefined threshold (addressed in Sect. 2.4). These form another candidate skill list, L_B.

Cross-Encoder. We combine L_A and L_B based on a predefined weighting strategy (addressed in Sect. 2.4), creating another candidate list, L_C. Subsequently, we perform a re-ranking operation on L_C. We measure the semantic similarity between the sentence and all skill descriptions from L_C again. In this step, instead of the Bi-Encoder, we use the Cross-Encoder from Sentence-BERT to predict the similarity score, taking the most similar K skills to form the final result, L_R. Note that we also set a threshold in this step (addressed in Sect. 2.4). It is important to note that the Cross-Encoder, while providing enhanced accuracy due to its fine-grained analysis, is less efficient with larger datasets. Therefore, its application is particularly advantageous for the re-ranking of shorter, pre-filtered candidate lists where its performance benefits can be realized.

Weighting and Thresholding. At steps 2 and 3 above, weights and thresholds are used to fuse and/or refine the output of the skills classifiers. The development of these weights and thresholds is different at each stage but in combination they draw upon input from domain experts, the ESCO skills hierarchy and the correspondence between skills and occupations in the ESCO taxonomy.

Employing BERT and Sentence-BERT for Skill Matching. The DBSSM classifier leverages BERT [6] and Sentence-BERT [17] architectures for skill matching, as shown in Fig. 2. The Bi-Encoder (Fig. 2A) generates embeddings for input sentences, facilitating efficient indexing and comparison. Sentence-BERT enhances embeddings' utility for skill matching through pooling. The Cross-Encoder (Fig. 2B) processes sentence pairs directly to produce a classification outcome, used for fine-grained comparisons [5].

2.2 Text Preprocessing

Given a resume denoted as $C = \{c_1, c_2, \ldots\}$, we clean and pre-process each sentence c_i:

(1) Noise removal: Unwanted characters including punctuation, accents, and special characters are removed; all words are converted to lowercase.
(2) Tokenization: Each sentence is split into tokens and converted into a list: $c_i = \{c_{i,1}, c_{i,2}, \ldots\}$.
(3) Stop words removal: Commonly appearing words are removed using a stopword list (NLTK's English stop words [12]).
(4) Lemmatization: Various forms of a word are reduced to their base or root using NLTK [12].

The same text preprocessing is applied to each skill label s_i' and its description d_i pair (s_i', d_i) from the ESCO skills taxonomy S, comprising $N = 13,485$ entries. Thus, S is represented as a set of pairs $\{s_i = (s_i', d_i) \mid i = 1, 2, \ldots, N\}$. For simplicity, we refer to the paired skills labels and descriptions as ESCO skills.

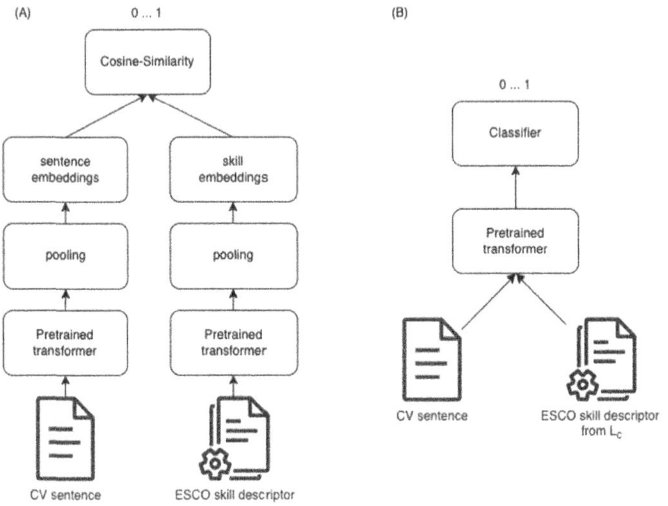

Fig. 2. BERT Bi-Encoder (Panel A) and Cross-Encoder (Panel B) for Skill Matching.

2.3 Generating Candidate Skills

Skill candidates are generated and refined for each resume sentence c_i from S using four complementary classification methods.

Keyword Search. Using FlashText [21], this method searches for keyword matches between resume sentences and skill labels. Skills in ESCO have preferred and alternative labels, mapped in the Trie dictionary of FlashText to identify matches, generating a candidate skill list, L_A.

Semantic Matching. A pre-trained Bi-Encoder assesses similarities between resume sentences c_i and every ESCO skill s_i, generating similarity scores. The network structure is shown in Fig. 2A (adapted from [17]). Each sentence-skill pair (c_i, s_i) is fed into a pretrained BERT model to derive their embedding representations. Cosine similarity scores above the threshold θ_s form the candidate skill list L_B.

Fusing Candidate Lists. Lists L_A and L_B are merged into a unified list L_C using a predefined weighting scheme. Each skill in L_C has two similarity scores: h_s from semantic matching and h_k from keyword search. We apply a predefined weight w_s (derived from the weighting strategy described in Sect. 2.4) to combine these two scores, yielding a new score h_f for each skill:

$$h_f = w_s h_s + (1 - w_s) h_k \tag{1}$$

This integration yields the refined skill candidate list $L_C = L_A \cup L_B$.

Re-generating Skill Candidates using the Cross-Encoder. The skills in L_C are reassessed using the Cross-Encoder for a finer-grained ranked list. For every candidate skill in L_C, each skill-resume sentence pair is processed by the Cross-Encoder to predict a match weights h_c (see Fig. 2B). Subsequently, we reweigh the skills in L_C using Eq. (2), to derive the weight h_r. Skills with similarity scores below the predefined threshold θ_r are eliminated. We then sort the remaining results to form a detected skill list L_R for each resume sentence. Although the Cross-Encoder is computationally inefficient, the number of skills in L_C has been significantly pruned from the entire skill list, making the re-ranking process feasible. The fine-tuned skill list $L_R = \{s \in L_C \mid h_r > \theta_r\}$ will have its own weights, calculated as:

$$h_r = (h_f + h_c)/2. \tag{2}$$

Re-Weighting Skill Candidates Using Occupational Information. The candidate skills generated by the Cross-Encoder are further refined using occupational information from the resume. ESCO links skills to occupations, allowing the use of occupational data to narrow down potential skill candidates. Specifically, we employed the Bi-Encoder described in Sect. 2.5 to identify five occupations from the resume, each achieving a similarity score above 0.3. If no occupation meets this threshold, no valid occupation is detected. Domain experts calibrated this threshold by analyzing the distribution of similarity scores between resume content and ESCO occupational profiles to ensure relevance. Essential skills for each identified occupation are selected from ESCO and amalgamated as potential skill candidates. These skills can increase the weight h_r by h_o, resulting in a new weight h'_r (see Eq. (3)), whenever the potential skill is present in L_R. The value of h_o, which also depends on the type of skill, is defined in Eq. (6).

$$h'_r = h_r + h_o \tag{3}$$

2.4 Developing Weights and Thresholds

At multiple steps in the development of the DBSSM, weights and thresholds are used to refine skill match candidates. Table 1 lists the thresholds and weights used at each stage, described below.

Fine-Tuning the Bi-Encoding Threshold. The first threshold to be fine-tuned is θ_s, filtering out skills matched using the Bi-Encoder. Lower values include more skills for downstream processing with the Cross-Encoder, increasing computational complexity but not affecting classification performance since a different threshold θ_r is used for the cross-encoding classifier. Domain experts determined the lowest viable threshold for an inclusive skill list L_B. An experiment (its setup is described in Sect. 3.2) was used to determine the optimal weighting. The range for the weight w_s, as defined in Eq. (1), was established between 0.6 and 0.9, with increments of 0.05. Lower values of w_s give less weight to the semantic match relative to the exact match output.

Fine-tuning Skill Thresholds for the Cross-Encoder Skills Matching. Skills in the ESCO taxonomy vary widely in specificity and commonality, classified into four categories: transversal, sector-specific, occupation-specific, and

Table 1. Thresholds and weights for fine-tuning

Thresholds & weights		Applied weights	Purpose
θ_s		h_s	Threshold for semantic matching results L_B
w_s		h_f	Weight for adjusting importance of semantic over keyword matching
θ_r	$\theta_r^{\text{transversal}}$	h_r	Threshold for re-ranked skill list $L_C^{\text{transversal}} \in L_C$
	$\theta_r^{\text{sector-specific}}$		Threshold for re-ranked skill list $L_C^{\text{sector-specific}} \in L_C$
	$\theta_r^{\text{occupation-specific}}$		Threshold for re-ranked skill list $L_C^{\text{occupation-specific}} \in L_C$
	$\theta_r^{\text{cross-sector}}$		Threshold for re-ranked skill list $L_C^{\text{cross-sector}} \in L_C$
h_o	$h_o^{\text{transversal}}$	see Eq. (3)	Adjusting weightings for skill type transversal
	$h_o^{\text{sector-specific}}$		Adjusting weightings for skill type sector-specific
	$h_o^{\text{occupation-specific}}$		Adjusting weightings for skill type occupation-specific
	$h_o^{\text{cross-sector}}$		Adjusting weightings for skill type cross-sector

cross-sector. Transversal skills (e.g., "work in teams") are meant to transfer to different disciplines and context [8] so they are broad in scope and are likely to appear frequently in resumes, requiring a higher threshold. In contrast, occupation-specific skills (e.g., "recognise fake goods") tend to be narrower and less common, needing a lower threshold for accurate matching. Therefore, using the same threshold for all Bi-Encoder skills is not appropriate. To fine-tune θ_r, we developed a dataset and consulted domain experts to evaluate matched skill candidates. This dataset guided the selection of optimal thresholds. A process flowchart (Fig. 1B), helped derive the ideal threshold $\theta_{r,i}$ for each skill s_i, aiming to optimize the F1 score (Eq. (5)). The optimal threshold θ_r^p for each skill category p (transversal, sector-specific, occupation-specific, and cross-sector categories) was calculated using Eq. (4), where L'^p_C denotes a subset of L^p_C that appears in the fine-tuning dataset.

$$\theta_r^p = \frac{1}{|L'^p_C|} \sum_{i=1}^{|L'^p_C|} \theta_{r,i}^p \tag{4}$$

$$\theta_{r,i} = \arg\max_{\theta} \mathrm{F1\_Score}(\theta, s_i) \tag{5}$$

This process established the thresholds for these four types of skills as $\theta_r^{\text{transversal}}$, $\theta_r^{\text{sector-specific}}$, $\theta_r^{\text{occupation-specific}}$, and $\theta_r^{\text{cross-sector}}$, respectively.

2.5 Weighting the Cross-Encoder Skills Matches Using ESCO Skills to Occupation Correspondence

After generating a final set of skill candidates using the cross-encoder, we refined the weightings attached to each skill match using ESCO skills to occupation correspondence. Resumes generally mention one or more occupations, and this information, along with the ESCO skills to occupation correspondence, enhances the accuracy of skill matches.

We used the Bi-Encoder to identify the top 5 occupations from a given resume and listed all corresponding required skills according to the ESCO skills to occupation correspondence. The weight h_r for corresponding skills in L_R, were then increased. For example, if a resume is identified as belonging to an audiologist who requires the skill "diagnose hearing impairment" according to ESCO, the confidence level of that skill from the resume will be increased by h_o to h'_r. If such a skill appears in L_C, it will be included in the final skill list L_R when the similarity score $h'_r > \theta_r$. The determination of h_o for each skill category p is given by Eq. (6). The value of $h_{o,i}$ is determined by Eq. (7), which optimizes the F1 score for skill detection while having h_r as the base weight.

$$h_o^p = \frac{1}{|L'^p_C|} \sum_{i=1}^{L'^p_C} h_{o,i}^p \tag{6}$$

$$h_{o,i} = h_r + \arg\max_{h_o} \mathrm{F1\_Score}(h_r + h_o, s_i) \tag{7}$$

3 Experimental Setup

3.1 Datasets

ESCO Taxonomy. We utilized ESCO V1.0.7 [4], encompassing 13,485 skills across four categories: transversal, cross-sector, occupation-specific, and sector-specific. Each skill has a preferred label, alternative labels, and a brief description. Additionally, ESCO offers an occupational taxonomy detailing required skills for 2,942 occupations[1].

Resume Data. Our DBSSM training and evaluation used a dataset of 11,082 anonymized resumes from www.postjobfree.com, provided by an industry collaborator. This dataset was divided into a cross-validation set (8,598 resumes), a fine-tuning set (1,988 resumes), and a test set (496 resumes). Human experts labeled the fine-tuning and test sets with ESCO skills, while the cross-validation set was automatically labeled using DBSSM.

3.2 Fine-Tuning Weights and Thresholds

The optimal weight w_s was determined using the fine-tuning dataset. The process involved classifying resume sentences and assigning F1 scores to evaluate the balance between semantic and exact matching. This balance was modulated by adjusting w_s. Each resume sentence in the fine-tuning dataset was labelled with a list of skills L_C with weights h_f (see Eq. (1)). Starting from $w_s = 0.6$, we computed individual F1 scores for each resume sentence, aggregating them to obtain an overall F1 score. Incrementing w_s by 0.05 in each iteration, we continued until $F1 > 0.9$. The weight that yielded the highest aggregate F1 score was deemed optimal. Details of this optimal weight are disclosed in Sect. 4.1.

For threshold fine-tuning, we selected resume sentences that positively or negatively predicted skill s_i, using a baseline threshold θ_r of 0.4. Predictions

[1] https://esco.ec.europa.eu/en/about-esco/publications/publication/skills-occupations-matrix-tables.

with weight h_r above θ_r are considered "valid predictions" (could be true positive or false positive), while those below are "invalid" (either true negative or false negative predictions). The F1 score, balancing precision and recall, was used as the optimization criterion. The initial value of $F1_{\text{max}}$ was set to 0. If the F1 score surpassed $F1_{\text{max}}$, $F1_{\text{max}}$ was updated. Afterwards, the threshold θ_r was incremented by 0.01 until it exceeded 0.8, establishing the optimal threshold $\theta_{r,i}$ for the skill s_i. Aggregating these individual thresholds resulted in the global thresholds $\theta_r^{\text{transversal}}$, $\theta_r^{\text{sector-specific}}$, $\theta_r^{\text{occupation-specific}}$, and $\theta_r^{\text{cross-sector}}$, respectively.

We began the fine-tuning of $h_{o,i}$ by an initial weight of 0.0. Using a method similar to θ_r fine-tuning, we incrementally adjusted $h_{o,i}$ and assessed model performance using the F1 score. This iterative process continued until $h_{o,i} > 0.2$. The weights $h_{o,i}$ were then aggregated across all the skill categories according to Eq. (6), determining the final category-specific weight h_o^p. The outcomes of this fine-tuning process are presented in Sect. 4.1.

3.3 Training the Extreme Multi-label Classifier

We used the cross-validation dataset to train an extreme multi-label classifier, with labels derived from our weighting and thresholding procedure. A five-fold cross-validation approach optimized hyperparameters and identified the best-performing models. The implementation was done using Amazon's Pecos framework [25], with hyperparameter tuning conducted on a separate validation set to ensure robustness and generalizability.

3.4 Evaluation

The held-out test set of 496 resumes was manually labeled by domain experts to provide ground-truth skills for each sentence. The performance of the DBSSM classifier was compared with candidate skills generators, including the current industry standard (exact match skills identification by Lightcast) and widely used open-access semantic matching classification tools [14,17].

- **EM (Exact Match)**: Detects skills in resume sentences if it matches an entry in our predefined ESCO skill dictionary. The comparison skill set L_A represents the baseline of exact lexical matches.
- **BE (Bi-Encoder)**: Converts resume sentences and skill definitions into fixed-size embeddings for direct comparison. The comparison skill set L_B is prioritized based on semantic weight h_s and threshold $\theta_s > 0.4$.
- **CE (Cross-Encoder)**: Re-evaluates similarity between each resume sentence and skill pair within L_C, assigning new weights h_c. The skills in L_C are then re-ranked according to these weights h_c for comparison.
- **EM+BE+CE**: Combines skill lists from EM, BE, and CE, re-ranked based on weight h_r. Skills with weightings above 0.5 are retained in the comparison set $L'_C = \{s \in L_C \mid h_r > 0.5\}$.

- **EM+BE+CE+fine-tuning**: Drawing upon the thresholding methodology outlined in Sect. 2.5, we refine the skill set L_C to obtain $L_R = \{s \in L_C \mid h_r > \theta_r\}$. The subset L_R, is then sequenced according to the weight h_r.
- **DBSSM (BE+CE+fine-tuning+occupational information)**: Re-weights skills from the EM+BE+CE+fine-tuning process using occupation information from resumes and ESCO. The skill set L'_R is prioritized according to $h'_r = h_r + h_o$, reflecting both generic skill relevance and occupational specificity.
- **XMLC**: An extreme multi-label classifier constructed from the dataset, leveraging the fine-tuned thresholding strategy from DBSSM. Skills in L_S are organized based on prediction confidence scores.

For a detailed discussion of the metrics used to evaluate DBSSM's prediction accuracy and alternative methods, see S1 Performance Metrics in the supplementary materials. Metrics include Mean Average Precision (MAP@K), precision, recall, and F1 scores, essential for assessing skill classification efficacy.

Table 2. The optimal weights and thresholds derived from the experiment

Thresholds	Values	Weightings	Values
θ_s	0.4	w_s	0.8
$\theta_r^{\text{transversal}}$	0.608	$h_o^{\text{transversal}}$	0.010
$\theta_r^{\text{sector-specific}}$	0.540	$h_o^{\text{sector-specific}}$	0.044
$\theta_r^{\text{occupation-specific}}$	0.535	$h_o^{\text{occupation-specific}}$	0.040
$\theta_r^{\text{cross-sector}}$	0.547	$h_o^{\text{cross-sector}}$	0.042

Table 3. Performance metrics for each skill classification method

	MAP@1	MAP@2	MAP@3	MAP@4	MAP@5	AP	AR	AF1
EM	0.094	0.07	0.049	0.04	0.033	0.091	0.04	0.045
BE	0.325	0.276	0.244	0.217	0.195	0.119	0.4	0.130
CE	0.345	0.297	0.253	0.224	0.201	0.231	0.254	0.179
EM+BE+CE	0.419	0.357	0.312	0.275	0.248	0.205	0.371	0.194
EM+BE+CE+fine-tuning	0.422	0.349	0.299	0.254	0.227	0.266	0.306	0.205
DBSSM	0.413	0.34	0.304	0.27	0.237	0.253	0.323	0.206
XMLC	0.332	0.312	0.291	0.265	0.24	0.222	0.318	0.217

Note: For each metric (i.e., column), dark green indicates the best performance and dark red the worst performance, while other colours indicates performance in between. The ranking is relative within each metric column.

4 Results

4.1 Weighting and Thresholding Analysis

We optimized the weighting coefficient w_s to achieve the highest F1 score, arriving at a weight of 0.80. For thresholding θ_r, we confirmed that positive samples received higher weights than negative counterparts using a paired T-test across all four skill categories (see Figure S2 in the supplementary materials). Following the procedure in Sects. 2.4 and 2.5, we established an optimal thresholds (Table 2). In the occupation-specific weight h_o adjustments, sector-specific, occupation-specific, and cross-sector skills received greater weight increments compared to transversal skills. These results support the need for tailored thresholds and weights in skills matching.

4.2 Performance Metrics and Analysis

Table 3 compares the performance metrics (MAP@k, AP, AR, and AF1) of the DBSSM and alternative skills classification methods. Although DBSSM did not consistently outperform others, it achieved high performance across all metrics. The "EM+BE+CE" method excelled in several metrics, especially MAP@2 to MAP@5. The "EM+BE+CE+fine-tuning" variant had the highest AP score, BE led in AR, and XMLC excelled in AF1 but performed poorly on other metrics. Despite being trained on a larger dataset, XMLC did not show a marked advantage, challenging the notion that more data always improves performance.

The DBSSM demonstrates a significant advantage in terms of efficiency. Unlike traditional models that often necessitate extensive datasets to refine their parameters, the DBSSM method achieves superior performance with substantially less data, making it particularly effective for applications with limited available data. This efficiency is likely attributable to the method's sophisticated thresholding strategy, which leverages domain expertise and a nuanced understanding of the skill matching task to fine-tune its parameters precisely and effectively.

Section S2 in the supplementary materials provides detailed insights into the DBSSM method's development and performance, analyzing skill matches at each developmental stage with dataset examples illustrating various methods' advantages and limitations.

5 Conclusion

This study explores advanced techniques for extracting and mapping skills from text documents into a standardized skills taxonomy, addressing semantic interpretation and thresholding complexities. Traditional methods like exact matching lacked nuance, while advanced algorithms like the Bi-Encoder (BE) showed high recall but low precision. The Cross-Encoder (CE) improved precision by filtering out irrelevant skills but sometimes missed relevant ones. The hybrid

EM+BE+CE approach, and further fine-tuning, balanced precision and recall by applying stricter thresholds and reducing noise. The DBSSM classifier proved the most effective, capturing accurate skills with few irrelevant entries and navigating the extensive ESCO taxonomy.

The findings affirm the potential of automated skills classification for fine-grained insights and big data analysis in the labor market. DBSSM's performance suggests its utility for efficient resume screening. Future research should test DBSSM's adaptability for other text documents, like job advertisements and training course descriptions. With further validation, DBSSM could support workforce research, training updates, job searches, and policy decisions [13].

Acknowledgments. We would like to express our gratitude to those who supported this research. Special thanks to Reejig, a recruitment services provider, for their partial funding of this study. Their contribution has been invaluable to the progression and success of our work.

Disclosure of Interests. The authors have no competing interests to declare that are relevant to the content of this article.

References

1. Acemoglu, D., et al.: Artificial intelligence and jobs: evidence from online vacancies. J. Labor Econ. **40**(S1), S293–S340 (2022)
2. Bastian, M., et al.: Linked in skills: large-scale topic extraction and inference. In: RecSys 2014 - Proceedings Of The 8th ACM Conference On Recommender Systems (2014)
3. Chiarello, F., et al.: Towards ESCO 4.0 - Is the European classification of skills in line with Industry 4.0? A text mining approach. Technol. Forecast. Soc. Change. **173**, 121177 (2021)
4. European Commission, ESCO Handbook: European Skills, Competences, Qualifications and Occupations. (European Commission Brussels,2019). https://op.europa.eu/en/publication-detail/-/publication/ce3a7e56-de27-11e7-a506-01aa75ed71a1
5. Deming, D., Kahn, L.: Skill requirements across firms and labor markets: evidence from job postings for professionals. J. Labor Econ. **36**(S1), S337–S369 (2018)
6. Devlin, J., et al.: BERT: pre-training of deep bidirectional transformers for language understanding. In: NAACL HLT 2019 - 2019 Conference Of The North American Chapter Of The Association For Computational Linguistics: Human Language Technologies - Proceedings Of The Conference, vol. 1 (2019)
7. Gugnani, A., Misra, H.: Implicit skills extraction using document embedding and its use in job recommendation. In: Proceedings Of The 32nd Innovative Applications Of Artificial Intelligence Conference, IAAI 2020 (2020)
8. Konstantinou, C., et al.: ATS2020 Developing and assessing transversal skills in primary and lower secondary education. In: ICERI2017 Proceedings. vol. 1, pp. 4416-4423 (2017,11)
9. Liu, X., et al.: Label Disentanglement in Partition-based Extreme Multilabel Classification. ArXiv Preprint ArXiv:2106.12751 (2021)

10. Liu, Y., et al.: RoBERTa: a robustly optimized BERT pretraining approach. ArXiv Preprint ArXiv:1907.11692. (2019)
11. LiveCareer Resume Check. https://www.livecareer.com/resume/check Accessed 18 March 2024
12. Loper, E., Bird, S.: NLTK: the natural language toolkit. ArXiv Preprint arxiv:Cs/0205028 (2002)
13. Mason, C., et al.: Illustrating the application of a skills taxonomy, machine learning and online data to inform career and training decisions. Int. J. Inf. Learn. Technol. **40**(4), 353–371 (2023)
14. Minaee, S., et al.: Deep learning-based text classification: a comprehensive review. ACM Comput. Surv. (CSUR). **54**, 1–40 (2021)
15. Oracle Australia Taleo. https://www.oracle.com/au/human-capital-management/taleo/ Accessed 18 March 2024
16. Peterson, N., et al.: Understanding work using the occupational information network (O*NET): implications for practice and research. Pers. Psychol. **54**(2), 451–492 (2001)
17. Reimers, N., Gurevych, I.: Sentence-BERT: sentence embeddings using siamese BERT-networks. arXiv preprint arXiv:1908.10084 (2019)
18. Resume Worded Free CV Checker. https://resumeworded.com/cv-checker Accessed 18 March 2024
19. Sayfullina, L., et al.: Learning representations for soft skill matching. Lecture Notes In Computer Science (including Subseries Lecture Notes In Artificial Intelligence And Lecture Notes In Bioinformatics). **11179 LNCS** (2018)
20. Shi, B., et al.: Salience and market-aware skill extraction for job targeting. In: Proceedings Of The 26th ACM SIGKDD International Conference On Knowledge Discovery & Data Mining, pp. 2871–2879 (2020,8)
21. Singh, V., Co, B.: Replace or Retrieve Keywords In Documents at Scale (2017,10)
22. Smedt, J., Vrang, M., Papantoniou, A.: ESCO: towards a semantic web for the European labor market. In: CEUR Workshop Proceedings, vol. 1409 (2015)
23. Vaswani, A., et al.: Attention is all you need. Adv. Neural Inf. Process. Syst. (2017-December) (2017)
24. Wowczko, I.: Skills and vacancy analysis with data mining techniques. Inf. **2** (2015)
25. Yu, H., et al.: PECOS: prediction for enormous and correlated output spaces. In: Proceedings Of The ACM SIGKDD International Conference On Knowledge Discovery And Data Mining (2022)
26. Zhao, M., et al.: SKILL: a system for skill identification and normalization. In: Proceedings Of The National Conference On Artificial Intelligence, Vol. 29, pp. 5 (2015)

Designing an Adaptive AI System for Operation on Board the SpIRIT Nano-Satellite

Zaher Joukhadar(✉), Jonathan Morgan, Christopher Bayliss, Miguel Ortiz del Castillo, Jack McRobbie, Robert Mearns, Krista A. Ehinger, Benjamin I. P. Rubinstein, Richard O. Sinnott, Michele Trenti, and James Bailey

University of Melbourne, Melbourne, Australia
{zjoukhadar,morgan.j,christopher.bayliss,miguel.ortizdelcastillo, mcrobbiej,robert.mearns,kris.ehinger,benjamin.rubinstein,rsinnott, michele.trenti,baileyj}@unimelb.edu.au

Abstract. This paper presents a detailed account of the adaptive AI system deployed on the recently launched SpIRIT nano-satellite. SpIRIT (Space Industry Responsive Intelligent Thermal) is a 6U CubeSat, an Australia-Italy mission led by the University of Melbourne. Launched in December 2023, SpIRIT is equipped with a payload designed to perform advanced computer vision experiments. The nano-satellite was launched with a pre-trained model for cloud detection, featuring the ability for onboard in-orbit fine-tuning. This paper highlights the software and system design choices made to address the unique challenges of running AI in space. The system is engineered for autonomy, robustness, and fault tolerance, allowing it to operate under the extreme conditions of space, including limited power, restricted communications, domain shifts, and constrained computing resources. The system consists of two distinct subsystems: one designed for space operations and the other for Earth-based activities. The space-bound subsystem handles inference and onboard fine-tuning of the model, while the Earth-based subsystem provides oversight and manages ground-truth labeling through a custom-built software called the Ground Truth Factory.

Keywords: AI in Space · Autonomy · CubeSat · Nano-satellite · Earth observation (EO)

1 Introduction

Miniature satellites, such as nano-satellites, have enhanced space-based computing capabilities. Weighing less than 20kg and often the size of a shoebox, they use commercial off-the-shelf components. The CubeSat is one example of such a nano-satellite. Such satellites are far less expensive than regular satellites

M. Gong et al. (Eds.): AI 2024, LNAI 15442, pp. 329–341, 2025.
https://doi.org/10.1007/978-981-96-0348-0_24

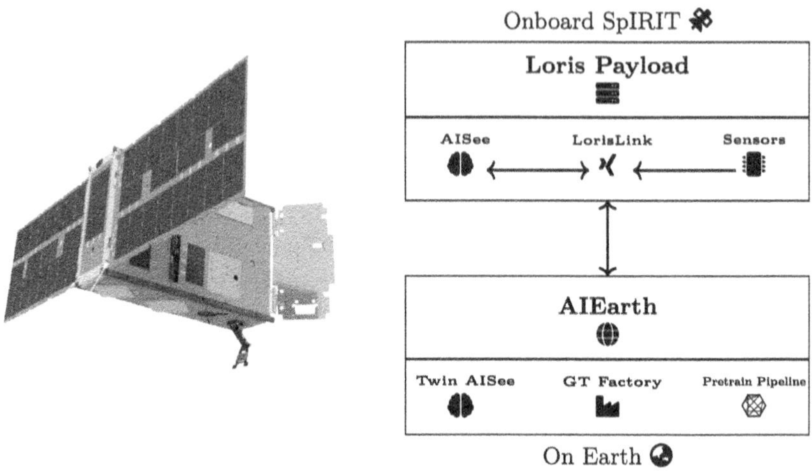

Fig. 1. Left: Computer-generated render of SpIRIT. Right: System Overview: Our system comprises components on the nano-satellite in orbit and their counterparts on Earth. The Ground Truth Factory receives image metadata (e.g., geographical coordinates, image acquisition time) and queries Earth-based data to create ground truth labels, which are uplinked to SpIRIT for onboard fine-tuning.

and enable new opportunities for onboard AI computations, Earth and space observations, and agile mission deployments.

Despite the potential, delivering AI on board a nano-satellite faces several challenges. AI models must be able to operate in a hostile and limited resource environment with reduced availability of power, memory and compute resources. This restricts both the complexity of AI models and the opportunities to retrain and dynamically tune them. [7,15] provide a comprehensive overview of these challenges. Furthermore, AI software must be resilient to unexpected conditions and shifts in data distribution, as well as have a high degree of autonomy to cope with extremely limited uplink/downlink communications to Earth (both in bit rate and length of transmission window).

Pre-training models with Earth-based data and deploying them on nano-satellites can enhance in-orbit AI capabilities. However, due to variations in sensors, orbits, and environmental factors, which cause domain shifts in data - meaning the data distribution changes significantly from the training data - pre-trained models need adaptation for different satellites and sensors. While [4,13] address domain shifts in space-acquired data, these experiments are typically conducted offline using downloaded data. Our paper presents a unique system offering onboard domain adaptation, an area not covered by previous works except for [1,16]. The former focuses on hardware aspects, and the latter performs in-orbit training in an unsupervised manner without ground truth labels. In contrast, our approach involves fine-tuning AI models directly onboard, providing adaptive capabilities that respond to evolving in-orbit conditions in real time.

Fine-tuning AI models is a well-known method in the AI-in-space research used to address domain adaptation. Usually, it is done through offline experiments on downlinked data. The unique aspect of our mission is that we perform onboard fine-tuning while in orbit, an emerging field at the intersection of space and AI sciences.

In collaboration with national and international partners, the University of Melbourne recently launched SpIRIT [18], a nano-satellite featuring a specialized payload named Loris. Loris is equipped with NVIDIA Jetson Nano hardware and provides a unique opportunity for us to deploy and test our adaptive AI system directly on an operational nano-satellite in orbit.

SpIRIT is currently in the calibration phase, and no live AI experiments have been conducted in orbit yet. This paper focuses on presenting the AI system and software designed for onboard fine-tuning and computer vision experiments. Future papers will detail actual AI experiments conducted once the calibration phase is complete. However, we see great benefit in sharing the learnings and insights we acquired while designing our system, as this can serve as a valuable building block for other researchers planning AI systems on a nano-satellite.

2 Related Work

Several projects have implemented machine learning on board nano-satellites with various end-goal aims. For instance, autonomous scheduling and planning are planned to be achieved by using AI to process data onboard and autonomously decide on operational schedules on CogniSAT-6 [14]. Additionally, other projects have utilized AI for selective data downlinking, such as in [5,8]. In the CubeSat ASRTU project, machine learning models were deployed to detect clouds [19]. The HEPCATS project [11] demonstrated the use of a pre-trained deep neural network for detection of auroral substorms. These projects highlight the potential of machine learning to be deployed on nano-satellites and the variety of tasks for which machine learning may be used.

Other works have explored the technical aspects of implementing AI in space. For example, [12] demonstrated deploying convolutional neural networks on small satellite platforms using TensorFlow. [6] proposed an efficient software approach for spacecraft AI with lock-free algorithms and data pipelining. [10] conducted AI experiments on image classification, image clustering, and fault detection on the European Space Agency's CubeSat (OPS-SAT) using platform they built called SmartApp platform.

Among the few previous projects that ran AI experiments on in-orbit nano-satellites, [1,16], and [10] are the closest to our work. Additionally, [17] proposed a concept system using a pre-trained model, but it was not run on an actual in-orbit CubeSat. [1,16], and [10] all shipped or uplinked pre-trained models to their nano-satellites for computer vision experiments. While these works conducted or plan to conduct AI experiments onboard in-orbit CubeSats, none provided a comprehensive overview of their AI pipeline or addressed space challenges in detail, except for [1], which focuses mostly on the hardware aspects of the mission.

Currently, there is a notable lack of documented research on the design of space-deployed AI systems, particularly with regards to software and system aspects. While some works address specific elements of system design [6,20] and others discuss hardware challenges [1], a comprehensive overview of software and system design strategies for the space environment is still missing. This gap in the literature motivated us to write this paper, focusing on the software and system aspects of designing AI systems for space missions.

3 System Design

3.1 Host Hardware

We first briefly summarise the hardware architecture underpinning the Loris system (see Fig. 1). This consists of:

- **Payload Hardware (Sensors)**: The Loris payload includes six visible light cameras, three infrared cameras, a camera control board, and a Jetson Nano system-on-module running Linux. The Jetson Nano, the central component, features a 128-core Maxwell GPU, Quad-core ARM A57 processor, 4GB LPDDR4 RAM, and 16GB eMMC storage. With Linux installed, about 2GB of RAM and 2GB of storage are available for operations. The cameras use Sony IMX219 sensors, offering 3820×2464 pixels and a ground resolution of approximately 200 m per pixel.
- **LorisLink**: Loris includes a Python-based software module called LorisLink, designed to interface with both SpIRIT and its onboard cameras. LorisLink receives commands from Earth, executes them on SpIRIT, and sends acknowledgment messages back to Earth, indicating the success or failure of the executed commands.

3.2 Software Architecture

3.3 AISee

AISee is our *in situ* deployed computer vision pipeline, written in Python utilizing PyTorch and TensorRT. It processes and analyzes data directly on the satellite, leveraging the onboard GPU for data processing, inference, and fine-tuning. The pipeline is compact, robust, and efficient, and has been designed to withstand the harsh conditions of space. AISee addresses several key challenges. The following sections discuss these challenges and our strategies to overcome them.

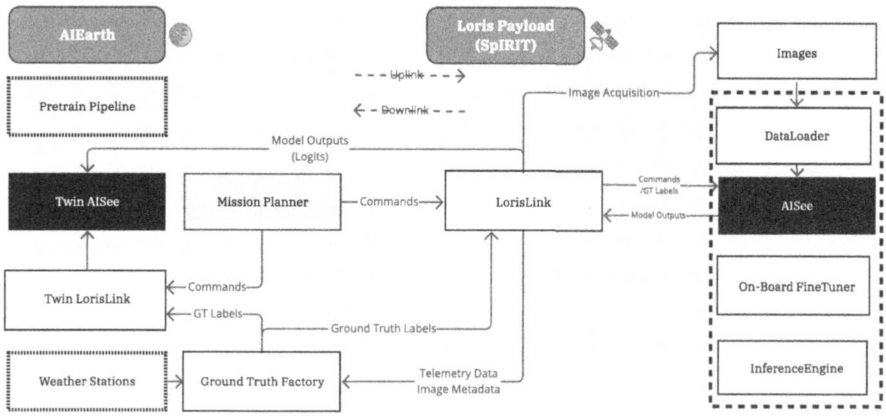

Fig. 2. Overview of AISee and AIEarth components, illustrating the interaction between the components onboard SpIRIT in orbit and their counterparts on Earth.

Extremely Limited Transmission Bandwidth. The main challenge with nano-satellites is limited transmission bandwidth. Loris is designed with an uplink capacity of 100KB per day (20% duty cycle, shared with other payloads) and a downlink capacity of 1MB per day. For context, a single compressed image acquired onboard SpIRIT ranges from 200KB-400KB.

Given these constraints, we chose a simple proof-of-concept task: *Cloud Detection*, requiring binary labels (no cloud = 0, cloud = 1). These 1-bit labels allow frequent model training and fine-tuning by uplinking many labels. This strategy enables us to build a large fine-tuning image set, acquired and processed entirely on SpIRIT without downloading the images.

Limitations of Onboard Image Processing. The Loris System lacks significant software or hardware for onboard post-processing of images, such as radiometric correction or cartographic geometry. Consequently, the curvature of the Earth may sometimes appear in the images because no cartographic geometry is applied. The acquired images are raw Level-0 data, and AISee must work with these raw images.

Data Variability and Model Adaptation. Due to bandwidth limitations, AISee avoids the downlink and uplink of images from SpIRIT. This constraint complicates monitoring data domain shifts and updating models. To address this, AISee includes an onboard fine-tuning module that uses ground truth labels from Earth via the Ground Truth Factory and downlinks model outputs (logits) for monitoring and evaluation.

Limited Compute Resources. AISee is comprised of three Python modules that perform data loading, inference, and fine-tuning. The *DataLoader* module interacts with both the *InferenceEngine* and *FineTuner* modules. Each of these modules is further divided into submodules dedicated to specific tasks, ensuring clear separation of their inner workings. An overview of AISee modules and their interactions with other system components is provided in Fig. 2.

This modular approach allows for dynamic loading and unloading of modules based on current needs, optimizing resource usage and improving performance. For instance, the FineTuner module can be loaded only when needed and unloaded afterward, conserving memory and processing power while enhancing adaptability to varying operational demands in space.

Key AISee Design Decisions and Strategies: The AISee solution's design and data organization incorporate strategies to address space environment challenges and facilitate Earth interaction. These include:

1. AISee uses non-recyclable unsigned integers for naming acquired images and fine-tuned models. This reduces transmission data size on SpIRIT because these integers have a fixed length and compact representation, eliminating the need for additional metadata. Additionally, it enhances pipeline robustness by ensuring unique identifiers without the risk of collisions.
2. AISee logs errors in a small file containing only code numbers corresponding to specific errors. We maintain a dictionary of these codes and their descriptions on Earth, allowing us to identify any errors after the file is transferred back following each command execution
3. Image acquisition is done by the imaging system on SpIRIT. This is initialized by a command sent from AIEarth (Sect. 3.4). Images are placed in a single location that serves as single source of truth for image data. AISee loads images from that location and does not re-store or place them anywhere else in the pipeline.
4. Due to transmission limitations, we do not downlink images or uplink trained models. Instead, we only download minimal image metadata, such as acquisition time and SpIRIT location at the time of acquisition. We also downlink fine-tuning performance metrics and inference outputs. For uplink, we transmit only the minimal data required for fine-tuning, such as ground truth labels in JSON format.
5. We downlink the raw model output (logits) instead of the final output, providing richer information and flexibility for analysis, albeit at the cost of much larger data compared to only downlinking the labels. (Listing 1.1).
6. AISee includes comprehensive and fully controlled capabilities for deleting model files and images, aiding in resource management on SpIRIT.
7. AISee utilizes TensorRT which is a machine learning inference framework developed by NVIDIA and it is highly optimized to run on NVIDIA GPUs.

Listing 1.1. Example of dowlinked inference output


```
{
    "imageID": "3",
    "model_name": "9",
    "inference_stime": "2023-07-14T00:56:27.782Z",
    "inference_etime": "2023-07-14T00:56:40.158Z",
    "inference_output": "-0.3567668497562408 4"
}
```


3.4 AIEarth

AIEarth is a suite of software programs and frameworks built to complement our AISee pipeline onboard SpIRIT and assist in mission operations. It includes components designed to function as counterparts to the nano-satellite's systems. This section explains the various components that make up AIEarth:

Mission Planner: is a lightweight Python program designed to manage our mission commands and planning. It is responsible for sending identical commands to both LorisLink and Twin LorisLink, such as image acquisition, running inference, or initiating fine-tuning processes.

Twin AISee and Twin LorisLink: Twin LorisLink is an exact copy of the onboard LorisLink, and Twin AISee is an exact copy of the onboard AISee pipeline, which we maintain on a replica (i.e. engineering model) of the Loris payload on Earth. The aim of Twin AISee is to mirror the state and operation of the AISee pipeline onboard SpIRIT as much as possible. This means that every command sent to SpIRIT is also sent to the Twin LorisLink to be executed on the Twin AISee.

When Mission Planner sends an image acquisition command and receives a success acknowledgment from SpIRIT, Twin LorisLink generates mock images with matching IDs. This allows us to monitor the image data status without downloading the actual data.

Similarly, Twin AISee generates mock fine-tuned models and inference outputs to mirror the data on SpIRIT. When the actual outputs are downlinked, these mock files are updated with real metrics from the satellite, ensuring an accurate reflection of onboard data. Twin AISee and Twin LorisLink also execute the same deletion and cleanup commands sent to SpIRIT, aiding in resource management.

Ground Truth Factory: is a critical component in the overall design of the system. It is a Python-based system that sources ground truth labels from Earth-based data sources like weather station APIs. For our initial model, it provides binary labels (cloud = 1, no cloud = 0) from ground-based weather datasets and APIs.

The Ground Truth Factory receives geographical coordinates and timestamps of images from SpIRIT (Fig. 3), derived from telemetry data. It sources ground truth labels from weather datasets and APIs, then sends these labels back to the satellite. This allows AISee to fine-tune the model in orbit in a supervised manner. The Ground Truth Factory is designed to be extensible. Initially, it focuses on ground truth labels for cloud detection but can be extended to include labels for ship locations, forest coverage, urban areas, deserts, coastlines, and more.

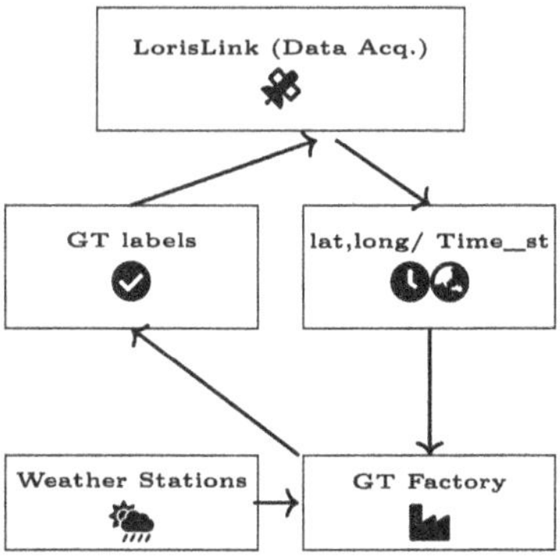

Fig. 3. Ground-Truth Factory

The modular design of our system make it adaptable to other satellite-based applications, such as land use classification, vegetation monitoring, or disaster management

Pre-train Pipeline: is a full-scale machine learning pipeline utilizing high-performance computing (HPC) and large-scale datasets for pre-training computer vision models. This pipeline exists exclusively on Earth. We used it to pre-train a cloud detection model called *Model0*, which we then deployed on SpIRIT. *Model0* serves as our baseline model, which will be fine-tuned in orbit to adapt to the image data acquired by SpIRIT.

3.5 Pre-trained Model and Fine Tuning Experiment

In this section, we describe the pre-trained model shipped with SpIRIT and the fine-tuning experiments conducted to simulate the expected domain shift and to test the onboard fine-tuning process pre-launch. We pre-trained our model using a curated subset of Sentinel-2 imagery [3]. For fine-tuning, we needed images vastly different from Sentinel-2 as well as those similar to those that SpIRIT will capture. Sourcing such images was challenging due to the lack of publicly available options, but one of our partners with a similar nano-satellite provided a suitable set, which we call the AdaptTest Set.

Sentinel-2 Set: This set consists of 76,800 tiles, each sized 224×224. Ground truth labels (cloud = 1, no-cloud = 0) were established using the Sentinel-2 Level-2A product [3], which includes a fourth channel for pixel classification. We focused on pixels classified as `CLOUD_MEDIUM_PROBABILITY`, `THIN_CIRRUS`, and `CLOUD_HIGH_PROBABILITY`. An image was labeled as cloudy (cloud = 1) if 20% or more of its pixels fell into these classifications. The 20% threshold aligns with meteorological standards using the oktas system, which divides the sky into eight parts. A cloud cover of 2 oktas (25%) is considered cloudy. The class breakdown (cloud, no-cloud) is shown in Table 1. Examples from the dataset are shown in Fig. 4.

Table 1. Dataset Breakdown by Cloud Coverage for Sentinel-2 and AdaptTest

	Sentinel-2		AdaptTest	
Label	Count (tiles)	Percentage	Count (tiles)	Percentage
No Cloud	18,646	24.23%	166	58.61%
Cloud	58,154	75.77%	117	41.39%

AdaptTest Set: This data set consists of 283 images (Table 1) acquired from our partner's nano-satellite currently operating in space. This nano-satellite is equipped with a similar imaging sensor to the IMX219 onboard the SpIRIT. Ground truth labels for AdaptTest images were produced manually also using the oktas system to label the Sentinel-2 Images. Several image examples of the new set are shown in Fig. 5.

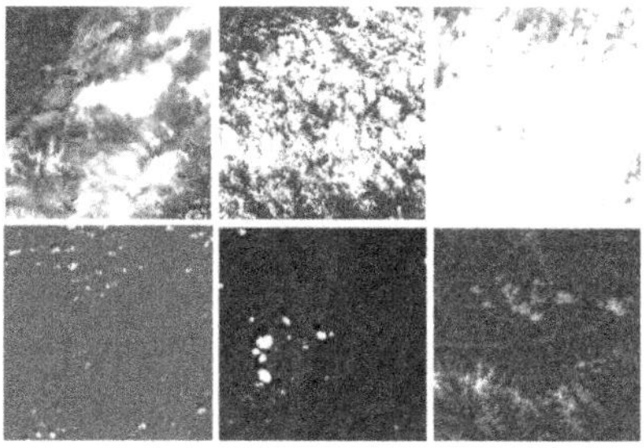

Fig. 4. Samples from Sentinel-2

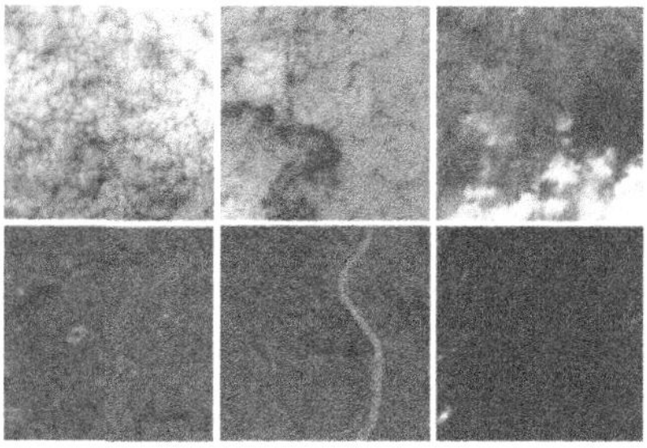

Fig. 5. Samples from AdaptTest

Comparison Between Sentinel-2 and AdaptTest Image Sets. To demonstrate how the fine-tuning experiment simulates the domain shift expected on SpIRIT, we must examine the differences between *Sentinel-2 set* and *AdaptTest set*. This involves comparing the imaging sensors, operating altitudes, and post-processing methods of each satellite. Table 2 compares the imaging sensor of Sentinel-2 with that of SpIRIT. Although it would be ideal to compare Sentinel-2 with our partner's nano-satellite - which directly speaks to the fine-tuning experiment presented here - we are unable to disclose the characteristics of our partner's nano-satellite. However, SpIRIT has similar sensors and hardware specifications to our partner's nano-satellite. Both satellites orbit at the same altitude and do not perform any post-processing on the acquired images.

Table 2 highlights significant differences between the two imaging sensors. Sentinel-2 uses a push broom sensor, capturing each line sequentially as the

Table 2. Comparison of imaging sensors used by Sentinel-2 and SpIRIT

	Sentinel-2 (visible bands) [2]	SpIRIT
Type of Sensor	MSI Push Broom	CMOS (IMX219)
Ground Sample Distance	10 m/pixel	200 m/pixel
Swath Width	290 km	541 km
Orbit Altitude	786 km	450 km
Post-processing	Level 2A Product	Raw Images

satellite moves over the Earth's surface, resulting in high-quality, low-distortion images with high spatial resolution. The CMOS sensors on our partner's and our nano-satellites capture the entire frame simultaneously, which can reduce distortion and allow for high-speed imaging but might introduce rolling shutter artifacts in fast-moving scenes or when the platform is moving such as on a nano-satellite. Additionally, the coverage widths differ: Sentinel-2 captures a 290 km swath in a single pass, while SpIRIT covers approximately 541 km. Sentinel-2 images are post-processed Level 2A product [3], which includes radiometric corrections and cartographic geometry, while SpIRIT images are raw and unprocessed. This comparison clearly shows that *AdaptTest Set* and *Sentinel-2 Set* are vastly different. *AdaptTest Set* is likely to be similar to the images that will be acquired by SpIRIT.

It is evident that SpIRIT uses simpler, less expensive hardware compared to Sentinel-2, which is not a CubeSat and can carry more sophisticated equipment. The primary mission of Sentinel-2 is high-quality imagery, while image acquisition and AI processing are supplementary functions for SpIRIT.

Pre-Trained Model *(Model0)*: *Model0* is our pre-trained cloud detection model, trained using the *AIEarth Pipeline* on a curated dataset from Sentinel-2 [3] and fine-tuned on AdaptTest.

For pre-training, we developed a cloud detection CNN model based on the MobileNetV3Small architecture [9], optimized for edge devices with limited resources. This model, with 22 layers and 2.5 million parameters, was trained both from scratch and using ImageNet weights, with the latter performing better. We froze all but the last 8 layers of the MobileNet initialized with ImageNet weights. We selected MobileNetV3Small for its optimized architecture, designed specifically for edge devices with limited computational resources, making it ideal for onboard nano-satellite processing.

We reserved 30% of the *Sentinel-2 Set* as a test set and used the rest for training. Using 5-fold cross-validation on the training set, each fold ran for 25 epochs with a batch size of 32, and we evaluated the model on the test set at the end of each fold. The test metrics at K=5 (end of the final fold) are presented in Table 3 (left column).

Table 3. Performance metrics of the pre-trained model on the Sentinel 2 test set and both the pre-trained and fine-tuned models on the AdaptTest whole and test sets, respectively.

Metric	Pre-trained model (Sentinel 2 Test Set)	Pre-trained model (AdaptTest Whole Set)	Fine-tuned model (AdaptTest Test Set)
Accuracy	94.00%	50.20%	84.50%
Precision	96.60%	55.60%	83.00%
Recall	95.30%	42.20%	91.70%
F1Score	95.90%	48.00%	87.10%
AUROC	98.20%	46.40%	95.30%
AveragePrecision	99.40%	57.60%	97.00%

Fine-Tuning Addresses Domain Shifts: Using the replica payload, we tested our pre-trained model on all AdaptTest images, as shown in Table 3 (middle column). As expected, the performance was poor due to significant differences between the AdaptTest images and the Sentinel-2 images used for pre-training. While an exact replication of the domain shift expected on SpIRIT is impossible, this offers an approximation.

To simulate fine-tuning, we set aside 30% of AdaptTest images (84 images) as a test set, with 30 labeled as cloudy and 54 as clear. The remaining images were used for fine-tuning. We fine-tuned the pre-trained model using 5-fold cross-validation and evaluated it on the test set at the end of each fold. Table 3 (right column) shows the improved metrics, demonstrating successful adaptation despite the domain shift.

Fine-tuning is a well-established method for addressing domain shift. However in space-related applications it is mostly done offline on Earth using downlinked data. Our project will perform this process in-orbit, which is novel. This experiment suggests it should work on SpIRIT, as the AdaptTest images are likely to be similar to those expected to be acquired. Although we cannot be certain of the exact domain shift, this provides a good direction. The fine-tuned model, *Model0*, was deployed on SpIRIT prior to launch.

4 Conclusion

In this paper, we detailed the design of an AI system recently deployed on SpIRIT, highlighting key design areas for building robust and adaptable AI systems for space. We emphasized addressing the data variability and domain gap between Earth and space data when training AI models. Our proposed solution includes onboard fine-tuning and a ground truth factory for sourcing labels from ground-based sources. Adaptable AI models are crucial for space applications, and future validation of our framework is anticipated from in-orbit operations during the main mission of the SpIRIT hosting AISee.

Acknowledgments. This research was supported by funding from the Office of National Intelligence under the National Intelligence and Security Discovery Research Grants program (Project NI220100072)

References

1. del Castillo, M.O., et al.: Mitigating challenges of the space environment for onboard artificial intelligence: Design overview of the imaging payload on SpIRIT. In: Proceedings of the IEEE/CVF Conference on Computer Vision and Pattern Recognition (2024)
2. ESA: Sentinel-2 user handbook (2015). https://sentinel.esa.int/documents/247904/685211/sentinel-2_user_handbook
3. European Space Agency: Level-2A algorithm overview - sentinel-2 MSI technical guide (2023). https://sentinels.copernicus.eu/web/sentinel/technical-guides/sentinel-2-msi/level-2a/algorithm-overview
4. Fanizza, V., Rijlaarsdam, D., González, P.T.T., Espinosa-Aranda, J.L.: Transfer learning for on-orbit ship segmentation. In: Computer Vision ECCV 2022 Workshops: Tel Aviv, Israel, October 23-27, 2022, Proceedings, Part I, pp. 21–36. Springer, Berlin, Heidelberg (2023). https://doi.org/10.1007/978-3-031-25056-9_2
5. Furano, G., et al.: Towards the use of artificial intelligence on the edge in space systems: challenges and opportunities. IEEE Aerosp. Electron. Syst. Mag. **35**, 44–56 (2020)
6. Ghiglino, P., Harshe, M.: A low power and high performance software approach to artificial intelligence onboard. In: 2023 IEEE Space Computing Conference (SCC), pp. 63–70 (2023).https://doi.org/10.1109/SCC57168.2023.00019
7. Ghiglione, M., Serra, V.: Opportunities and challenges of ai on satellite processing units. In: Proceedings of the 19th ACM International Conference on Computing Frontiers, pp. 221–224. CF '22, Association for Computing Machinery, New York, NY, USA (2022). https://doi.org/10.1145/3528416.3530985
8. Giuffrida, G., et al.: The ϕ-sat-1 mission: the first onboard deep neural network demonstrator for satellite earth observation. IEEE Trans. Geosci. Remote Sens. **60**, 1–14 (2022). https://doi.org/10.1109/TGRS.2021.3125567
9. Howard, A., et al.: Searching for MobileNetV3 (2019)
10. Labrèche, G., et al.: OPS-SAT spacecraft autonomy with TensorFlow lite, unsupervised learning, and online machine learning. In: 2022 IEEE Aerospace Conference (AERO), pp. 1–17 (2022). https://doi.org/10.1109/AERO53065.2022.9843402
11. Lesser, V., Peercy, C., Siva, V., Sullivan, C.: Automatic detection of auroral substorms from a CubeSat platform using machine learning. In: AIAA SciTech 2020 Forum (2020)
12. Manning, J., et al.: Machine learning space applications on SmallSat platforms with TensorFlow. In: Proceedings of the Small Satellite Conference (2018). https://digitalcommons.usu.edu/smallsat/2018/all2018/458/
13. Mateo-García, G., Laparra, V., López-Puigdollers, D., Gómez-Chova, L.: Transferring deep learning models for cloud detection between Landsat-8 and Proba-v. ISPRS J. Photogramm. Remote. Sens. **160**, 1–17 (2020). https://doi.org/10.1016/j.isprsjprs.2019.11.024

14. Rijlaarsdam, D., et al.: Autonomous operational scheduling on CogniSAT-6 based on onboard artificial intelligence. In: Proceedings of the 16th Symposium on Advanced Space Technologies in Robotics and Automation (ASTRA 2023), European Space Agency (ESA) (2023)
15. Russo, A., Lax, G.: Using artificial intelligence for space challenges: a survey. Appl. Sci. **12**(10) (2022). https://doi.org/10.3390/app12105106
16. Rika, V., et al.: Fast model inference and training onboard of satellites. In: IGARSS 2023 - 2023 IEEE International Geoscience and Remote Sensing Symposium, pp. 2002–2005 (2023). https://doi.org/10.1109/IGARSS52108.2023.10282715
17. Salazar, C., et al.: Cloud detection autonomous system based on machine learning and cots components onboard small satellites. Remote Sens. **14**(21), 5597 (2022). https://doi.org/10.3390/rs14215597
18. Trenti, M., et al.: SpIRIT mission: in-orbit results and technology demonstrations. arXiv preprint arXiv:2407.14034 (2024)
19. Zhang, Z., Xu, G., Song, J.: CubeSat cloud detection based on JPEG2000 compression and deep learning. Adv. Mech. Eng. **10**(10), 1687814018808178 (2018). https://doi.org/10.1177/1687814018808178
20. Zhou, H., Shyam, R.B.A., Rathinam, A., Yang, G.: Intelligent spacecraft visual GNC architecture with the state-of-the-art AI components for on-orbit manipulation. Front. Rob. AI **8** (2021). https://doi.org/10.3389/frobt.2021.639327

LSTM Autoencoder-Based Deep Neural Networks for Barley Genotype-to-Phenotype Prediction

Guanjin Wang[1(✉)], Junyu Xuan[2], Penghao Wang[3], Chengdao Li[4,5,6], and Jie Lu[2]

[1] School of Information Technology, Murdoch University, Perth, Australia
guanjin.wang@murdoch.edu.au
[2] Australia Artificial Intelligence Institute, Faculty of Engineering and Information Technology, University of Technology Sydney, Sydney, Australia
[3] School of Medical, Molecular and Forensic Sciences, Murdoch University, Perth, Australia
[4] Western Crop Genetics Alliance, Murdoch University, Perth, Australia
[5] Western Australian State Agricultural Biotechnology Centre, Perth, Australia
[6] Department of Primary Industries and Regional Development, Perth, Australia

Abstract. Artificial Intelligence (AI) has emerged as a key driver of precision agriculture, facilitating enhanced crop productivity, optimized resource management, and sustainable farming practices. Also, the expansion of genome sequencing technology has greatly increased crop genomic resources, offering deeper insights into genetic variation and enhancing desirable crop traits for better performance across various environments. Machine learning (ML) and deep learning (DL) algorithms are gaining traction for genotype-to-phenotype prediction, due to their excellence in capturing complex interactions within large, high-dimensional datasets. In this work, we present a new LSTM autoencoder-based model for barley genotype-to-phenotype prediction, specifically targeting flowering time and grain yield estimation. Our model outperformed the other baseline methods, highlighting its effectiveness in handling complex, high-dimensional agricultural datasets and enhancing the accuracy of crop phenotype prediction predictions. This approach has the potential to optimize crop yields and improve management practices.

Keywords: deep learning · barley phenotyping · phenotype prediction

1 Introduction

AI, particularly ML and DL, has been widely applied across various industries in recent years [1,2]. In particular, it has become a key enabler of precision agriculture, also known as smart farming, which has transformed modern farming practices [3–5]. As the global population grows and climate change intensifies, the need for sustainable farming practices has become increasingly critical [6].

M. Gong et al. (Eds.): AI 2024, LNAI 15442, pp. 342–353, 2025.
https://doi.org/10.1007/978-981-96-0348-0_25

AI techniques have demonstrated significant potential in enhancing agricultural productivity and efficiency, optimizing resource use, and making farming operations more sustainable and profitable. These techniques also enable farmers and other stakeholders to make more informed decisions [6,7]. One of the significant applications of precision agriculture is predicting crop phenotypes from genotypes. Thanks to the expansion of genome sequencing technology, crop genomic resources have greatly increased, deepening our understanding of genetic variation and enhancing desirable plant traits to optimize performance in various environments [8].

In this work, we focus on barley (*Hordeum vulgare* L.), a crucial cereal crop both globally and nationally. Barley is cultivated in highly productive agricultural areas as well as in marginal environments subject to adverse conditions [9]. Known for its resilience compared to other cereals like wheat and rice, barley can adapt to various biotic and abiotic stresses, making it essential for maintaining and increasing production in marginal areas to ensure food security [10]. In Western Australia, for example, barley ranks as the second-largest cereal crop, contributing 25% of the state's total grain production and generating over $1 billion annually in export earnings from barley grain and malt. About 30% of this barley is classified as malting grade for the international beer industry, while the remaining 70% is feed grade, primarily exported to the Middle East [11].

To maximize yield and minimize exposure to environmental stresses such as frost, heat, and drought during the growing season, it is crucial for barley to flower within a specific time window [12]. Also genes that control flowering time, often overlap with those related to grain yield [13]. Understanding the genetic data and their association with flowering time prediction and grain yield is vital for advancing barley improvement to meet future food and feed demands, enhance crop quality, and optimize management practices, including pest and disease control and harvesting schedules. Many previous studies in this area have utilized traditional statistical methods, but recent years have seen a growing interest in ML and DL algorithms for genotype-to-phenotype prediction due to their advanced learning capabilities [8]. These algorithms excel at capturing complex, higher-order interactions and achieving higher predictability with high-dimensional datasets, making them highly effective at linking plant genotypes with phenotypes [8,14]. Some existing studies has demonstrated the success of ML and DL models such as ensemble learning methods, kernel-based methods, and deep neural networks in predicting a wide range of agronomic traits by capturing the intricate interactions between genotype, phenotype, and the environment [8,15–17], showcasing their significant potential. In this work, we propose a new Long Short-Term Memory (LSTM) autoencoder-based deep nueral network model for crop genotype-to-phenotype prediction to enhance predictive performance on the complex, high-dimensional datasets, with a specific application in predicting the barley flowering time and grain yield. We use a real barley dataset that includes multi-environment field trials conducted over five diverse geographical locations across two years in Western Australia, encompassing high-dimensional genomic, phenotypic, and environmental information.

The remainder of this paper is organized as follows: Sect. 2 provides a brief review of crop genotype-to-phenotype prediction using AI and relevant techniques. Section 3 details the methodologies of our proposed model. Section 4 discusses the adopted dataset, experimental setup, and results. Finally, Sect. 5 concludes the paper.

2 Previous Work

This section first briefly introduces different crop genotype-to-phenotype modelling methods. Then, it reviews the LSTM model, which is used as a main component of our proposed model.

2.1 Crop Genotype-to-Phenotype Prediction

Understanding plant genotype-to-phenotype relationships is crucial for improving crop performance and resilience, food security and sustainability. Linear modeling approaches such as Genomic Best Linear Unbiased Prediction (GBLUP) [18] and Bayesian systems [19] have traditionally been used in genomic selection and genotype-to-phenotype prediction. However, these methods may face performance limitations due to the high dimensionality of marker data and the complex patterns within. There is growing interest in utilizing ML and DL techniques to predict plant phenotypes, as these methods can capture nuanced relationships among variables and efficiently handle large datasets, leading to improved predictive accuracy. For instance, Ma et al. [20] demonstrated the effectiveness of Convolutional Neural Networks (CNNs) in extracting informative genomic features, thereby improving selection accuracy in plant breeding programs. Kick et al. [21] examined optimized Deep Neural Network (DNN) models, which produced more consistent maize yield estimates despite having a slightly higher average error than the best BLUP model. These results show the DNN's promise for complementing existing models in crop selection and improvement. Wu et al. [22] investigated the use of Transformer-based DNNs for genomic prediction, introducing a new model named GPformer. GPformer integrates information from all relevant SNPs, irrespective of their physical distance, to achieve a holistic understanding. Extensive experiments across five diverse crop datasets demonstrated that GPformer consistently outperformed traditional methods such as ridge regression-based linear unbiased prediction (RR-BLUP), support vector regression (SVR), light gradient boosting machine (LightGBM), and deep neural network genomic prediction (DNNGP) in terms of reducing mean absolute error. Kkut et al. [23] reviewed major DL approaches, including fully connected DNNs, Recurrent Neural Networks (RNNs), CNNs, and Long Short-Term Memory (LSTM) networks, as well as various variations of these architectures for complex trait genomic prediction. However, the use of DL architectures such as RNNs still remains largely unexplored in genotype-to-phenotype predictions [8], despite presenting a potential alternative to traditional statistical methods [24,25].

2.2 LSTM

The LSTM network, introduced by Hochreiter and Schmidhuber [26], is a type of RNN model [27,28] that specifically addresses the vanishing gradient problem found in traditional RNNs. LSTM networks are designed with special units called memory cells that can maintain information from multiple previous layers and pass them through the network as needed, allowing them to effectively find and utilize relationships and patterns within the data. Unlike standard RNNs, LSTM networks have a hidden layer with additional units to manage the flow of information to and from the memory cells. These units include the input unit, which determines what information should be added to the memory cell based on its high activation levels; the forget unit, which clears the memory cell when its activation is high, effectively 'forgetting' unnecessary information; and the output unit, which transfers information from the memory cell to the next neuron if it has high activation. The mathematical functions governing these units' operations are formulated as follows [29]:

$$f_t = \sigma(W_{xf} \cdot x_t + W_{hf} \cdot h_{t-1} + b_f) \tag{1}$$

$$i_t = \sigma(W_{xi} \cdot x_t + W_{hi} \cdot h_{t-1} + b_i) \tag{2}$$

$$o_t = \sigma(W_{xo} \cdot x_t + W_{ho} \cdot h_{t-1} + b_o) \tag{3}$$

$$c'_t = \tanh(W_{xg} \cdot x_t + W_{hg} \cdot h_{t-1} + b_g) \tag{4}$$

$$c_t = f_t \odot c_{t-1} + i_t \odot c'_t \tag{5}$$

$$h_t = o_t \odot \tanh(c_t) \tag{6}$$

where x_t represents the input at time step t, W are the weight parameter matrices, and b are the bias vectors. Equation (5) denotes c_t, the cell state at time step t, while Eq. (6) denotes h_t, the hidden state at time step t. Here, $\cdot$ indicates standard matrix multiplication, $\odot$ represents the elementwise product, and σ is the Sigmoid function. The weights and biases remain consistent across all time steps. Equations (1)-(3) describe the three gates: the input gate i, forget gate f, and output gate o. These gates control the flow of information within the cell by generating values between 0 and 1 to write to the internal memory c_t, reset the memory, or read from the memory, respectively.

3 Methodology

In this section, we propose a LSTM autoencoder-based DNN architecture for genotype-to-phenotype prediction, as illustrated in Fig. 1. It is composed of two main components: one for genomic data encoding and the other for genotype-to-phenotype prediction. Next, we are going to explain these two components in detail.

3.1 Genomic Data Encoding

Crop genomic data encodes critical traits such as disease resistance, drought tolerance, and yield potential, making it essential for genotype-to-phenotype

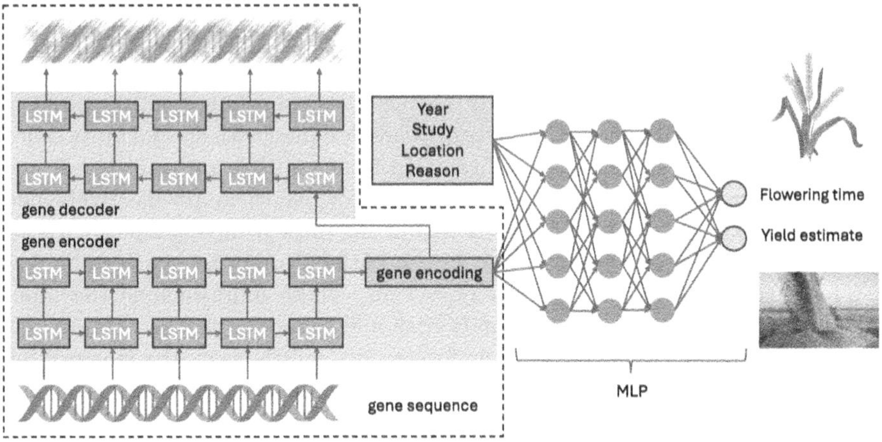

Fig. 1. Our LSTM autoencoder-based deep neural network framework

prediction. Our prior experiments indicated that naively using genomic data directly as features for ML models is insufficient. In this work, we propose using an LSTM as the encoder to learn the hidden representation of high-dimensional, large-sized genomic data for each crop variant, thereby reducing the reliance on traditional feature engineering and processing. We follow the classical setting[1] to stack two layers of LSTM rather than one to obtain enhanced information abstraction and increase the capability to capture more complex hidden feature representation information.

To further improve the performance, we propose to pretrain the LSTM using an autoencoder structure [30] before the phenotype prediction. In particular, a corresponding LSTM decoder is constructed to decode the gene from its latent representation, as shown in Fig. 1, using the following loss function

$$\mathcal{L} = \ell(x_{\text{gene}}, f^{\text{dec}}(f^{\text{enc}}(x_{\text{gene}}))) \tag{7}$$

where x_{gene} represents the genomic data, f^{enc} and f^{dec} denote the LSTM encoder and decoder. The advantage of this pretraining is that it does not need the labels (e.g., flowering time and grain yield) so we can use the large-scale genomic data to obtain a reasonably good encoder before the training of the predictor. Furthermore, since x_{gene} is high-dimensional, the gradient vanishing problem may still happen to LSTM even though it is much better than other RNNs. We propose to segment the original high-dimensional input data into several same-size frames,

$$x_{\text{gene}} = \left[x_{\text{gene}}^{(0)}, x_{\text{gene}}^{(1)}, x_{\text{gene}}^{(2)}, \ldots, x_{\text{gene}}^{(L)}\right] \tag{8}$$

then the encoding of these frames would be trained using the LSTM autoencoder, and the aggregation of frame encodings is finally used as the encoding of the

[1] https://github.com/fabiozappo/LSTM-Autoencoder-Time-Series.

gene as

$$z_{\text{gene}} = f^{\text{enc}}(x^{(0)}_{\text{gene}}) \oplus f^{\text{enc}}(x^{(1)}_{\text{gene}}) \oplus \cdots \oplus f^{\text{enc}}(x^{(L)}_{\text{gene}}). \tag{9}$$

3.2 Genotype-To-Phenotype Prediction

With the genomic data encoding from the above component, we propose to learn the unknown relationship between genomic data encoding, additional external variables (i.e., Year, Study, Location, and Season), and phenotypes via a DNN. We use the DNN architecture in this component for two main reasons: 1) as guaranteed by the universal approximation theorem [31], a deep neural network with a sufficient number of neurons and the appropriate nonlinear activation function can approximate any continuous function; and 2) there is no explicit knowledge about the targeted relationship, making a DNN a less-biased choice. We tested both multilayer perceptron (MLP) and CNN architectures, and found that the MLP outperformed the CNN (results can be found in the following section).

4 Experiments

4.1 Barley Dataset

We adopted a barley genotype data provided by the Western Crop Genetics Alliance (WCGA) at Murdoch University for modelling the barley genotype-to-phenotype prediction. A total of 894 barley accessions were genotyped using Next-Generation Sequencing. After filtering for heterozygosity, a mapping quality of 20, and a minor allele frequency (MAF) of 0.01, we obtained 30,543 high-quality single nucleotide polymorphism (SNP) markers, which were used as the genotype data. The average density of these genetic markers is approximately 150 kb.

Additionally, we included various environmental variables to account for the growing conditions. These environmental variables encompass location data (five different geographical sites across Western Australia), temporal data (years 2015 and 2016), light conditions (an extended light exposure trial conducted in 2016 at the South Perth site under 18 h of artificial lighting versus natural light), and agricultural practices (an irrigation trial at Merredin comparing irrigated and non-irrigated conditions). The target phenotype variables include 'ZS49' and 'GrYld(kg/ha)'. Table 1 describes the external environmental variables recorded, which are used along with the genotype and phenotype data to build predictive models for barley performance.

Necessary preprocessing steps were performed based on the data types and specific data problems encountered. The genotype data includes 30,543 SNP markers for distinct barley varieties, with alleles 'A', 'C', 'G', and 'T' encoded numerically, and missing or unavailable alleles assigned a value of -1. Missing data in environmental variables were assigned a value of -1. The resulting processed dataset consists of 4,203 records and 30,554 variables, encompassing genotype, environmental, and phenotype information.

Table 1. Description of external environmental variables and genotype variables in the adopted dataset

Variables	Descriptions
Year	The year in which data was recorded, ranging from 2014 to 2016
Study	Various conditions to which the crop was exposed, including 18 hrs of artificial lighting, natural lighting, 2 hrs of irrigation, no irrigation, and 1 hr of natural light
Location	The Western Australia location where the experiments were conducted: Geraldton, Merredin, South Perth, Katanning, and Esperance
ZS49	The flowering time of barley, measured in days, a crucial parameter for understanding and predicting crop performance.
GrYld(kg/ha)	Grain yield measured in kilograms per hectare to quantify the amount of grain produced by a crop per unit area of land.

4.2 Experiment Settings

For the LSTM autoencoder, we used two LSTM layers. The lower layer has an input dimension of 1 and a hidden dimension of 128, while the higher layer has an input dimension of 128 and a hidden dimension of 64. The pretraining epoch number is set to 500. The default dimension segment is 100, with different values tested and shown in Fig. 4. The default genomic data embedding dimension is 10, with results for different values shown in Fig. 3. For the MLP, the network width is set to 1,000 with a default depth of 3. Other values are also tested, and the results are shown in Fig. 2. The activation function used is ReLU. Batch normalization and dropout (with a probability of 0.2) are added between two linear layers. Note that the results in Figs 2, 3, and 4 are all from the MLP with the LSTM encoder (without pretraining). The epoch number for model training is 1,000. In the experiments, the dataset is randomly split into training (80%) and testing (20%) sets. The training set is used for model training, and the testing set is used for performance evaluation. We report the average testing results and standard deviations from five runs with five random seeds (123, 124, 125, 126, and 127).

5 Results

We evaluated our model's performance (termed as preLSTMMLP) on a processed barley dataset using two prediction tasks: predicting barley flowering time ('ZS49') and grain yield ('GrYld (kg/ha)'). The average Mean Absolute Error (MAE), Root Mean Squared Error (RMSE), and standard deviation results on the testing set are presented in Tables 2 and 3, respectively. Our model achieved the lowest MAE and RMSE in both prediction tasks, with an MAE of 7.55 and RMSE of 10.70 for predicting flowering time, and an MAE of 647.36 and RMSE of 843.24 for predicting grain yield. We compared our model against several baseline models, including XGBoost (eXtreme Gradient Boosting)[2], which is a

[2] https://github.com/dmlc/XGBoost.

widely used ensemble learning method for its high performance and scalability to large datasets [8]. XGBoost trained on the full processed dataset resulted in higher MAE and RMSE for both tasks compared to our model.

We also tested an MLP neural network trained directly on the same full processed dataset and without genomic data (termed as MLP(w/o gene)). The results indicated that including genomic data directly in the MLP without the genomic data encoding component led to decreased performance compared to our model in both tasks. Additionally, MLP(w/o gene) performed better than MLP trained on the full processed dataset. This decrease in performance is likely due to the curse of dimensionality, as the genomic data has over $3 * 10^4$ dimensions. We also tested our model without pretraining (termed as LSTMMLP) trained on the same full processed dataset. This model performed better than MLP, but not as well as our model with pretraining. Finally, after adding the autoencoder pretraining mechanism, our model's performance significantly improved for both ZS49 and GrYld in terms of MAE and RMSE, demonstrating that the pretraining in the genomic data encoding is crucial as expected. Additionally, we tested a CNN on the same dataset, with results shown in Table 4. The CNN's performance was inferior to that of the MLP in Table 3. This may be because MLPs tend to perform well on relatively small datasets, whereas CNNs might overfit due to their specialized architecture and thus require more data to learn meaningful features.

Table 2. Predictive results (MAE) using MLP in predicting flowering time ('*ZS49*') and grain yield ('*GrYld*')

Outputs	XGBoost	MLP(w/o gene)	MLP	LSTMMLP	preLSTMMLP
ZS49	16.30 ± 0.5	8.10 ± 0.1	11.57 ± 0.2	8.05 ± 0.2	**7.55** ± **0.3**
GrYld	727.98 ± 20.7	694.60 ± 6.9	733.74 ± 18.4	710.61 ± 24.9	**647.36** ± **8.0**

Table 3. Average predictive results (RMSE) using MLP in predicting flowering time ('*ZS49*') and grain yield ('*GrYld*')

Outputs	XGBoost	MLP(w/o gene)	MLP	LSTMMLP	preLSTMMLP
ZS49	23.61 ± 0.9	11.32 ± 0.3	15.13 ± 0.3	11.30 ± 0.3	**10.70** ± **0.4**
GrYld	954.84 ± 30.7	904.02 ± 14.5	941.79 ± 30.9	911.12 ± 45.5	**843.24** ± **8.9**

We further illustrate the impact of parameters including MLP depth, gene embedding dimension, and dimension segment length on prediction performance in Figs. 2, 3 and 4, respectively. Our findings indicate that an MLP depth of 4 achieved the lowest MAE for both prediction tasks, striking a balance between model complexity and predictive performance. For the gene embedding dimension, a value of 15 yielded the lowest MAE for both outcomes. However, increasing the dimension to 20 significantly increased the MAE for both tasks, while

Table 4. Average predictive results (RMSE) using CNN in predicting flowering time ('*ZS49*') and grain yield ('*GrYld*')

Outputs	CNN(w/o gene)	CNN	LSTMCNN	preLSTMCNN
ZS49	17.73 ± 0.8	18.98 ± 2.9	16.56 ± 0.7	17.13 ± 0.9
GrYld	974.50 ± 14.0	1071.81 ± 57.0	955.39 ± 12.9	1020.15 ± 106.3

reducing it to 10 also led to higher MAE values. This suggests a sensitive selection is necessary for this parameter to optimize model performance. In terms of dimension segment length, a length of 500 resulted in the lowest MAE for both outcomes, compared to shorter lengths ranging from 10 to 100. This indicates that the LSTM encoder architecture more effectively captures relationships within the higher dimension data.

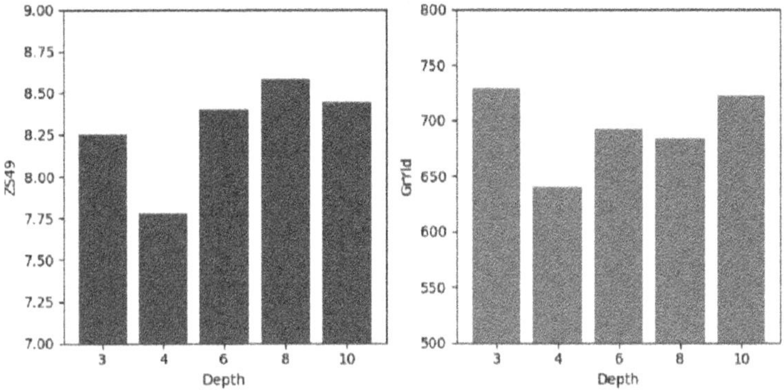

Fig. 2. Impact of MLP Depth on predictive results (MAE): Left - ZS49, Right - GrYld

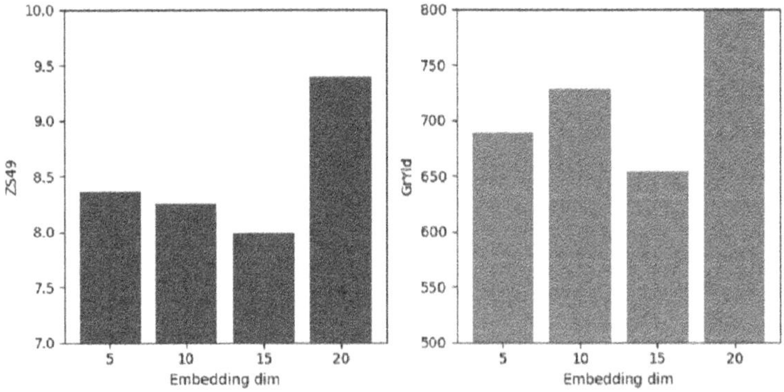

Fig. 3. The impact of gene embedding dimension on predictive results (MAE): Left - ZS49, Right - GrYld

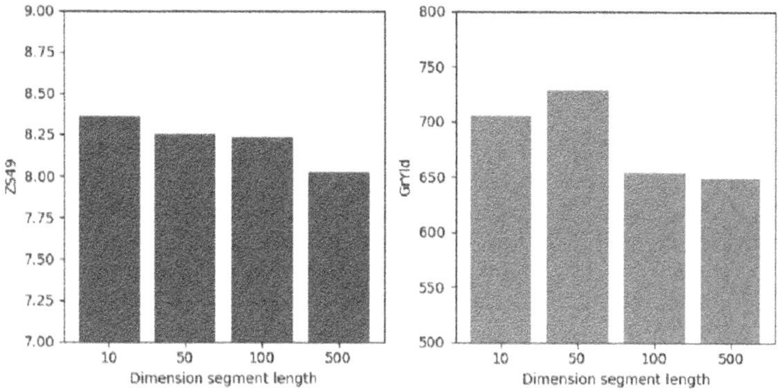

Fig. 4. The impact of dimension segment length on predictive results (MAE): Left - ZS49, Right - GrYld

6 Conclusion

We proposed a new LSTM autoencoder-based DNN model for crop genotype-to-phenotype prediction, and applied it to predicting barley genotype to flowering time and grain yield with improved performance. Specifically, we introduced genomic data encoding by pretraining the LSTM using an autoencoder structure before phenotype prediction to extract latent feature representations from the complex high-dimensional genomic data. Our model achieved the lowest MAE and RMSE in both prediction tasks compared to other baseline models, demonstrating its potential to enhance predictive power in handling complex datasets encompassing genotype, phenotype, and environmental data in the agriculture context. In the future, we plan to include time series environmental variables such as soil temperature and rainfall based on locations to further enhance the model's prediction performance. We will also expand the comprehensive testing of our models and compare them with other state-of-the-art methods on various crop types.

Acknowledgements. The work was supported by the MU School of IT Emerging Researcher Grant and the 2024 UTS FEIT Blue Sky Research Grant.

References

1. Eli-Chukwu, N.C.: Applications of artificial intelligence in agriculture: a review. Eng. Technol. Appl. Sci. Res. **9**(4) (2019)
2. Kwok, W.H., Zhang, Y., Wang, G.: Artificial intelligence in perinatal mental health research: a scoping review. Comput. Biol. Med. **177**, 108685 (2024)
3. Sharma, A., Jain, A., Gupta, P., Chowdary, V.: Machine learning applications for precision agriculture: a comprehensive review. IEEE Access **9**, 4843–4873 (2020)
4. Bhat, S.A., Huang, N.F.: Big data and AI revolution in precision agriculture: survey and challenges. IEEE Access **9**, 110209–110222 (2021)

5. Adewusi, A.O., Asuzu, O.F., Olorunsogo, T., Iwuanyanwu, C., Adaga, E., Daraojimba, D.O.: AI in precision agriculture: a review of technologies for sustainable farming practices. World J. Adv. Res. Rev. **21**(1), 2276–2285 (2024)
6. Mana, A.A., Allouhi, A., Hamrani, A., Rahman, S., el Jamaoui, I., Jayachandran, K.: Sustainable AI-based production agriculture: exploring AI applications and implications in agricultural practices. Smart Agric. Technol. **7**, 100416 (2024)
7. Pandey, D.K., Mishra, R.: Towards sustainable agriculture: harnessing AI for global food security. Artif. Intell. Agric. **12**, 72–84 (2024)
8. Danilevicz, M.F., et al.: Plant genotype to phenotype prediction using machine learning. Front. Genet. **13**, 822173 (2022)
9. Baum, M., et al.: Molecular approaches and breeding strategies for drought tolerance in Barley. In: Genomics-Assisted Crop Improvement: Vol 2: Genomics Applications in Crops, pp. 51–79 (2007)
10. Tester, M., Langridge, P.: Breeding technologies to increase crop production in a changing world. Science **327**(5967), 818–822 (2010)
11. Western Australian Department of Primary Industries and Regional Development: Barley. https://www.agric.wa.gov.au/crops/grains/barley. Accessed 30 June 2024
12. Maurer, A., et al.: Modelling the genetic architecture of flowering time control in barley through nested association mapping. BMC Genomics **16**, 1–12 (2015)
13. Hill, C.B., et al.: Hybridisation-based target enrichment of phenology genes to dissect the genetic basis of yield and adaptation in Barley. Plant Biotechnol. J. **17**(5), 932–944 (2019)
14. LeCun, Y., Bengio, Y., Hinton, G.: Deep learning. Nature **521**(7553), 436–444 (2015)
15. CrossaJ, J., et al.: Deep kernel and deep learning for genome-based prediction of single traits in multienvironment breeding trials. Front. Genet. **10**, 1168 (2019)
16. Grinberg, N.F., Orhobor, O.I., King, R.D.: An evaluation of machine-learning for predicting phenotype: studies in yeast, rice, and wheat. Mach. Learn. **109**(2), 251–277 (2020)
17. Khaki, S., Wang, L., Archontoulis, S.V.: A CNN-RNN framework for crop yield prediction. Front. Plant Sci. **10**, 1750 (2020)
18. Clark, S.A., van der Werf, J.: Genomic best linear unbiased prediction (gBLUP) for the estimation of genomic breeding values. In: Genome-Wide Association Studies and Genomic Prediction, pp. 321–330 (2013)
19. Tong, H., Nikoloski, Z.: Machine learning approaches for crop improvement: leveraging phenotypic and genotypic big data. J. Plant Physiol. **257**, 153354 (2021)
20. Ma, W., et al.: A deep convolutional neural network approach for predicting phenotypes from genotypes. Planta **248**, 1307–1318 (2018)
21. Kick, D.R., et al.: Yield prediction through integration of genetic, environment, and management data through deep learning. G3 Genes Genomes Genet. **13**(4), jkad006 (2023)
22. Wu, C., et al.: A transformer-based genomic prediction method fused with knowledge-guided module. Briefings Bioinf. **25**(1), bbad438 (2024)
23. Okut, H.: Deep learning algorithms for complex traits genomic prediction. Hayvan Bilimi ve Ürünleri Dergisi **4**(2), 225–239 (2021)
24. Muneeb, M., Henschel, A.: Eye-color and type-2 diabetes phenotype prediction from genotype data using deep learning methods. BMC Bioinf. **22**, 1–26 (2021)
25. Zhang, G., Zhang, Z., Kou, X., Chen, Y.: FF-LSTM: phenotype prediction based on feature fusion. In: Third International Conference on Electronic Information Engineering and Data Processing (EIEDP 2024), vol. 13184, pp. 1629–1635. SPIE (2024)

26. Hochreiter, S., Schmidhuber, J.: Long short-term memory. Neural Comput. **9**(8), 1735–1780 (1997)
27. Grossberg, S.: Recurrent neural networks. Scholarpedia **8**(2), 1888 (2013)
28. Weerakody, P.B., Wong, K.W., Wang, G., Ela, W.: A review of irregular time series data handling with gated recurrent neural networks. Neurocomputing **441**, 161–178 (2021)
29. Yu, Y., Si, X., Hu, C., Zhang, J.: A review of recurrent neural networks: LSTM cells and network architectures. Neural Comput. **31**(7), 1235–1270 (2019)
30. Srivastava, N., Mansimov, E., Salakhudinov, R.: Unsupervised learning of video representations using LSTMs. In: International Conference on Machine Learning, pp. 843–852. PMLR (2015)
31. Nishijima, T.: Universal approximation theorem for neural networks. arXiv preprint arXiv:2102.10993 (2021)

An Improved Prescriptive Tree-Based Model for Stochastic Parallel Machine Scheduling

Siping Chen[1(✉)], Debiao Li[2], Nasimul Noman[1], Kyle Harrison[1], and Raymond Chiong[1,3]

[1] University of Newcastle, University Drive, NSW, Australia
siping.chen@uon.edu.au
[2] Fuzhou University, University Town, Fuzhou, China
[3] University of New England, Elm Avenue, Armidale, Australia

Abstract. Machine scheduling serves as a vital function for industrial and service operations, and uncertainties always pose a significant challenge in real-world scheduling practices. In this paper, we propose to solve the stochastic machine scheduling problems with uncertain processing times by an improved prescriptive tree-based (IPTB) model. Our approach includes a novel way of combining historical processing time data with current scheduling constraints to strengthen the quality of historical decisions. We apply these improved historical decisions and incorporate an improved model for calculating the optimisation loss and accelerate the training of our IPTB model. Our trained model can directly prescribe downstream scheduling solutions with high robustness in the face of uncertainties. We evaluate the proposed IPTB method on a stochastic parallel machine scheduling problem originating from printed circuit board assembly lines. Through a series of comparative experiments, our findings demonstrate the IPTB method's superior accuracy and robustness, highlighting its resilience in noisy data environments. Additionally, we interpret the model through feature importance analysis and examine the model's behaviours under noisy conditions.

Keywords: Prescriptive analytics · Machine learning · Machine scheduling problem

1 Introduction

Machine scheduling is essential in both industrial and service sectors. An effective schedule facilitates optimal resource utilization, leading to enhanced efficiency and often resulting in significant cost savings. Uncertainties are key factors that impact machine scheduling problems, for instance, unpredictable task durations can significantly affect productivity goals [15]. Similarly, unexpected resource breakdowns might undermine the objectives of resource utilisation [12]. Without

M. Gong et al. (Eds.): AI 2024, LNAI 15442, pp. 354–365, 2025.
https://doi.org/10.1007/978-981-96-0348-0_26

strong scheduling methods, these uncertainties can disrupt scheduling goals and may affect the practicality of the schedules.

Traditional approaches, such as stochastic optimisation [19] and robust optimization techniques [21], are commonly used for handling the various uncertainties in machine scheduling. However, these methods have either distribution or uncertainty set assumptions for scheduling problems. In real-world applications, obtaining a high-quality distribution or uncertainty set is often unrealistic. Additionally, the effectiveness of the schedule solutions may be compromised if the model inaccurately captures the significance of the uncertainty. In contrast, when high-quality data is available, the 'predict, then optimise' (PO) framework emerges as an effective tool. It combines the predictive ability of machine learning (ML) models for forecasting with optimization techniques for decision-making. The PO approach has been successfully implemented in various scheduling contexts [1, 14], showing promising results in practical scenarios.

Despite its success, the PO solution can be prone to optimisation loss caused by prediction errors [13]. To improve decision quality, an effective approach is integrating prediction with optimisation. Bertsimas et al. [4] proposed a prescriptive ML method by directly 'prescribing' the solutions rather than simply providing the prediction results, thereby shifting the predictive decision-making process to a prescriptive one. In another research stream, Elmachtoub et al. [6] proposed a novel method called smart 'predict, then optimise' (SPO), which uses the true optimisation loss of operational research problems as the loss function for predictive model training. Both methods have proven asymptotic optimality for general decision-making problems under structural restrictions. However, in scheduling problems, this condition is usually conflicted, making the application of these integrations challenging. Additionally, they require finding the optimal solution for optimisation problems during training. This is a challenging task for scheduling problems due to their NP-hard nature. To the best of our knowledge, there's no research applying these methodologies with true optimisation loss to solve optimising problems with uncertainties, especially in a complex optimisation environment like machine scheduling.

In this paper, we solve a machine scheduling problem with uncertain processing time by introducing an improved prescriptive tree-based (IPTB) model. In contrast to existing ML models, we train our IPTB model using scheduling loss to improve its accuracy. The primary challenge for this problem is the significant computational cost, as it involves solving machine scheduling problems - typically NP-hard - numerous times during model training. To overcome this, we propose a novel methodology for constructing datasets, which remarkably improves the effectiveness of historical decision data. We also provide an approach to pre-calculate an objective value matrix for accelerating the training process. To scale up the training efficiently, we apply a surrogate decision to guarantee both the effectiveness and efficiency of the model. Our model performance is tested in a real-world machine scheduling scenario: a machine scheduling problem of printed circuit board assembly lines. Through a series of comparative experiments, our findings demonstrate the IPTB method's superior accuracy and robustness, highlighting its resilience in noisy data environments.

This paper is organised as follows. In Sect. 2, we review the related literature. In Sect. 3, we describe the machine scheduling problem with uncertainty and its methodologies. In Sect. 4, we present the experimental results and analysis. In Sect. 5, the conclusion and future works are presented.

2 Related Work

Multiple techniques have been studied in the literature to deal with uncertainty in the machine scheduling problem. A traditional way is to formulate scheduling problems with uncertainty as stochastic optimisation problems considering the probabilistic distribution of uncertainty. This method optimises the expectation of scheduling objectives to obtain its solution. Cai and Zhou proposed a weighted expected shortest processing time method for single machine scheduling problems with exponentially distributed processing time [5]. Soroush studied a stochastic single-machine scheduling problem with uncertain processing time and proposed a heuristic to generate near-optimal sequences [18]. In addition, there are multiple studies on stochastic machine scheduling problems, e.g., flow shop problems [9] and job shop problems [10].

The integration of ML models and optimisation problems is a recently-proposed data-driven method. With this approach, forecast tasks are selected or evaluated based on their contribution to the downstream cost of the optimisation problem, rather than relying on different measures of forecast quality, such as mean absolute error. Elmachtoub et al. [8] developed a specialised algorithm to build decision trees directly for the actual optimisation target. In a separate study, Elmachtoub and Grigas [6] introduced a differentiable surrogate of the optimisation loss function. Mandi et al. [13] further applied this research stream to more real-world operational research problems and proposed techniques to efficiently learn the model. Bertsimas et al. [4] proposed a prescriptive optimisation method that combines ideas from ML and operational research.

However, current research studies general operational research problems with strong assumptions on the problem structure, which could be invalid when applied to scheduling. All of these applications only considered uncertainties in the strictly linear optimisation objective, while in a more complex decision-making environment, the problem could be non-linear and the uncertainties usually affect the feasibility of the decision. Despite there being research on scheduling problems with SPO loss function [13], it is still a strongly linear problem. Besides, solution algorithms in the current landscape lean heavily towards exact solving techniques, which look for globally optimal solutions during model training [4,6]. Yet, real-world scheduling challenges, such as the machine scheduling problems studied in this paper, present intricate complexities and are usually NP-hard. This is further exacerbated when training ML models that require solving the problem multiple times. Further research should be conducted in this field.

3 Methodology

In this section, we introduce the proposed IPTB model for solving stochastic machine scheduling problems. We first formulate the stochastic machine scheduling problem in a data-driven manner. Then, our data is constructed with procedures including historical instance reformulation and data pre-processing. The proposed IPTB model is trained by a surrogate loss function and can be applied directly for prescribing scheduling solutions.

3.1 Data-Driven Formulation of Machine Scheduling Problem

To provide an intuitive start, we first present the formulation of stochastic machine scheduling problems. Consider a scenario with n_j jobs to be processed on n_m machines. In this problem, we focus on uncertain processing time y_j for each job, which originates from a multivariate distribution $\mathcal{Y}$. We formulated this stochastic problem as $\arg\min_{z \in \mathcal{L}} \mathbb{E}(\zeta(z, \mathcal{Y}))$. Here, $\mathcal{L}$ is the constraint space of the machine scheduling problem, encompassing all valid constraints for the specific machine scheduling problem, such as ensuring each job is processed only once on one machine; z is a schedule solution and the function $\zeta(z, \mathcal{Y})$ is the stochastic objective function for this problem, representing the goal of minimizing the expected value of the scheduling outcome.

However, obtaining the full probability distribution $\mathcal{Y}$ is unrealistic in practical applications. Therefore, instead of relying on the stochastic formulation and its distributional assumptions, we shift to a data-driven formulation of the problem. This approach leverages historical scheduling data, which includes observations of job-related features X (such as order quantity and factors related to job complexity) and historical processing time data Y (including historical values of y_j). In this context, we consider n_s scenarios in the historical data. More specifically, Y has $[\boldsymbol{y}_1, ..., \boldsymbol{y}_{n_s}]$ processing time arrays for each historical machine scheduling problem. Similarly, X has the form $X = [\boldsymbol{x}_1, ..., \boldsymbol{x}_{n_s}]$ with n_s historical observation data. Bertsimas [4] proposed a formulation to obtain the optimal decision given only the observation data:

$$z^P(\boldsymbol{x}) = \arg\min_{z \in \mathcal{L}} \sum_{i=1}^{n_s} w_i(\boldsymbol{x}) \mu(z, \boldsymbol{y}_i) \tag{1}$$

Here, the z^P is calculated by an weighted sample averaged approximation (SAA) procedure controlled by an ML model, which is used to estimate weight functions of all instances from historical scenarios, $[w_1(), ..., w_{n_s}()]$.

3.2 Data Construction

To begin, we construct a dataset specified for training the IPTB model. Suppose we have n_s historical scheduling solutions, i.e., n_s job processing time arrays $\boldsymbol{y}$. Rather than using these historical scheduling solutions directly, we reformulate new schedule instances by combining each historical processing time array $\boldsymbol{y}_i$

with the parameters of the current machine scheduling problem. For example, we assume the problem is a parallel machine scheduling problem with precedent constraints P, and its objective is minimizing the makespan. Then, instead of using historical scheduling solutions, which might not be a feasible solution under current precedent constraints $\mathcal{L}(P)$, we combine the historical processing time array $\boldsymbol{y}_i$ with $\mathcal{L}(P)$ to form new scheduling instances. We solve all these instances to gain new solutions, which are all feasible scheduling solutions to the current machine scheduling problem. We record the solutions of these instances as new solutions data $Z = z_1, ..., z_N$. Subsequently, each solution z_i from this new dataset Z is applied to other reformulated instances to calculate the objective function values $\mu(z_i, \boldsymbol{y_k})$, for all $k = 1 \dots n_s$, where $k \neq i$. We compile these calculated objective values into a matrix U. This matrix U is used later for calculating the optimisation loss in the context of Equation (1), thus allowing us to assess and refine our model's outcomes.

The calculation of U can be done by using the exact solver to obtain all optimal solutions. However, finding optimal solutions for machine scheduling problems, particularly those with large sizes, is challenging due to their NP-hard nature. To circumvent this, we employ weaker solutions - suboptimal solutions that are easier to compute but still reflect the basic pattern of the instance, provided the algorithm maintains a bounded ratio to the optimal solution. While it might seem that using weak oracles could reduce optimisation results, similar approaches are applied in a stochastic energy-aware scheduling problem during the training of an SPO model [13], and, according to their empirical results, the relaxed model has a similar performance to an optimal one.

3.3 Improved Prescriptive Tree-Based Model

As present in Equation (1), the major task is to train the weight function of each instance w_i. We use tree-based model in this paper. Its weight function w_i^{DT} has the form $w_i^{DT}(\boldsymbol{x}) = \frac{\mathcal{I}(R(x)=R(x_i))}{|\{j:R(x_j)=R(x)\}|}$, where $\mathcal{I}(.)$ denotes an indicator function that is equal to 1 if the expression within is true, and 0 otherwise; $|.|$ represents the number of elements of input sets; and R is the leaf-splitting function of the trained decision tree. In typical ML model training, the training target is minimizing the prediction loss of objective values. As discussed earlier, this training method has a major drawback in that it can be misled by two similar objective values. Instead of prediction loss, we use the optimisation loss from matrix U to guide the model training. We reformulate the true optimisation loss for the decision tree model [8] in a prescriptive manner, as shown here:

$$\min_{j,s} \frac{1}{N} \Big(\sum_{i \in R_{1(j,s)}} (\mu(z^P(Y_l), \boldsymbol{y_i}) - \mu_i^*) + \sum_{i \in R_{2(j,s)}} (\mu(z^P(Y_r), \boldsymbol{y_i}) - \mu_i^*) \Big) \tag{2}$$

where $R_1(j,s)$ and $R_2(j,s)$ denote the left and right split functions, which split the observation data on their jth feature by this feature's sth possible value; Y_l and Y_r are the within-leaf sample set obtained by $R_1(j,s)$ and $R_2(j,s)$, respectively.

This formulation can produce a robust decision by identifying appropriate SAA samples for each leaf node in the decision tree. However, training by this function requires countless computations of $z^{SAA}(Y_l)$, which can hardly be achieved due to the NP-hardness of the machine scheduling problem. Instead, we applied the pre-calculated value in U as a surrogate of the loss. We propose Equation (3) to obtain a decision z^{SP} from historical decisions:

$$z^{SP} = \arg\min_{z \in Z} \sum_{\boldsymbol{y}_i \in Y} \mu(z, \boldsymbol{y}_i) \tag{3}$$

Here, each of the values of $\mu(z, \boldsymbol{y}_i)$ has been pre-calculated in U. According to [4], this decision is asymptotically optimal with several ML models, and z^P empirically converges to the optimal when the observation number is large enough. For our surrogate solution, if its solution quality is bounded in a specific ratio, then it will have similar convergence performance with a sufficient amount of data. Consequently, we propose Algorithm 1 to train the IPTB model.

Algorithm 1. An IPTB training algorithm

Input: X, Y, Z, U, n_d and n_l
  Initialise decision tree, layer count $l = 1$
  while $l \leq n_d$ **do**
    for each split in layer $l - 1$ **do**
      for each input feature j in X **do**
        Obtain candidate splits set S by splitting function, $\hat{\boldsymbol{z}} = \{\}$
        for s in S **do**
          $\hat{z}_l = \arg\min_{z \in Z_l} \sum_{i \in Y_l} \mu(z, \boldsymbol{y}_i)$
          $\hat{z}_r = \arg\min_{z \in Z_r} \sum_{i \in Y_r} \mu(z, \boldsymbol{y}_i)$
          $\hat{\boldsymbol{z}}[j, s] = (\hat{z}_l, \hat{z}_r)$
        end for
        $s = \arg\min_{s \in S} L_{SP}(\hat{\boldsymbol{z}}[j, s])$
      end for
      Update layer l, $l = l + 1$.
    end for
  end while
Output decision tree model

Initially, we input the observation data X, uncertainty data Y, historical decisions Z, the maximum depth of the decision tree n_d, and IPTB depth n_l into the initialised decision tree model. During the training of each layer l, we applied the decision tree's splitting algorithm [2] to generate candidate splits. For each candidate split, decision z^{SP} was applied for both the left split decision $\hat{z}_l$ and the right split decision $\hat{z}_r$. The candidate split that minimises the L_{SP} was chosen and added to the decision tree, where L_{SP} is the Eq. 2 using z^{SP} as decision. In this stage, most data samples with similar scheduling patterns were retained within each split. Further, the model calculates the loss with a pre-calculated objective value matrix U, which avoids solving a significant number of SAA problems with a high number of scenarios. This can reduce the optimisation times during training and improve the algorithm's training speed.

After we gain the trained decision tree model, it can be applied to 'prescribe' a decision given any observation array $\boldsymbol{x}$. It first gets the leaf-splitting function according to its structure. Then, it calculates the weight value of each historical sample i given observation $\boldsymbol{x}$. Instead of using the surrogate decision used during the training, the decision Z^P are calculated according to Equation (1) and are outputted as the final decision. This final scheduling solution is an optimal SAA solution, thus ensuring model performance.

To enhance decision quality, we train an ensemble of trees using the random forest (RF) training method. RF is known to exhibit less variance compared with individual decision trees, and it thus improves model stability [17]. To construct an RF model, n_t tree models are trained on bootstrapped samples of the training dataset, where n_t represents the number of desired trees in the forest. According to the prescriptive RF model proposed by [4], the final decision is the weighted average of SAA samples from selected leaves in each tree. The RF model's performance is sensitive to the hyper-parameter settings of the number of decision trees n_t and n_d.

4 Experiment and Analysis

In the experiment, we consider a parallel machine scheduling problem with precedent constraints and try to minimise its makespan under uncertain processing times. This problem originated from a real-world machine scheduling problem of printed circuit board (PCB) assembly lines.

4.1 PCB Assembly Lines Scheduling Problem

In this problem, there are n_j PCB assembly jobs to be processed on n_m identical parallel surface mounting technology machines, and each job has an uncertain processing time y, which is identical on all the machines. The exact value of $\boldsymbol{y} = [y_1, ...y_{n_j}]$ will not be revealed during the decision-making process. However, there were three different data related to PCB assembly jobs prior to our decision-making: historical observation data X, historical uncertainty data Y, and current observation array $\boldsymbol{x}$.

The problem can be formulated as a parallel machine scheduling problem with precedent constraint and makespan minimisation objective. Each machine can only process one job at any time during the production. There is no preemption during the processing of each job, so each job should continue processing during its processing time period. Some PCB orders need to be processed twice on any machine of the production line: firstly on its top side and then on its bottom side. We assume the bottom-side job can start processing right after the top-side job is finished, therefore, the bottom-side job's starting time is larger or equal to the top-side's completion time. We assume the setup time of each job is included in the processing time based on the given observation data.

4.2 Data Generation

We first generated the processing time data by a multivariate distribution for PCB assembly lines [11]. There are four crucial features that impact the processing time data: simulated cycle time x_1, number of orders x_2, warm-up penalty x_3, and warm-up threshold x_4. Simulated cycle time (x_1) is the estimated production time of each piece of the PCB. This estimation is conducted by the surface mounting technology machine and it can be utilised as a measurement of PCB assembly job complexity. The number of orders (x_2) is the exact number of PCBs for this job. Ideally, these two features (x_1 and x_2) can determine the processing time of each job by simply multiplying them. However, in the real production scenario of PCB, the machine can not achieve the simulated speed x_1 at the beginning of production as it requires a 'warm-up' procedure, producing some pieces of PCB at a lower speed and accelerating to simulated speed after x_4 PCBs have been produced. We assume that in this 'warm-up' procedure, the processing time is fluctuating by a distribution related to x_3. To simulate this process, the cycle time CT satisfies a truncated normal distribution $CT = Trun(\mu, \sigma, \mu, +\inf)$, where the mean value $\mu = x_1$ and the standard deviation $\sigma = \frac{x_3 * x_1}{1+e^{x_4 - x_2}}$ (a sigmoid function to make σ higher if $x_2 < x_4$, i.e., the number of PCBs is not enough for warm-up). After having the real cycle time value, the processing time y is calculated by $y = CT \cdot x_2$.

For the PCB assembly scheduling problem, we set $n_j = 20$ and $n_m = 3$, which has been proven to be NP-hard when $n_m > 2$ without precedent constraint. Among all the jobs, 5 pairs of jobs have the precedent constraint, i.e., these paired jobs are the first and bottom side jobs. Since the processing time is uncertain during the decision-making process, we sort the jobs by a non-increasing sequence of $x_1 \cdot x_2$ as a prior knowledge of job complexity.

4.3 Comparison with Different Data-Driven Optimisation Methods

In this experiment, we compare 4 different data-driven optimisation methods to the proposed method for evaluating its performance on machine scheduling problems. The first method is the SAA method with all the historical samples applied to the decision-making. This method can be regarded as a model-free baseline for this optimisation problem. Then, we applied the PO method with an RF model as the predictor (PO-RF). By comparing PO to the proposed method, we could find out the improvement after applying prescriptive methodologies. We then used two different methods that included the basic elements of the proposed IPTB model for further comparison. We applied the prescriptive RF model as its weighting ML model to find historical data for SAA, denoted by PSAA-RF. We also made a comparison with the SPO-forest model proposed by [6], which has been applied to a cargo inspection problem [20]. We tested four different sizes

Table 1. Comparison among data-driven optimisation methods in terms of average regret

N	SAA	PO-RF	PSAA-RF	SPO-RF	IPTB-RF
400	20.60%	19.30%	19.73%	19.30%	**18.85%**
800	22.21%	6.65%	6.52%	8.80%	**5.23%**
1,600	22.54%	4.77%	4.88%	8.59%	**3.78%**
4,000	21.68%	**1.86%**	3.19%	8.79%	3.37%

of training data: $N \in \{400, 800, 1600, 4000\}$, and generated 10 testing instances $Y_{test} = [\boldsymbol{y}_1 ... \boldsymbol{y}_{10}]$. The final results are compared in terms of average regret calculated by the following equation: $regret = \sum_{\boldsymbol{y} \in Y_{test}} \frac{\mu(\hat{z}, \boldsymbol{y}) - \mu(z^*(\boldsymbol{y}), \boldsymbol{y})}{\mu(z^*(\boldsymbol{y}), \boldsymbol{y})}$ where $z^*(\boldsymbol{y})$ is the full-information optimal of testing sample $\boldsymbol{y}$, and $\hat{z}$ is the schedule solution from 5 compared methods.

Table 1 showcases the outcomes of data-driven optimisation models. At $N = 400$, the sample number is relatively small. All the methods have a similar performance to the model-free baseline SAA method, but IPTB-RF turns out to be the best. The proposed IPTB-RF keeps its top ranking on $N = 800$ and $N = 1600$. In these settings, the SPO-RF method did not exhibit continuous improvement as observed in other methods. As discussed earlier, SPO-RF's condition is conflicted with the machine scheduling problem since its uncertainty affects the constraint space. This causes diminished performance gain with sufficient data. Conversely, the PSAA-RF method can improve its performance with sufficient samples, indicating that it finds suitable samples for conducting SAA. However, without training by true optimisation loss like IPTB-RF, its performance still could not surpass PO-RF when $N = 1600$. For PO-RF, a rich dataset facilitates high-quality predictions of processing times, securing it the second position at $N = 1600$ and the prime spot at $N = 4000$. It is important to note that this does not necessarily imply IPTB-RF's inability to outperform PO-RF in more complicated distributed problem settings. The output mean absolute percentage error for PO-RF fell below 8% at $N = 4000$, suggesting that the model's high-precision prediction capability is under this particular uncertainty. In this context, the problem is well-specified, i.e. the predictive model can capture sufficient features of the uncertainty [7]. Therefore, PO can perform better than another method.

4.4 Analysis Using Noisy Data and Feature Importance

To validate the robustness of the IPTB model in a more complex environment, we introduced a noise term into the processing time distribution and retested the model's performance. This noise term $\gamma \sim U(0.5, 1.5)$ was directly multiplied by the cycle time CT. It significantly varied the final processing time, as the processing time will be calculated by $y = \gamma \cdot CT \cdot x_2$. This makes the problem

Table 2. Comparison among data-driven optimisation methods using noise data

N	SAA	PO-RF	PSAA-RF	SPO-RF	IPTB-RF
400	20.38%	17.35%	15.03%	18.81%	**12.09%**
800	19.14%	18.66%	13.55%	17.34%	**11.04%**
1600	18.11%	17.08%	13.98%	17.10%	**11.14%**
4000	19.54%	19.54%	12.44%	17.04%	**10.79%**

worse-specified than the previous case: the included features can not reflect the distribution of uncertainty.

With the introduction of a noise term in the data, the results exhibit a new pattern across the different data-driven optimisation models as presented in Table 2. At the smallest sample size of $N = 400$, SAA shows the highest regret at 20.38%, while IPTB outperforms all methods with the lowest error rate of 12.09%. As the sample size increases to $N = 800$, unlike previous results, PO-RF's performance decreased, indicating more optimization errors with additional noise. IPTB maintains its lead with the lowest error rate of 11.04%, further cementing its robustness against noise. While PSAA shows a greater improvement than IPTB at $N = 800$, its performance declines as the sample size increases to 1600, revealing its instability with larger datasets. Considering the largest sample size of $N = 4000$, IPTB-RF demonstrates the best performance with the lowest error rate of 10.79%. This outcome suggests that IPTB-RF becomes increasingly effective as sample sizes grow, highlighting its scalability and robustness in noisy environments.

Next, to enhance our understanding of the model's behaviour, we incorporated an irrelevant input feature $x_5 \sim U(-10.0, 10.0)$ distribution, into the input observation data X. This feature has no actual influence on the processing time and therefore should be assigned a lower weight during the RF model's training process. We employed two distinct sample sizes-800 and 4,000-for comparative purposes. The RF model inherently calculates feature weights during the bootstrapping training process. We evaluate these weights by their percentage contribution to the overall importance, providing insight into the model's feature prioritisation. It should be noted that the presented feature importance for each feature is the aggregate of the importance scores across all jobs that have that feature.

As depicted in Fig. 1, there are noticeable differences in feature importance between datasets with and without the noise term. In the absence of the noise term, the IPTB model successfully identifies the irrelevance of the feature x_5, assigning it minimal importance. At $N = 800$, this feature accounts for approximately 6% of the total importance, which further diminishes to less than 1% at $N = 4000$, indicating a refined feature selection process as sample size increases. However, when the noise term is introduced, the model's ability to diminish the importance of x_5 is compromised at both $N = 800$ and $N = 4000$. This suggests that the noise term interferes with the model's capacity to accurately assess

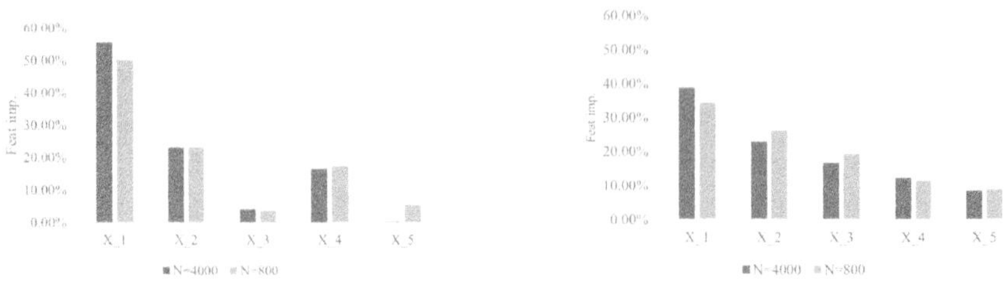

(a) Feature importance percentage of data without noisy term

(b) Feature importance percentage of data with noisy term

Fig. 1. Feature importance plot of two datasets

feature relevance, potentially leading to an overestimation of the importance of redundant attributes. This could result in a less efficient or less interpretable model, as the noise term introduces randomness when sample numbers are small, thereby reducing the model's predictive precision and interpretability.

5 Conclusion and Future Work

The prescriptive ML model has been shown to be able to 'prescribe' a high-quality decision given the observation data. The main challenge of applying this method to a machine scheduling problem comes from both its low accuracy after training by prediction loss and the computational complexity due to the problem's NP-hardness. In this paper, we applied several techniques to an IPTB model to solve this problem. The empirical evaluation conducted on scheduling problems from the printed circuit board assembly line industry underscores the IPTB model's accuracy and robustness. Notably, the model's performance in noisy data environments reveals its robustness, which is a crucial attribute for industrial applications where variability is the norm.

To extend this study, an interesting direction is to find the optimal structure of the tree-based model, which could be achieved by methods such as the branch and bound algorithm [3]. We also plan to investigate the model performance in the practical implementations. Finally, interpreting the scheduling solution 'prescribed' by the decision trees can produce more insights not only for data-driven modelling but also for management significance. The implementation of ML interpretation techniques like LIME [16] is promising for this analysis.

References

1. Ahmed, A., He, L., an Chou, C., Hamasha, M.M.: A prediction-optimization approach to surgery prioritization in operating room scheduling. J. Ind. Prod. Eng. **39**(5), 399–413 (2022)
2. Bartusch, M., Mohring, R.H., Radermacher, F.J.: Scheduling project networks with resource constraints and time windows. Ann. Oper. Res. **16**(1-4), 201–240 (1988)

3. Bertsimas, D., Dunn, J., Mundru, N.: Optimal prescriptive trees. INFORMS J. Optim. **1**(2), 164–183 (2019)
4. Bertsimas, D., Kallus, N.: From predictive to prescriptive analytics. Manage. Sci. **66**, 1025–1044 (2019)
5. Cai, X., Zhou, X.: Single-machine scheduling with exponential processing times and general stochastic cost functions. J. Global Optim. **31**, 317–332 (2005)
6. Elmachtoub, A.N., Grigas, P.: Smart predict, then optimize. Manage. Sci. **68**, 9–26 (2021)
7. Elmachtoub, A.N., Lam, H., Zhang, H., Zhao, Y.: Estimate-then-optimize versus integrated-estimation-optimization versus sample average approximation: a stochastic dominance perspective. arXiv preprint arXiv:2304.06833 (2023)
8. Elmachtoub, A.N., Liang, J.C.N., McNellis, R.: Decision trees for decision-making under the predict-then-optimize framework. In: Proceedings of the 37th International Conference on Machine Learning, vol. 119, pp. 2858–2867. Proceedings of Machine Learning Research (2020)
9. Framinan, J.M., Perez-Gonzalez, P.: On heuristic solutions for the stochastic flowshop scheduling problem. Euro. J. Oper. Res. **246**, 413–420 (2015)
10. Ghasemi, A., Ashoori, A., Heavey, C.: Evolutionary learning based simulation optimization for stochastic job shop scheduling problems. Appl. Soft Comput. **106** (2021)
11. Li, D., Chen, S., Chiong, R., Wang, L., Dhakal, S.: Predicting the printed circuit board cycle time of surface-mount-technology production lines using a symbiotic organism search-based support vector regression ensemble. Int. J. Prod. Res. **59**(23), 7246–7265 (2021)
12. Lu, Z., Cui, W., Han, X.: Integrated production and preventive maintenance scheduling for a single machine with failure uncertainty. Comput. Ind. Eng. **80**, 236–244 (2015)
13. Mandi, J., Demirovi, E., Stuckey, P.J., Guns, T.: Smart predict-and-optimize for hard combinatorial optimization problems. In: Proceedings of the AAAI Conference on Artificial Intelligence, vol. 34, pp. 1603–1610 (2020)
14. Meilanitasari, P., Shin, S.J.: A review of prediction and optimization for sequence-driven scheduling in job shop flexible manufacturing systems. Processes **9**(8) (2021)
15. Nesbitt, P., et al.: Underground mine scheduling under uncertainty. Eur. J. Oper. Res. **294**(1), 340–352 (2021)
16. Ribeiro, M.T., Singh, S., Guestrin, C.: Why should I trust you? Explaining the predictions of any classifier. In: Proceedings of the 22nd ACM SIGKDD International Conference on Knowledge Discovery and Data Mining, pp. 1135–1144. KDD '16, Association for Computing Machinery, New York, NY, USA (2016)
17. Salles, T., Rocha, L., Gonçalves, M.: A bias-variance analysis of state-of-the-art random forest text classifiers. Adv. Data Anal. Classif. **15**(2), 379–405 (2021)
18. Soroush, H.M.: Minimizing the weighted number of early and tardy jobs in a stochastic single machine scheduling problem. Euro. J. Oper. Res. **181**, 266–287 (2007)
19. Yaakoubi, Y., Dimitrakopoulos, R.: Learning to schedule heuristics for the simultaneous stochastic optimization of mining complexes. Comput. Oper. Res. **159**, 106349 (2023)
20. Yan, R., Wang, S., Fagerholt, K.: A semi- smart predict then optimize (Semi-SPO) method for efficient ship inspection. Transp. Res. Part B Methodol. **142**, 100–125 (2020)
21. Zhang, N., Zhang, Y., Song, S., Chen, C.L.P.: A review of robust machine scheduling. IEEE Trans. Autom. Sci. Eng. **21**(2), 1323–1334 (2024)

Economic Graph Lottery Ticket: A GNN Based Economic Forecasting Model

Htoo Wai Aung[1,2], Ying Guo[2(✉)], Jiaming Li[2], Geoffrey Lee[2], and Zili Zhu[2]

[1] University of Technology Sydney, Sydney, Australia
HtooWai.Aung@student.uts.edu.au
[2] Commonwealth Scientific Industrial and Research Organisation (CSIRO), Canberra, Australia
{HtooWai.Aung,Ying.Guo,Jiaming.Li,Geoffrey.Lee,Zili.Zhu}@data61.csiro.au

Abstract. In econometrics, conventional statistical methods often struggle with modelling complex, non-linear relationships among economic variables. The Wilkie Investment Model and its derivatives, such as the SUPA model, have limitations due to their assumptions of normal distribution and stationarity. Recent development in machine learning methodologies has produced more robust models such as the Graph Neural Network (GNN). In this paper, we propose a GNN based Economic Graph Lottery Ticket (EGLT) algorithm, a new method for economic forecasting based on GNN framework. The EGLT approach generates an optimal adjacency matrix without the need for prior knowledge of existing economic relationships. The algorithm identifies key interdependencies among the economic variables through an iterative process. This paper shows that the EGLT algorithm improves forecasting accuracy when compared to the well-established cascading stochastic economic model, SUPA model, calibrated by using eight major Australian economic variables. The EGLT method reduces the Root Mean Square Error (RMSE) significantly, highlighting its potential for better economic predictions. The results of the paper demonstrate that EGLT approach is more accurate and data-driven for modelling and analysing the dynamics of major economic variables.

Keywords: Economic Forecasting · Economic Forecasting · Economic Graph Lottery Ticket (EGLT) Algorithm

1 Introduction

The field of econometrics is constantly evolving due to on-going improvements in statistical methods, burgeoning data availability, and lately from rapid advances of machine learning techniques. The primary goal is to develop models that reveal the complex relationships among various economic variables, and enhance our understanding of the economic system.

M. Gong et al. (Eds.): AI 2024, LNAI 15442, pp. 366–380, 2025.
https://doi.org/10.1007/978-981-96-0348-0_27

Economic systems can be viewed as complex networks of interdependent relationships between variables or indicators such as inflation, wage price index, and equity returns. Simplifying these relationships between key economic variables can help create a logical network that facilitates a more intuitive understanding of the economic system.

The Wilkie Investment Model [14], a cascading stochastic economic model first developed in the actuary domain, has been extended to incorporate more economic variables, such as the SUPA model [4, 8] which is used for projecting accumulation and decumulation phases within the Australian superannuation system. However, the Wilkie model's assumptions of normally distributed asset returns and stationarity of economic variables may not always align with the real-world financial market dynamics [11]. Recently, there has been a shift towards using machine learning methods in economic modelling. For example, Scheidegger's framework combines Gaussian process regression with the active subspace method to address dynamic stochastic models [13].

The need to model non-linear relationships and to manage high-dimensional data has led to exploring the potential of Neural Networks in economic forecasting [10]. Despite criticism surrounding interpretability, these neural networks do have the potential to capture information hidden in large datasets. However, standard implementation of Neural Networks does not provide the mechanism to account for inherent structure and connectivity in economic data and therefore cannot encapsulate complex interactions between these input economic variables. Addressing this limitation, Graph Neural Networks (GNNs) have been developed [12]. By incorporating data structure into their learning process, GNNs excel at uncovering complex patterns in structured data, such as networks of economic variables. Traditionally, these interrelationships need to be delineated via an adjacency matrix before applying GNNs. However, establishing the adjacency matrix requires prior knowledge of the unknown relationships between the economic variables, the challenge is to utilise the delineation process to establish the complex relationship between the key economic variables without any preconceived knowledge about the relationship [15].

In this paper, we propose the Economic Graph Lottery Ticket (EGLT) algorithm, which is a novel GNN-based method that, without the prior knowledge of the inter-relationships between the economic variables, we can generate the optimal adjacency matrix as a by-product of the GNN training process from historical data only. The trained EGLT model is proven to provide more accurate economic forecasting. Compared with the SUPA model, the prediction RMSE reduced significantly for eight major Australian economic variables, highlighting the efficacy of EGLT in economic prediction.

2 Economic Graph Lottery Ticket Approach

As mentioned in Sect. 1, the EGLT algorithm extends the Graph Neural Network (GNN) framework offering a more effective approach to find the optimal adjacency matrix automatically without requiring the prior knowledge to construct an adjacency matrix. The EGLT algorithm iteratively reduces the connections between economic variables, identifying the key interdependencies that drive economic dynamics. In this section, we will first introduce the traditional GNN framework, followed by a detailed introduction of the novel EGLT approach.

2.1 Graph Neural Network

Graph Neural Networks (GNNs) are a crucial development in machine learning, offering novel ways to process and understand non-Euclidean data, setting them apart from traditional Euclidean-based networks. GNNs can be primarily classified into two types: spectral domain-based methods [2] and spatial domain-based methods [5], each bringing unique strategies to the analysis of graph data. Non-Euclidean data, which does not conform to the standard Euclidean geometries, is typically represented as complex relational network structures, posing significant challenges for traditional neural networks. While these networks handle Euclidean data like images and texts adeptly, they fall short when working with the complexities of non-Euclidean data [1], necessitating the role of GNNs due to their inherent capability to handle such data forms effectively. GNNs have found utility across a wide range of domains, largely due to their proficiency in decoding intricate patterns within networks. They have been employed in diverse areas such as road networks, social networks [7], and Soybean trade networks [9]. By interpreting and predicting complex, interconnected structures in these fields, GNNs have enhanced our comprehension of various phenomena.

Addressing the significant concern of computational efficiency in GNNs, strategies like the Unified GNN Sparsification (UGS) have been developed [3]. UGS seeks to identify specific sparsifications in the graph and weights of the GNN without compromising performance accuracy, primarily motivated by reducing computational costs. However, one limitation is that UGS requires a known network to sparsify initially, which is the information sought. Despite this, UGS manages to simplify complex networks into more manageable structures while preserving the original network's essential attributes. A critical component in the successful deployment of GNNs is the graph adjacency matrix, a record of connections between nodes in a graph. Without this matrix, GNNs would lack the necessary relational context, thereby hindering their ability to perform effective analyses.

2.2 Graph Representation and Notations

Consider a directed weighted graph represented by $G = \{V, E\}$, where V denotes the set of nodes as $V = \{v_1, v_2, \ldots, v_{|V|}\}$, and E represents the set of edges with $|E|$ being the number of edges. Each node in the graph has F_{in} features, and the node feature matrix can be denoted as $X \in \mathbb{R}^{|V| \times F_{in}}$. The edges in the graph signify the relationships between pairs of nodes, given by $E = \{e_1, \ldots, e_{|E|}\}$.

The graph's overall topology can be captured using an adjacency matrix, denoted as $A \in \mathbb{R}^{|V| \times |V|}$. If node v_i influences v_j, there is an edge $(v_i, v_j) \in E$, and the corresponding element in the adjacency matrix satisfies $A[i,j] \neq 0$; otherwise, $A[i,j] = 0$.

The two-layer Graph Convolutional Neural Network model [6] can be represented as follows:

$$\hat{Y} = A.\ \sigma\left(A.\ X.\ \Theta^{(1)}\right).\Theta^{(2)} \tag{1}$$

The $\hat{Y} \in \mathbb{R}^{|V| \times K}$ represent the prediction generated by GNN, where $\Theta^{(1)}$ corresponds to the weight at layer-1, $\Theta^{(2)}$ represents the weight at layer-2, and A is the overall adjacency matrix, as detailed in Eq. 3. The activation function $\sigma(.)$ employed in this study is the Rectified Linear Unit ($ReLU$), defined as $ReLU(x) = max(0, x)$. Given that the objective is to address a regression problem, the average Mean Square Error (MSE) was chosen as the loss function, L as shown in the Eq. 2.

$$L = \frac{1}{N} \sum_{i=1}^{N} \sum_{j=1}^{K} \left(Y_j^{(i)} - \hat{Y}_j^{(i)}\right)^2 \tag{2}$$

where $Y \in \mathbb{R}^{|V| \times K}$ is the real valued vector of target values, N denotes the number of training instances and K signifies the number of output features.

2.3 Economic Graph Lottery Ticket Algorithm

In the proposed EGLT approach, the key difference is that we introduce a trainable adjacency matrix mask into the forward pass of the GNN. This would enable the adjacency matrix to be adaptively optimised. Such flexibility ensures that the model captures the most significant relationships without being constrained by preconceived notions. The EGLT algorithm also integrates the structure of economic data into the learning process, allowing for more accurate predictions.

The use of the adjacency matrix in GNNs is crucial for accurately modelling the relationships between nodes. By dynamically adjusting this matrix, the EGLT algorithm can better capture the complexities of economic interactions. This adaptability is particularly important in economic forecasting, where relationships between variables can change over time due to various factors such as policy changes, and market conditions or events.

More details of the EGLT approach are as follows. In the analysis, the forward pass of the Graph Neural Network (GNN) function $f(., \Theta)$ was performed on a given graph $G = \{A, X\}$, where A represents the overall adjacency matrix, as defined in Eq. 3. The original adjacency matrix $A_{original} \in \mathbb{1}^{|V| \times |V|}$ remained fixed as a unit matrix and was not subject to training. To modify the graph structure, a trainable adjacency matrix mask, denoted as $m_g \in \mathbb{R}^{|V| \times |V|}$, was introduced. At the start of Algorithm 1, the initial adjacency matrix mask is defined as $m_g^0 = A_{original}$.

$$A = A_{original} \odot m_g \tag{3}$$

The GNN forward pass trained variables m_g, $\Theta^{(1)}$ and $\Theta^{(2)}$ across N_{ep} epochs for each adjacency matrix mask density during the training process. The lowest parts of m_g's absolute values at s^{th} density were pruned, leaving the rest set to 1, guided by the minimal MSE loss from the validation set. This cycle persisted until the m_g density fell below a pre-set minimum, κ_{min}, with all edges in the m_g at this lowest density deemed essential. The adjacency matrix density at the sth pruned level, κ^s is defined in Eq. 4, where m_g^s represents the trainable adjacency matrix mask at the sth pruned level.

$$\kappa^s = \frac{||m_g^s||_0}{||A_{original}||_0} \tag{4}$$

The state of m_g producing the smallest validation set MSE loss at the ith iteration and lowest adjacency mask s^{th} density level during the GNN forward pass was defined as the Economic Graph Lottery Ticket (EGLT), with ties resolved by selecting the lower m_g density state.

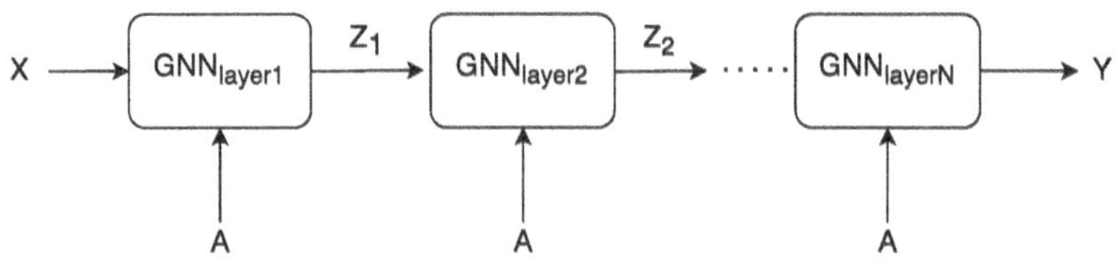

Fig. 1. Flowchart of a multiple layer EGLT approach.

The flowchart of a multiple layer GNN is shown in Fig. 1. While the number of layers increases, the computational cost and complexity of the algorithm increases. Further details on the training EGLT algorithm are in Algorithm 1. All edges in the m_{g_EGLT} were necessary for the best predictions.

3 Real Data Study: Implementation and Evaluation of the EGLT Approach

The EGLT algorithm provides a novel approach to economic forecasting by optimising the adjacency matrix with GNNs. In this section, we will detail the implementation of the EGLT model on real-world data, followed by an evaluation of its performance compared to the traditional SUPA model.

We implemented the EGLT approach to a comprehensive selection of eight Australian economic variables. These variables encompassed a wide range, including the Consumer Price Index (CPI), Wage Price Index (Wage), Short-Term Interest Rate (r_short), Long-Term Interest Rate (r_long), Domestic Bond Index (Dom_BI), International Bond Index (Int_BI), Unemployment Rate (Unemploy) and House Price Index (HPI).

Our analysis utilised monthly data, transforming daily variables into monthly averages and interpolating quarterly variables to match this frequency. Spanning

Algorithm 1. Finding Economic Graph Lottery Ticket

Input: Graph $G = \{A, X\}$, GNN $f(G, \Theta)$, GNN initialisation $\Theta_0^{(1)}$ and $\Theta_0^{(2)}$, $A_{original} \in \mathbb{1}^{|V| \times |V|}$, initial Adjacency Matrix Mask $m_g^0 = A_{original}$, set learning rate η, set pruning rate p_g, pre-defined lowest Graph Density Level κ_{min}.

Output: Economic Graph Lottery Ticket $(m_{g\_EGLT}) - m_g^{s,i}$ at the lowest MSE loss with the highest sparsity possible.

```
while κ^s ≥ κ_min do
    for for iteration i = 0, 1, 2, ..., N_ep do
        Forward f(., Θ_i) with G_s = {m_g^{s,i} ⊙ A_original, X} to compute MSE Loss, L
        Backpropagate to update Θ_{i+1} ← Θ_i − η∇_{Θ_i} L
        Update m_g^{s,i+1} ← m_g^{s,i} − η∇_{m_g^{s,i}} L
    end for
    Record m_g^{s,i} with the lowest MSE loss in validation set during the N_ep iteration
    Pre set p_g of the lowest absolute magnitude values in m_g^s to 0 and the others
    to 1, then obtain a new m_g^{s+1,0}
end while
```

20 years (July 2000-June 2020), the dataset provided a substantial 240 months of data for exploration. The study used these variables' monthly values as input features. Since some variables are recorded quarterly, consecutive months weren't used as input features. For instance, data from months like January, April, and July 2008 were used to predict October 2008 values in our study. This approach helps capture seasonal trends.

Given their varying value scales, these data required normalisation via z-score normalisation to ensure a mean of zero and a standard deviation of one. This step was crucial to prevent larger-scale features from dominating the learning process. We divided the dataset into a training set that includes data from July 2000 to December 2017, and a testing set that includes data from January 2018 to June 2020. Ten percent of the data in the training set was reserved for continuous performance monitoring as a validation set.

3.1 Model Initialisation and Hyper-parameter Settings

Our study employed a two-layer GNN model to optimise graph characteristics, as shown in Fig. 2. The node features are represented by $F_{in} \in \mathbb{R}^3$ and the number of GNN hidden nodes, $M = 128$. We aimed to predict a single output month, thus $F_{out} \in \mathbb{R}^1$. Our research considered various scenarios, encompassing different input month and GNN weight combinations. The implementation details are lised in Table 1.

We used the PyTorch deep learning package to construct the two-layer GNN, with the training process running over 500 epochs (N_{ep}) and including a 0.5

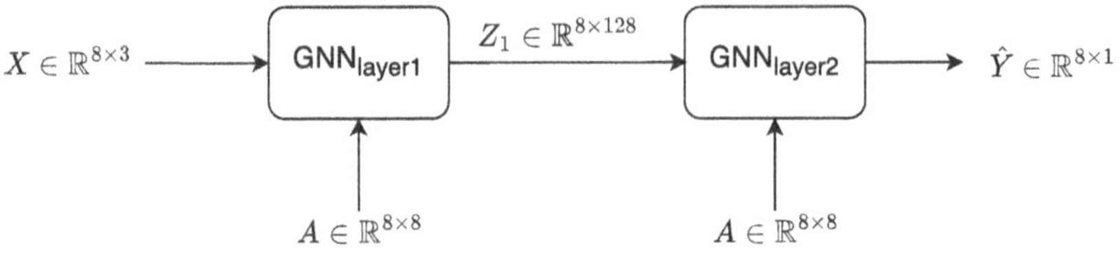

Fig. 2. The implemented two-layer EGLT model for eight economic variables

Table 1. Implementation Details of the 2-layer EGLT Model

Layer	Type	Input Size	Weight	Output	Number of parameters
Input	Input graph features	$\|V\| \times F_{in}$	-	-	-
H1	First graph convolution layer	$\|V\| \times F_{in}$	$\Theta^{(1)} \in \mathbb{R}^{F_{in}\times M}$	$\|V\| \times M$	$F_{in} * M$
H2	Second graph convolution layer	$\|V\| \times M$	$\Theta^{(2)} \in \mathbb{R}^{M\times F_{out}}$	$\|V\| \times F_{out}$	$M * F_{out}$
$\widehat{Y}$	Predicted Output	$\|V\| \times F_{out}$	-	-	-

dropout rate to reduce overfitting. The model was optimised using the Adam optimiser with a learning rate (η) of 0.01. The adjacency matrix mask m_g was set to prune at a rate (p_g) of 10% at each m_g density, down to a minimum density of 10.71%. The average MSE metric was used to assess the model's regression problem performance.

3.2 EGLT Model Implementation

Using the Algorithm 1, the RMSE losses corresponding to different adjacency matrix densities, κ^s, were obtained as shown in Fig. 3. The optimal adjacency matrix, m_{g_EGLT}, was found at a density of 21.88%, as it yielded the lowest RMSE loss. Figure 4 (top) shows the heat map illustrating the interrelationships among the variables. Figure 4 (bottom) presents the nodes-edges graph, providing a visual understanding of how these economic factors are interconnected and their potential impact on the overall economic system. For example, the Unemployment Rate (Unemploy) directly affects the Wage Price Index (Wage) and House Price Index (HPI). This relationship aligns with economic observations where a high unemployment rate impacts the affordability of mortgages, hence influencing house prices.

In addition to the direct relationships, the heat map and nodes-edges graph reveal some indirect connections. For instance, the Domestic Bond Index (Dom_BI) shows interrelation with both short-term (r_short) and long-term interest rates (r_long). This reflects the real-world scenario where bond prices are sensitive to changes in interest rates. Similarly, the Consumer Price Index (CPI) and International Bond Index (Int_BI) have connections, indicating the influence of international economic conditions on domestic inflation.

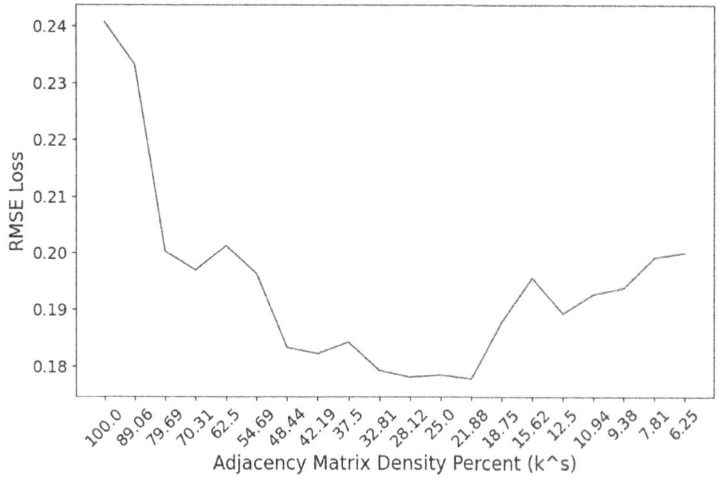

Fig. 3. RMSE losses at different adjacency matrix densities, κ^s

The graphical representation of the adjacency matrix (Fig. 4 Bottom) not only provides a clear depiction of the essential connections but also highlights the complexity of the economic network.

3.3 Prediction Analysis

The graphical representation of the optimised adjacency matrix shows the necessary connections between the provided economic variables. The EGLT algorithm identifies these critical relationships without prior knowledge. Such direct and indirect interactions can be crucial for accurate predictions. The performance of the trained EGLT model is validated against the current state-of-the-art SUPA model using data from January 2018 to June 2020, evaluated in terms of RMSE values. Table 2 lists the comparison across multiple variables. The average RMSE loss of the EGLT model is 0.1778, which is about 40% lower than the RMSE loss of the SUPA model.

We also show all variables' prediction results in more details as time series in Figs. 5, 6, 7, and 8. In each figure, the green line represents the real value, the orange line the EGLT model prediction, and the blue line the SUPA model prediction.

For the Consumer Price Index (CPI) and Wage Price Index (Fig. 5), both datasets are linear and monotonically increasing. Here, the SUPA model predic-

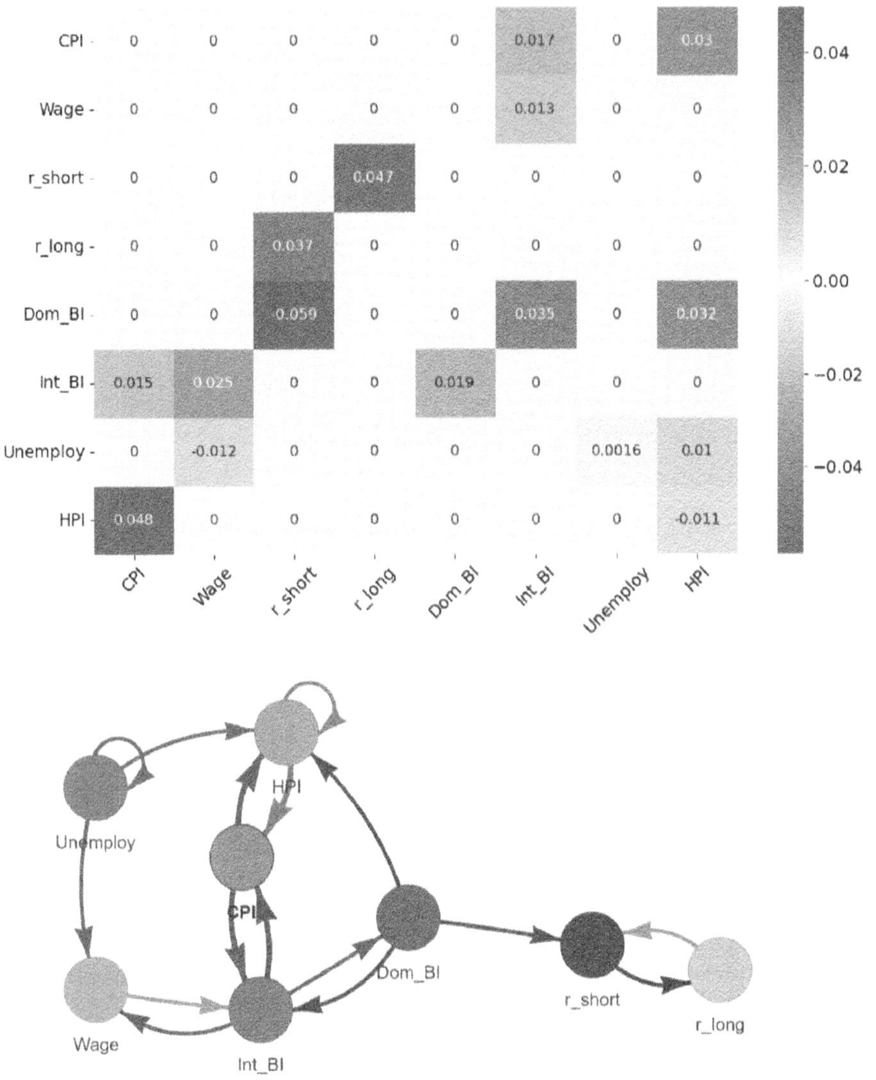

Fig. 4. The optimal adjacency matrix of EGLT model, m_{g_EGLT} is found at the 21.88% density with Necessary Edges only. Top: Adjacency Matrix in Heatmap, Bottom: Adjacency Matrix in Directed Nodes-Edges Graph

tions are better than the EGLT model results. The SUPA model closely follows the actual values, indicating its effectiveness in handling straightforward, linear trends.

In Fig. 6, the Short-Term Interest Rate (r_short) and Long-Term Interest Rate (r_long) prediction results are depicted. The EGLT model outperforms the SUPA model, especially in capturing the rapid fluctuations in r_short. The SUPA

Table 2. Comparison of the RMSE losses for the SUPA and the EGLT algorithm across various economic variables.

Variable	SUPA	EGLT
CPI	0.0204	0.0186
Wage	0.0186	0.0361
r_short	0.6170	0.0962
r_long	0.5303	0.3420
Dom_BI	0.1433	0.0795
Int_BI	0.5579	0.5536
Unemploy	0.3319	0.2108
HPI	0.0869	0.0852
Overall	**0.2883**	**0.1778**

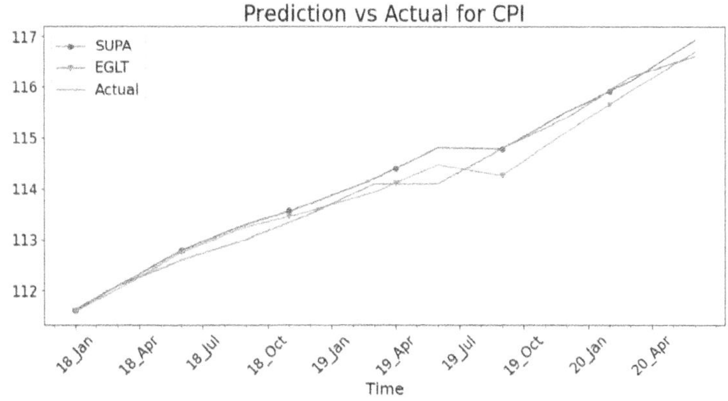

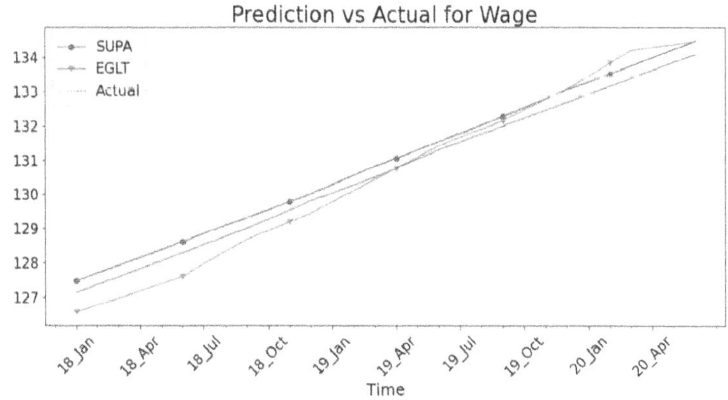

Fig. 5. Top: CPI prediction result, and Bottom: Wage Price Index prediction result

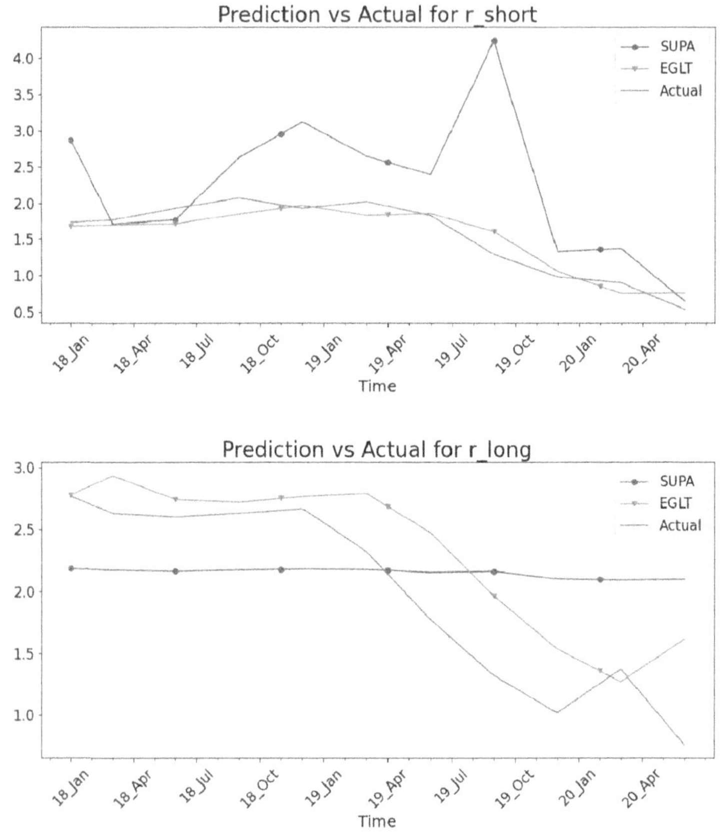

Fig. 6. Top: Short-Term Interest Rate prediction result, and Bottom: Long-Term Interest Rate prediction result

model shows significant lag in adjusting to these changes, while EGLT adapts more quickly, highlighting its suitability for volatile datasets.

Figure 7 shows the Domestic Bond Index (Dom_BI) and International Bond Index (Int_BI) prediction results. The EGLT model again provides better predictions, particularly for Dom_BI. The EGLT model captures both minor and major fluctuations more accurately than the SUPA model, demonstrating its robustness in dealing with the non-linear relationship.

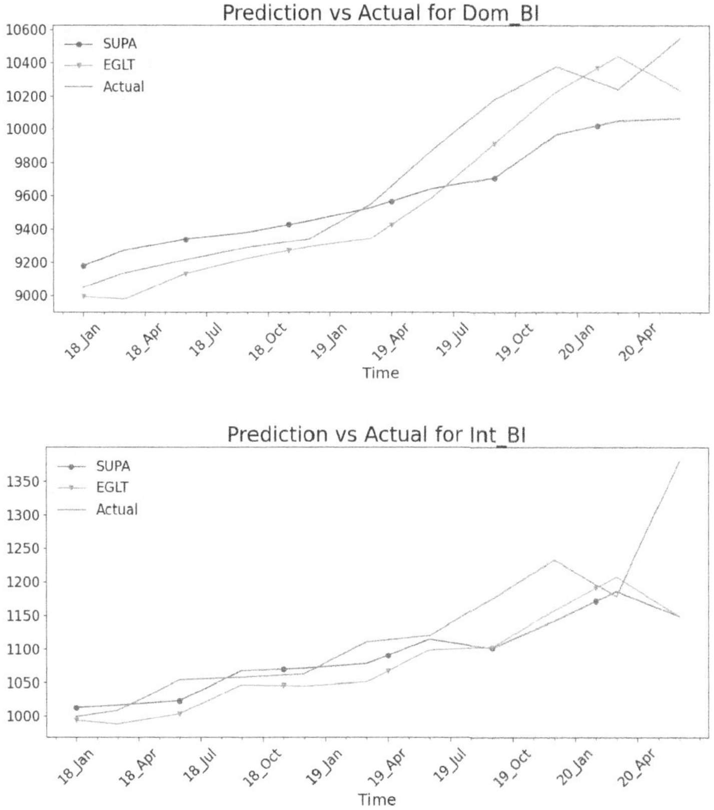

Fig. 7. Top: Domestic Bond Index prediction result, and Bottom: International Bond Index prediction result

The Unemployment Rate (Unemploy) and House Price Index (HPI) prediction results are shown in Fig. 8. The EGLT model significantly outperforms the SUPA model in predicting the Unemployment Rate. For the HPI, both models perform similarly, but EGLT has a slight edge in following the upward trends more closely.

Given the current small datasets, all the EGLT models are tested on single-step forecasting. Later on, when there are larger datasets covering longer time periods, the EGLT model can also be trained and evaluated for multiple-step long-term forecasting as well. The flexibility of the EGLT algorithm will allow it to adapt to different data conditions, making it a versatile tool for economic forecasting.

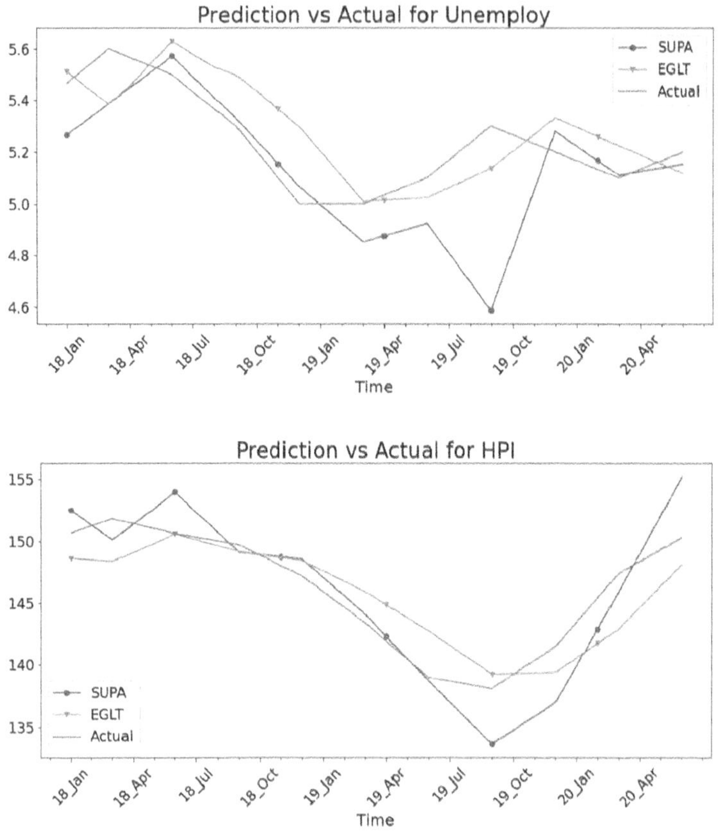

Fig. 8. Top: Unemployment Rate prediction result, and Bottom: House Price Index prediction result

4 Conclusion

In this paper, within a Graph Neural Network (GNN) framework, we have presented the Economic Graph Lottery Ticket (EGLT) algorithm as more accurate method for economic forecasting. The EGLT algorithm automatically generates an optimal adjacency matrix without requiring prior knowledge of existing economic relationships. The EGLT algorithm can iteratively converge and reveal the most parsimonious inter-connecting relationships by minimising the Root Mean Square Error for the dataset and therefore improving the forecasting accuracy.

The forecasting accuracy of the EGLT model was compared directly with the state-of-the-art cascading SUPA model for eight major economic variables in Australia from January 2018 to June 2020. The comparison shows that the EGLT model consistently outperforms the SUPA model across these economic variables, with a reduction in RMSE. Specifically, the EGLT model's average RMSE is 40% lower than that of the SUPA model.

In this paper, our study has focused on single-step forecasting. However, for applications in major decision making under uncertainty such as climate uncertainty, long-term multiple-step forecasting is required. In our subsequent study on EGLT algorithm, focus will be on extending the algorithm to the generation of uncertainty estimate for long-term forecasting.

The EGLT algorithm can be readily applied to the task of identifying inter-dependency relationship structure of large number of time-series variables. In this paper, we have used the EGLT algorithm to generate inter-dependency structure for eight major economic variables in Australia. For future work, an interesting test case would be for using the EGLT algorithm to quantify and reveal the complex inter-dependency relationship network among major commodity prices.

References

1. Bronstein, M.M., Bruna, J., LeCun, Y., Szlam, A., Vandergheynst, P.: Geometric deep learning: going beyond Euclidean data. IEEE Sig. Process. Mag. **34**(4), 18–42 (2017)
2. Bruna, J., Zaremba, W., Szlam, A., LeCun, Y.: Spectral networks and locally connected networks on graphs. arXiv preprint arXiv:1312.6203 (2013)
3. Chen, T., Sui, Y., Chen, X., Zhang, A., Wang, Z.: A unified lottery ticket hypothesis for graph neural networks. In: International Conference on Machine Learning, pp. 1695–1706. PMLR (2021)
4. Chen, W., et al.: Using a stochastic economic scenario generator to analyse uncertain superannuation and retirement outcomes. Ann. Actuarial Sci. **15**(3), 549–566 (2021). https://doi.org/10.1017/S1748499520000305
5. Hamilton, W., Ying, Z., Leskovec, J.: Inductive representation learning on large graphs. Adv. Neural Inf. Process. Syst. **30** (2017)
6. Kipf, T.N., Welling, M.: Semi-supervised classification with graph convolutional networks. arXiv preprint arXiv:1609.02907 (2016)
7. Kumar, S., Mallik, A., Khetarpal, A., Panda, B.: Influence maximization in social networks using graph embedding and graph neural network. Inf. Sci. **607**, 1617–1636 (2022)
8. Minney, A., Zhu, Z., Guo, Y., Li, J., Toscas, P., Koo, B., Pantelous, A.A.: Using the pension multiple to measure retirement outcomes. Financ. Res. Lett. **49**, 103149 (2022)
9. Monken, A., Haberkorn, F., Gopinath, M., Freeman, L., Batarseh, F.A.: Graph neural networks for modeling causality in international trade. In: The International FLAIRS Conference Proceedings, vol. 34 (2021)
10. Moshiri, S., Cameron, N.: Neural network versus econometric models in forecasting inflation. J. Forecast. **19**(3), 201–217 (2000)
11. Sahin, S., Cairns, A., Kleinow, T.: Revisiting the Wilkie investment model. In: 18th International AFIR Colloquium, Rome, pp. 1–24 (2008)
12. Scarselli, F., Gori, M., Tsoi, A.C., Hagenbuchner, M., Monfardini, G.: The graph neural network model. IEEE Trans. Neural Netw. **20**(1), 61–80 (2008)

13. Scheidegger, S., Bilionis, I.: Machine learning for high-dimensional dynamic stochastic economies. J. Comput. Sci. **33**, 68–82 (2019)
14. Wilkie, A.D.: A stochastic investment model for actuarial use. Trans. Fac. Actuaries **39**, 341–403 (1984)
15. Wu, Z., Pan, S., Chen, F., Long, G., Zhang, C., Philip, S.Y.: A comprehensive survey on graph neural networks. IEEE Trans. Neural Netw. Learn. Syst. **32**(1), 4–24 (2020)

Pattern-Based Trading by Continual Learning of Price and Volume Patterns

Patrick Liston[1](✉), Charles Gretton[1], and Artem Lensky[2,3]

[1] The Australian National University, College of Engineering, Computing and Cybernetics, Canberra, Australia
patrick.liston@anu.edu.au

[2] School of Engineering and Technology, The University of New South Wales, Canberra, ACT, Australia

[3] School of Biomedical Engineering, Faculty of Engineering, The University of Sydney, Sydney, NSW, Australia

Abstract. Automating trading decisions has been a pursuit of researchers and practitioners alike for decades. We contribute to the literature focusing on "pattern based" strategies. *Dynamic time warping* is used to group similar patterns into a representative category, while the method of *continual learning augmentation* is used to maintain the set of patterns used for decision-making. Thus, we implement a novel approach to pattern-based trading, utilising adaptive memory structures to enable adaptability of agent decision making and overall agent performance. Two new online pattern-based trading agents are introduced and tested on two-sets of historical cryptocurrency data, for the BTCUSDT pair over the periods of 2017–2023 and 2023–2024. We compare our newly formulated agents against an established baseline of rule-based agents, thereby comparing the relative profit generating abilities of a wide range of agents.

1 Introduction and Background

Algorithmic trading—defined as the automation of trading decisions, order submission, and management through computer algorithms [10]—has spurred the development of numerous systems aimed at achieving this automation [6,9]. These systems use trading rules to generate trading decisions. They operate using inputs such as market sentiment and technical factors [6,20]. Also using trading rules, discretionary traders have been known to employ charting techniques to identify price patterns within market signals that are said to indicate *reversals*, *breakouts*, and subsequent trends. Illustrations of such patterns are shown in Fig. 1. There exists a large assortment of tools and technical indicators available to practitioners, including strategies that aim to capitalise on this idea. Analysis tools employing technical indicators are proposed in [12,22], candlestick formations are used by [18], and of particular relevance to us, chart patterns are used in [26,27]. Integration of chart patterns into automated decision making is

M. Gong et al. (Eds.): AI 2024, LNAI 15442, pp. 381–393, 2025.
https://doi.org/10.1007/978-981-96-0348-0_28

less straightforward than use of other features, such as moving averages, Moving Average Convergence/Divergence (MACD), and the Relative Strength Index (RSI). At first blush patterns, for example "Head-and-shoulders" and "Flags", are seemingly more subjective however can be integrated in an algorithmic trading agent.

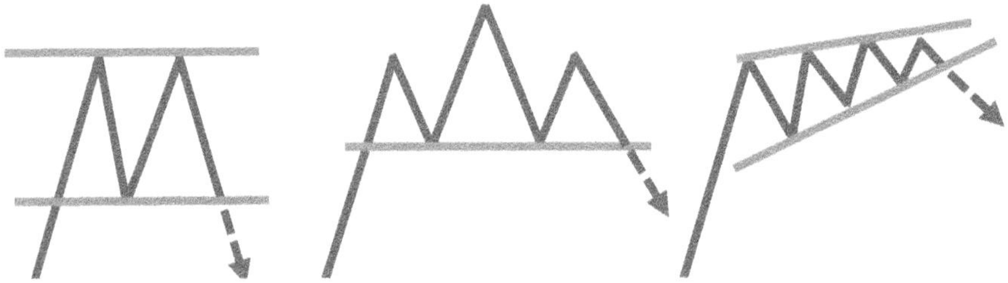

Fig. 1. Example of chart patterns: (left) Double top, (mid) Head and shoulders, (right) Rising wedge.

The recent work in [27] assessed pattern charting trading from a generic pattern based perspective. They used an algorithmic method, based on subsequence Dynamic Time Warping (DTW) in which bullish and bearish classes were assigned to a set of query patterns. Using previous price behaviour of similar historical sub-sequences of price and volume pairs, to inform trading decisions. The authors argued that the proposed algorithm could capture common, generic technical principles, while remaining free of technical specifications of particular technical patterns.

We develop new systems that trade according to market price patterns, and in a sense these patterns indicate *financial market regimes*. Regimes have been widely shown to be present within financial markets [19]. Regimes describe consistent behaviour—price movements, volatility, and trading volumes—over a given period of time, and can be largely influential when it comes to determining trading strategies and asset allocation [1]. Thus, regime changes can impact which strategies are performant and effective [2]. Regimes may be complex, covering a range of market features, however, some studies, such as [2], show that even simple regime constructions can be effective in segmenting market signals. Given that market regimes have been shown to be present within cryptocurrency markets [14], this lends support for the notion that this method of pattern detection can serve as a proxy for market regime identification and in turn serve as an important input for automated trading systems.

Building upon the works [25–27], in this paper we utilise DTW, a method of signal comparison to identify and aid in classifying chart patterns. DTW is often used to minimise the effect of time distortion between two time-series signals allowing for a flexible comparison based on the underlying shape of the data. Initially gaining recognition for its use in the field of speech recognition [24], DTW has since gained popularity for its applications to health data [23,28], and finance [25]. In [27] the authors show the ability of DTW based strategies to generate profits for stocks, and in [13] authors show that a pattern matching trading system built using DTW is able to produce a stable and effective return in low trading frequency environments. DTW-based agents alone have been shown to be profitable. Further, some authors [7,8,16] suggest deep learning-based agents may outperform traditional models for financial forecasting tasks, such as volatility prediction. However, one major disadvantage of these models is their lack of adaptability, hence it remains a compelling direction to investigate how to create dynamic agents that can continually adapt to novel market regimes.

We combine DTW with ideas in [21], in which authors develop a method of Continual Learning Augmentation (CLA). This is used to create a trading system that is able to adapt to, and learn from, changing market environments. Introduced as a regression approach and applied as a sliding window, CLA steps forward through time over a set of input data, typically one or more time-series. Of particular relevance to this paper, the approach is initialised with an empty memory structure, M, and some base model, such as a recurrent neural network (RNN) – e.g. long short-term memory (LSTM) network. The base model is first trained on a small subsection of data, and as new data becomes available the model is updated iteratively. The memory structure preserves essential information from previous time steps while continuously adapting to new data. The iterative process allows the model to dynamically adjust its parameters to account for changes and patterns in the data distribution.

We develop and evaluate a novel pair of agents that capitalise on ideas from previously designed pattern based agents, using DTW and a CLA inspired memory structure to create agents which use dynamic states and table structures with an aim to improve trading performance. Our agents are shown to have relatively good performance compared to baseline agents. Our evaluation is across two-sets of historical cryptocurrency data, for the BTCUSDT pair over the periods of 2017–2023 and 2023–2024.

2 Model Design

We introduce two new agent types: a Markov agent and an action-value agent, following a distinct approach to decision-making and trade execution. As shown in Fig. 2, these agents operate within a structured framework that involves: state classification, prediction/action selection, and finally trade execution.

Importantly, we note that agents used within this paper rely solely on historical backtesting processes, hence they do not influence market (price) trajectory. As a consequence, transition probabilities observed historically remain unchanged by the agents. This is of particular relevance, as agents within this paper deviate from typical reinforcement or Q-learning agents, as they update their observations and expectations based only on observed and idealised outcomes, and not by reference to rewards or returns generated by their own actions. Further, supported by [3,11], we note all agents rely on the Markov assumption—only the previous state has an impact on the future state—to construct all transition and action-value tables.

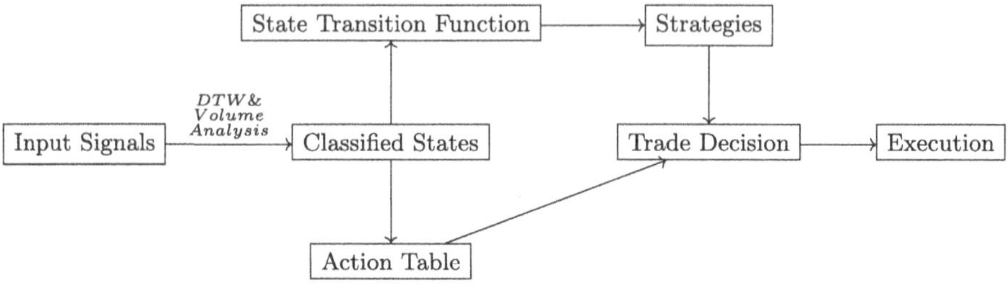

Fig. 2. High level design of our agents that employ DTW and CLA.

2.1 Input Signal and State Assignment

Agents determine their state with reference to the current status of the market. All agents within this paper utilise the same state assignment mechanisms. These agents refer to two input signals, historical price and volume, which are viewed as windows of length n, specifically in this paper we use $n = 24$. These price and volume signals are analysed to create separate *classes*, that when combined create an overall state.

DTW is used to analyse the price signal curve, and determine the similarity of the given pattern against all other price patterns within the *pattern dictionary*. Also characterising the agent's state, the sum of traded volume from the input signals is placed into a quintile.

The pattern dictionary is initially populated with a sample of 10 representative price patterns, drawn from [5], commonly cited by technical analysis practitioners. State assignment in this model is characterized by two critical features. First, ϵ measures the unsigned distance between the current market pattern and the closest database pattern to that. A margin constant θ is used to ensure that the agent does not act when the observed price signal is too dissimilar to the closest matching pattern. Given ϵ and θ, agents only act when the DTW distance satisfies the condition: $\epsilon \leq \theta$.

Second, α measures the DTW distance between the current pattern and all patterns within the dictionary. The maximum distance from this process is then used to determine if or when a pattern within the dictionary should be replaced, as described in Sect. 2.2. Agents patterns are initialised so that the strategy indicated is *buy* and *sell* equally often in the initial dictionary – i.e. for 50% of patterns the *ideal strategy* is *buy*, while for the other 50% the agent should *sell*.

The volume quintiles, used to characterize states, are derived by analyzing the entire trading history for each period. We consider the sum of volume in each 24 hour period, ranking these and creating equally sized quintiles. An agent will consider the volume traded within the given analysis period, and locate the appropriate quintile. That is then used in combination with the matched chart pattern from the agent's dictionary to determine the overall state of the market.

2.2 Updating Beliefs and Tables

Markov Agent. Given the Markov assumption and states described as above, we represent the state transition function explicitly in a tabular form (e.g., Table 1), with the transition probabilities corresponding to the empirical state transition probabilities. To act the agent then greedily selects the pattern which is mostly likely to occur in the following trading period according to the transition function. Based on this pattern, drawn from the pattern dictionary, the agent applies the strategy indicated by the successor function, as defined by Eq. 1, where P is the price at a given time.

$$Action = \begin{cases} Buy, & \text{if } P_{end} - P_{start} \geq 0 \\ Sell, & \text{if } P_{start} - P_{end} < 0 \end{cases} \quad (1)$$

Action-Value Agent. In contrast to the Markov agent this agent does not predict the next pattern type, but instead attempts to infer the next appropriate action directly from an action-value table, as illustrated in Table 2. This action-value table contains the empirical expected value of completing an action for each given state. The Action-Value agent considers the current state and under a purely exploitative strategy, executes a trade of constant size S in the direction associated with the highest reward.

Table 1. Example Structure of a Markov Transition Table

	S_1	S_2	S_3	...	S_n
S_1	0.20	0.15	0.05	...	0.55
S_2	0.15	0.30	0.20	...	0.05
S_3	0.55	0.20	0.10	...	0.10
S_n	0.03	0.00	0.65	...	0.15

Table 2. Example Structure of an Action-Value agent's decision table

	Buy	Sell
S_1	475.24	70.20
S_2	1256.50	587.10
S_3	-56.20	1875.21
S_n	23.10	75.23

Pattern Replacement. An important feature of the newly designed agents is their ability to update their memory structures – including their pattern dictionary, transition table and Action-value decision table. This update occurs on the basis of DTW price signal comparison.

Consider the agent encounters a new pattern with distance value ϵ wrt the current dictionary. If ϵ exceeds a maximum value α—i.e., the maximum DTW distance value found when the current pattern is compared to every pattern within the pattern dictionary, exceeds the predefined threshold α—the pattern within the current dictionary that is most similar to the observed pattern is replaced with the new pattern. More formally, after the current signal is compared to all patterns within the pattern dictionary, the minimum ϵ value is found and if $\epsilon \geq \alpha$ (and additionally $\epsilon \leq \theta$), the pattern within the dictionary that has the lowest ϵ when compared to the current pattern, is replaced.

When a pattern is replaced, this pattern no longer exists within the pattern dictionary and thus the state space of the agent (be it the Markov agent or either of the Action-Value agents) is altered. To address this the agent revisits the observed history to the current point, recalculating transition probabilities and expected rewards and updating the relevant tables as required. Proceeding this change the agent continues trading as required. (Note: All previous trades the agent has made remain unaltered, only future actions are effected by this change).

3 Results and Discussion

We perform a backtesting experimental evaluation using candle data for the BTCUSDT trading pair, collected from Cryptocurrency exchange Binance. We evaluate using trials, each comprising 1 h candles with the *Open* being considered as the *true* price. Trials were run for a training/test period of 01/01/2017 - 01/01/2023. We employ a "burn in", with all agents observing 1,000 results from notional trades, to inform the starting pattern dictionary and related information, before it executes trades within the market. In other words, ≈ 500 days of trades are seen by the agent, with trading results being recorded from March of 2018 onwards. 100 trials were conducted, with each trial beginning at a different time. After each trial the agent parameters (transition matrix and Action-Value-tables) were saved. Saved strategies were further evaluated during the *test* period of 01/01/2023 - 01/01/2024, a period which was not part of the above trials. Here, we evaluate how well the agent performs in an unseen period given strategies developed for a historical period.

All agents utilised a prediction window of $w = 12$, such that the price movement was predicted 12 hours ahead of the current time. Agents had an observation window (pattern length) of 24 hours. Further, the dictionary of patterns consisted of 10 comparison patterns. Initial experiments did not consider friction (trading fees).

Our introduced agents were compared against a sample of less sophisticated agents, called *naive* here, common within the literature.

1. **Buy and Hold** - Agent buys at t_1 and sells at the end of a trial t_{SimEnd}.
2. **Random** - Chooses the next pattern type (and associated action) with equal probability from all possibilities listed in the dictionary of patterns.
3. **Most Common** - Selects the next pattern and action, by picking the most frequently occurring pattern in the observed history.
4. **Previous** - Considers only the immediate past (pattern and action) and assumes this will continue to be optimal in the next time period.

Table 3. Mean results for various agent types. Return - Multiple, Pattern class prediction accuracy (%), Action prediction accuracy (%), Drawdown (%)

Agent Type	Return	Pattern%	Action%	Drawdown
Updating Phase - 2017-2023				
Buy & Hold	0.396	-	-	-0.358
Random	-0.055	9.59	49.81	-100.0
Previous	0.44	15.21	51.24	-35.10
Most Common	0.010	**21.92**	49.93	-100.0
Markov	0.194	19.35	50.83	-55.06
Action-Value	**4.021**	-	**56.69**	**-5.993**
Test Phase - 2023-2024				
Buy & Hold	0.355	-	-	-0.000984
Random	0.002	9.80	50.15	-22.92
Previous	0.260	15.41	50.56	-30.61
Most Common	-0.396	**21.37**	49.45	-44.0
Markov	0.223	20.11	50.53	-26.39
Action-Value	**2.774**	-	**59.60**	**-1.36**

3.1 Pattern Prediction

First, in Table 3, Column 3 (Pattern%), for agents that operate according to a chosen pattern we measure how well agents' choices predict patterns realised in the market. We set the dictionary size to be 10, thus expect an agent acting uniformly at random to have success 10% of the time. The *Random* agent achieves expected performance (mean 9.59% and 9.80% for respective evaluated periods). The performance of naive agents here reflect regular structures in the market. Our more sophisticated Markov agent achieves good performance, i.e., superior to random, and we shall see later is able to generate better returns according to its choices compared to all naive agents.

3.2 Action Selection

We now assess strategies for predicting the most profitable action, (*buy* or *sell*), at timestep t_{n+w}. We have a balanced problem, because our analysis of data reveals that *buy* actions are optimal 50.3% (*sell* Optimal 49.7%) of the time, across trials. Table 3 summarises the mean accuracy of action predictions as a

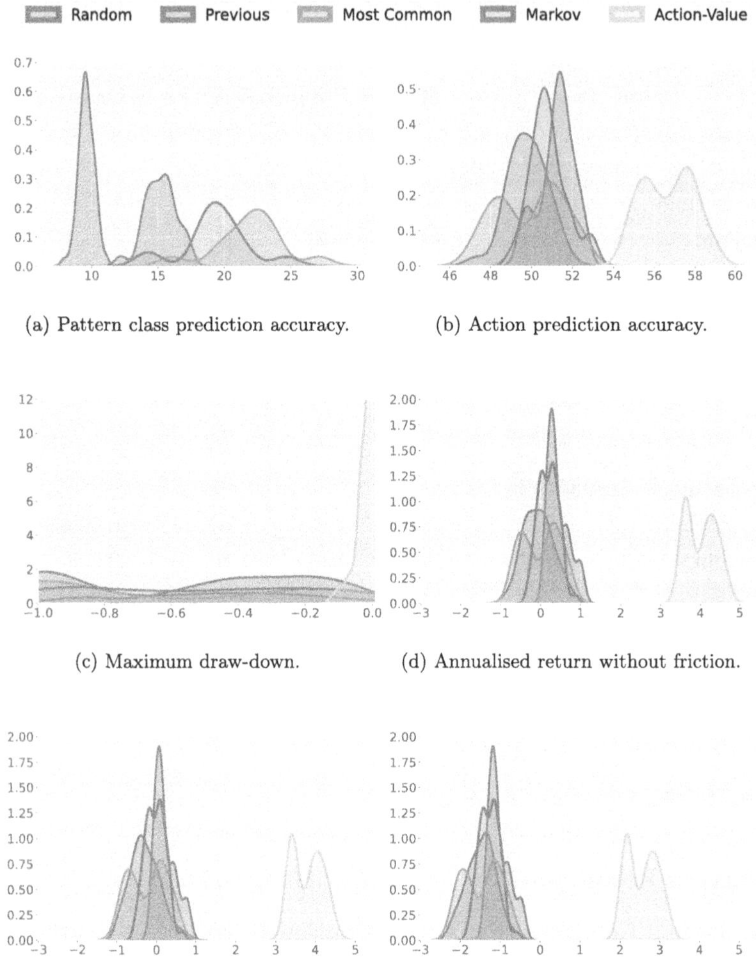

(a) Pattern class prediction accuracy.

(b) Action prediction accuracy.

(c) Maximum draw-down.

(d) Annualised return without friction.

(e) Annualised return with 0.015% friction

(f) Annualised return with 0.1% friction

Fig. 3. Distributions of various metrics for the period 2017 – 2023 each consisting of the 100 trials.

percentage, while Fig. 3b illustrates the distribution of these predictions across 100 trials.

While Fig. 3a suggests the *Markov* agent excels in pattern prediction compared to the *Previous* agent, Fig. 3b indicates the reverse for action prediction. This discrepancy may imply the *Markov* agent's trading strategy is overly simplistic. Alternatively, this could be a symptom of DTW, where its time distortion minimisation effect might result in a pattern closely matching a predictive pattern, yet potentially losing crucial time-dependent gradients necessary for precise trading decisions.

The *Action-Value* agent is able to more accurately select the correct future action. The *Action-Value* agent outperforms other agents with an accuracy of 56.69%, compared to its nearest competitor at 51.24%. This again calls into question the method the *Markov* agent uses to devise strategies for action selection. But shows the potential of the *Action-Value* agent and its use of an adaptive pattern memory structure to improve upon naive methods.

3.3 Returns

We examine the annualised returns obtained by our agents. After an initial observation period, agents began executing trades within the prediction window pw, opening positions at t_1 and closing them at t_pw – the final time-step in the prediction window. All agents had a trade size S of 1 BTC, while profits are measured in excess return, as a percentage of the initial price (price at t_1).

BTCUSDT experienced substantial price fluctuations during both back-testing periods. To mitigate the influence of these fluctuations, profits were measured on a trade-by-trade basis as excess profit. Thus excluding gains from Bitcoin's value changes. This approach ensures that all returns reflect strategy performance **in addition to** the baseline *Buy and Hold* returns shown in Table 3.

Simple strategies like *Random*, *Most common*, and *Previous* all achieved negative or negligible returns, as did the *Markov* agent despite its earlier success in predicting trading directions. The *Action-Value* agent on the other hand is able to capitalise on its superior performance in predicting the next action, and hence generated significant gains.

Interestingly, returns in the testing period where the agents dictionary is static are lower compared to those in the period preceding 2023. This decrease likely stemmed from reduced market volatility in 2023 relative to 2017-2023, where Bitcoin's price increase resulted in larger dollar-value price deviations but smaller relative changes. Consequently, while the *Action-Value* agent produced larger profits in USDT (109, 496.89 vs. 72, 691.99), annualised relative returns were lower at 2.774 in 2023 compared to 4.021 over 2017 – 2023.

3.4 Drawdowns

A crucial aspect of any trading agent is avoiding catastrophic drawdowns that could entirely deplete its capital thus stopping trading. Drawdowns were analysed assuming the initial portfolio value equaled twice the initial BTC price. Maximum drawdowns, shown in Table 3, are capped at −100%, indicating insolvency if reached.

From Table 3 it is clear that *Most Common* and *Random* agents inevitably face −100% drawdowns, leading to failure. In contrast, *Action-value* experiences minor drawdowns, with a maximum of 11.8% (5.93% on average). This underscores the reliability of the *Action-value* agent. The *Markov* agent also demonstrates lower drawdowns compared to other baseline agents, highlighting the positive impact of the pattern dictionary, but more importantly the importance of the transition matrix, and Action-Value table on performance.

Additionally, Table 3 shows varied drawdowns between the 2017-2023 and 2023-2024 backtest periods, with lower drawdowns in 2023-2024 likely due to the shorter period. Despite negative action prediction rates for *Random*, *Most Common*, and *Markov* agents, extended testing might further reduce their drawdown percentages over time.

3.5 Fees

Trading fees can significantly impact a strategy's returns. To examine the effect of fees upon our agents, three fee structures were tested: no fee (0bps), 0.015% (1.5bps), and 0.1% (10bps). The results from the 2017-2023 test period are detailed in Table 4. We omit the *Buy & Hold* agent from this analysis due to its lack of trading frequency. Given the small size of returns from some positive actions, these results motivate future work that would consider an agent that can hold if projected profits are marginal. Similarly, we only considered agents with actions that make fixed-size trades. Future work should consider more versatile agents that can trade dynamic quantities.

Table 4. Annual Return for various fee conditions during the initial 2017 – 2023 backtest period.

Fee Percentage	0%	0.015%	0.1%
Random	-0.055	-0.278	-1.470
Previous	0.440	0.221	-1.020
Most Common	0.010	-0.209	-1.451
Markov Model	0.194	-0.025	-1.266
Action-Value	**4.021**	**3.802**	**2.560**

4 Conclusions and Future Work

4.1 Future Work

Currently agents use a fixed-size pattern dictionary. Future research could explore expanding or dynamically adjusting the dictionary size to include more diverse and representative price patterns. Although, finding the optimal dictionary size would involve balancing specificity of accurate and frequent actions with generalisability across different market conditions, which remains a challenging task.

To this end a focus on integrating anomaly detection mechanisms into the agent, allowing it to identify and react to outlier deviations from expected patterns could be beneficial to the agent. Anomalies caused by large market events or liquidation cascades as seen in [17] may be modelled prior to developing such agents and be used to create a more robust memory and agent for real-life trading. Hence, we plan to integrate this pattern based trading agent into synthetic market simulations and observe its; performance, effect within the market and if such training leads to performance gains in real markets.

While the construction of these agents is reminiscent of reinforcement learning (RL), we do not present an RL agent in a true sense. The use of RL is becoming more popular and has previously be shown to be effective in trading and portfolio allocation tasks [4,15,29]. With this in mind, adapting our proposed agents, as well as utilising our proposed pattern dictionary and memory structure methods to inform an RL agents dictionary of features and state selection, opens an interesting line of research and may lead to enhanced performance in trading tasks. However, an arguably more intriguing pursuit for practitioners of charting methods, would be to investigate if there are discoverable relationships between the states as defined within our paper within financial markets.

4.2 Conclusions

We combine ideas from pattern-based trading literature, notably DTW, with concepts from the field of continual learning to create two new online agents. With these two online DTW-based strategies—one using a Markov model and another an Action-Value table—we demonstrate significant improvement over baseline agents in predicting the next pattern class and selecting a profitable trading action. Moreover, the Action-Value based agent yields significantly higher profit when applied to trading tasks. While the Markov-based strategy achieves greater profitability than some baseline agents, it also exhibits large draw-downs, suggesting a trade-off between performance and risk. The Action-Value agent's ability to offer high returns coupled with limited drawdowns indicates its potential for use in competitive trading environments. These findings encourage continued interest in pattern-based trading agents and support to the effectiveness of incorporating historical pattern information and adaptive memory structures to characterise trading strategies.

References

1. Ammann, M., Verhofen, M.: The effect of market regimes on style allocation. Financ. Mark. Portfolio Manag. **20**(3), 309–337 (2006)
2. Ang, A., Bekaert, G.: How regimes affect asset allocation. Financ. Anal. J. **60**(2), 86–99 (2004)
3. Ang, A., Timmermann, A.: Regime changes and financial markets. Annu. Rev. Financ. Econ. **4**(1), 313–337 (2012)
4. Carapuço, J., Neves, R., Horta, N.: Reinforcement learning applied to forex trading. Appl. Soft Comput. **73**, 783–794 (2018)
5. CMC Markets: 11 most essential stock chart patterns (2024). https://www.cmcmarkets.com/en/trading-guides/stock-chart-patterns
6. Feuerriegel, S., Prendinger, H.: News-based trading strategies. Decis. Support Syst. **90**, 65–74 (2016)

7. Ge, W., Lalbakhsh, P., Isai, L., Lensky, A., Suominen, H.: Comparing deep learning models for the task of volatility prediction using multivariate data. arXiv preprint arXiv:2306.12446 (2023)
8. Ge, W., Lalbakhsh, P., Isai, L., Lenskiy, A., Suominen, H.: Neural network-based financial volatility forecasting: a systematic review. ACM Comput. Surv. **55**(1) (2022). https://doi.org/10.1145/3483596
9. Geva, T., Zahavi, J.: Empirical evaluation of an automated intraday stock recommendation system incorporating both market data and textual news. Decis. Support Syst. **57**, 212–223 (2014)
10. Hendershott, T., Jones, C.M., Menkveld, A.J.: Does algorithmic trading improve liquidity? J. Financ. **66**(1), 1–33 (2011)
11. Jacobson, L.: Warne: growth, saving, financial markets, and Markov switching regimes. Stud. Nonlinear Dyn. Econometrics **5**(4), 99–110 (2002)
12. Kim, K.J., Han, I.: The extraction of trading rules from stock market data using rough sets. Expert Syst. **18**(4), 194–202 (2001)
13. Kim, S.H., Lee, H.S., Ko, H.J., Jeong, S.H., Byun, H.W., Oh, K.J.: Pattern matching trading system based on the dynamic time warping algorithm. Sustainability **10**(12), 4641 (2018)
14. Kumar, A.: Empirical investigation of herding in cryptocurrency market under different market regimes. Rev. Behav. Fin. **13**(3), 297–308 (2020)
15. Leem, J., Kim, H.Y.: Action-specialized expert ensemble trading system with extended discrete action space using deep reinforcement learning. PLoS ONE **15**(7), e0236178 (2020)
16. Lensky, A., Hao, M.: Learning to predict short-term volatility with order flow image representation. In: 2024 IEEE Conference on Artificial Intelligence (CAI) (2024)
17. Liston, P., Gretton, C., Lensky, A.: The role of stop-loss orders in market efficiency and stability: an agent-based study. In: ICAART (1), pp. 280–288 (2024)
18. Lu, T.H., Shiu, Y.M.: Can 1-day candlestick patterns be profitable on the 30 component stocks of the DJIA? Appl. Econ. **48**(35), 3345–3354 (2016)
19. Maringer, D., Ramtohul, M.: Regime-switching recurrent reinforcement learning for investment decision making. CMS **9**(1), 89–107 (2012)
20. Nguyen, T.H., Shirai, K., Velcin, J.: Sentiment analysis on social media for stock movement prediction. Expert Syst. Appl. **42**(24), 9603–9611 (2015)
21. Philps, D.: A temporal continual learning framework for investment decisions, Ph.D. thesis, City Research Online, University of London (2020)
22. Pätäri, E., Vilska, M.: Performance of moving average trading strategies over varying stock market conditions: the Finnish evidence. Appl. Econ. **46**(24), 2851–2872 (2014)
23. Raghavendra, B., Bera, D., Bopardikar, A.S., Narayanan, R.: Cardiac arrhythmia detection using dynamic time warping of ECG beats in e-healthcare systems. In: 2011 IEEE International Symposium on World Wireless Mobile Multimedia Networks, pp. 1–6 (2011)
24. Sakoe, H., Chiba, S.: Dynamic programming algorithm optimization for spoken word recognition. IEEE Trans. Acoust. Speech Sig. Process. **26**(1), 43–49 (1978)
25. Tsinaslanidis, P., Alexandridis, G., Zapranis, A., Livanis, E.: Dynamic time warping as a similarity measure: applications in finance. J. Fin. Data Sci. (2014)
26. Tsinaslanidis, P., Guijarro, F.: What makes trading strategies based on chart pattern recognition profitable? Expert. Syst. **38**(5), e12596 (2021)
27. Tsinaslanidis, P.E.: Subsequence dynamic time warping for charting: bullish and bearish class predictions for NYSE stocks. Expert Syst. Appl. **94**, 193–204 (2018)

28. Vullings, H., Verhaegen, M., Verbruggen, H.: Automated ECG segmentation with dynamic time warping. In: Proceedings of the 20th Annual International Conference of the IEEE Engineering in Medicine and Biology Society, pp. 163–166 (1998)
29. Zhang, Z., Zohren, S., Roberts, S.: Deep reinforcement learning for trading. arXiv preprint arXiv:1911.10107 (2019)

An Experimental Study on Decomposition-Based Deep Ensemble Learning for Traffic Flow Forecasting

Qiyuan Zhu[1], A. K. Qin[1(✉)], Hussein Dia[1], Adriana-Simona Mihaita[2], and Hanna Grzybowska[3]

[1] Swinburne University of Technology, Melbourne, Australia
{qiyuanzhu,kqin,hdia}@swin.edu.au
[2] University of Technology Sydney, Sydney, Australia
adriana-simona.mihaita@uts.edu.au
[3] Data61 CSIRO, Sydney, Australia
hanna.grzybowska@data61.csiro.au

Abstract. Traffic flow forecasting is a crucial task in intelligent transport systems. Deep learning offers an effective solution, capturing complex patterns in time-series traffic flow data to enable the accurate prediction. However, deep learning models are prone to overfitting the intricate details of flow data, leading to poor generalisation. Recent studies suggest that decomposition-based deep ensemble learning methods may address this issue by breaking down a time series into multiple simpler signals, upon which deep learning models are built and ensembled to generate the final prediction. However, few studies have compared the performance of decomposition-based ensemble methods with non-decomposition-based ones which directly utilise raw time-series data. This work compares several decomposition-based and non-decomposition-based deep ensemble learning methods. Experimental results on three traffic datasets demonstrate the superiority of decomposition-based ensemble methods, while also revealing their sensitivity to aggregation strategies and forecasting horizons.

Keywords: Ensemble learning · Deep learning · Decomposition · Traffic flow forecasting

1 Introduction

Traffic flow forecasting is one of its most important tasks for intelligent transport systems (ITS) in daily traffic management and operations [23]. Several operations, such as incident management, require reliable flow forecasting for a short horizon in future to support decision-making. However, accurate forecasting remains challenging due to the complex patterns in time-series traffic data. Factors like road congestion, vehicle breakdowns and traffic signal timing

M. Gong et al. (Eds.): AI 2024, LNAI 15442, pp. 394–407, 2025.
https://doi.org/10.1007/978-981-96-0348-0_29

[15,17] contribute to irregular and unpredictable patterns in traffic data, making accurate forecasting difficult to achieve.

Traditional statistical and shallow machine learning techniques often struggle to capture the complex patterns within this data, rendering them less effective in this context. In contrast, deep learning techniques are better suited to adapt to the intricate nature of time-series traffic data [21]. However, these methods can be biased, overfitting the intricate details of flow data, leading to poor generalisation.

Ensemble learning [7] is a potential solution to mitigate the limitations of deep learning by combining the outputs of multiple models [25]. Recent advances in ensemble learning have promoted various kinds of deep ensemble learning approaches, including using conventional ensemble methods [14,28] and time-based ensemble methods [3]. Among these ensemble learning methods, the decomposition-based ensemble is a less explored category that transforms time-series into simple components for modelling [20]. The components extracted from decomposition-based methods may reduce the complexity of the data, and thus yield a more robust solution. However, there is a lack of comparative studies on whether this approach can better benefit deep learning models than non-decomposition-based ensemble methods. Currently, the comparison studies are mostly restricted to certain types of methods [9,12,20].

This paper compares several decomposition-based ensemble methods with conventional bagging and time domain multi-resolution ensemble methods [3] under the traffic flow forecasting tasks. The main contributions of this paper are:

1. An empirical study is conducted to assess the helpfulness of decomposition-based ensemble methods for deep learning models in traffic flow forecasting tasks. Results yield that decomposition-based methods better enhance the performance of deep learning models than baseline methods.
2. We explored the effectiveness of optimised aggregation for decomposition methods. The results show that decomposition methods are sensitive to the aggregation methods.
3. We investigated the impact of inputs and forecasting horizons for decomposition methods. Results indicate that these methods are sensitive to the input and do not always benefit from extensive data.

2 Background and Related Works

This section briefly reviews the existing decomposition-based ensemble methods and time-domain multi-resolution ensemble approaches.

2.1 Decomposition-Based Ensemble Methods

Most research on decomposition-based ensemble methods follows a divide-and-conquer concept, which transforms complex original time-series data into a set of simple components [20]. The first category, including Fourier Transform (FT) [4] and Wavelet Transform (WT) [6], uses use predefined basis functions. However, predefined basis functions suffer from frequency resolution issues, which means they lack or have a limited ability (due to predefined function) to distinguish between closely spaced frequency components in a signal. To address this gap, the second category avoids predefined functions, extracting signals based on local time scales and, as such, distinguishing closely spaced frequency components into Intrinsic Mode Functions (IMFs). Notable models in this category include Empirical Mode Decomposition (EMD), Ensemble Empirical Mode Decomposition (EEMD), and Complete Ensemble Empirical Mode Decomposition with Adaptive Noise (CEEMDAN) [20]. Previous studies [12,24] show that these methods outperform direct raw data predictions. Thus, we selected EMD, EEMD, and CEEMDAN for this study.

2.2 Time Domain Multi-resolution Ensemble

The time interval (i.e., temporal resolution) of recorded time-series data is connected to forecasting performance, making it a beneficial approach for ensemble learning [3,27]. The challenge of overfitting in deep learning methods due to complex time-series data can be mitigated by considering outcomes of other deep learning models trained on aggregated time-series data that focus on different trends. Previous research suggests that such a multi-resolution ensemble may outperform the single model in terms of performance and robustness [3,11].

Frequency resolution refers to the ability to distinguish between closely spaced frequency components in a signal.

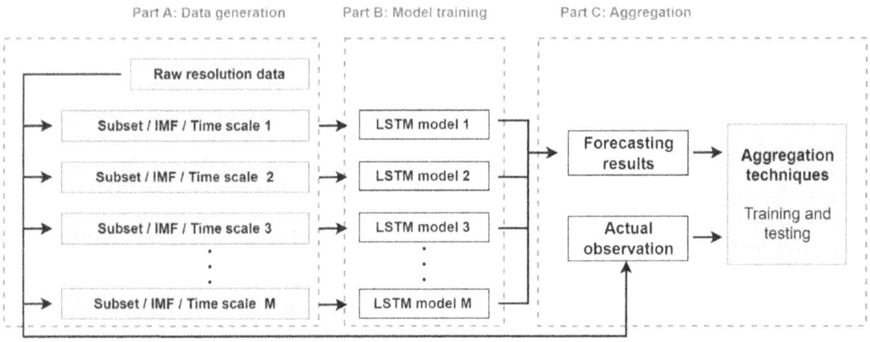

Fig. 1. The comparison design with three modules: (a) The data generation module is responsible for extracting the subsets of time patterns from the original time-series. (b) The training module is responsible for training base learners. (c) The aggregating module trains the final learner based on meta-data extracted from base learners.

3 Comparative Design

In this work, we attempt to compare and evaluate the deep ensemble methods on traffic flow forecasting tasks. The proposed comparison design is illustrated in Fig. 1.

3.1 Data Generation Module

Firstly, the data generation module extracts the subsets of time patterns from the original time-series for each selected ensemble method. The generated subsets are adjusted according to the specific ensemble learning method.

For Multi-resolution Ensemble and Bagging. The data generation for time domain multi-resolution ensemble is an addition to simple data aggregation, representing views based on aggregation levels [1]. Firstly, the original time-series will be sliced into a set of input-output pairs, denoted by set $X = \{x_1, x_2, ..., x_i\}$ and set $Y = \{y_1, y_2, ..., y_i\}$. Each x_i contains I element and each y_i is the sum of T elements following I. $X^M = \{X^1, X^2, ..., X^m\}$ is obtained by sum aggregation on X based on each time interval. Assuming the original time-series is recorded in 1 min time interval, input I is 60 min, the timer-series input size for x_1^1 in X^1 and x_1^{10} in X^{10} are $60/1 = 60$ and $60/10 = 6$, respectively. For conventional bagging, the X^M and Y^M are generated by taking 90% elements from X and Y with replacement in their original form, and the number of generated series M is equal to 25 as suggested by [5], representing subsets of the original data.

For Decomposition-Based Methods. Since decomposition will alter the feature characteristics of the original time-series data, the input and output are combined into a single sequence s and are sliced after decomposition. The first step is to prepare the raw set $S = \{s_1, s_2, ..., s_i\}$ and each sequence s_i contains $I+T$ elements. $S = \{s_1, s_2, ..., s_i\}$ will be decomposed to M sets $\widehat{S^M} = \{\widehat{S^1}, \widehat{S^2}, ..., \widehat{S^m}\}$ with each $\widehat{S^m} = \{\widehat{s_1^m}, \widehat{s_2^m}, ..., \widehat{s_i^m}\}$.

After decomposition, we need to extract the input and output from $\widehat{S^M}$. Assuming I and T are 60 min and 10 min, arbitrary $\widehat{s_i^m}$ in $\widehat{S^m}$ can be sliced into $\widehat{x_i^m} = \{\widehat{s_{1,i}^m}, ..., \widehat{x_{60,i}^m}\}$ and $\widehat{y_i^m} = \{\widehat{s_{61,i}^m}, ..., \widehat{s_{70,i}^m}\}$ and the task is to forecast $\widehat{y_i^m}$ using corresponding $\widehat{x_i^m}$. The final prediction y_i' will be aggregated from predicted $\{\widehat{y_i^{1'}}, \widehat{y_i^{2'}}, ..., \widehat{y_i^{m'}}\}$ and compared with grand truth y_i. The details of each method are explained below.

EMD Method was developed to adaptively decompose complex signals into multiple simpler time-series called Intrinsic Mode Functions (IMFs), providing a better ability to distinguish frequency components compared with the Fourier Transform. We use $s(t)$ for the value in time t in sequence s for simplification. Its process can be described as following steps:

Step (1): local maxima and minima for a sequence s will be determined over time. All local maxima compose an upper envelope $u(t)$ (usually by cubic interpolation), and local minima points will form a lower envelope $l(t)$ in the same way. We calculate the mean value of the upper and lower envelops for each t by:

$$m(t) = (u(t) + l(t))/2. \tag{1}$$

Step (2): The $m(t)$ will create a new time-series $m_{envelop}$, and we can get the value of the candidate IMF in each t by

$$IMF_{candidate}(t) = s(t) - m(t) \tag{2}$$

Step (3): we need to verify this candidate IMF by checking two conditions: (1) the number of extreme points and the number of zero-crossing points must be equal or differ at most by 1, and (2) the $m_{envelop}$ must be 0 everywhere [10]. If these two conditions are not satisfied, steps (1)-(2) must repeatedly apply to this $IMF_{candidate}$ until it meets the criteria.

When an IMF is found, we will collect it, and the remaining signal becomes:

$$r(t) = s(t) - IMF(t) \tag{3}$$

where the sequence r obtained from $r(t)$ will replace the s as input for steps (1)-(3) to extract the next IMF components until the total number of local maxima and minima for residue sequence r can no longer exceed two. The final EMD output is $\{IMF_1, IMF_2, ..., IMF_m\}$ for given s, while the number m of IMFs is not predefined but discovered from the signal.

EEMD Method. Some signal components can occasionally appear and disappear within a signal at irregular intervals. The IMF generated by EMD under such a signal can be misleading: (1) several IMFs can be similar, and (2) a single IMF can falsely contain components of different frequencies. Instead of calculating IMF directly, EEMD computes each IMF based on an ensemble of multiple noise-added IMFs. The added white noise provides a uniform reference in the time-frequency space, making the method less sensitive to small variations in the signal that might cause misleading IMFs in standard EMD. Much like the EMD, its process can be described as following steps:

Step (1): Add different Gaussian noise $\omega^k(t), k = 1, 2, ..., K$ and corresponding noise coefficient ϵ to the sequence s to generate the noise-added data $s^k(t) = s(t) + \epsilon * \omega^k(t), k = 1, 2, ..., K$, where K is the number of trials pre-defined.

Step (2): Each sequence $s^k, k = 1, 2, ..., K$ is decomposed by EMD to extract all IMFs, i.e., IMF_m^k refers to m_{th} IMF for s^k.

Step (3): Get each final IMF for s by ensemble IMFs extracted from $\{s^1, s^2, .., s^k\}$:

$$\overline{IMF_m(t)} = \frac{1}{K} \sum_{k=1}^{K} IMF_m^k(t) \tag{4}$$

CEEMDAN Method. The noise EEMD used improves the IMF quality and hampers the reconstruction ability. Thus, the EEMD does not guarantee an accurate reconstruction of s, which can be important. CEEMDAN changes Gaussian noise to adaptive noise, which adds adaptive noise at each decomposition stage and helps to reduce the noise in the final IMFs [22]. It ensures a complete decomposition, where the sum of all IMFs and the residual exactly reconstructs the original signal. The quality of IMF is also improved due to an optimal solution for noise addition. Its process can be described as following steps:

Step (1): Add different Gaussian noise $\omega^k(t), k = 1, 2, ..., K$ and corresponding noise coefficient ϵ_0 to the sequence s to generate the noise-added data $s^k(t) = s(t) + \epsilon_0 * \omega^k(t), k = 1, 2, ..., K$, where K is the number of trials predefined.

Step (2): Each $s^k, k = 1, 2, ..., K$ is decomposed by EMD to extract the first IMF, i.e., IMF_1^k.

Step (3): Calculate the $\overline{IMF_1}$ of s by ensemble the IMF_1^k using mean values as Eq. 4.

Step (4): Calculate the first residual sequence r_1 like Eq. 3 by:

$$r_1(t) = s(t) - \overline{IMF_1(t)} \tag{5}$$

Step (5): we define operator $E(.)$ as an EMD process, and $E_1(.)(t)$ represents the value of EMD process's first IMF in time t, Then:

$$\overline{IMF_2(t)} = \frac{1}{K}\sum_{k=1}^{K} E_1(r_1 + \epsilon_1 * E_1(\omega^k))(t) \tag{6}$$

Step (6): we further define $E_m(.)$, representing EMD processs's m_{th} IMF, then we can acquire all IMFs by:

$$\overline{IMF_{m+1}(t)} = \frac{1}{K}\sum_{k=1}^{K} E_1(r_m + \epsilon_m * E_m(\omega^k))(t) \tag{7}$$

The stop condition is much like the EMD: until the total number of local maxima and minima for residue sequence r can no longer exceed two. Although ϵ_m allows optimisation at each stage of the decomposition, some studies suggest the constant value is also practical [22].

3.2 Modelling Module

Long short-term memory (LSTM) is a recurrent neural network type that stores information across the time stamp to model the long-term dependencies [8] effectively. We apply LSTM as the base learner in the modelling module for its good time-series learning capacity. Each base learner will be trained in sequential order under the corresponding input-output pair until the maximum training iteration is reached or the early stopping rule on the validation set is violated. A hyperparameter tuning process is responsible for optimising the base learner. To maintain a fair comparison, the input and forecasting horizons are the same for all ensemble learning methods regardless of the data generation process.

3.3 Aggregation Module

In the aggregation module, the outputs of base learners will be extracted and utilised using an optimised final learner or baseline aggregation strategy. Mean aggregation is the baseline strategy for time domain multi-resolution ensemble and conventional bagging. Due to the nature of decomposition, sum aggregation is the baseline strategy for decomposition-based methods. We further apply linear regression and MLP as the optimised final learners to utilise the predictions generated from base learners.

4 Experimental Study

To evaluate the performance of decomposition-based deep ensemble learning methods, we conducted a series of experiments on Melbourne traffic datasets and PEMS datasets to investigate:

1. RQ1: Whether decomposition-based methods better benefit the deep learning models than other ensemble methods on forecasting tasks?
2. RQ2: Is the performance of decomposition-based ensemble learning sensitive to aggregation methods?
3. RQ3: Whether the answer of RQ1 and RQ2 depends on input and forecasting horizon?

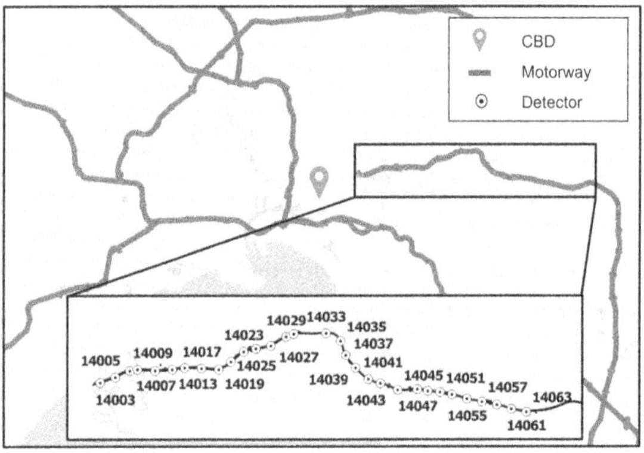

Fig. 2. The location of sensors in eastern residential suburbs, Melbourne CBD.

4.1 Data

In this study, we conducted the experiments on 3 datasets: the Melbourne data set, the PEMSD4 data set and the Portland data set. The Melbourne data set are corridors stretching from Melbourne CBD to the eastern residential suburbs; see Fig. 2. The traffic flow, speed and occupancy are recorded in 1-minute intervals from 1st Jul 2016 to 30th Sep 2016. In this study, 5 sites that do not have large maintenance (e.g., road construction and fixing) are selected regardless of the weekday and weekend.

The collected PEMS data set records the highway traffic flow of the San Francisco Bay Area in California, USA [2] from 1st Jan 2018 to 28th Feb 2018 in 30-second intervals. The collected Portland data set comes from the FHWA research project [19] in Portland I-205N from September 15, 2011 to November 15, 2011. In each dataset, 5 sites are selected using the same criteria as the Melbourne dataset. For all datasets, the train-val-test set is split on 70%/10%/20%. The validation set is only used for hyperparameter tuning.

4.2 The Hyper-Parameter Setting

To generate IMFs for decomposition-based methods, we use the implementation from the pyEMD package. The Spline method is set to Akima for complex signal processing for all decomposition-based methods. EMD adaptively decompose the signal without preset parameters. For EEMD and CEEMDAN, all noise coefficients ϵ are set to 0.2, and the number of trials is set to 25. We use the top 5 IMF components in modelling.

The proposed base learners' model structure is shown in Table 1. LSTM first completes the time-series feature extraction, and the returned sequence is then passed to fully connected layers. The proposed model is implemented using tensorflow-gpu (version 2.6.0), and the hyper-parameter tuning is conducted by keras-tunner (version 1.1.3).

Table 1. Hyper-parameter tuning of LSTM model (deep base learner)

Model layers	Hidden units
LSTM	[8,16,32,64,128]
Dense	[8,16,32,64,128]
Dropout	[0, 0.1, 0.2, 0.3, 0.4, 0.5]
Dense	[8,16,32,64,128]
Dropout	[0, 0.1, 0.2, 0.3, 0.4, 0.5]
Dense	10
Dense	Output

The proposed MLP final learners' model structure has three dense layers and two dropouts in between. The search space for the dense layer's hidden unit is [8,16,32,64], and the search space for the dropout rate is [0, 0.1, 0.2, 0.3, 0.4, 0.5]. The input and output of the final learner are both 2-dimensional arrays.

4.3 Evaluation Matrix

This research uses root mean squared error (RMSE) for performance evaluation. RMSE is used to measure the fitness between the predictions and the real observation of traffic flows:

Table 2. Results of the ensemble learning methods using baseline aggregation against the baseline methods in Melbourne data. The method(s) highlighted in bold if its performance is statistically the best one(s).

	site14005	site14049	site14059	site14061	site14063
Model	RMSE	RMSE	RMSE	RMSE	RMSE
I = 120 min					
LSTM	59.571	90.093	57.492	56.638	47.474
Bagging + mean	48.641	61.146	45.595	47.933	36.349
Multi-reso + mean	49.031	65.619	51.342	53.779	39.157
EMD + sum	**25.495**	41.054	36.986	35.786	33.445
EEMD + sum	34.242	**30.374**	**33.360**	46.221	**27.455**
CEEMDAN + sum	28.040	37.329	44.193	**35.428**	29.592
I = 240 min					
LSTM	54.201	72.146	55.417	55.770	39.333
Bagging + mean	49.042	55.783	47.451	49.639	36.745
Multi-reso + mean	47.799	60.911	49.050	49.571	37.377
EMD + sum	32.267	46.931	36.186	33.227	31.250
EEMD + sum	**31.110**	**29.354**	**22.904**	**30.757**	**15.993**
CEEMDAN + sum	31.853	40.098	38.695	30.779	27.626
I = 360 min					
LSTM	58.276	76.819	54.426	54.874	44.023
Bagging + mean	49.079	56.300	46.259	48.500	37.524
Multi-reso + mean	49.396	60.222	46.547	50.556	38.976
EMD + sum	**23.204**	48.403	35.376	31.975	**24.332**
EEMD + sum	27.552	**30.033**	**31.024**	**27.598**	25.395
CEEMDAN + sum	46.705	48.293	45.486	47.839	26.736

$$RMSE = \sqrt{\frac{1}{n} * \sum_{i=1}^{n} (tfp_i - y_i)^2}$$

where y_i is the actual value of traffic flows, tfp_i represents the forecasting value corresponding to each y_i, and n denotes the number of observed values.

4.4 Overall Comparison

To assess the effectiveness of decomposition-based ensemble methods, we compared EMD, EEMD, and CEEMAND against conventional bagging and multi-resolution ensemble, all trained on time-series recorded in 1-minute intervals over 10 repeated runs. A standalone LSTM model, trained on the aggregated

Table 3. Results of the decomposition-based ensemble learning methods using optimised strategy against the baselines aggregation in Melbourne data. The method(s) highlighted in bold if its performance is statistically the best one(s).

	site14005	site14049	site14059	site14061	site14063
Model	RMSE	RMSE	RMSE	RMSE	RMSE
I = 120 min					
EMD + sum	25.495	41.054	36.986	35.786	33.445
EMD + linear	24.207	33.555	31.979	31.470	28.858
EMD + MLP	27.685	38.204	34.679	36.811	35.994
EEMD + sum	34.242	30.374	33.360	46.221	27.455
EEMD + linear	**22.142**	**26.747**	**22.730**	**21.193**	**17.567**
EEMD + MLP	23.975	27.544	23.122	24.551	22.321
CEEMDAN + sum	28.040	37.329	44.193	35.428	29.592
CEEMDAN + linear	26.990	31.893	35.267	33.211	27.943
CEEMDAN + MLP	27.632	29.299	40.161	32.815	28.029
I = 240 min					
EMD + sum	32.267	46.931	36.186	33.227	31.250
EMD + linear	26.184	31.717	31.878	29.021	30.491
EMD + MLP	37.696	33.864	35.806	33.640	31.292
EEMD + sum	31.110	29.354	22.904	30.757	15.993
EEMD + linear	**20.579**	**25.382**	**21.272**	**22.127**	**15.363**
EEMD + MLP	23.927	25.726	21.505	23.093	25.374
CEEMDAN + sum	31.853	40.098	38.695	30.779	27.626
CEEMDAN + linear	28.842	31.542	30.484	29.071	23.270
CEEMDAN + MLP	30.957	35.395	29.155	37.400	30.473
I = 360 min					
EMD + sum	23.204	48.403	35.376	31.975	24.332
EMD + linear	19.965	33.480	28.697	30.292	22.381
EMD + MLP	24.914	43.897	35.089	33.223	22.920
EEMD + sum	27.552	30.033	31.024	27.598	25.395
EEMD + linear	**18.072**	**26.038**	**21.379**	**22.792**	**18.165**
EEMD + MLP	23.655	28.645	22.166	25.464	18.232
CEEMDAN + sum	46.705	48.293	45.486	47.839	26.736
CEEMDAN + linear	28.026	38.748	40.342	37.731	20.284
CEEMDAN + MLP	43.151	39.521	45.883	42.264	29.699

time-series data with an equal time interval and forecasting horizon, was also compared with the aforementioned methods as an additional baseline, as suggested by [1]. The input horizon consists of 120, 240, and 360 min, with a forecasting horizon of 10 min for this test. The results for the Melbourne dataset are reported in Table 2, while the other results are reported in the supplementary material [1] Based on RMSE measurements, EEMD is the best ensemble in the Melbourne dataset, outperforming other methods in 10 out of 15 test scenarios. In the case of the PEMS and Portland datasets, EEMD outperformed other methods in 13 and 14 test scenarios, respectively. Overall, EEMD methods outperform the conventional bagging and multi-resolution ensemble methods. In all cases, decomposition-based methods outperformed non-decomposition-based methods.

4.5 The Ablation Study of Aggregation Strategy

We compare linear regression and MLP methods against baseline aggregation across the ensemble methods to assess the sensitivity of decomposition-based methods to aggregation strategy. The test setup is the same as the overall comparison. The results for the Melbourne dataset are reported in Table 3, and the other results are reported in the supplementary material. Based on RMSE measurement, linear is the best ensemble in the Melbourne dataset. In the case of the PEMS and Portland datasets, the best strategy is also linear.

Furthermore, linear regression and MLP outperform the baseline aggregation strategy in 127 and 94 cases, respectively. This might be because our progress didn't add residue to modelling, and some uncovered IMFs remain out of the top 5 IMFs, which are further utilised by linear regression and MLP methods.

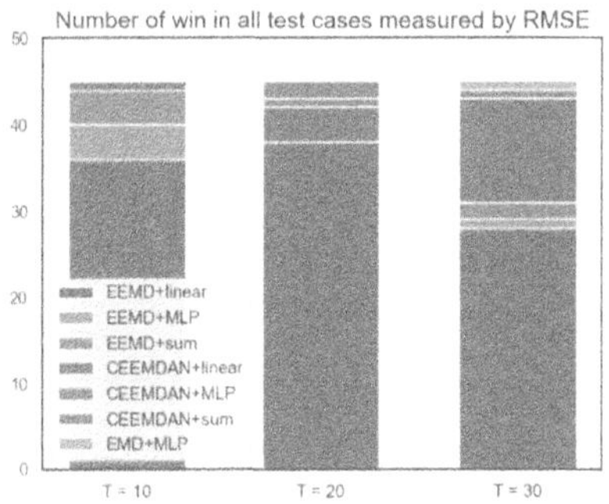

(a) Number of wins for all methods in all test cases measured by RMSE.

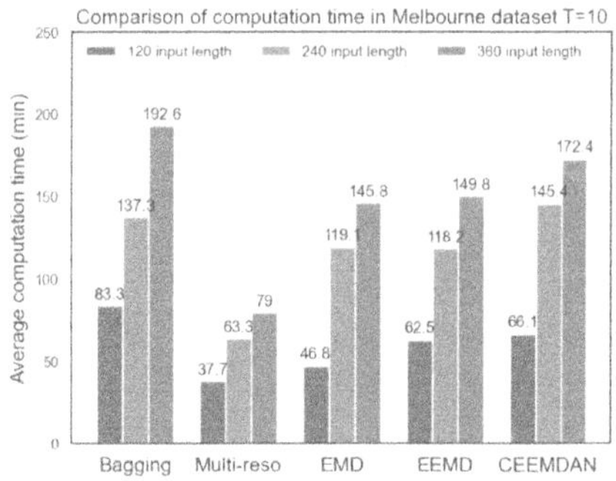

(b) Comparison of computation time (in minutes) of conventional bagging, multi-resolution ensemble, EMD, EEMD, and CEEMDAN methods in Melbourne dataset, target horizon 10.

Fig. 3. The computation time analysis and sensitivity analysis.

[1] https://alexkaiqin.org/doc/AJCAI2024DDELSupp.pdf.

4.6 Sensitivity of 20-Min and 30-Min Forecasting Horizons

We further evaluate the performance of all combinations of ensemble and aggregation methods for the forecasting tasks with 20-min and 30-min forecasting horizons. The rest of the test setup is the same as the overall comparison. The results are summarised in Fig 3a. We evaluated the results and noticed that 360-min input would improve the general performance of non-decomposition-based methods. This is consistent with the previous study [16] that longer inputs generally improve forecasting accuracy. However, decomposition-based are not sensitive to specific setups, and the CEEMDAN method tends to perform better when the output size is large. Figure 3b. compares each ensemble's computation time (in minutes) paired with linear regression for different inputs. It can be observed that decomposition-based methods have higher computation time due to the IMF construction progress, whereas CEEMDAN is more computationally expensive because of the adaptive noise strategy.

5 Conclusions and Future Work

In this paper, we compare three decomposition-based deep ensemble learning methods with two common non-decomposition-based ones, including bagging and multi-resolution ensemble, for their performance in solving traffic flow forecasting tasks. Experimental results on several traffic flow datasets demonstrate the superiority of decomposition-based methods, with the EEMD-based method outperforming others in most test scenarios. Future work includes the exploration of advanced ensemble learning strategies based on multi-task optimisation [26] and the incorporation of other types of base learners [18]. We also plan to evaluate the decomposition-based methods in other application scenarios involving time-series forecasting missions [13].

Acknowledgments.. This work was supported by the Australian Research Council (ARC) under Grant No. LP180100114. During the preparation of this work the authors used ChatGPT in order to improve language. After using this tool, the authors reviewed and edited the content as needed and take full responsibility for the content of the publication.

References

1. Abdulhai, B., Porwal, H., Recker, W.: Short-term traffic flow prediction using neuro-genetic algorithms. Intell. Transport. Syst. J. **7**(1), 3–41 (2002). https://doi.org/10.1080/10248070212011
2. Bai, L., Yao, L., Li, C., Wang, X., Wang, C.: Adaptive graph convolutional recurrent network for traffic forecasting. ArXiv **abs/2007.02842** (2020). https://api.semanticscholar.org/CorpusID:220363737
3. Chen, C., Liu, H.: Medium-term wind power forecasting based on multi-resolution multi-learner ensemble and adaptive model selection. Energy Convers. Manage. **206**, 112492 (2020)

4. Chen, L., Zheng, L., Yang, J., Xia, D., Liu, W.: Short-term traffic flow prediction: From the perspective of traffic flow decomposition. Neurocomputing **413**, 444–456 (2020)
5. Chen, L., Chen, C.P.: Ensemble learning approach for freeway short-term traffic flow prediction. In: 2007 IEEE International Conference on System of Systems Engineering, pp. 1–6. IEEE (2007)
6. Chen, S., Wang, W.: Traffic volume forecasting based on wavelet transform and neural networks. In: Wang, J., Yi, Z., Zurada, J.M., Lu, B.-L., Yin, H. (eds.) Advances in Neural Networks - ISNN 2006, pp. 1–7. Springer Berlin Heidelberg, Berlin, Heidelberg (2006). https://doi.org/10.1007/11760191_1
7. Hansen, L.K., Salamon, P.: Neural network ensembles. IEEE Trans. Pattern Anal. Mach. Intell. **12**(10), 993–1001 (1990)
8. Hochreiter, S., Schmidhuber, J.: Long short-term memory. Neural Comput. **9**(8), 1735–1780 (1997). https://doi.org/10.1162/neco.1997.9.8.1735
9. Huang, H., Chen, J., Huo, X., Qiao, Y., Ma, L.: Effect of multi-scale decomposition on performance of neural networks in short-term traffic flow prediction. IEEE Access **9**, 50994–51004 (2021)
10. Huang, N.E., et al.: The empirical mode decomposition and the hilbert spectrum for nonlinear and non-stationary time series analysis. Proc. Royal Society London. Series A: Math., Phys. Eng. Sci. **454**(1971), 903–995 (1998)
11. Liu, H., Duan, Z., Chen, C.: Wind speed big data forecasting using time-variant multi-resolution ensemble model with clustering auto-encoder. Appl. Energy **280**, 115975 (2020). https://doi.org/10.1016/j.apenergy.2020.115975
12. Lu, W., Rui, Y., Yi, Z., Ran, B., Gu, Y.: A hybrid model for lane-level traffic flow forecasting based on complete ensemble empirical mode decomposition and extreme gradient boosting. IEEE Access **8**, 42042–42054 (2020)
13. Mistry, S., Bouguettaya, A., Dong, H., Qin, A.K.: Metaheuristic optimization for long-term iaas service composition. IEEE Trans. Serv. Comput. **11**(1), 131–143 (2018)
14. Moretti, F., Pizzuti, S., Panzieri, S., Annunziato, M.: Urban traffic flow forecasting through statistical and neural network bagging ensemble hybrid modeling. Neurocomputing **167**, 3–7 (2015)
15. Okutani, I., Stephanedes, Y.J.: Dynamic prediction of traffic volume through kalman filtering theory. Transp. Res. Part B: Methodol. **18**(1), 1–11 (1984)
16. Petelin, G., Hribar, R., Papa, G.: Models for forecasting the traffic flow within the city of ljubljana. Eur. Transp. Res. Rev. **15**(1), 30 (2023)
17. Qiao, F., Yang, H., Lam, W.H.: Intelligent simulation and prediction of traffic flow dispersion. Transp. Res. Part B: Methodol. **35**(9), 843–863 (2001)
18. Qin, A.K., Suganthan, P.N.: Initialization insensitive LVQ algorithm based on cost-function adaptation. Pattern Recogn. **38**(5), 773–776 (2005)
19. RESEARCH, T., CENTER, E.: Multimodal transportation data research. https://trec.pdx.edu/transportation-data-research. Accessed 26 Apr 2024
20. Shi, J., Leau, Y.B., Li, K., Park, Y.J., Yan, Z.: Optimization and decomposition methods in network traffic prediction model: a review and discussion. IEEE Access **8**, 202858–202871 (2020)
21. Tedjopurnomo, D.A., Bao, Z., Zheng, B., Choudhury, F.M., Qin, A.K.: A survey on modern deep neural network for traffic prediction: trends, methods and challenges. IEEE Trans. Knowl. Data Eng. **34**(4), 1544–1561 (2020)
22. Torres, M.E., Colominas, M.A., Schlotthauer, G., Flandrin, P.: A complete ensemble empirical mode decomposition with adaptive noise. In: 2011 IEEE International

Conference on Acoustics, Speech and Signal Processing (ICASSP), pp. 4144–4147. IEEE (2011)
23. Vlahogianni, E.I., Karlaftis, M.G., Golias, J.C.: Short-term traffic forecasting: where we are and where we're going. Transp. Res. Part C: Emerg. Technol. **43**, 3–19 (2014)
24. Wang, H., Liu, L., Dong, S., Qian, Z., Wei, H.: A novel work zone short-term vehicle-type specific traffic speed prediction model through the hybrid emd-arima framework. Transp. B: Transp. Dyn. **4**(3), 159–186 (2016)
25. Wang, Z.j., Liu, H.x., Qiu, S., Fang, J.p., Wang, T.: The predictability of short-term urban rail demand: choice of time resolution and methodology. Sustainability **11**(21), 6173 (2019)
26. Xu, H., Qin, A.K., Xia, S.: Evolutionary multitask optimization with adaptive knowledge transfer. IEEE Trans. Evol. Comput. **26**(2), 290–303 (2022)
27. Zhong, C., Batty, M., Manley, E., Wang, J., Wang, Z., Chen, F., Schmitt, G.: Variability in regularity: mining temporal mobility patterns in London, Singapore and beijing using smart-card data. PLoS ONE **11**(2), e0149222 (2016)
28. Zhou, T., et al.: δ-agree adaboost stacked autoencoder for short-term traffic flow forecasting. Neurocomputing **247**, 31–38 (2017)

Author Index

M. Gong et al. (Eds.): AI 2024, LNAI 15442, pp. 409–412, 2025.
https://doi.org/10.1007/978-981-96-0348-0

The manufacturer's authorised representative in the EU is Springer Nature Customer Service Centre GmbH, Europaplatz 3, 69115 Heidelberg, Germany. If you have any concerns regarding our products, please contact ProductSafety@springernature.com

Printed and bound by CPI Group (UK) Ltd, Croydon, CR0 4YY
06/05/2026
02104369-0003